LONG-TIME PREDICTION IN DYNAMICS

NONEQUILIBRIUM PROBLEMS IN THE PHYSICAL SCIENCES AND BIOLOGY

Editors: **I. Prigogine and G. Nicolis**

Université Libre de Bruxelles
Brussels, Belgium

Volume I: **G. NICOLIS, G. DEWEL, AND J. W. TURNER,**
Order and Fluctuations in Equilibrium and Nonequilibrium Statistical Mechanics: XVIIth International Solvay Conference on Physics

Volume II: **C. W. HORTON, JR., L. E. REICHL, AND V. G. SZEBEHELY,**
Long-Time Prediction in Dynamics

LONG-TIME PREDICTION IN DYNAMICS

Edited by

C. W. HORTON, JR., L. E. REICHL, and V. G. SZEBEHELY

A WILEY-INTERSCIENCE PUBLICATION

JOHN WILEY & SONS

New York · Chichester · Brisbane · Toronto · Singapore

Library of Congress Cataloging in Publication Data:

Main entry unter title:

Long-time prediction in dynamics.

(Nonequilibrium problems in the physical sciences and biology, ISSN 0275-9292; v. 2)

"A Wiley-Interscience publication."

Presentations from the Workshop on Long-time Predictions in Nonlinear Conservative Dynamical Systems held in Mar. 1981 in Lakeway, Tex, under the auspices of the Center for Studies in Statistical Mechanics, the Center for Orbital Dynamics, and the Institute for Fusion Studies at the University of Texas at Austin.

Includes index.

1. Statistical mechanics—Congresses. 2. Dynamics—Congresses. 3. Plasma dynamics—Congresses. 4. Nonlinear theories—Congresses. 5. Many-body problem—Congresses. I. Horton, C. W. (Claude Wendell), Jr., 1942- . II. Reichl, L. E. III. Szebehely, Victor G., 1921- . IV. Workshop on Long-time Predictions in Nonlinear Conservative Dynamical Systems (1981: Lakeway, Tex.) V. University of Texas at Austin. Center for Studies in Statistical Mechanics. VI. University of Texas at Austin. Center for Orbital Dynamics. VII. University of Texas at Austin. Institute for Fusion Studies.

QC174.7.L66 1982 530.1′3 82-17609

ISBN 0-471-86447-1

Printed in the United States of America

10 9 8 7 6 5 4 3 2 1

CONTRIBUTORS

TASSOS C. BOUNTIS	Department of Mathematics, and Computer Science, Clarkson College of Technology, Potsdam, New York 13676
DANIEL H. E. DUBIN	Plasma Physics Laboratory, Princeton University, P.O. Box 451, Princeton, New Jersey 08544
CHARLES R. EMINHIZER	Physical Dynamics, P.O. Box 1883, La Jolla, California 92038
D. F. ESCANDE	Laboratoire de Physique des, Milieux Ionisés, Groupe de Recherche No 29, du Centre National de la Recherche Scientifique, École Polytechnique, 91128 Palaiseau Cedex, France
JOSEPH FORD	School of Physics, Georgia Institute of Technology, Atlanta, Georgia 30332
J. P. FREIDBERG	Nuclear Engineering Department and Plasma Fusion Center, Massachusetts Institute of Technology, Cambridge, Massachusetts 02139
SHELDON GOLDSTEIN	Department of Mathematics, Rutgers University, New Brunswick, New Jersey 08903
HAROLD GRAD	Courant Institute of Mathematical Sciences, New York University, 251 Mercer Street, New York, New York 10012
CELSO GREBOGI	Lawrence Berkeley Laboratory and Department of Physics, University of California, Berkeley, California 94720
JOHN M. GREENE	Plasma Physics Laboratory, Princeton University, Princeton, New Jersey 08544
ROBERT H. G. HELLEMAN	Department of Theoretical Physics, Twente University of Technology, 7500 AE Enschede, The Netherlands
S. P. HIRSHMAN	Oak Ridge National Laboratory, Oak Ridge, Tennessee 37830
C. W. HORTON, JR.	Institute for Fusion Studies, University of Texas, Austin, Texas 78712
J. M. HYMAN	Research Group T-7, Los Alamos Scientific Laboratory, Los Alamos, New Mexico 87545
YOSHI H. ICHIKAWA	Institute of Plasma Physics, Nagoya University, Nagoya 464, Japan
C. F. F. KARNEY	Plasma Physics Laboratory, Princeton University, Princeton, New Jersey 08540

ALLAN N. KAUFMAN Lawrence Berkeley Laboratory and Department of Physics, University of California, Berkeley, California 94720

S. KHEIFETS Stanford Linear Accelerator Center, Stanford University, Stanford, California 94305

KIMIAKI KONNO Department of Physics, College of Science and Technology, Nihon University, Tokyo 101, Japan

JOHN A. KROMMES Plasma Physics Laboratory, Princeton University, P.O. Box 451, Princeton, New Jersey 08544

JOEL L. LEBOWITZ Department of Mathematics and Physics, Rutgers University, New Brunswick, New Jersey 08903

M. A. LIEBERMAN Department of Electrical Engineering and Computer Sciences, University of California, Berkeley, California 94720

ROBERT S. MACKAY Plasma Physics Laboratory, Princeton University, P.O. Box 451, Princeton, New Jersey 08544

D. W. MCLAUGHLIN Research Group T-7, Los Alamos Scientific Laboratory, Los Alamos, New Mexico 87545

B. MISRA Faculté des Sciences, Université Libre de Bruxelles, 1050 Brussels, Belgium

K. MOLVIG Nuclear Engineering Department and Plasma Fusion Center, Massachusetts Institute of Technology, Cambridge, Massachusetts 02139

EDWARD OTT Laboratory for Plasma and Fusion Energy Studies, University of Maryland, College Park, Maryland 20742

R. POTOK Nuclear Engineering Department and Plasma Fusion Center, Massachusetts Institute of Technology, Cambridge, Massachusetts 02139

I. PRIGOGINE Faculté des Sciences, Université Libre de Bruxelles, 1050 Brussels, Belgium and Center for Studies in Statistical Mechanics, University of Texas, Austin, Texas 78712

A. B. RECHESTER Plasma Fusion Center, Massachusetts Institute of Technology, Cambridge, Massachusetts 02139

M. N. ROSENBLUTH Institute for Fusion Studies, University of Texas at Austin, Austin, Texas 78712

A. SALAT Max-Planck-Institut für Plasmaphysik, 8046 Garching, Federal Republic of Germany

A. C. SCOTT Department of Electrical and Computer Engineering, University of Wisconsin, Madison, Wisconsin 53706

V. G. SZEBEHELY Department of Aerospace Engineering, University of Texas, Austin, Texas 78712

T. TAJIMA Institute for Fusion Studies, University of Texas, Austin, Texas 78712

J. TATARONIS Department of Electrical and Computer Engineering, University of Wisconsin, Madison, Wisconsin 53706

JEFFREY L. TENNYSON Electronic Research Laboratory, University of California, Berkeley, California 94720

R. O. VICENTE Department of Applied Mathematics, Faculty of Sciences, Lisbon, Portugal

MIKI WADATI Institute of Physics, College of General Education, University of Tokyo, Tokyo 153, Japan

R. B. WHITE Plasma Physics Laboratory, Princeton University, Princeton, New Jersey 08540

J. C. WHITSON Oak Ridge National Laboratory, Oak Ridge, Tennessee 37830

PREFACE

The subject of this book is the unsolved and important problem of long-time prediction in nonlinear conservative systems. This ubiquitous problem is of concern in modern physics, chemistry, and engineering. Traditionally, research in this area has centered around either integrable systems with a few degrees of freedom or the thermodynamic limit with an infinite number of degrees of freedom. We know that systems with an infinite number of degrees of freedom exhibit irreversibility by virtue of their continuous spectra. Thus "infinite" systems can be expected to approach a thermodynamic equilibrium state. However, for real systems with a finite number of degrees of freedom, the problem is more difficult. One would like to show that such systems are mixing, and yet real systems such as a system of molecules interacting via short-range attractive interactions or a system of ions interacting via the long-range Coulomb potential or even a system of planets interacting via the gravitational interaction are not mixing. Indeed, we know from the Kolmogorov–Arnol'd–Moser (KAM) theorem that the phase space of anharmonic oscillator systems exhibits ordered motion at low energy and *may* become chaotic on a global scale as the energy is raised. General conditions for the onset of chaotic behavior in nonlinear systems are only now being established by researchers in the particular areas that comprise this volume.

The problem of determining regions of instability in nonlinear systems and predicting the long-time behavior in such systems is central to many fields of natural science and engineering. For example, traditional statistical mechanics is based on the assumption that the many-body systems described are ergodic. Once the phase space of a system is sufficiently chaotic (i.e., mixing), the very concept of an individual trajectory loses meaning. In this regime of unstable trajectories it is natural to define operators acting on distribution functions. With this mathematical description it is found that the initial dynamic equations transform into a probabilistic kinetic equation.

In actuality many dynamic systems are not mixing. Still, there may exist classes of states that approach equilibrium. A sufficient condition for this to be true appears to be the validity of the classical Poincaré theorem. Poincaré's theorem implies an *infinite* number of resonances, whereas stochastic behavior may start already with the appearance of a small number of resonances.

The question of dynamic stability when only a few nonlinear resonances are present was recognized early as an essential problem in the confinement of plasmas. The existence of confined trajectories of the magnetic field lines in a torus, where the phase space is the actual three-dimensional configuration space of the laboratory, is predicted by the KAM theorem. The chaotic trajectories between the KAM surfaces are regions of lost confinement, since the electrons follow the chaotic magnetic field lines. The motion of a single ion in a magnetic bottle was an early stimulus to the development of the criterion of overlapping nonlinear resonance for the onset of unstable trajectories allowing the ion to escape the magnetic trap.

Next in complexity from the single-particle motion is the beam–beam interaction problem of accelerator physics. Area preserving maps provide the natural mathematical formulation of this physical problem. The rich variety of dynamics from this simplest of two-dimensional deterministic systems has only recently become fully appreciated, and the search for universal classes of behavior in the transition to chaos is most fully developed for these simple dynamical systems. Perhaps the simplest conservative nonlinear self-consistent field problem, and certainly one of central importance, is the plasma problem posed by the Vlasov–Poisson system of equations. As in statistical mechanics, the theory of trajectory instabilities provides a foundation for the development of a renormalized theory of plasma kinetic theory.

The question of long-time prediction also touches many other problems of considerable interest. The question of how energy is absorbed and distributed in large molecules is not well understood. There are also astrophysical problems of considerable interest. It is not known whether the solar system is stable. The problem of predicting the long-time behavior of natural and artificial satellites and planets continues to be of great importance. Some tools used in celestial mechanics to predict the long-time behavior of celestial objects include the use of integrals to establish regions of possible motions, series expansions that, however, are not uniformly convergent, numerical integration with limited-time duration, study of secular behaviors where unproved approximations must be used, the KAM theory (which unfortunately is not applicable to the solar system), and the Kuiper–Nacozy–Szebehely theory, which attempts to establish long-time behavior with increased perturbation strength and reduced duration of numerical integration. However, even with these techniques, predictions cannot be made very far into the future, and long-time prediction remains an ever-present problem.

The aim of the Workshop on Long-Time Predictions in Nonlinear Conservative Dynamical Systems was to bring together scientists from various areas of the natural and engineering sciences where the problem of long-time prediction is a central issue. The workshop was organized under the auspices of the Center for Studies in Statistical Mechanics, the Center for Orbital Dynamics, and the Institute for Fusion Studies at the University of Texas at Austin, and was made possible by funding from the National Science Foundation, the U.S. Department of Energy, and the University of Texas at Austin. Sessions were conducted on the origins of chaotic behavior in nonlinear dynamical systems, the foundations of statistical mechanics, the internal dynamics of molecules, the stability of the gravitational many-body problem, and particle dynamics in confined plasmas.

We are grateful for the support provided by the foregoing institutions and the excellent quality of the presentations made by participants at the workshop. We offer this volume to our colleagues in the hope that it is the beginning of closer cooperation between the workers of the fields considered.

C. W. HORTON, JR.
L. E. REICHL
V. G. SZEBEHELY

Austin, Texas
May 1982

CONTENTS

4. BEAM–BEAM INTERACTION

LONG-TIME PREDICTION IN DYNAMICS

Part 1

STATISTICAL MECHANICS

MICROSCOPIC DYNAMICS AND MACROSCOPIC LAWS

JOEL L. LEBOWITZ
Institute for Advanced Study, Princeton, New Jersey

Abstract. *Some thoughts about the relation between reversible microscopic dynamics and irreversible macroscopic laws are presented in somewhat sketchy form. It is argued that not only are they compatible but that one may (eventually) be able to derive the latter from the former. The problems and possible resolutions are illustrated (and hopefully illuminated) by means of a "gedanken-experiment" that is analyzed both heuristically and from the point of view of Lanford's rigorous derivation of the Boltzmann equation in a certain well-defined (Boltzmann–Grad) limit.*

1. INTRODUCTION

Time present and time past
Are both perhaps present in time future,
And time future contained in time past.
If all time is eternally present
All time is unredeemable.

T. S. Eliot, *Four Quartets*

I. Prigogine[1] has mentioned the perception of time in a way that recalls the opening lines of one of my favorite poems, quoted above. These lines should remind us that we do not comprehend the universe by logic alone. The poet appears to grapple with the paradox of simultaneous coexistence of past and future, as in a trajectory, despite the asymmetry of time's arrow. This *apparent* paradox is my subject (cf. Refs. 1–6).

There have been many references to coin tossings, roulette wheels, and other games of chance. It is well known that these games require differences of

---

Work supported in part by National Science Foundation grant PHY 78-15920.
Permanent address: Department of Mathematics and Physics
Rutgers University
New Brunswick, New Jersey

opinion—there would not be much betting at a horse race if everyone agreed on the horses. I do not know if anyone is making bets on the eventual resolution of the apparent paradoxes relating to the coexistence, in the description of the same phenomena, of both determinism and randomness, reversibility and time asymmetry, and so on. If there are people betting, however, I would be very happy to be the banker and keep the money until everyone has agreed on the answer.

Let me lay my cards on the table: I agree entirely with Prigogine's statement that it seems inconceivable that the deterministic, reversible microscopic laws (classical or quantum) do not hold for the evolution of the local density in an initially nonuniform fluid. Similarly, I too do not believe that irreversibility is due to approximations; that is, it is not just some small term missing from the diffusion equation that makes it irreversible. There remains, therefore, the question of how to derive the heat equation or the Boltzmann equation from microscopic dynamics. I say "derive" rather than "reconcile" because I do not believe that there is any contradiction, but there certainly is a need for a convincing mathematical derivation. Such a derivation would also help dispel some of the confusion surrounding the subject.

To illustrate, let us consider a somewhat idealized version of a typical time–asymmetric macroscopic event. It contains parts that may not be easy or even possible to achieve in practice (I consider an isolated system, use a classical description, etc.). The relevant question, however, is not simply what idealization can actually be carried out practically, but rather what is the right idealization for understanding rationally and being able to predict what will be observed at a particular level of precision under given circumstances. Misunderstanding of this question leads to statements like one that appeared some years ago in a respected popular magazine, namely, that Aristotle was right (hence Galileo wrong) in asserting that heavy objects fall faster than light ones under the action of gravity—just drop a feather and a penny together.

2. EXAMPLE

Figure 1 illustrates a box Λ, 10 cm on each side, divided into two equal parts connected by a channel. There are altogether $N \simeq 10^{21}$ atoms in the box, and the macroscopic experiment or observation consists in determining the numbers N_1 and N_2 of particles in the left and right parts, Λ_1 and Λ_2, $N_1 + N_2 = N$. Consider first just the left-hand side of the figure. The sequence represents qualitatively a macroscopic experiment carried out as follows. We start with the box with a plug in the hole and fill the left side Λ_1 with a gas, say helium, at room temperature and 0.01 atmospheric pressure. We then wait for a few minutes and at time $t = t_1$, the beginnings of our observations, we remove the plug. The configurations x_j are my imagined ones at the observations times t_j, $j = 1, 2, 3, 4, \ldots$. No surprise here as long as I tell you that $t_1 < t_2 < t_3 < t_4$. The system goes from a highly nonuniform density to a uniform one: precisely the kind of behavior we are used to seeing in such experiments. It is surely consistent with Hamilton's equations of motion for a system of particles interacting with a Lennard–Jones or hard sphere pair potentials. It is also true, and consistent with the microscopic equations, that the evolution of the density, kinetic energy, and other similar quantities can be described *for the times observed* very accurately by an irreversible kinetic equation (e.g., the Boltzmann equation). (If this does not seem credible, try a computer experiment.)

What *is inconsistent* is to say that the behavior of the system will be accurately described by the Boltzmann equation in *all* situations that can be imagined. Thus, if we imagine that at time t_2 we somehow put this system in the microscopic state y_2 obtained from x_2 by reversing all velocities, $y_2 = \bar{x}_2$, then for the time interval (t_2, t_3) we would not observe the density getting more uniform, as would be predicted by the Boltzmann equation. What should happen to such a system is shown on the right-hand side of Fig. 1 including, on top, the microscopic state at t_1, which would give y_2 at t_2 without further intervention. The appropriate question, then, is: Why is it typically or essentially always the case that Boltzmann's equation (or other irreversible equations) makes the right predictions? This has to do with what is going to happen (or is likely to happen) when we carry out experiments or observations of certain types on a *macroscopic* system prepared explicitly or implicitly in a certain way.

In terms of the example given above, why is it that when we observe a gas at t_2 looking as it does in the row x_2 we can quite safely predict that it will follow the course on the left rather than the one on the right? The apparent answer is that we have a prescription for constructing the pictures on the left, but we do not know how to construct the state y_2 or y_1. There are, of course, situations (e.g., the spin-echo experiments of Hahn[2]) in which a spin state "similar" to our y_2 is produced. The spin reversal is, however, considered to be something special. Also the "isolation" of the spin system is relatively low, and interactions with other degrees of freedom in the system, which have not been reversed, means that the effect is limited and of short duration. This has permitted us to accept this occasional "antiuniformization" behavior in spin systems without modifying our predictions about the course of events in general macroscopic systems such as in our example following the observation at time t_2.

Unfortunately, or fortunately, no one has succeeded in reversing all velocities in a macroscopic system, and all our experience corresponds to seeing the sequence on the left rather than the one on the right. We explain this behavior intuitively by saying that the amount of "phase space volume" consistent with a macroscopic observation of N_1 and N_2 and (energy E) increases in the left-hand sequence. But since *each* trajectory on the left has a counterpart trajectory on the right, the preparation of the system at t_1 must somehow be relevant in assigning appropriate probabilities to different configurations, and this is where we need more understanding.

3. SYSTEM ISOLATION

Our failure thus far to take into account outside perturbations makes the example of Section 2 incomplete. However, this is not very relevant for the time scales considered; that is, even if there were an isolated macroscopic system I do not believe that it would behave differently for the times considered. Certainly neither the Boltzmann nor other kinetic equations includes any terms due to walls, cosmic rays, or other outside interactions. It seems, therefore, inappropriate to invoke such outside perturbations in justifying them.

However, Poincaré's famous theorem about isolated systems should be mentioned here (see Ref. 3). The theorem states that if we surround the trajectory of an isolated Hamiltonian system by a tube of "diameter" δ then—since the energy surface has finite area—any point on the trajectory will be inside the tube infinitely often, no matter how small δ. This means if we keep *on* observing the system in Fig. 1 we will see it returning again and again to configurations "close" to x_1, to x_2, and so on. Boltzmann's equation or indeed any equation predicting an approach to uniformity without reversal cannot therefore be an even approximately valid description of an isolated system for *all* times. These "recurrence times" are, however, likely to increase very rapidly with the size of the system—being probably longer than the age of the universe for $N \simeq 10^{21}$. As Boltzmann is supposed to have told Zermelo, who raised this objection to Boltzmann's equation, "You should live so long."[4]

This extrapolation to *arbitrary times* is, therefore, invalid for isolated systems. But since no experiment lasts for a *very long* time, this need not worry us. In this case the idealization of an isolated system is not a useful one. Even small interactions with the "outside" world are likely to destroy completely this recurrence for dynamically unstable systems. Indeed, when we consider ensembles, this is not even a problem for an isolated system if it is at least mixing. For such systems ensembles can and do approach a uniform (coarse-grained) state.[5]

It should also be noted that the dynamical instability of a system's trajectory implies that if the reversal of velocities is not absolutely precise, the subsequent motion might behave very differently from the exactly reversed one. Doing a nonexact velocity reversal might then not have much observable effect on a real system. I shall come back to this point.

4. STATISTICAL MECHANICS

The discussion above is clearly far from conclusive, and to continue in this vein would be likely to add to the confusion. Therefore, we proceed to a more concrete analysis of the problem. The central question is how to describe, in a way amenable (at least in principle) to quantitative study, the observed "typical" behavior of macroscopic systems. The answer to this lies in our beloved statistical mechanics: the study of dynamics combined with probability. Here we replace the study of the microscopic trajectory of a macroscopic system (prepared initially by some macroscopic means) by the study of the time evolution of an initial ensemble. The question is now simpler: How are the appropriate ensembles to be characterized? This question is discussed in detail, although not entirely resolved, in a recent article by Penrose,[6] which I recommend highly. The article also contains a discussion of various approaches to the problem of irreversibility and a very extensive list of references.

Instead of considering the general problem, however, I discuss in more detail a further, more drastic idealization of the foregoing example for which one can actually *prove* some results. This involves consideration of a well-defined limit introduced by Harold Grad (see Ref. 7). It is the appropriate idealization of a dilute classical gas for which the Boltzmann equation ought to hold exactly and, therefore, *might* perhaps be proved rigorously. Grad's program was carried out brilliantly by Oscar Lanford[8] with certain limitations.

I now use the Lanford theorem to make the discussion above more precise, hence, I hope, more clear. The material that follows is from a joint paper with van Beijern, Lanford, and Spohn[9] in which the Lanford theorem is extended to all times for certain initial states corresponding to the motion of a test particle in a dilute gas in equilibrium. In the considerations here, however, I deal with the original theorem, as explicated in King's thesis,[10] so that to be strictly applicable the observation times t_j would have to be much sooner than is indicated in Fig. 1. (Truth is a necessary but not sufficient condition for a mathematical proof.)

5. LANFORD'S THEOREM

We consider a system of hard spheres of diameter ε and unit mass inside a box Λ. The spheres are elastically reflected among themselves and at the boundary of Λ. Let the state of the system be specified by the absolutely continuous distribution functions $\{\rho_n^\varepsilon | n > 0\}$. These satisfy the Bogolubov–Born–Green–Kirkwood–Yvon (BBGKY) equation for hard spheres.[11]

$$\frac{\partial}{\partial t}\rho_n^\varepsilon(x_1,\ldots,x_n,t) = H_n^\varepsilon\rho_n^\varepsilon(x_1,\ldots,x_n,t) + 2\sum_{j=1}^{n}\int_{R^3}dp_{n+1}\int_{S^2}d\omega\,\omega$$

$$\times\big((p_{n+1}-p_j)\rho_{n+1}^\varepsilon(x_1,\ldots,x_n)q_j+\varepsilon\omega,\,p_{n+1},\,t\big) \quad (1)$$

Here

$$x_i = (q_i,\,p_i) \in \Lambda \times R^3$$

ω is a unit vector in R^3 and $d\omega$ is the surface measure of the unit sphere S^2 in three dimensions; H_n^ε describes the evolution of n hard spheres of diameter ε

inside Λ. The solutions of the BBGKY hierarchy are denoted by

$$\rho_n^\varepsilon(x_1,\ldots,x_n,t) = (V_t^\varepsilon \rho^\varepsilon)_n(x_1,\ldots,x_n) \tag{2}$$

for the initial vector of distribution functions

$$\rho^\varepsilon = (\rho_1^\varepsilon, \rho_2^\varepsilon, \ldots)$$

We want to study the low density, Boltzmann–Grad limit of the solutions of the BBGKY hierarchy. This limit is obtained by letting the fraction of volume occupied by the particles $\sim \rho\varepsilon^3$, with ρ the average density, go to zero while keeping the mean free path of the hard spheres, $\sim 1/\varepsilon^2\rho$, constant. This requires that as ε approaches zero, the density is increased as ε^{-2}. Therefore for each value of the diameter ε one chooses an initial state with distribution functions ρ_n^ε such that $\rho_n^\varepsilon \sim \varepsilon^{-2n}$. With this in mind we define the rescaled distribution functions

$$r_n^\varepsilon(x_1,\ldots,x_n) = \varepsilon^{2n}\rho_n^\varepsilon(x_1,\ldots,x_n) \tag{3}$$

Regarding the sequence $\{r_n^\varepsilon | n \geqslant 0\}$ as the vector r^ε, one can write Eq. (1) compactly as

$$\frac{d}{dt} r^\varepsilon(t) = H^\varepsilon r^\varepsilon(t) + C^\varepsilon r^\varepsilon(t) \tag{4}$$

where H^ε is a diagonal matrix with entries H_n^ε, and C^ε is a matrix with entries $C_{n,\,n+1}^\varepsilon$ and zero otherwise.

For $t > 0$ the time evolution of $r_n^\varepsilon(t)$ is determined by backward streaming. Therefore it seems natural to replace, for a collision, the phase point

$$(x_1,\ldots,q_j,\, p_j,\ldots,q_j + \varepsilon\omega,\, p_{n+1})$$

with outgoing momenta by the phase point

$$(x_1,\ldots,q_j,\, p_j',\ldots,q_j + \varepsilon\omega,\, p_{n+1}')$$

with incoming momenta. (These are just two different representations of the same phase point.) This leads to

$$\begin{aligned}
\frac{\partial}{\partial t} r_n^\varepsilon(x_1,\ldots,x_n,t) &= H_n^\varepsilon r_n^\varepsilon(x_1,\ldots,x_n,t) \\
&\quad + \sum_{j=1}^{n} \int_+ dp_{n+1}\, d\omega\, \omega \cdot (p_j - p_{n+1}) \\
&\quad \times \left\{ r_{n+1}^\varepsilon(x_1,\ldots,q_j,\, p_j',\ldots,q_j' - \varepsilon\omega,\, p_{n+1}',\, t) \right. \\
&\qquad \left. - r_{n+1}^\varepsilon(x_1,\ldots,q_j,\, p_j,\ldots,q_j + \varepsilon\omega,\, p_{n+1},\, t) \right\}
\end{aligned} \tag{5}$$

where $\int_+$ indicates that the integration over ω is restricted to the upper hemisphere $\omega \cdot (p_j - p_{n+1}) \geqslant 0$.

Formally, the limiting form of Eq. (5) that might be satisfied for $t \geqslant 0$ by the distribution functions $r(t) = \lim_{\varepsilon\to 0} r^\varepsilon(t)$ is obtained by simply setting $\varepsilon = 0$ in Eq. (5),

$$\begin{aligned}\frac{\partial}{\partial t} r_n(x_1,\ldots,x_n,t) = {} & \sum_{j=1}^{n} p_j \frac{\partial}{\partial q_j} r_n(x_1,\ldots,x_1,t) \\ & + \sum_{j=1}^{n} \int_+ dp_{n+1}\, d\omega\, \omega \cdot (p_j - p_{n+1}) \\ & \times \{ r_{n+1}(x_1,\ldots,q_j,p_j',\ldots,q_j,p_{n+1}',t) \\ & \quad - r_{n+1}(x_1,\ldots,q_j,p_j,\ldots,q_j,p_{n+1},t) \}\end{aligned} \tag{6}$$

(Implicitly, the free motion $-\Sigma_{j=1}^n p_j\, \partial/\partial q_j$ includes the specular reflection at the boundaries of Λ.)

For $t < 0$ the time evolution of $r_n^\varepsilon(t)$ is determined by forward streaming. In that case, for a collision, the phase point with incoming momenta should be replaced by the phase point with outgoing momenta. The formal limit of the resulting equation is then again Eq. (6) but with the sign of the collision term reversed.

Equation (6) for $t \geqslant 0$ (and with the sign of the collision term reversed for $t \leqslant 0$) is *called* the Boltzmann hierarchy, which can be written in the form

$$\frac{d}{dt} r(t) = Hr(t) + Cr(t) \tag{7}$$

Let $r^\varepsilon(t) \equiv V^\varepsilon(t) r^\varepsilon(0)$ and $r(t) \equiv V(t) r(0)$ be the solution of Eqs. (4) and (7), respectively, as defined, for example, by the Dyson series with $r(0) \equiv \lim_{\varepsilon\to 0} r^\varepsilon(0)$.

To prove that $r^\varepsilon(t)$ converges to $r(t)$, for $t \neq 0$, as $\varepsilon \to 0$, we need two conditions.

First, the initial distributions $r^\varepsilon(0)$ must be uniformly bounded in ε. This guarantees the uniform convergence of the Dyson series solution for some interval $|t| < t_0$. If h_β denotes the normalized Maxwellian at inverse temperature β, a suitable choice for this bound is as follows:

Condition 1. There exist a pair (z, β) such that

$$r_n^\varepsilon(x_1,\ldots,x_n) \leqslant Mz^n \prod_{j=1}^{n} h_\beta(p_j) \tag{8}$$

for all $\varepsilon < \varepsilon_0$ with a positive constant M independent of ε.

Second, $r_n^\varepsilon(0)$ must converge to $r_n(0)$ in such a way that the Dyson series for $r^\varepsilon(t)$ converges term by term to the series for $r(t)$. For the initial phase point $x^{(n)} = (x_1,\ldots,x_n) \in (\Lambda \times R^3)^n$, let $q_j(t, x^{(n)})$, $j = 1,\ldots,n$, be the position of the jth point particle at time t under the free motion. Then

$$\Gamma_n(t) = \left\{x^{(n)} = x_1,\ldots,x_n \in \left(\Lambda \times R^3\right)^n | q_i(s, x^{(n)}) \neq q_j(s, x^{(n)})\right\} \quad (9)$$

for $i \neq j = 1,\ldots,n$ and $-t \leqslant s \leqslant 0$ if $t \geqslant 0$, $0 \leqslant s \leqslant -t$ if $t \leqslant 0$.

In words, $\Gamma_n(t)$ is the restriction of the n-particle phase space to the set of phase points that under free backward streaming over a time t, if t is positive (or free forward streaming over a time $|t|$, if t is negative) do not lead to a collision between any pair of particles, regarded as point particles. By this restriction only a set of Lebesgue measure zero is excluded from $(\Lambda \times R^3)^n$.

Note that (i) $\Gamma_n(t)$ depends only on the free motion, (ii) $\Gamma_n(t)\ \Gamma_n(t')$ for $t' = \alpha t$, $\alpha \leqslant 1$, (iii) $\Gamma_n(t) \neq \Gamma_n(-t)$, and (iv) $x^{(n)} \in \Gamma_n(t)$ is equivalent to $\bar{x}^{(n)} \in \Gamma_n(-t)$, where $\bar{x}^{(n)} \equiv Rx^{(n)}$ is the phase point obtained from $x^{(n)}$ under the reversal $p_j \to -p_j$. In particular $\Gamma_n(t)$ is not invariant under reversal of velocities.

The suitable choice of convergence is then as follows:

Condition 2. There exists a continuous function r_n on $(\Lambda \times R^3)^n$ such that

$$\lim_{\varepsilon\to 0} \varepsilon^{2n}\rho_n^\varepsilon = \lim_{\varepsilon\to 0} r_n^\varepsilon = r_n \quad (10)$$

uniformly on all compact sets of $\Gamma_n(s)$ for some $s > 0$.

Theorem (Lanford). *Let $\{\rho_n^\varepsilon | n \geqslant 0\}$ be a sequence of initial distribution functions of a fluid of hard spheres of diameter ε inside a region Λ and let the sequence $\{r_n^\varepsilon | n \geqslant 0\}$ of rescaled distribution functions satisfy Conditions* 1 *and* 2. *Let $r_n^\varepsilon(t)$ be the solution of the BBGKY hierarchy with initial conditions r_n^ε, and let $r_n(t)$ be the solution of the Boltzmann hierarchy with initial conditions r_n.*

Then there exists a $t_0(z, \beta) > 0$ such that for $0 \leqslant t \leqslant t_0(z, \beta)$ the Dyson series for Eqs. (4) *and* (7) *converge and such that $r_n(t)$ satisfies a bound of the form* (*Condition* 1) *with $z' > z$ and $\beta' < \beta$. Furthermore,*

$$\lim_{\varepsilon\to 0} r_n^\varepsilon(t) = r_n(t) \quad (11)$$

uniformly ***on compact sets of*** $\Gamma_n(s + t)$.

For $-t_0(x, \beta) \leqslant t \leqslant 0$, Eq. (11) *holds, provided in Condition* 2 $s \leqslant 0$ *and in the Boltzmann hierarchy the collision term $C_{n,n+1}$ is replaced by $-C_{n,n+1}$.*

Remark. An interesting property of the Boltzmann hierarchy is the well-known "propagation of chaos": if the initial conditions of the Boltzmann hierarchy factorize,

$$r_n(x_1,\dots,x_n) = \prod_{j=1}^{n} f(x_j) \tag{12}$$

then the solutions with this initial condition stay factorized,

$$r_n(x_1,\dots,x_n,t) = \prod_{j=1}^{n} f(x_j,t) \tag{13}$$

$f(x, t)$ is the solution of the Boltzmann equation

$$\frac{\partial}{\partial t} f(q,p,t) = -p\frac{\partial}{\partial q} f(q,p,t) + \int_{+} dp_1\, d\omega\, \omega\cdot(p-p_1) \times \{f(q,p',t)f(q,p'_1,t) - f(q,p,t)f(q,p_1,t)\} \tag{14}$$

with initial condition $f(q, p)$.

Note, however, that even when Eq. (12) is not satisfied, the solutions $r_n(t)$ of the Boltzmann hierarchy are *not* reversible, whereas the $r_n^\varepsilon(t)$ are—so what has happened?

The answer lies in the fact that the set on which the $r_n^\varepsilon(t)$ converge to $r_n(t)$ gets smaller and smaller as t increases, $\Gamma_n(s+t_2) < \Gamma_n(s+t_1)$ for $t_2 > t_1 \geqslant 0$ and $\Gamma(t) \neq \Gamma(-t)$.

When I get confused at this point (this happens at least four out of five times), the following picture is sometimes helpful.

Let $\Gamma(0) = U_{n>0}\Gamma_n(0)$ be represented schematically by the upper quadrant of the plane $x \geqslant 0$, $y \geqslant 0$. Let $\Gamma(t) = U_n\Gamma_n(t)$ correspond to the set $x \geqslant t$, $y \geqslant 0$ for $t > 0$ and to the set $x \geqslant 0$, $y \geqslant t$ for $t < 0$. Then for the case of Condition 2 satisfied on $\Gamma_n(0)$, the convergence at $t = t_2 > 0$ holds on the set $\Gamma(t_2) = \{x > t_2, y > 0\}$ but may not hold on the complementary set $\Gamma^c(t_2) = \{x, y | x < t_2, y > 0\}$. If we now reverse *all* velocities at $t = t_2$, the convergence of the *new* r_n^ε at t_2 will be on $\Gamma(-t_2)$ but will no longer hold on $\Gamma(s) = \{x, y | x > s, y > 0\}$, for *any* $s \geqslant 0$. Therefore since Condition 2 is no longer satisfied for any $s > 0$, the Lanford theorem need not hold for any time $t_2 + \tau$, $\tau > 0$.

Armed with these mathematical weapons, let us now return to our example.

6. EXAMPLE REVISITED

Let us now consider Fig. 1 from the point of view of ensembles. The initial state, at $t = 0$, corresponds to a canonical Gibbs state of N hard spheres of diameter ε all in the left-hand half-box Λ_1. It is clear that since the initial state is invariant to reversal of velocities, its distribution functions $\rho^\varepsilon = (\rho_1^\varepsilon, \varepsilon_2^\varepsilon, \dots)$

satisfy the equality

$$V_t^\varepsilon \rho^\varepsilon = R V_{-t}^\varepsilon \rho^\varepsilon \tag{15}$$

where

$$(R\rho)_n(q_1, p_1, \ldots, q_n, p_n) = \rho_n(q_1, -p_1, \ldots, q_n, -p_n) \tag{16}$$

Furthermore,

$$V_t^\varepsilon(R V_t^\varepsilon \rho^\varepsilon) = \rho^\varepsilon \tag{17}$$

while

$$V_t^\varepsilon(V_t^\varepsilon \rho^\varepsilon) = V_{2t}^\varepsilon \rho^\varepsilon \tag{18}$$

Equation (17) states that if at time t we reverse all velocities, the system, after another time interval t, will return to its initial state in which all the particles are in Λ_1.

Consider now the sequence of initial states with distribution functions ρ^ε in which as ε approaches 0 the number of particles inside Λ_1 increases with fixed $N\varepsilon^2 = z$. Then

$$\lim_{\varepsilon\to 0} \varepsilon^{2n} \rho_n^\varepsilon(x_1, \ldots, x_n) = \lim_{\varepsilon\to 0} r_n^\varepsilon(x_1, \ldots, x_n) = r_n(x_1, \ldots, x_n)$$

$$= \prod_{j=1}^{n} \{x_{\Lambda_1}(q_j) z h_\beta(p_j)\} \tag{19}$$

on $\Gamma_n(0)$, where X_{Λ_1} is the characteristic function of the set Λ_1, and, since Conditions 1 and 2 are satisfied, by Lanford's theorem

$$\lim_{\varepsilon\to 0} \varepsilon^{2n} (V_t^\varepsilon \rho^\varepsilon)_n(x_1, \ldots, x_n) = (V_t r)_n(x_1, \ldots, x_n) = \prod_{j=1}^{n} \{f(x_j, p_j, t)\} \tag{20}$$

on $\Gamma_n(t)$ for $|t| < t_0(z, \beta)$, where $f(x, t)$ is the solution of the Boltzmann equation with initial conditions $f(q, p) = X_{\Lambda_1}(q) z h_\beta(p)$.

Let us now reverse the velocities at time t, $0 < t < t_0/2$, and let us consider $R V_t^\varepsilon \rho^\varepsilon$ as the new initial state. Clearly

$$V_t(R V_t r) \neq r = \lim_{\varepsilon\to 0} V_t^\varepsilon(R V_t^\varepsilon r^\varepsilon) \tag{21}$$

according to Eq. (17), so the limiting r do not have the time reversibility of the r^ε. Indeed, the Boltzmann H-function decreases up to t, remains unchanged by R, and continues to decrease as $R V_t r$ is evolved for a time interval t.

At first sight this seems to contradict Lanford's theorem, which appears to assert that the right-hand side of Eq. (21) should indeed equal the left-hand side. There is, however, no such contradiction, for although

$$\lim_{\varepsilon \to 0} \varepsilon^{2n}(V_t^\varepsilon \rho^\varepsilon)_n = (V_t r)_n \qquad \text{on} \quad \Gamma_n(t) \tag{22}$$

we also have

$$\lim_{\varepsilon \to 0} \varepsilon^{2n}(RV_t^\varepsilon \rho^\varepsilon)_n = (RV_t r)_n \quad \text{on} \quad \Gamma_n(-t) \neq \Gamma_n(t+s), \qquad Vs \geqslant 0 \tag{23}$$

Therefore, continuing in the same time direction as before, the reversal of velocities, $RV_t^\varepsilon \rho^\varepsilon$ no longer satisfies Condition 2 of Lanford's theorem. The theorem asserts nothing about the convergence of $\varepsilon^{2n}(V_t^\varepsilon(RV_t^\varepsilon \rho^\varepsilon))_n$ as $\varepsilon \to 0$. Of course, by Eq. (17) we can say something about this limit. The point is that we cannot conclude from Lanford's theorem that the limit is $(V_t(RV_t r))_n$, since Condition 2 is violated. For the theorem to be applicable with the initial condition at time t, one must consider either $V_t^\varepsilon(V_t^\varepsilon \rho^\varepsilon)$ or $V_{-t}^\varepsilon(RV_t^\varepsilon \rho^\varepsilon)$. In both cases the system evolves further toward equilibrium (e.g., in Fig. 1*a*, right, or Fig. 1*c*, left).

7. CONCLUDING REMARKS

1. Lanford's theorem deals with correlations that are absolutely continuous with respect to Lebesgue measure; see Eqs. (8) and (10). This singling out of Lebesgue measure, though "intuitively" very reasonable, cannot be justified on mathematical grounds alone. However, it may not be essential—there may be other physical conditions that rule out "bad" initial configurations of macroscopic systems prepared in the laboratory or found in nature.

2. The irreversible Boltzmann hierarchy is consistent with the reversible BBGKY hierarchy, since the approximation by the Boltzmann hierarchy is valid only for a particular class of initial states. Condition 2 excludes initial states such as the one just constructed by reversal of velocities. Although the question how to generally characterize good initial states for systems other than very dilute gases remains, this example does illustrate what form such an answer might take, that is, Condition 2. This is time asymmetric in just the right way[6]: the property is preserved under *forward* time evolution V_t^ε but is not invariant under R.

More precisely if Condition 2 is satisfied for some $s > 0$ (say 1 hour), the state evolves for some t, $t > 0$, Condition 2 is still satisfied on the smaller set $\Gamma(s+t)$. If we, however, do a reflection at t then Condition 2 is no longer satisfied for forward times. The same statements are true for $s < 0$, $s < t < 0$. The initial state considered in the example consists of (1) is symmetric under R and (2) has Condition 2 satisfied with $s = 0$. We can, therefore, derive either the forward or backward Boltzmann equation (both leading to identical uniformization of the density as $|t|$ increases), *but* we cannot go first in one direction and then "backtrack" by using R. It is this restriction that permits derivation of irreversible equations from reversible dynamics and symmetric initial conditions. Which way we actually go physically is determined by the fact that we *first* prepare the system (i.e., select the state), *then* observe it.

3. How would our conception of the "arrow of time" change if a method were found for actually reversing velocities in a fluid (or changing appropriate quantum phases)? It is not suggested that all the velocities in the universe be reversed, but only in systems like that of our example—something modest that would permit the right-hand side of Fig. 1 to represent the result of an actual experiment in which the system was isolated beginning with t_1. Would this be just like the spin-echo experiment,[2] which is now almost forgotten, or would this substantially change our concept of time's arrow. Put differently, do the laws of nature, as we understand them at present, exclude the possibility of ever observing the right-hand side of Fig. 1 in a real-life experiment? The instability of trajectories may be relevant here—precluding sufficiently exact

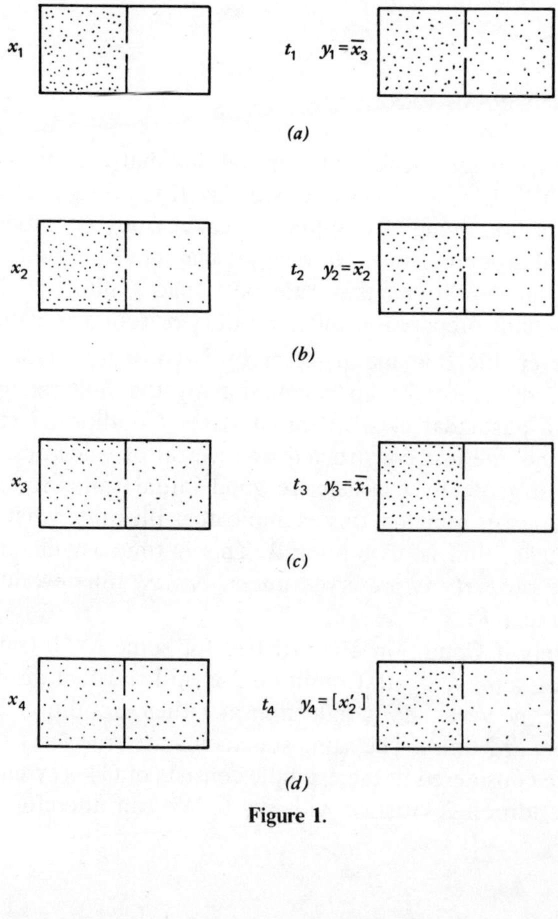

Figure 1.

reversal of velocities to give a macroscopically observable effect—like Maxwell's demon, it would cost more (in entropy) than it would gain.

This instability of trajectories relates to the good ergodic properties that physical systems are believed to possess. Thus despite having played down the role of ergodicity in providing the key to understanding irreversible macroscopic behavior, I nevertheless believe that macroscopic systems generally have good ergodic properties. Their absence would, I think, lead to effects that have not been observed.

Note however that Lanford's theorem never uses the ergodic properties of hard spheres—it is equally valid for cubes with the appropriate modification of the collision kernel. It is thus macroscopic size (remember $N \to \infty$ in the limit) that is essential for Lanford's derivation of the Boltzmann equation.

4. To emphasize the importance of the macroscopic size of the system in the observation of irreversible behavior, note that the example in Fig. 1 would not make much sense if there were only three particles in the system. Systems with few degrees of freedom can certainly exhibit instabilities in their trajectories. They can even be Bernoulli systems, like the baker's transformation, or the point particle moving among fixed convex scatterers—Sinai's billiard. In this case *certain types of initial* ensembles will behave irreversibly but a single trajectory *will not* exhibit irreversible behavior. For a macroscopic system, however, a simple trajectory can give observational results, which we would call irreversible as in the example above.

Thus the "stochastic-type" behavior of trajectories of nonlinear dynamical systems with a few degrees of freedom can serve as only one ingredient in the derivation of kinetic equations describing the time evolution of real macroscopic variables. This is true even though the study of the consequences of good ergodic properties on the behavior of measures absolutely continuous to a given stationary measure may be relevant directly to the behavior of ensembles for macroscopic systems. The situation here is similar to, but much less well understood than, the situation in equilibrium. The use of equilibrium Gibbs ensembles is formally similar for systems of few or many particles, but the relation between ensemble averages and observations is quite different in the two cases; it is only for macroscopic size systems that these can be expected to (approximately) coincide. Also, certain interesting behavior (e.g., phase transitions) shows up in large systems only.

5. Consider again the second row in Fig. 1. Suppose we measure, at time t_2, the numbers N_1, N_2 and also the total energy E of the system. We then want to *predict the future* behavior of this system without knowing "anything else" about its past or future history. Being statistical mechanicians, we would construct an ensemble to represent the initial state at t_2 and use Liouville's equation (which is equivalent to the Hamiltonian equations of motion) for the time evolution of this ensemble. It would seem appropriate to use an initial ensemble that is symmetric under velocity reversal. It is also reasonable that the ensemble be a "smooth" function on the energy surface E.

One such ensemble is $\mu(dx) = \chi(x|N_1, N_2)\,dx$, where dx is the Liouville measure projected on the energy surface S_E, $H(x) = E$ on S_E, and $\chi(x|N_1, N_2)$ is the characteristic function of the set in which there are N_1 particles on the left-hand side and N_2 particles on the right-hand side [i.e., $\chi(x|N_1, N_2)$ is 1 or 0 depending on whether the phase point is consistent with the observation]. This is the so-called generalized microcanonical ensemble, for the use of which (or of its relatives, the generalized canonical or grand canonical ensembles) many "justifications" have been given.[6] None of the arguments is entirely convincing on logical grounds *alone*—but then perhaps neither are the arguments for equilibrium Gibbs ensembles. The important question is how well this will predict the outcome of measurements. It seems clear on the basis of phase space volume arguments that the predictions would favor the left-hand sequence over the right-hand one, at least in a qualitative way. I, furthermore, think that for simple macroscopic systems (e.g., an inert fluid), the prescription would work also quantitatively, after some "short" transient time necessary for the system to establish its own quasi-steady state consistent with the macroscopic constraints. In our example this would presumably be something close to a product state with a one-particle distribution given by the Chapman–Enskog solution of the Boltzmann equation. What the appropriate quasi-steady state ensemble looks like for a more general system, even just a dense gas or an anharmonic crystal, is an open question. It is the *big* question in nonequilibrium statistical mechanics at the present time.[12]

6. Finally, the Boltzmann hierarchy, Eq. (7), does not have underlying it any flow in the phase space. Thus unlike the BBGKY hierarchy, Eq. (4), where the evolution of the states having these correlations can be implemented via the evolution of the phase points, there is no point transformation in the phase space that yields the time evolution of the correlation functions given by Eq. (7). This is yet another manifestation of the "loss of information" resulting from the use of the Boltzmann–Grad limit—a loss necessary to make dissipative macroscopic laws consistent with reversible microscopic dynamics.

ACKNOWLEDGMENTS

It is a great pleasure to thank P. G. Bergmann, S. Goldstein, O. Lanford, O. Penrose, I. Prigogine, the late P. Résibois, and H. Spohn for many useful discussion and arguments about the subject of irreversibility. A paper closely related to this work has appeared in Annals of the New York Academy of Sciences, **373**, 220 (1981).

APPENDIX

The main theme of the Workshop on Long-Time Prediction in Nonlinear Conservative Dynamical Systems appears to have been that in most cases such

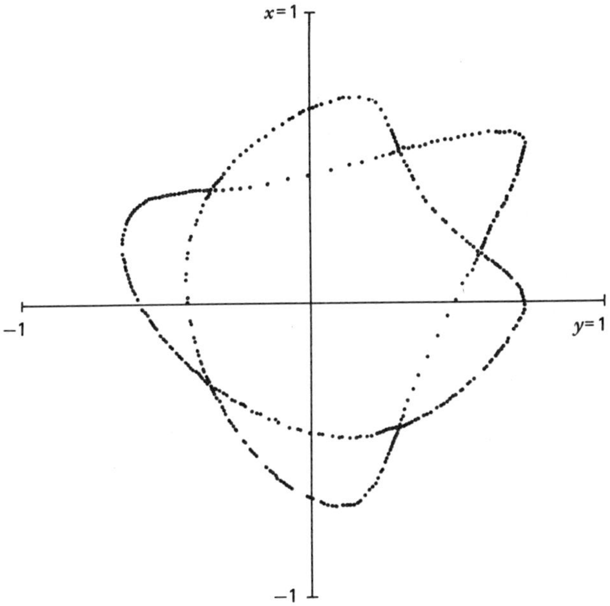

(a)

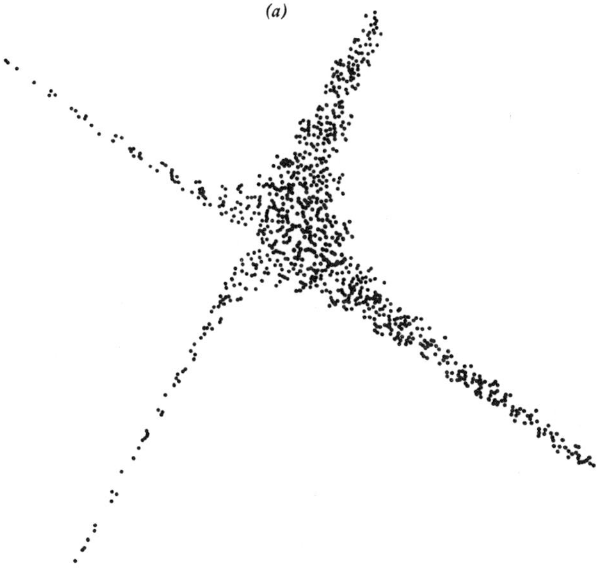

(b)

Figure A1.

prediction is impossible. Unlike the questions about irreversibility, there is general consensus here. The reasons for, or the names given to, this phenomenon are sometimes "instabilities," at other times "sensitivity to initial conditions." Often the condition is simply referred to as "stochasticity of the motion." Therefore, I mention here briefly (or simply record in pictures) a particularly striking example of such an erratic trajectory astutely tracked by Channon[13] in his study of stochasticity in the Hénon area preserving quadratic map.[14] The still life picture of regular and stochastic trajectories of this system, familiar to statistical mechanicians, is presented again in Fig. A1. What I want to show, however, is a "moving" picture, which I think is quite impressive.

The discrete time area preserving evolution in the plane with phase point $x = (q, p)$ is $x_{n+1} = Tx_n$. It is given by

$$q_{n+1} = (\cos\alpha)q_n - (\sin\alpha)q_n + q_n^2(\sin\alpha)$$

$$p_{n+1} = (\sin\alpha)q_n + (\cos\alpha)p_n - q_n^2(\cos\alpha)$$

that is, just a rotation with some shear. Figure A2 shows the trajectory, in steps of T^5, passing through the point $q_0 = 0.718$, $p_0 = 0$ for $\cos\alpha = 0.24$, $\alpha \simeq 2\pi/5$ (the "canonical" stochastic value). The calculations were done with a precision

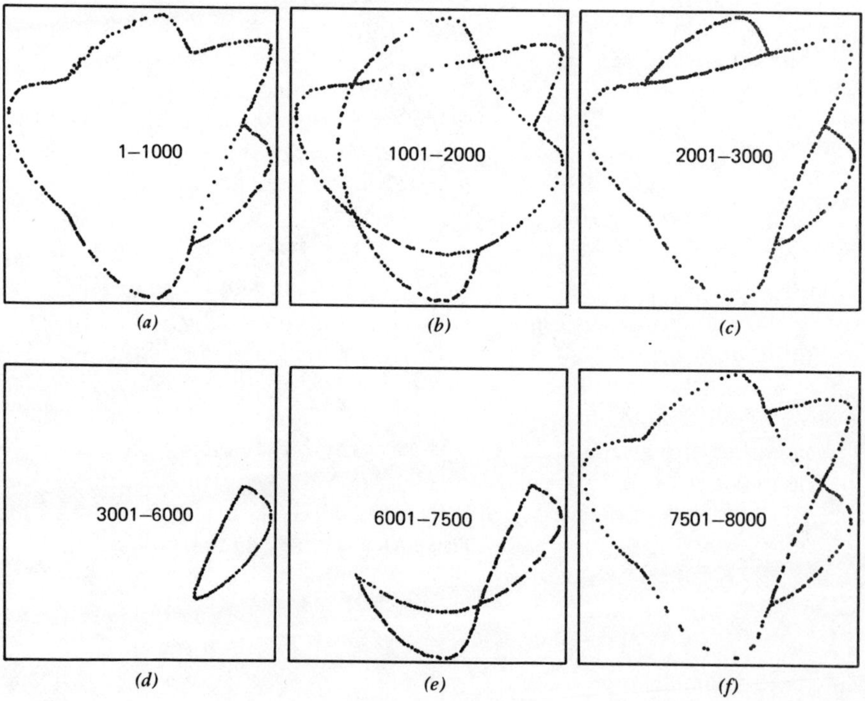

Figure A2.

of 358 decimal digits (this provides reliability for about 8000 steps of T^5). Surely this trajectory is erratic, stochastic, and difficult to predict. More information about this exciting system is provided in the thesis of Channon.[15]

REFERENCES AND NOTES

1. I. Prigogine, lecture at The 1981 Workshop on Long-Time Predictions in Nonlinear Conservative Dynamical Systems (see chapter with B. Misra, this volume); see also *From Being to Becoming* (Freeman, San Francisco: 1980).
2. E. L. Hahn, *Phys. Rev.* **80**, 580 (1950); S. Hartmann and E. L. Hahn, *Phys. Rev.* **128**, 2024 (1962).
3. P. Ehrenfest and T. Ehrenfest, *Encykl. Mat. Wiss.* IV 2.II (Taubner, Leipzig, 1911); English translation, *The Conceptual Approach in Mechanics* (Cornell University Press, Ithaca, NY 1959).
4. For a history of the subject, see S. G. Brush, *The Kind of Motion We Call Heat*, Vols. 1 and 2 (North-Holland, New York, 1976).
5. J. L. Lebowitz and O. Penrose, *Phys. Today*, p. 23 (February 1973).
6. O. Penrose, *Rep. Prog. Phys.* **42**, 1937 (1979).
7. H. Grad, *Principles of the Kinetic Theory of Gases*, in *Handbuch der Physik*, Vol. 12, S. Flugge, Ed. (Springer-Verlag, Berlin, 1958).
8. O. E. Lanford, "Time Evolution of Large Classical Systems," in *Dynamical Systems, Theory and Applications*, J. Moser, Ed., Lecture Notes on Physics, Vol. 38 (Springer-Verlag, New York, 1975).
9. H. van Beyern, O. E. Lanford III, J. L. Lebowitz, and H. Spohn, *J. Stat. Phys.* **22**, 237 (1980).
10. F. King, Ph.D. thesis, Department of Mathematics, University of California, Berkeley (1975).
11. The derivation of Eq. (1) requires some careful analysis, cf. Refs. 8 and 10.
12. Cf. J. L. Lebowitz, *Prog. Theor. Phys. Suppl.* **64**, 35 (1978).
13. S. Channon and J. L. Lebowitz, *Ann. N.Y. Acad. Sci.*, **357**, 108 (1980).
14. M. Henon, *Q. Appl. Math.* **27**, 291 (1969).
15. S. Channon, Ph.D. thesis, Rutgers University, New Brunswick, NJ, 1982.

TIME, PROBABILITY, AND DYNAMICS

B. MISRA AND I. PRIGOGINE

Faculté des Sciences
Université Libre des Bruxelles
Brussels, Belgium

Abstract. *It is shown that a nonunitary equivalence exists between the dynamic and stochastic evolutions of unstable conservative dynamical systems. This leads to a theory of irreversibility and the "arrow of time." The physical origin of these is seen to reside in a restriction on the class of physically realizable states.*

1. INTRODUCTION

In classical (as well as quantum) mechanics time appears in the most elementary form as a numerical parameter serving to label the states occurring in the course of dynamical evolution. Moreover, the laws of dynamical evolution (with the single possible exception of the law relating to the decay of K-meson, which is not directly relevant in the context of this discussion) are themselves invariant with respect to the transformation $t \mapsto -t$ of the time parameter. The dynamical laws are also strictly deterministic. Classical (as well as quantum) mechanics thus envisage a timeless reality in which there is no true change and no intrinsic distinction between past and future.

By contrast, both our conscious experience of time and (thermodynamically) irreversible physical phenomena seem to point to a far richer and subtler concept of time that is endowed with an intrinsic direction, "time's arrow," in Eddington's phrase. Furthermore, it is, in general, meaningless in both classical and quantum mechanics to say that one state is more "aged" than the other. As opposed to this, it is the very essence of the second law governing irreversible thermodynamic evolution to enable one to attribute such "age" to physical states on the basis of their entropy. Time in thermodynamics thus

---

Combined version of the talks presented by the authors at the Workshop on Long-Time Prediction in Nonlinear Conservative Dynamical Systems, March 1981.

I. Prigogine is also the Director of the Center for Studies in Statistical Mechanics, University of Texas, Austin, Texas.

seems to refer to a concept of "internal time" associated with a certain class of physical systems to which the second law applies. Does this concept of oriented and internal time represent a fundamental feature of the physical world, or is it merely a phenomenological concept that is ultimately subjective in origin?

The conventional viewpoint in this respect is that the fundamental laws of physics being symmetric with respect to the time-reversal transformation, there can be no fundamental physical basis for oriented time. The appearance of irreversibility in physical processes and the related "arrow of time" are only the result of statistical averaging (or "coarse graining"), which is necessitated not by any objective aspect of physical phenomena but simply to take into account our ignorance (or lack of interest) of the *exact* dynamical state of the system. Thus Born, for example, asserts that "irreversibility is a consequence of the explicit introduction of ignorance into the fundamental laws."[1]

Recent developments in physics and chemistry, however, make it increasingly difficult to maintain such a view of irreversibility. The progress toward a unified theory of particle interaction, which seems to indicate that all matter is unstable, the discovery of the background black-body radiation and the associated "big bang" theory of the origin of the universe, theories indicating that the so-called black holes are subject to thermodynamic laws, and finally the study of dissipative structures that can emerge in far-from-equilibrium situations, all testify to a growing recognition of the essential and constructive role of irreversibility in fundamental physical processes. In view of these developments, it becomes necessary to reconsider the relation between the dynamical description with its reversible and deterministic laws of evolution on the one hand and the thermodynamic description with its law of monotonic increase of entropy on the other.

The problem, obviously, is that of the existence of a suitable "mechanism" for breaking the time-reversal symmetry of dynamics. Not all forms of violation of time-reversal invariance can, however, lead to the irreversibility expressed in the second law. For this one needs a special form of symmetry breaking that entails a transition from the unitary group describing the reversible dynamical evolution to a semigroup describing an irreversible evolution to which one can associate an H-function or a Lyapunov functional. As is well known, probabilistic Markov processes (with an invariant measure) provide suitable models of such irreversible evolution admitting an H-function. Indeed, if the evolution $\tilde{\rho}_0 \mapsto \tilde{\rho}_t$ (of distribution functions $\tilde{\rho}$) is induced from such a Markov process, then the usual expression for (negative) entropy

$$\int_\Gamma \tilde{\rho}_t \log \tilde{\rho}_t \, d\mu$$

(as well as any other convex functional of $\tilde{\rho}_t$) is an H-function.[2] The important question, thus, is how, and for what class of dynamical systems, can a symmetry breaking transition from dynamics to probabilistic Markov processes occur? We consider this question in the light of modern developments in the theory of (classical) dynamical systems.

In this connection, let us recall that it is in Boltzmann's work that one first sees the deep link between irreversibility and probability. The infusion of probabilistic concepts into dynamics plays so central a role in Boltzmann's work that Gibbs remarked that in reading Boltzmann "We seem rather to be reading in the theory of probabilities."[3] But in Boltzmann's time, dynamics and probability theory were not sufficiently advanced to support Boltzmann's physical insight and, as is well known, the validity of Boltzmann's approach was soon the subject of violent controversy. We need not follow the course of this debate here. Instead, let us turn our attention to the recent developments in the theory of classical dynamical systems and probabilistic processes that have made the situation somewhat clearer than it could be in the late nineteenth century.

One of the most important outcomes of recent advances in the theory of dynamical systems is a growing recognition of the limited scope of the concept of phase space trajectories. Classical mechanics starts by *idealizing* temporal evolution of dynamical systems in terms of the deterministic motion of phase points along phase space trajectories. Implicit in this idealization is the supposition that phase points in initial conditions can be determined with *infinite* precision. This idealization is unobjectionable as long as one deals only with simple dynamical motion (the periodic planetary motion, etc.) for which the phase space trajectories depend on the initial conditions in a continuous manner. In this situation any initial imprecision about initial phase points does not get amplified in an unbounded manner in the course of time. However, as the present symposium testifies, interest has shifted now to dynamical systems exhibiting strong trajectory instability. For such systems *arbitrarily close* initial conditions can give rise to exponentially diverging or qualitatively distinct types of trajectory. Obviously, in this situation the very concept of deterministic motion along phase space trajectories ceases to be a physically meaningful idealization.

In our opinion, this limitation of the concept of phase space trajectories implied by instability of motion is not just a practical limitation but is of conceptual character. This makes it *conceptually* necessary to give up the unphysical idealization of phase points and trajectories and to go to a probabilistic description of physical states in terms of open regions of phase space, or, more generally, Gibbs distribution functions.

Obviously, such a step beyond phase points and trajectories to a probabilistic description of physical states in terms of distribution functions is *necessary* for irreversibility implied in the second law to occur. This by itself is, however, not sufficient. The evolution $\rho_0 \mapsto \rho_t$ of distribution functions is governed by a (unitary) group U_t (or in the infinitesimal form by the Liouville equation) that is induced by the dynamical motion of the phase points. And, of course, no symmetry breaking can occur in this transition from the Hamiltonian description of dynamics in terms of phase space trajectories to the Liouvillian description in terms of the evolution of distribution functions. The usual expressions for (negative) entropy such as

$$\int_{\Gamma} \rho_t \log \rho_t \, d\mu$$

remains constant in the course of the time evolution of the system.

The passage from the Hamiltonian description to the Liouvillian description is, however, not without certain important gains. First, it permits the formulation of concepts in ergodic theory that do express at least some aspects of irreversible thermodynamical evolution. For instance, the concept of mixing dynamical systems expresses that initial conditions are forgotten with time going on. Moreover, Gibbs, who was the first to have arrived at the concept of mixing, has also observed that the very use of the probability concept in the description of physical states implies a sort of time asymmetry. Despite the reversibility, in a mathematical sense, of this evolution of distribution functions, as Gibbs says, "It should not be forgotten when ensembles are chosen to illustrate the probabilities of events in the real world, that while the probabilities of subsequent events may often be determined from the probabilities of prior events, it is rarely the case that probabilities of prior events can be determined from those of subsequent events, for we are rarely justified in excluding the considerations of the antecedent probability of the prior events."[3]

Nevertheless, concepts of ergodic theory and the above-mentioned time asymmetry implicit in the use of the probability concept do not provide the dynamical basis for the characteristic feature of irreversible thermodynamic evolution (i.e., the *monotonic* character of the evolution expressed in the second law). As said before, this aspect can be described most naturally in terms of a symmetry breaking that causes the unitary group U_t of dynamics to be "realized" as a dissipative semigroup associated with a probabilistic Markov process.

Such a symmetry breaking would occur, for example, if for some reason not all distribution functions, but only a certain proper subset of them, could be physically observed or realized. Then the physically realized evolution would be described not by the unitary group U_t, but by the "restriction" or projection of U_t to this subset. Furthermore, if the subset of physically realizable distributions were asymmetric with respect to time-reversal transformation in an appropriate manner, one might expect the projected evolution to reduce, in one time-direction [say $(t \geqslant 0)$], to the semigroup evolution associated with a probabilistic Markov process. Mathematically, the question of the possibility of such a symmetry breaking is the question of the existence of a suitable operator Λ such that the evolution $\Lambda\rho_0 \equiv \hat{\rho}_0 \mapsto \Lambda U_t \rho_0 \equiv \hat{\rho}_t$ of the transformed states is described for $t \geqslant 0$ by a strongly irreversible Markov semigroup W_t^* (for details see Section 2).

The idea of a possible nonunitary equivalence between dynamic and stochastic evolutions has been introduced earlier by one of us and his co-workers in the study of kinetic theory. This study implies an N-body problem and the taking of the thermodynamic limit $N \to \infty$, $V \to \infty$, which leads to complicated mathematical problems. Therefore it is important that a new approach to the problem summarized in this chapter establish this possibility rigorously for a well-defined class of dynamical systems.

The symmetry breaking can take two different forms according to the assumption made on the operator Λ that relates ρ to $\hat{\rho}$. First, Λ may be invertible, in which case the dynamical group U_t is *nonunitarily equivalent* or *similar* to a strongly irreversible Markov semigroup. In the other case Λ is a projection, say, P_0 and the projected evolution $P_0 U_t P_0$ is a strongly irreversible Markov semigroup for $t \geqslant 0$.

At this point let us emphasize that the symmetry breaking projection operator that we are considering is distinct from the projection operator employed in the usual scheme in nonequilibrium statistical mechanics for deriving the so-called generalized master equation from dynamics. This latter projection is taken to correspond to an operator of "coarse graining" or contraction of description that is *independent of dynamics and does not break the time-reversal symmetry*. The resulting projected evolution then obeys the so-called generalized master equation, which is, in general, non-Markovian. To restore the Markovian character of the evolution, one needs to resort to further asymptotic approximation schemes (the weak coupling limit, etc.).

In contrast, the symmetry breaking mechanism we are discussing asks for a projection operation that itself breaks the time-reversal symmetry and leads directly to the master equation of a Markov process without the necessity of resorting to special approximation schemes. In this sense, we are considering the possibility of deriving an *exact* Markovian master equation that does not depend on special approximation schemes.

Naturally, one must ask whether such a procedure of symmetry breaking (either through an invertible Λ or a projection) leading from deterministic dynamics to probabilistic Markov processes is at all possible for conservative dynamical systems. As already mentioned, the possibility of a nonunitary equivalence or similarity between dynamical groups and dissipative semigroups associated with irreversible processes was already considered in the "subdynamics" approach to the foundation of kinetic theory.[4] A rigorous proof of this possibility was given, however, only recently for a class of dynamical systems called the K-flows in modern ergodic theory.[5,6] Subsequently it was found that the class of dynamical systems for which a symmetry breaking *projection* leading to an *exact* master equation (for all $\rho \in L^2_\mu \cap L^1_\mu$) exists is identical with the class of K-flows.[7,8] The K-flows are known to include several systems of physical interest, such as the system of hard spheres within a box, the Lorentz gas model, and geodesic flow on a manifold of negative curvature.[9–11]

The existence of a link between instability of dynamical motion and irreversibility has been intuitively grasped in several early works, notably that of the Russian physicist Krylov.[12] The results mentioned above make this link precise and are of obvious interest in nonequilibrium statistical mechanics. They show how (thermodynamic) irreversibility can be the manifestation of a special form of symmetry breaking entailed by limitations on realizable physical states of the dynamical system.

The possibility of symmetry breaking as described here is based on the existence of an internal time operator* introduced in Ref. 13. Briefly, this is an operator acting on the distribution functions of the phase space that is canonically conjugate to the generator of motion (Liouvillian). Its existence permits one to attribute (average) internal time (or age) to individual distribution functions in such a manner that advance in internal age corresponds to decrease in H-function of the system (or increase of entropy). The internal time operator may thus serve as a microscopic model of the phenomenological concept of thermodynamic time (for a discussion of thermodynamic time, see Ref. 14).

Independently of the "thermodynamic" context, the significance of these results seems to extend beyond their application in statistical mechanics. As is evident from chapters in this volume, there is at present a growing interest in what may be called randomness of dynamical origin. Our result on nonunitary equivalence between dynamical motion and stochastic Markov processes contributes toward an understanding of the precise nature of this "dynamic randomness." It shows that in the presence of suitably strong forms of instability (expressed, e.g., by the K-flow condition) the (deterministic) dynamical motion can be transformed into the stochastic evolution of a Markov process simply through a "change of representation" that involves no loss of information. Such systems may hence be said to be *intrinsically random*. The demonstration of the existence of intrinsically random dynamical systems in this sense shows that the appearance of the probability concept in the description of physical phenomena need not involve the "introduction of ignorance into the fundamental laws."

Finally, let us briefly turn to the question of the fundamental physical basis for the so-called arrow of time. The second law of thermodynamics has two aspects. First, it implies a symmetry breaking transition from dynamics to irreversible probabilistic processes with which one can associate an H-function. The theory of symmetry breaking described in this chapter fully accounts for this aspect of the second law. But the second law also endows time with a "preferred" direction. This aspect of the second law or the existence of a "preferred" direction of time is, however, not described by the existence of a symmetry breaking transition from dynamics to Markov processes, for the very formulation of the idea of symmetry breaking as described here assumes the "correct" or "preferred" direction of time being given. Note that we require the symmetry breaking P or Λ to be such that a strongly irreversible

*In contrast, the symmetry breaking as described in the subdynamics approach is based on the study of the analytical continuation of the resolvent of the Liouville operator.

Markov process results for $t \geqslant 0$. If one wished, one could also consider a *necessarily different* symmetry breaking operation that would yield an *exact* master equation for the reverse direction of time. The reversibility of the dynamical group implies that if symmetry breaking is possible for one direction of time, it is also possible for the reverse direction.

In this respect the situation is exactly similar in the usual approach to deriving the Boltzmann equation or a master equation. In the usual approach to deriving the master equation, the symmetry breaking occurs at the stage of a special asymptotic limit (e.g., the so-called $\lambda^2 t$ limit); and one could consider a suitable limit to yield a master equation in either the "forward" direction ($t \geqslant 0$) or the "backward" direction ($t \leqslant 0$) of time. Thus neither the usual approach nor the theory of symmetry breaking presented here answers the question of the "preferred" direction of time or "time's arrow." This fundamental question is related to the problem of *intrinsic* physical distinction (if any) between the two possibilities of symmetry breaking corresponding to the opposite directions of time. We shall not discuss this question in detail here, only offering a few brief remarks in Section 4.

We now proceed to a more precise and detailed formulation of the ideas described above.

2. FORMULATION OF THE PROBLEM

Consider an abstract dynamical system $(\Gamma, \mathfrak{B}; \mu, S_t)$. Here Γ denotes the phase space of the system equipped with a σ-algebra $\mathfrak{B}$ of measurable subsets, and S_t a group of measurable transformations mapping onto itself and preserving the measure μ. For example, Γ could be the energy surface of a classical dynamical system, S_t the group of dynamical evolution and μ the invariant measure whose existence is assured by Liouville's theorem. For convenience we assume the measure μ to be normalized: $\mu(\Gamma) = 1$. As is well known, the evolution $\rho \mapsto \rho_t$ of density functions under the given deterministic dynamics is described by the unitary group U_t induced by S_t:

$$\rho_t(\omega) \equiv (U_t \rho)(\omega) = \rho(S_{-t}\omega)$$

The generator L of the unitary group U_t is called Liouvillian operator of the system $U_t = e^{-iLt}$. It is given by

$$L_\rho = i[H, \rho]_{\text{P.B.}}$$

if the evolution is generated by the Hamiltonian function H. Here $[\ ,\ ]_{\text{P.B.}}$ denotes the usual Poisson bracket.

Every measure preserving deterministic evolution S_t thus defines a unitary group. Conversely [under certain mild assumptions about the measure space $(\Gamma, \mathfrak{B}, \mu)$], every unitary group that preserves positivity (i.e., maps nonnegative functions to nonnegative functions) and leaves the constant functions unchanged is induced by a group S_t of measure preserving transformations on Γ.[15]

On the other hand, stochastic Markov processes on the state space Γ, preserving μ, are associated with contraction semigroups of $L^2_\mu(\Gamma)$.[16] In fact, let $p(t, \omega, \Delta)$ denote the probability of transition from the point $\omega \in \Gamma$ to the region Δ in time t. Then the operators W_t defined by:

$$(W_t f)(\omega) = \int_\Gamma f(\omega')p(t, \omega, d\omega')$$

form a contraction semigroup for $t \geqslant 0$.

Moreover, W_t has the following properties:

Condition i. W_t preserving positivity (i.e., $f \geqslant 0$ implies $W_t f \geqslant 0$ for $t \geqslant 0$).

Condition ii. $W_t \cdot 1 = 1$.

The evolution of the distribution functions $\tilde{\rho}$ under the Markov process is described now by the adjoint semigroup W_t^*, which also preserves positivity because W_t does: $\tilde{\rho}_0 \mapsto \tilde{\rho}_t \equiv W_t^* \tilde{\rho}_0$. Since the measure μ is an invariant measure for the process (or equivalently, the *microcanonical distribution function* 1 is the equilibrium state of the process), we also have:

Condition iii. $W_t^* \cdot 1 = 1$.

Every Markov process on Γ with stationary measure μ is thus associated with a contraction semigroup satisfying Conditions 1–3. Conversely every contraction semigroup W_t on L^2_μ satisfying these conditions comes from a stochastic Markov process, the transition probabilities $p(t, \omega, \Delta)$, being given by

$$p(t, \omega, \Delta) = (W_t \varphi_\Delta)(\omega)$$

where φ_Δ denotes the characteristic (or indicator) function of the set Δ.

In the following we are interested in a special class of Markov processes whose semigroups W_t satisfy, in addition to Conditions i–iii the final condition:

Condition iv. $\|W_t^* \tilde{\rho} - 1\|^2$ decreases strictly monotonically to zero as $t \to +\infty$, for all states $\tilde{\rho}$ (i.e., all nonnegative distribution functions $\tilde{\rho}$ with $\int_\Gamma \tilde{\rho}\, d\mu = 1$). This condition expresses the requirement that any initial state $\tilde{\rho}$ tends *strictly monotonically* in time to the equilibrium distribution 1. For such processes the functional

$$\int_\Gamma \tilde{\rho}_t \log \tilde{\rho}_t \, d\mu; \qquad \tilde{\rho}_t \equiv W_t^* \tilde{\rho}_0$$

and indeed any other convex functional of $\tilde{\rho}_t$ is an H-function.

Semigroups satisfying the Conditions i–iv are called strongly irreversible Markov semigroups.

The problem before us is to determine the class of dynamical systems for which one can construct a bounded operator Λ having the following properties:

Condition i. Λ preserves positivity.

Condition ii. $\Lambda \cdot 1 = 1$

Condition iii. $\int_{\Gamma} \Lambda \rho \, d\mu = \int_{\Gamma} \rho \, d\mu$

Condition iv. The dynamical group $U_t = e^{-iLt}$ satisfies the intertwining relation

$$\Lambda U_t = W_t^* \Lambda \qquad t \geqslant 0$$

with a *strongly irreversible Markov semigroup* W_t.

The meaning of Conditions i–iii is obvious: expression of the requirements that the transformation $\rho \mapsto \Lambda\rho$ map states to states and leave the equilibrium state unchanged. The intertwining condition (iv), on the other hand, expresses that the transformation $U_t\rho \equiv \rho_t \mapsto \Lambda\rho_t \equiv \tilde{\rho}_t$ must bring about the desired form of symmetry breaking: the transformed evolution $\tilde{\rho}_0 \mapsto \tilde{\rho}_t$ obeys for $t \geqslant 0$ the master equation corresponding to a strongly irreversible Markov semigroup W_t.

We consider two cases. In the first case, Λ has a densely defined inverse Λ^{-1}. Existence of such a Λ means that the dynamical group U_t is *similar* (nonunitarily equivalent) to a strongly irreversible Markov semigroup: $\Lambda U_t \Lambda^{-1} \equiv W_t^*$ for $t \geqslant 0$. Dynamical systems admitting such a Λ may be said to be *intrinsically random*. For such systems a "change of representation" of dynamics $\rho_t \mapsto \Lambda\rho_t = \tilde{\rho}_t$ involving "no loss of information" (expressed by the invertibility of Λ) can convert the dynamical motion into that of a stochastic Markov process.

In the other case we consider, Λ is a projection operator P. The transformation $\rho_t \mapsto P\rho_t \equiv \tilde{\rho}_t$ may now be considered to be a generalized form of "coarse graining" that eliminates from ρ_t physically unobservable or "uncontrollable correlations." The intertwining condition now is equivalent to the requirement that the restriction PU_tP of the dynamical group U_t to the subspace of physically observable states (of the form $P\rho$) is a strongly irreversible Markov semigroup for $t \geqslant 0$. Let us reemphasize that we are considering not the usual form of "coarse graining" employed in the derivation of, say, non-Markovian generalized master equations but the existence of a symmetry breaking projection that leads directly to an *exact* master equation.

The time reversibility of the dynamical motion U_t implies, of course, that if there exists a symmetry breaking projection P_+ (or a similarity Λ_+) such that $W_t^{(+)} \equiv P_+ U_t P_+$ (or $\Lambda_+ U_t \Lambda_+^{-1}$) is a strongly irreversible Markov semigroup for the "forward" direction of time ($t \geqslant 0$), there also exists another projection P_- (or a similarity Λ_-) for which $W_t^{(-)} \equiv P_- U_t P_-$ (or $\Lambda_- U_t \Lambda_-^{-1}$) will be a strongly irreversible semigroup for the "backward" direction $t \leqslant 0$. The important point, however, is that the symmetry breaking projections P_+ and P_- (or Λ_+ and Λ_-) corresponding to the two directions of time must necessarily be distinct. In fact, it can be easily verified that if $P_+ U_t P_+$ is a strongly irreversible Markov semigroup for $t \geqslant 0$, then $P_+ U_t P_+$ is unitary (in the subspace of P_+) for $t < 0$, hence $P_+ U_t P_+$ cannot represent, in the backward direction of time, an irreversible stochastic evolution. Similarly, if $\Lambda_+ U_t \Lambda_+^{-1}$ represents the irreversible stochastic evolution of a Markov process for $t \geqslant 0$, then $\Lambda_+ U_t \Lambda_+^{-1}$ cannot even be positivity preserving, hence cannot represent any kind of physical evolution for $t < 0$.[15]

The symmetry between the two directions of time is thus broken by the *choice* of a symmetry breaking projection or similarity. But as we have seen, this choice between P_+ and P_- (or between Λ_+ or Λ_-) and the corresponding semigroups $W_t^{(+)}$ and $W_t^{(-)}$ presupposes that the "correct" direction of time is known. Thus, while the existence of a symmetry breaking projection P (or similarity Λ) expresses, at the dynamical level, the characteristic features of thermodynamic irreversibility, it cannot serve to define an intrinsic direction of time or the "arrow of time." As we have said, this is the further question of *intrinsic* physical distinction (if any) between P_+ and P_-.

3. INTERNAL TIME AND SYMMETRY BREAKING

If the projected evolution PU_tP to the subspace of states of the form $P\rho$ is to behave asymmetrically with respect to the two directions of time, one expects that the states $P\rho$ themselves must, in some suitable sense, contain the time asymmetry. To give meaning to such a notion of time-asymmetric states, one needs the notion of an "internal time" operator that permits the attribution (averaging) of "age" or "internal time" to individual states. In this section we define this concept and discuss the close connection between the existence of a symmetry breaking transition from dynamics to probabilistic processes and the existence of an internal time operator for the system. For further details see Refs. 5–8, and 13.

Let $\mathcal{H}_{-\infty}$ denote the one-dimensional subspace of constant functions on the phase space Γ, $P_{-\infty}$ the projection onto $\mathcal{H}_{-\infty}$, and $\mathcal{H}_{-\infty}^{\perp}$ the subspace of $L_\mu^2(\Gamma)$ that is orthogonal to $\mathcal{H}_{-\infty}$. By an internal time operator T of the (abstract) dynamical system with dynamical group U_t, we mean a self-adjoint operator T on $\mathcal{H}_{-\infty}^{\perp}$ satisfying the two following conditions:

Condition a. $U_t^* T U_t = T + tI$
(The infinitesimal form of this relation is the familiar canonical commutation relation $[T, L] = iI$ (on $\mathcal{H}_{-\infty}^{\perp}$) between the Liouvillian L and T.)
Condition b. The projections $P_\lambda = E_\lambda + P_{-\infty}$ of L_μ^2 preserve the positivity of functions: that is, $P_\lambda \rho \geqslant 0$ a.e. if $\rho \geqslant 0$ a.e.*. Here E_λ (λ real) denotes the

*a.e. = almost everywhere

spectral projections of T, that is, E_λ is an increasing family of projections such that: (i) $E_\lambda \to 0$ $(\lambda \to -\infty)$ (ii) $E_\lambda \to I_{\mathcal{H}^\perp_{-\infty}}$ $(\lambda \to +\infty)$, and (iii) given functions f and g (in the domain of T)

$$\langle f, Tg\rangle = \int_{-\infty}^{\infty} \lambda d\langle f, E_\lambda g\rangle$$

We have used here, as elsewhere in this chapter, the usual inner product notation that denotes, for example, the integral $\int_\Gamma \bar{f} g\, d_\mu$ by $\langle f, g\rangle$.

If such an internal time operator T exists, one can consistently interpret the quantity

$$\langle T\rangle_\rho \equiv \frac{\langle \bar{\rho}, T\bar{\rho}\rangle}{\langle \bar{\rho}, \bar{\rho}\rangle}$$

(with $\bar{\rho} = \rho - 1$ being the departure from microcanonical equilibrium ensemble 1), as the average internal time or "age" of the (Gibbsian) ensemble 1. Condition a on T then expresses the desirable consistency requirement that in the course of dynamical evolution the system's *average age* advance in step with the increase in the external time parameter t: that is, if $\rho_t = U_t \rho_0$ then $\langle T\rangle_{\rho_t} = \langle T\rangle_{\rho_0} + t$.

As in the case of quantum mechanical observables, the time operator T does not in general permit us to attribute a *definite* "age" to a distribution function. Definite "ages" can be attributed only to the eigenfunctions of T. Corresponding to the fact that a distribution function ρ is, in general, the "superposition" of several eigenfunctions of T, there will be associated with ρ not a definite "age" but a statistical distribution of possible "ages," the average being given by $\langle T\rangle_\rho$. The projection operator $P_\lambda = E_\lambda + P_{-\infty}$ associated with T may be interpreted as the projection to subspace of states, all whose possible "ages" lie in the interval $(-\infty, \lambda]$. The projection P_λ may, thus, be called the "projection to the past of λ." In particular, the projection P_0 will simply be called the projection to the past.

The positivity condition on P_λ (Condition b) means that the projection operation to this past maps states (i.e., nonnegative and normalized distribution) to states. Besides being a natural requirement, this condition is indispensable if P_λ is to be used for constructing symmetry breaking transformations Λ or P with properties stated in the preceding section.

Condition b is independent of condition a. In fact, the existence of a T satisfying condition a alone is known to be equivalent to a spectral property of the Liouvillian L: namely, that L (restricted to $\mathcal{H}^\perp_{-\infty}$) have absolutely continuous spectrum of uniform degeneracy extending over the entire real line. Condition b, on the other hand, cannot follow from merely *spectral properties* of L. The further restriction on dynamics that is imposed by the existence of T satisfying both Conditions a and b is discussed below. Here, let us study the connection between the possibility of constructing an internal time operator and the existence of symmetry breaking transitions from dynamics to stochastic processes. This connection is described by the following theorem.

4. THEOREM

1. Let the dynamical group admit a symmetry breaking projection P (i.e., a projection P mapping states to states and such that PU_tP is a strongly irreversible Markov semigroup for $t \geqslant 0$). Assume further that P is sufficiently large in the sense that given a state ρ and $\varepsilon > 0$ there exists a state ρ' and $t > 0$ such that $\| \rho - U_t P \rho' \| < \varepsilon$. In other words, the set of states in the subspace of P evolves in time to give rise to all possible states.) Then the dynamical system admits an internal time operator T whose "projection operator to the past" coincides with P.

2. Conversely, if the dynamical system U_t admits an internal time operator T, the associated "projection operator to the past" P_0 is a symmetry breaking projection for U_t.

3. Moreover, if an internal time operator T exists, the system is *intrinsically* random; that is, there exists a state preserving operator Λ with densely defined inverse Λ^{-1} such that $\Lambda U_t \Lambda^{-1}$ is a strongly irreversible Markov semigroup for $t \geqslant 0$.

A proof of part 3 of this theorem is essentially contained in Ref. 13. Let us only mention that the symmetry breaking invertible transformation can be constructed as an operator function of T. More specifically, Λ is of the form:

$$\Lambda = \int_{-\infty}^{\infty} h(\lambda)\, dE_\lambda + P_{-\infty}$$

where E_λ denotes the spectral projections of T and $h(\lambda)$ is a function satisfying the following conditions:

Condition i. $h(\lambda)$ is strictly monotonically decreasing with $\lim_{\lambda \to +\infty} h(\lambda) = 0$ and $\lim_{\lambda \to -\infty} h(\lambda) = 1$.

Condition ii. $[h(\lambda + s)]/[h(\lambda)] \equiv \tilde{h}_s(\lambda)$ is monotonically decreasing function of λ for every $s \geqslant 0$. The function $h(\lambda)$ can, for example, be the function $e^{-(e^\lambda)}$.

The proofs of parts 1 and 2 of the theorem are fairly straightforward. As an illustration, let us consider the proof of part 2. Let P_0 be the "projection to the past" associated with the internal time operator T. We must show that $P_0 U_t P_0$ has the semigroup property for $t \geqslant 0$ and satisfies the strong irreversibility condition $\| P_0 U_t P_0 \rho - 1 \|^2 \to 0$ as t approaches infinity. (The positivity preserving property of $P_0 U_t P_0$ follows automatically from the fact that P_0 and U_t

do so.) Obviously $P_0U_tP_0$ will form a semigroup for $t \geqslant 0$ if $P_0U_tQ_0 = 0$ for $t \geqslant 0$, where $Q_0 = I - P_0$. Consider now $P_0U_tQ_0U_t^*$:

$$P_0U_tQ_0U_t^* = P_0 - P_0U_tP_0U_t^*$$

At this point one may use the equivalence of the relation $U_t^*TU_t = T + tI$ (on $\mathcal{H}_{-\infty}^{\perp}$) between T and U_t to the following so-called *imprimitivity relation*

$$U_tP_\lambda U_t^* = P_{\lambda+t}$$

between the "projections to the past of λ" P_λ associated with T. Thus $P_0U_tP_0U_t^* = P_0P_t$. Since $P_t \geqslant P_0$ if $t \geqslant 0$, it follows that $P_0P_t = P_0$ for $t \geqslant 0$. Thus $P_0U_tQ_0U_t^*$, hence also $P_0U_tQ_0$ vanish for $t \geqslant 0$. Similarly, the strong irreversibility condition follows because

$$\begin{aligned}\|P_0U_tP_0\rho - 1\|^2 &= \|U_t^*(P_0U_tP_0\rho - 1)\|^2\\ &= \|P_{-t}P_0\rho - 1\|^2\\ &= \|P_{-t}\rho - 1\|^2\end{aligned}$$

where the last equality follows because $P_{-t}P_0 = P_{-t}$ for $t \geqslant 0$. The last term tends montonically to zero with $t \to \infty$ because P_{-t} decreases monotonically to the projection $P_{-\infty}$ on to the equilibrium state 1 as t approaches infinity.

The proof of part 1 of the theorem is essentially a reversal of the steps of the preceding argument.

Before discussing the properties of dynamical systems admitting an internal time operator, let us briefly comment on the possible physical meaning of internal time. We remarked in the introduction that the internal time operator seems to have the right properties to serve as a microscopic model for the phenomenological concept of thermodynamic time. Moreover, one can show that in a certain suitable sense, internal times associated with states ρ reflect the "degree of indistinguishability" of ρ from the equilibrium state. The more advanced the "age," the more indistinguishable is the state from equilibrium state. But one may still raise the question of the *microscopic* meaning of statistically distributed possible internal times or "ages" that are associated with individual states.

Now the description of physical states in terms of a (Gibbs) distribution function ρ is a mathematical shorthand for two types of information: the information about the velocity distribution of the particles (in our kinetic theory this is called the "vacuum of correlations") and the hierarchy of correlations between the particles, which depends *both* on the positions and the velocities.

We have therefore to expect some relation between the internal time and the correlations that may be the result of past events ("postcollisional" correlations) or the cause for future events ("precollisional" correlations). For instance, it is to be expected that the *projection* operation P_0 to the *past* eliminates all *future-directed* ("precollisional") correlations that would cause future "collisions." We return to this question in Section 5, where we shall find, at least in a qualitative manner, such a link.

5. K-FLOWS AND SYMMETRY BREAKING TRANSITIONS TO MARKOV PROCESSES

It can be shown that the necessary and sufficient condition for the existence of an internal time operator is that the dynamical system be a K-flow. In view of the theorem given in Section 4, this means that the K-flow condition is also both necessary and sufficient for the existence of a symmetry breaking projection leading to an exact master equation. The existence of an internal time operator for K-flows also implies that K-flows are *intrinsically random*; that is, the dynamical group U_t induced from a K-flow is (for $t \geqslant 0$) nonunitarily equivalent, through a state preserving and invertible similarity transformation, to the semigroup associated with a strong irreversible Markov process.

We shall not stop here to supply the proofs of the statements above (the interested reader may see Refs. 5–8 and 13). Let us only describe briefly the concept of K-flow[17] and the construction of the symmetry breaking projection in terms of associated K-partitions.

A K-flow is by definition a dynamical system $(\Gamma, \mathfrak{B}, \mu, S_t)$ for which there exists a distinguished (measurable) partition ξ_0 of the phase space into disjoint cells such that:

Condition i. $S_t\xi_0 \equiv \xi_t \geqslant \xi_s$ if $t \geqslant s$

Here ξ_t is the partition into which the original partition ξ_0 is transformed under dynamical evolution in time t. The notation $\xi_t \geqslant \xi_s$ signifies that the partition ξ_t is finer than ξ_s (i.e., every cell of ξ_t is entirely contained in some single cell of ξ_s).

Condition ii. The (least fine) partition $V_{t=-\infty}^{\infty}\xi_t$, which is finer than each ξ_t $(-\infty < t < \infty)$, is identical with the partition of the phase space into individual phase points.

Condition iii. The (finest) partition $\cap_{t=-\infty}^{\infty}\xi_t$ that is *less fine* than each ξ_t is the trivial partition consisting of a cell of full measure 1. (A partition ξ_0 with the stated properties is called a K-partition.)

Another mathematically equivalent characterization of K-flows is that they have completely positive Kolmogorov entropy. Without defining Kolmogorov entropy, we simply say that complete positivity of Kolmogorov entropy means that the knowledge about the past history of the system obtained from an infinite repetition of any realistic measurement that corresponds to a partition

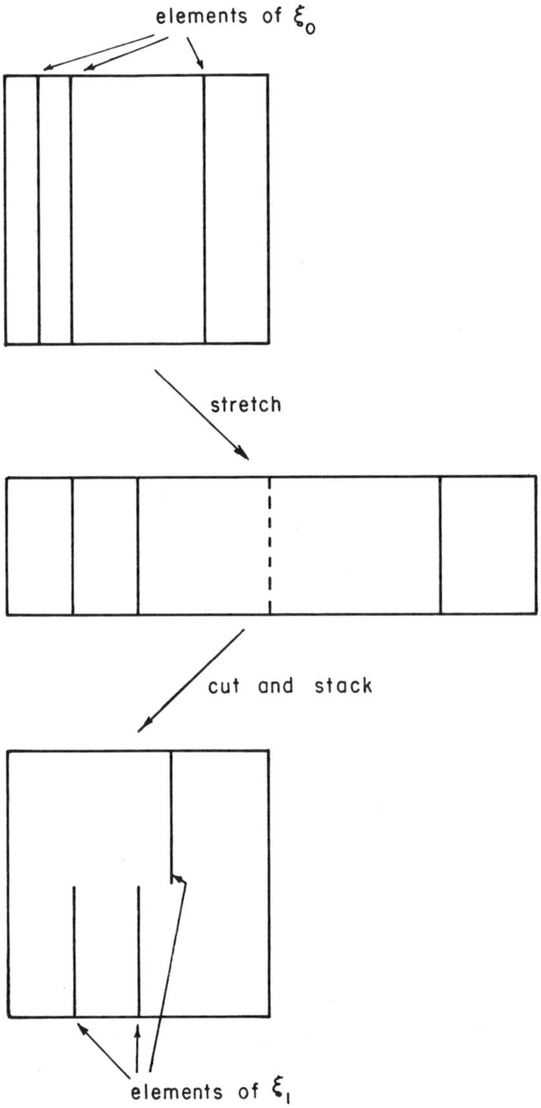

Figure 1.

of the phase space into a *finite* number of disjoint cells is insufficient to predict the future outcome of the same experiment. In this sense, the *observed* behavior of K-flows is nondeterministic.

The simplest example of K-flow, or rather a K-system (discretized time), is the so-called baker's transformation. It may be described as the transformation B of the unit square onto itself that is the result of two successive operations (Fig. 1): (a) first, the unit square is squeezed in the vertical direction to half its width, and it is at the same time stretched along the horizontal direction to double the length; (*b*) next, the resulting rectangle is cut in the middle and the right half is stacked on the left half.

The iterates B^n of the baker's transformation can be considered to describe the evolution of an *abstract* dynamical system at unit intervals of time. It is clear that the partition of the unit squares into vertical lines is a K-partition for the baker's system. In fact, under the operation of B^n $(n \geqslant 0)$, the partition in question becomes successively finer and the reader can easily verify that the defining Conditions i–iii of K-partitions hold.

Besides this rather artificial example, many systems of physical interest, such as the system of Lorentz gas and hard spheres in a box,[9] and the geodesic flow on a compact Riemann manifold of negative curvature[10, 11] are known to be K-flows. Although the existence of K-partitions for these more realistic systems has been established, their construction (and description) cannot yet be given analytically with a sufficient degree of explicitness.

As an illustration, and also for future comment, let us geometrically describe the elements (individual cells) of a K-partition for the (two-dimensional) Lorentz model of gas. In the model we have a fixed configuration of disks (scatterers) and light point particles that do not interact among themselves but move between the scatterers freely with constant velocity and on reaching a scatterer are reflected elastically. Since the interaction between the light point particles is neglected, the study of the behavior of a beam of such particles reduces to the study of the motion of a distribution function on the phase space of the one-particle system: the fixed convex scatterers and a light particle. This system is known to be a K-flow.[9]

The cells of the K-partition associated with this system may be obtained as follows (see Fig. 2). Let ω_0 be an initial point in the phase space and ω_t the point after a time t. If we vary the direction of the particle (but not the spatial position) around ω_t and trace back the motion to $t = 0$ to determine the initial phase point from which the particle would have started, we obtain a curve Σ_t containing the given point ω_0. The curve Σ_t will be smooth at least if one considers a reasonably small neighborhood of ω_0. Now one can construct Σ_t, in the manner described above, for various values of t (but for the same initial points).

A typical element (cell) of the K-partition associated with the system will be the limiting curve Σ_∞ to which the family of curves converge as t approaches infinity. In other terms, the elements of K-partition for the system under consideration are curves in phase space such that all particles having initial conditions on one such curve are so intricately correlated that after being scattered repeatedly by the scatterer, they all converge toward a single position as t goes to infinity. Thus, their correlation is similar to the correlations obtained in an incoming wave front that will converge in the infinitely distant future to a single point. Later, we shall make use of this feature of K-partitions to suggest a possible distinction between the two semigroups (corresponding to

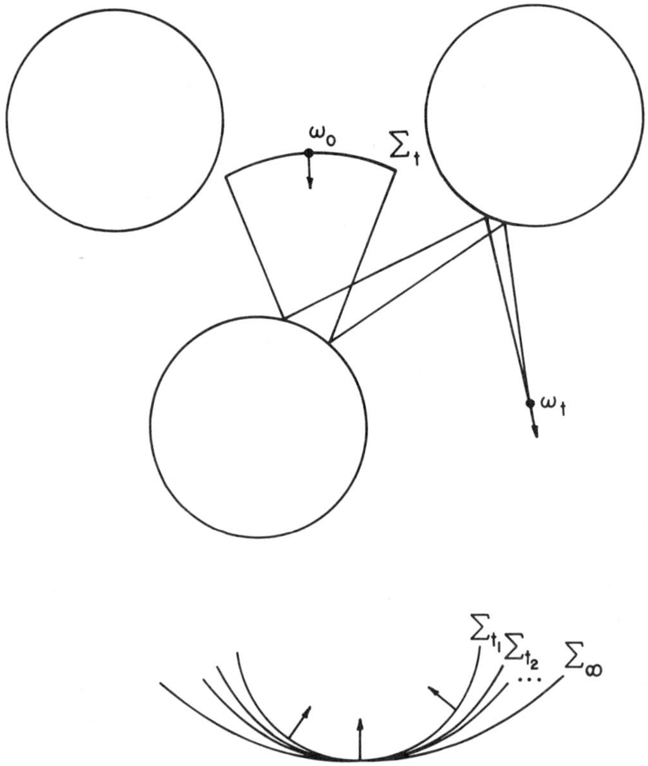

Figure 2.

the two directions of time) that result from the symmetry breaking mechanism discussed in this chapter.

The interest in explicit construction of K-partitions comes from the following property: namely, for K-flows the symmetry breaking projections leading to exact master equations (and equivalently the projection to the past associated with the internal time operator of the K-flow) can be constructed as the *projection operation of "coarse graining" with respect to K-partitions*. More explicitly, if ξ_0 is a K-partition associated with the K-flow $(\Gamma, \mathcal{B}, \mu, S_t)$, the orthogonal projection P_0 of $L^2_\mu(\Gamma)$ onto the subspace $L^2(\mathcal{A}(\xi_0), \mu)$ is a symmetry breaking projection; that is, $P_0 U_t P_0$ is a strongly irreversible Markov semigroup for $t \geqslant 0$. Here we have denoted by $\mathcal{A}(\xi_0)$ the σ-subalgebra of $\mathcal{B}$ formed by only the (measurable) subsets Δ *that are unions of elements (cells) of the partition* ξ_0. And $L^2(\mathcal{A}(\xi_0), \mu)$ denotes the subspace of $L^2_\mu(\Gamma)$ that consists of functions that are measurable with respect to $\mathcal{A}(\xi_0)$. For instance, in the example of the baker's transformation, $\mathcal{A}(\xi_0)$ is the σ-algebra of subsets formed by all *vertical* rectangular strips only.

The projection P_0 onto $L^2(\mathscr{A}(\xi_0), \mu)$ has the following properties:

Condition i. $f \geqslant 0 \Rightarrow P_0 f \geqslant 0$

Condition ii. $\int_\Delta f\, d\mu = \int_\Delta (P_0 f)\, d\mu$ for all $\Delta \in \mathscr{A}(\xi_0)$ and $f \in L^2_\mu(\Gamma)$.

Condition iii. $P_0 f$ [being measurable with respect to $\mathscr{A}(\xi_0)$] assumes constant values on each element (individual cell) of the partition ξ_0.

In the language of probability theory, the function $P_0 f$ is the *conditional expectation* of the function (random variable) f, given (the outcomes of the measurement represented by) the partition ξ_0. The symmetry breaking projections for K-flows may thus be viewed as representing a generalized process of "coarse graining" which, in a sense, averages over the cells of the K-partition. To put it differently, the symmetry breaking projection P_0 eliminates from dynamical considerations the degree of freedom and the "correlations" associated with the cells of K-partitions.

When the K-partition ξ_0 is of simple structure, one can give an explicit analytic formula for the symmetry breaking projection P_0 onto $L^2(\mathscr{A}(\xi_0), \mu)$ and display in explicit form the *exact* master equation that results from symmetry breaking. Such explicit calculations have recently been made for the infinite ideal gas[18] using the explicit form of *K*-partitions described in Ref. 19.

To conclude this section, let us briefly return to the question of the choice between the two semigroups (corresponding to the two different directions of time) that can result from symmetry breaking. As discussed above, one semigroup (for $t \geqslant 0$) comes from "coarse graining" with respect to the K-partition. The other semigroup for $t \leqslant 0$ would correspondingly come from the projection associated with the "time-reversed" K-partition, and such a partition, of course, exists because the originally given dynamics is assumed to be reversible. (The "*time-reversed*" K-partition is a partition ξ'_0 such that the partition $\xi'_t \equiv S_t \xi'_0$ becomes *less fine* as t increases: $\xi'_t \leqslant \xi'_s$ if $t \geqslant s$.)

The question then is whether there exists any physical distinction between the two semigroups to recommend one over the other as the physically realized temporal evolution. It seems to us that such a distinction can be made on the basis of the type of "correlations" that are eliminated by the two projections respectively. As an illustration, "coarse graining" projections associated with the K-partition of the Lorentz gas model would eliminate from the description of dynamical states precisely the future-directed correlations of the type present in "incoming waves" that would converge to a point in the *infinite future*. The projection associated with the "time-inverted" K-partition would, on the other hand, eliminate the "outgoing wave" type of correlation but would retain the "incoming wave" type of correlation. It is the physical unrealizability of correlations of the "incoming wave" type that recommends the former semigroup arising from the elimination of such unrealizable correlations as the physically preferred semigroup over the other.

These remarks, although qualitative, indicate how an intrinsic asymmetry between the two time directions can be sought in terms of the question of physical realizability and unrealizability, respectively, of correlations of the "outgoing wave" and "incoming wave" types.

Similar considerations can also be advanced in the "subdynamics" approach to the problem of irreversibility. In brief, in the subdynamics approach the distinction between the semigroups can be formulated as follows: between particles situated at large distances there may exist only postcollisional correlations but no precollisional correlations. Collisions can give rise to correlations and correlations to collisions, but the relation between them is not a symmetrical one. However, these considerations are developed elsewhere.

6. INTERNAL TIME AND GEODESIC INSTABILITY IN COSMOLOGY

An interesting application of the foregoing ideas occurs in certain simple cosmological models described below. In his interesting chapter in Part 2 of this volume, V. G. Szebehely exhibits in a striking manner the stochasticity of motion in the restricted three-body problem. It is interesting that in the cosmological model we consider, stochasticity appears already at the level of a one-body problem.

The model we consider is the usual model of an expanding universe (with Robertson–Walker metric) where three-dimensional spatial hypersurfaces of simultaneity are assumed to be of negative curvature. Furthermore, we suppose that the spatial hypersurfaces are compactified by suitable identification of points. It is known that this is possible to do without changing the metric structure of the space.

A free test particle in the universe moves, of course, along the four-dimensional space–time geodesics. Because of the symmetry of the Robertson–Walker metric, however, one can show that the *three* spatial coordinates (in the comoving frame) of the freely moving test particles follow geodesic curves in (any given) three-dimensional spatial hypersurface of simultaneity. In other terms, the four-dimensional geodesic motion of test particles when projected to a three-dimensional spatial hypersurface defines a geodesic flow in the three-space. The important point to note is that this affine parameter λ of the projected geodesic flow on the three-space turns out not to be the proper time of the test particle nor the *cosmic time parameter* t, but a nonlinear function $\lambda(t)$ of the form:

$$\lambda(t) = \lambda(t_0) + A\int_{t_0}^{t} \frac{ds}{R(s)}$$

for massless particles, and

$$\lambda(t) = \lambda(t_0) + A\int_{t_0}^{t} \frac{ds}{R(s)[\alpha^2 + R^2(s)]^{1/2}}$$

for massive test particles. Here $R(t)$ denotes the scale factor or expansion parameter of the universe. It is to be determined, of course, from the Einstein equation in conjunction with an equation of state for the cosmological fluid and the equation of conservation of energy. The quantity α is a constant related to the speed of the particle. The above-mentioned reduction of the four-dimensional geodesic motion to geodesic flow in three-space is possible in any model with Robertson–Walker metric irrespective of the sign of spatial curvature. But it is only in the case of models with negative spatial curvature and compactified three-space that this possibility has interesting implications.

This is because in this case the reduced geodesic flow in three-space associated with free motion of test particles is known to be K-flow (and, in fact, Bernoulli flow), hence a fortiori mixing and ergodic. Thus in such a universe arbitrary initial beams of test particles distributed in space and direction (or more precisely, beams corresponding to square-integrable distribution functions on the space of line elements of three-space) will tend toward the uniform (microcanonical) distribution as time progresses. There is thus a natural mechanism leading to homogeneity and (local) isotropy in such a universe. In particular, photons from different regions of the early universe would form the isotropic background radiation after a sufficient lapse of time.

In the cosmological model of the universe under consideration, distributions or beams of massive test particles behave somewhat differently from beams of massless test particles. Though the "mixing mechanism" is present for massive particles also, an initial distribution of massive particles may not be able to reach the uniform distribution. This difference in behavior between photons and massive test particles exists because the parameter $\lambda(t)$ for massive test particles, in contrast to that of photons, stays bounded even as t approaches infinity: the physically admissible values of λ (for massive test particles) cannot be made arbitrarily large, whereas the mixing property of three-dimensional geodesic flow would lead to a uniform distribution in general only in the asymptotic limit $\lambda \to \infty$.

Let us now turn to the internal time operator T associated with the geodesic motion of test particles in the cosmological model under consideration. Its existence is assured because the "projected" geodesic flow on the fixed three-dimensional hypersurface of simultaneity is a K-flow. The internal time operator T under discussion will thus be an operator acting on the distribution functions on the space of line elements of the three-dimensional hypersurface and will satisfy the relation

$$U_\lambda^* T U_\lambda = T + \lambda(t) I$$

Here U_λ denotes, of course, the unitary group induced by the projected three-dimensional flow.

The important point to note is that the "time parameter" of the projected flow is not the cosmic time parameter t but is a nonlinear function $\lambda(t)$ given in the preceding section, and it is $\lambda(t)$ rather than t that must occur in the defining relation of T. As a result, the (average) age of distribution functions on the space of line elements changes as the test particles move freely, keeping step not with t, but with $\lambda(t)$. The time scale defined by internal time is thus distinct from the cosmic time scale and corresponds to that of $\lambda(t)$. It is interesting that (see the expression for $d\lambda/dt$ given in the preceding section) in very early epochs of the universe the internal time flows more rapidly compared with the cosmic time: $d\lambda/dt \to 1/R(t) \to +\infty$ as $t \to 0$. Similarly, as the universe ages the internal time scale gets dilated relative to the cosmic time: $d\lambda/dt \to 0$ as $t \to +\infty$. The physical meaning of these relative rates of flow of internal time and cosmic time is that the mixing rate (i.e., the rate of approach to equilibrium) *with respect to change in t* approaches infinity as one nears the singularity, whereas there is practically no mixing in a sufficiently aged universe.

The existence of an internal time operator T in the cosmological model under discussion has important implications. As discussed in this chapter, it allows one to associate a Lyapunov variable or H-function with the geodesic motion of test particles. Moreover, the existence of T implies that the free geodesic motion of test particles are intrinsically random. This illustrates how irreversibility and randomness could emerge as essential features of dynamical systems embedded in a suitable cosmological model. A further interesting feature of the cosmological model under discussion follows from the uncertainty relation

$$(\Delta T)_\rho (\Delta L)_\rho \geqslant \tfrac{1}{2}$$

which is a consequence of the canonical commutation relation

$$[T, L] = iI$$

between the Liouvillian L and T. Now $(\Delta L)_\rho$ represents the dispersion of frequencies of periodic components into which the (Gibbs) distribution ρ can be decomposed. The uncertainty relation above thus shows that for a ρ in the early epochs of the universe (when $(\Delta T)_\rho$ is necessarily small), the distribution function ρ must have a wide dispersion of frequencies; hence its motion must have been very chaotic.

More details on these questions will be provided in a forthcoming publication.[20]

It is amusing to recall that Einstein cherished the belief that "God does not play dice." A serious challenge to this point of view comes from quantum mechanics. In our opinion, an equally important challenge comes from the recent studies of classical systems exhibiting strong forms of trajectory instability. As was said before, such systems are intrinsically random, and we find here that Einstein's own theory allows cosmological models in which the simplest and most fundamental of all motions, the geodesic motion of test particles, has this feature of nondeterminism.

7. CONCLUDING REMARKS

The preceding considerations thus lead us to the viewpoint that irreversibility expressed in the second law results from a special form of symmetry breaking at the dynamical level that causes the dynamical group to be "realized" as a dissipative semigroup associated with a probabilistic process admitting an H-function. The physical origin of the symmetry breaking in question is a limitation on physically observable states. Such a limitation comes, in the first place, from (strong) instability of dynamical motion as a consequence of which the concept of phase space trajectories ceases to be physically meaningful and the physically realizable states of the system need to be described in terms of (Gibbs) distribution functions. But the existence of symmetry breaking under consideration is the expression of a further limitation: not all distributions but only a suitable proper subset of them can correspond to physically realizable states. We have presented arguments indicating that this second limitation is a consequence of the fact that certain types of "future-directed" correlations cannot exist in physical systems so that only the distributions that do not contain such "future-directed" correlations can represent physically realizable states.

Thus the second law, which implies at the macroscopic level a limitation on the possibilities of "manipulation" of matter (e.g., the impossibility of perpetual machines of the second kind), implies a limit to our manipulation also at the *microscopic level*. To put it differently, the second law makes explicit on the macroscopic level a basic structure referring to the microscopic level. It expresses an essential new element foreign to the laws of dynamics but, of course, compatible with them. The analogy with quantum statistics may perhaps clarify what we want to say. The limitation to symmetrical (or antisymmetrical) wave functions is of course not a consequence of the Schrödinger equation. However, once a restriction on the symmetry of wave functions is formulated, it is propagated by the laws of quantum mechanics.

In the present cultural context, entropy plays a considerable role from basic physics to economics and political thought. It is therefore gratifying that we begin to understand somewhat better the microscopic meaning of the second law. Boltzmann's famous conclusion that increase of entropy means the evolution to the "most probable" state remains basically correct. However, it hides some rather complex features. Indeed, we have first to express the dynamical conditions that lead to a Markov process, since it is only for such processes that Boltzmann's statement is correct. Moreover, we must find physical reasons to introduce a selection principle that would permit us to choose the right semigroup. Anyway, the wide gap that existed between the far-ranging macroscopic applications of dissipativity and the microscopic theory of irreversible processes is beginning to narrow, and further progress can be expected in the near future.

ACKNOWLEDGMENTS

We thank Dr. C. M. Lockhart for his help in preparing this chapter. This work was partly supported by the Robert A. Welch Foundation of Houston, Texas, and the Instituts Internationaux de Physique et Chimie (Solvay), Brussels.

REFERENCES

1. M. Born, *Natural Philosophy of Cause and Chance* (Clarendon Press, Oxford, 1949).
2. K. Yosida, *Functional Analysis* (Springer-Verlag, New York, 1974).
3. J. W. Gibbs, *Elementary Principles in Statistical Mechanics* (Dover, New York, 1960).
4. I. Prigogine, C. George, F. Henin, and L. Rosenfeld, *Chem. Scripta* **4**, 5–32 (1973).
5. B. Misra, I. Prigogine, and M. Courbage, *Physica A* **98**, 1–26 (1979).
6. S. Goldstein, B. Misra, and M. Courbage, *J. Stat. Phys.* **25**, 111–126 (1981).
7. B. Misra, in *Proceedings of the Special Session in Mathematical Physics of the American Mathematical Society Meeting* (March 1980, Boulder, Colorado), (Plenum Press, New York, 1981).
8. B. Misra and I. Prigogine, *Prog. Theor. Phys. Suppl.* **69**, 101–110 (1980).
9. Ya. G. Sinai, *Usp. Mat. Nauk* **27**, 137 (1972).
10. Ya. G. Sinai, *Sov. Math. Dokl.* **1**, 335–339 (1960).
11. D. Anosov, *Proc. Steklov Inst.* No. 90 (1967).
12. N. S. Krylov, *Works on the Foundations of Statistical Physics* (Princeton University Press, Princeton, NJ, 1979).
13. B. Misra, *Proc. Natl. Acad. Sci. U.S.A.* **75**, 1627–1631 (1978).
14. I. Prigogine, *Études Thermodynamiques des Phénomènes Irreversibles* (Dunod, Paris, 1947).
15. K. Goodrich, K. Gustafson, and B. Misra, *Physica A* **102**, 379–388 (1980).
16. E. B. Dynkin, *Markov Processes* (Springer-Verlag, New York, 1965).
17. V. I. Arnol'd and A. Avez, *Ergodic Problems of Classical Mechanics* (Benjamin, New York, 1968).
18. M. Theodosopulu and C. Coutsomitros (to appear).
19. Ya. G. Sinai, *Funct. Anal. Appl.* **6**, 35 (1972).
20. C. Lockhart, B. Misra, and I. Prigogine, "Geodesic Instability and Internal Time in Relativistic Cosmology" (to appear).

CORRELATIONS, FLUCTUATIONS, AND TURBULENCE IN A RAREFIED GAS

HAROLD GRAD
Courant Institute of Mathematical Sciences
New York University
New York, New York

Abstract. *This chapter brings up to date an investigation*[19] *of the validity of long-term predictions from fluid dynamic and Boltzmann equations and the relation of this problem to molecular chaos and long-range correlations in rarefied gases. Also discussed are the connections between kinetic theory and turbulence, the propagation of correlations by nonchaotic Boltzmann equations, and the relation between ascending and descending hierarchies (the former commonplace in fluid turbulence and the latter in statistical mechanics). For a linear, determinate evolution equation, the governing equation for any higher order correlation is precisely determinate; the hierarchy requires no closure or truncation assumption. In a nonlinearly unstable situation (as in turbulence) both the Boltzmann and the Navier–Stokes equations are, in principle, invalid (in the sense that pointwise solutions of the two models disagree with each other and with those of Liouville's equation); in a rarefied gas one must solve at least the nonlinear* Boltzmann hierarchy *rather than the Boltzmann or Navier-Stokes equations. The postulate that* statistical *(as distinguished from pointwise) properties of the various models may be comparable has not yet been given qualitative or quantitative justification.*

This work was supported by the U.S. Department of Energy, contract number DE-ACO2-76ER-03077 and the Air Foorce Office of Scientific Research, contract number AFOSR-81-0020.

1. INTRODUCTION AND SUMMARY

One can consider three sources of fluctuations and stochastic behavior in a gas:

1. Inherent graininess of molecular matter.
2. Imperfect control of measurable experimental conditions.
3. Instability, for example, turbulence or strange attraction.

In statistical mechanics, situation 1 leads one to introduce probability into the description. In fluid dynamics, 2 or 3 leads to probability in the model. In basic conception, 1 and 3 can be grouped together as two aspects (microscopic and macroscopic) of loss of continuous dependence on initial data [cf. Refs. 12, 17). In practice, 2 and 3 are grouped together as being macroscopic, whereas 1 is microscopic; (the respective literatures have been, for many years, almost nonoverlapping). The microscopic and macroscopic descriptions *may* differ in the use of mathematical densities that are sums of δ-functions or smooth, respectively; but this is not an essential difference. A more significant factor in the grouping of 2 with 3 is that from the point of view of statistical mechanics, each involves the introduction of a second independent stochastic structure.[19]

For example, in fluid dynamics the Navier–Stokes or Euler equations are assumed to be exact for a specific dynamical system, yielding an exact solution of, say, the Navier–Stokes equations. A probability structure is introduced either casually by operating with an undefined expectation operator $\langle \; \rangle$, or precisely[21] by considering a sample space of solutions of the Navier–Stokes equations with an infinite-dimensional measure that satisfies something like a generalized Liouville equation. However, if the Navier–Stokes equations are considered to have a kinetic origin, an a priori probability has already been introduced into the n-particle dynamics (Liouville equation), and a second (Hopf) structure is redundant, incompatible, or (most likely) unduly enlarges the space. This possibility of discrepancy is compounded when (as is always necessary) approximations are introduced into one or the other probability structure.

Note that the usual Navier–Stokes correlation formalism, introducing successive equations for the average one-point velocity $\langle u(x) \rangle$, two-point velocity $\langle u(x_1)u(x_2) \rangle$, and so on, gives an *ascending* infinite, coupled hierarchy of equations, starting from a determinate equation for a fluctuating $u(x)$. This is in direct contrast to statistical mechanics, where a *descending* hierarchy is obtained starting from a determinate n-particle density satisfying Liouville's equation. The truncation or closure questions for both types of hierarchy appear to be somewhat similar, but this may mask inherently different logical structures.

As a second example, the conventional Boltzmann equation governs a one (or small number) particle density *expectation* in a certain continuum limit[9,23] in which the particle number density becomes infinite. A *fluctuating* Boltzmann equation can be introduced in three ways:

- **a.** Inherently, by looking at the fine structure in the approach to the Boltzmann limit (related to case 1 above, cf. Ref. 29, also the f_1^σ discussion later in this section).
- **b.** Expediently, by superimposing a second probability structure, representing uncertainty in initial or boundary data (case 2), or the enhancement of this by instability (case 3), after the Boltzmann limit is reached.[19]
- **c.** Also expediently, by adding an ad hoc stochastic term to a postulated Boltzmann equation.[6]

The mental compartmentalization of case 1 versus cases 2 and 3 is typified by the phenomenon of fluid turbulence in an ideal gas. In fluid dynamics, turbulence is always described in terms of long-range, lasting correlation; in the theory of the Boltzmann equation (which is clearly applicable to turbulent flow), it has frequently been stated (and sometimes "proved") that no such correlations can exist (e.g., that long-range correlations arise only in a dense gas or are quickly destroyed by collisions). Here the fluid viewpoint offers a correction to the common kinetic position. In the opposite direction, kinetic theory has something significant to offer to the theory of fluid turbulence, simply from its distinct logical structure of nonlinearity and truncations and not from intrusion of the mean-free-path (see Ref. 19 and Section 9, below, and references cited there.)

Thermal fluctuations vanish in the Boltzmann limit; this allows us to separate out cases 2 and 3 above. Consider an experiment with measurable imperfections (such as small ripples in a wind tunnel) or turbulence. If the variable macroscopic (or one-particle Boltzmann) state is taken into account in the initial a priori Liouville probability, an infinite *Boltzmann hierarchy* results by taking the Boltzmann limit (implicit in Refs. 9 and 27, with more precision in Refs. 23 and 30). Alternatively, if the initial state is assumed to be chaotic, then the n-particle probability is compatible with an essentially unique macroscopic and one-particle state, and the determinate *Boltzmann equation* is obtained in the limit.[9,23] One can subsequently, as in fluid dynamics, superpose a second probability structure, describing the experimental imperfections, and construct from the determinate Boltzmann equation an ascending hierarchy of two-point, three point, and so on, correlations.

The two infinite systems, descending (direct from Liouville) and ascending (two successive probability structures) are shown to be identical in Section 8 [also see Ref. 30]. This answers one of the questions raised in Ref. 19. But the problem of compatibility between two superposed probability structures reappears as soon as one tries to solve specific problems by introducing approximations, truncations, and so on, in these abstract infinite systems.

Return now to the question of incorporating thermal fluctuations, case 1. The Boltzmann equation can be given two distinct physical interpretations.[9,32] One is as the *expected* number density (relative to an a priori probability), and the other as a *fluctuating* density, relating to a specific dynamic state. The elementary formal derivation of the mathematically more accessible first form of the Boltzmann equation as given in Ref. 9 depends on a trick, namely, to introduce f_1^σ, the one-particle phase space density of the molecules which are not undergoing collisions at a given moment (measured by any neighbor closer than σ). This density evidently changes discontinuously as pairs, triplets, and so on are formed and split up following Liouville's equation. In the Boltzmann limit (Section 4), the number of colliding pairs at any instant becomes infinite, but the fraction of particles undergoing collision at any time goes to zero (the occupied volume, $n\sigma^3$, approaches zero). The Boltzmann limit involves a two-space scale and two-time scale process. The collision duration and interaction distance go to zero. But since $n\sigma^2$ (the reciprocal mean free path) remains finite, the time evolution of $f_1^\sigma \to f_1$ remains finite.

All the *formal* operations relating f_1 and f_1^σ at fixed n are equally valid for f_1 as a sum of δ-functions and for smooth f_1 and f_n. The reason is that Liouville's equation is linear allowing distributions as solutions; the integration operations to obtain $\partial f_1/\partial t$, $\partial f_2/\partial t$, and so on, are also linear; and the operations that define f_1^σ in terms of $f_1, f_2, \ldots$ are also linear. (To be precise, the standard distribution space, dual to C_∞ and compact support, is inappropriate, but the relevant calculus formulas can be confirmed.)

The hierarchy of equations $\partial f_1^\sigma/\partial t, \partial f_2^\sigma/\partial t, \ldots$ for finite n as obtained by Grad[9] are therefore exact, without approximation, for both f_r and f_r^σ in terms of δ-functions as well as smooth arguments. There is one important distinction, however: for δ-function densities (cf. Section 3), $f_2(x_1, x_2)$ is automatically a product $f_1(x_1)f_1(x_2)$, $f_3(x_1, x_2, x_3) = f_1(x_1)f_1(x_2)f_1(x_3)$, and so on, within correction terms $0(\sigma) = 0(n^{-1/2})$. In other words, making the identifications $f_1^\sigma \sim f_1$, and so on [also, formally, with an error $0(\sigma)$], we obtain for an individual (fluctuating) dynamical system the determinate Boltzmann equation, automatically including molecular chaos (in contrast to the case of a smooth expectation f_1). Of course, the error $0(n^{-1/2})$ is exactly comparable to thermal fluctuations, which are not properly treated by this very easy derivation of the Boltzmann equation (rather than hierarchy) for an individual dynamical system.

Since a large fraction of the literature on statistical mechanics fluctuations is concerned with thermal noise, and relatively little is devoted to macroscopic fluctuations, we shall, in the sequel, assume that thermal fluctuations are negligible. Until recently, fluctuations have been usually included in the Boltzmann equation as ad hoc, stochastic inhomogeneous terms[6] (and similarly in fluid dynamics[22]). A more careful treatment, carrying error terms $0(n^{-1/2})$ in the Boltzmann limit is given by Spohn.[29]

One important point is that the insertion of inhomogeneous stochastic terms (such as Gaussian noise) vitiates the equivalence of the ascending and descending hierarchies. Another important semantic point (cf. Ref. 19), is the widespread confusion in the literature between the conceptually distinct *multiple collisions* and *multipoint correlations*. On the one hand, the virial-type expansions for higher density deviations from the Boltzmann equation successively requires two-particle collisions, three-particle collisions, and so on, and, in consequence, successively higher correlations. On the other hand, the strictly rarefied gas (Boltzmann limit) hierarchy describes higher order *correlations* without higher order *collisions*.

The real difference between the two concepts lies in scale length. Expansion in density, away from an ideal gas, produces a concentration on molecular dimensions (i.e., small compared to a mean free path) over which any correlations, not sustained by recollisions, decay very rapidly. On the other hand, correlations on distances longer than a mean free path are preserved for macroscopically long times. In principle, a statistical mechanics calculation that incorporates both length scales can simultaneously account for thermal fluctuation and macroscopic fluid correlations.

The key to understanding the distinction between the determinate Boltzmann equation and the infinite Boltzmann hierarchy lies in a study of molecular chaos (Sections 2, 3, 5, and 6). The miracle of the transfer of determinism from Liouville's equation to fluid equations or to the Boltzmann equation (but probably to no other higher order kinetic equations) is briefly discussed in Section 4. The important distinction between a linear ascending hierarchy (which is strictly determinate at every level) and the much more subtle nonlinear hierarchy is discussed in Sections 7 and 8. The Boltzmann hierarchy for *small amplitude fluctuations* is shown to be essentially determinate (this formal argument is probably susceptible to proof using standard Boltzmann equation theory). The relation of stability to the determinate solution of fluid or Boltzmann equations, and the related loss of molecular chaos with time, are treated in Section 9.

It is worth pointing out that the classical use of the word *chaos* as in *molecular chaos* is quite distinct (and in some respects opposite) to its use to describe random behavior as chaotic. For example, instability destroys molecular chaos while creating the possibility of chaotic (i.e., random) behavior.

2. A THOUGHT EXPERIMENT

The relation between the kinetic theory concept of molecular chaos and macroscopic fluid correlations can be illustrated by the following shock tube experiment. Consider an automated shock tube, namely, a tube with a barrier (Fig. 1), and a button that initiates the experiment by causing air to be pumped into the left-hand chamber (experiment 1) and, after a suitable delay, causing the barrier to be broken. A second button (experiment 2) starts the mirror image experiment by pumping air into the right-hand chamber. The macroscopic solution to experiment 1 consists of a shock wave moving to the right

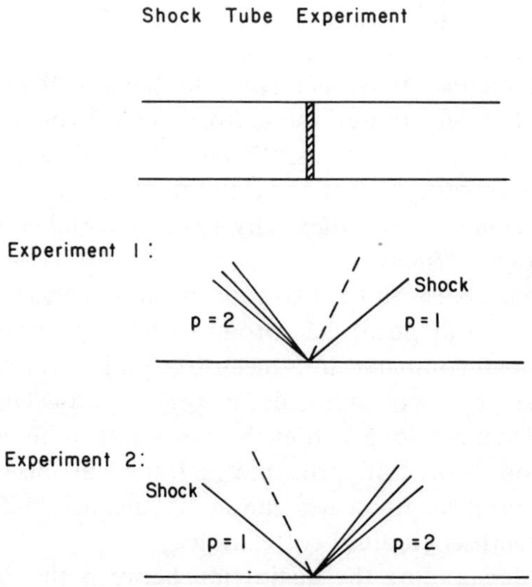

Figure 1.

and a rarefaction wave to the left, leaving a contact front (temperature discontinuity) between. For shock waves of medium strength, the Navier–Stokes solution and the Boltzmann solution both represent the flow accurately. We have every reason to believe that the Liouville solution, if obtainable, would also agree. The a priori initial probability distribution in n-particle space (the initial value for Liouville's equation) is determined, in some way, by the behavior of the pumps and by the subsequent approximate equilibration in the two chambers to the "initial" state for the shock problem (an approximate thermodynamic equilibrium in each chamber).

In experiment 3 we toss a coin to decide whether to push button 1 or 2. For the composite experiment, the a priori initial condition for Liouville's equation is the arithmetic mean of the two initial conditions for experiments 1 and 2, respectively. Since Liouville's equation is strictly linear, the solution at any later time is the arithmetic mean of the two mirror image solutions (experiments 1 and 2) at that time. The one-particle distribution function (argument of the Boltzmann equation) and the macroscopic fluid state (argument of the Navier–Stokes equations) are obtained by appropriate integration of the Liouville solution at any time. These are all linear operations.* Therefore the one-particle distribution and fluid states as predicted by Liouville's equation for experiment 3 are exactly the averages of the corresponding quantities for experiments 1 and 2. These macroscopic and one-particle descriptions of experiment 3 from Liouville's equation do not satisfy the Boltzmann equation

*To be precise, ρ, ρu, and $\rho(\frac{3}{2}RT + \frac{1}{2}u^2)$ are obtained linearly.

or the Navier–Stokes equations, which are nonlinear. As a matter of fact, the initial state (as measured by the macroscopic or one fluid state) is approximately a uniform equilibrium state across the entire shock tube and from this, by solving Liouville's equation, there develops in time a nonuniform macroscopic state.

Of course, the initial two-particle distribution is far from thermodynamic equilibrium and, in particular, molecular chaos is not satisfied. The initial state for experiment 3 is highly correlated: if a pressure measurement on the left gives approximately $p = 2$, then on the right it will almost certainly give $p = 1$, and vice versa. This high degree of correlation persists for a macroscopically long time. We have here an example of a solution of Liouville's equation that does not approximate either the solution of Boltzmann's equation or any macroscopic system. Molecular chaos can be violated in a simple, realistic experiment.

A more subtle example in which chaos is not satisfied (or *is* satisfied, depending on precisely how the problem is formulated) is turbulent flow. Any macroscopically unstable flow gives an example in which molecular chaos can be lost after a period of time.

The two successive stochastic structures referred to in the last section arise very naturally in the "mixed state" of experiment 3 above. Ignoring the fact that the one-particle distribution that evolves according to the Boltzmann equation is already a probability density (or expectation), one superposes a discrete probability space with two states and probabilities $(\frac{1}{2}, \frac{1}{2})$. In this simple case, there is no question of compatibility of the two successive probability structures. But it seems to be less trivial to verify, for example, that the Hopf introduction of an infinite-dimensional probability on the space of Navier–Stokes solutions is compatible with the structure introduced by a double limit of Liouville problems as n goes to infinity and the mean free path goes to zero.

3. AN ELEMENTARY IDENTITY

Let x represent a coordinate in the physical or phase space of a molecule (e.g., 3 or 6 or any other fixed dimension), x_i the coordinate of molecule i, $i = 1 \ldots n$, $X = (x_1, \ldots, x_n)$, and $f(X)$ a normalized probability density that is symmetric under permutation of $(x_1, \ldots, x_n)$. For a suitable function $\phi(X)$, $\langle \phi \rangle = \int \phi f \, dX$ is the expectation of ϕ.

Let D' and D'' represent two possibly intersecting domains (Fig. 2) in the space of a representative molecule x, and $D' \cap D'' = D_0$. Let $\phi'(x_i) = 1$ if $x_i \in D'$, $\phi'(x_i) = 0$ otherwise, and similarly for ϕ'' and ϕ_0. The number of molecules in D' is $N'(X) = \Sigma \phi'(x_i)$ and similarly for N'' and N_0. Define $f^{(r)} = \int f \, dx_{r+1} \cdots dx_n$ and introduce the notation $f^{(2)}(x_3, x_5) = f_{35}$, $f^{(1)}(x_2) = f_2$, and so on. Also define $F^{(r)} = n(n-1) \cdots (n-r+1) f^{(r)}$.

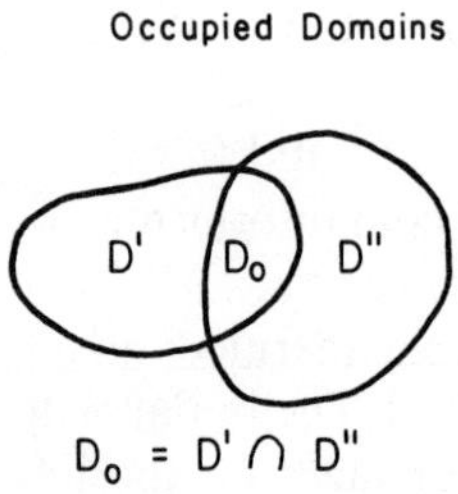

Figure 2.

A trivial computation (counting) yields

$$\langle N' \rangle = n \int_{D'} f^{(1)} = \int_{D'} F^{(1)} \tag{3.1}$$

$$\langle N'N'' \rangle = \langle N_0 \rangle + \int_{D' \times D''} F_{12} \tag{3.2}$$

$$\langle (N' - \langle N' \rangle)(N'' - \langle N'' \rangle) \rangle = \int_{D_0} F_1 + \int_{D' \times D''} (F_{12} - F_1 F_2) \tag{3.3}$$

There is very much useful information contained in the elementary, exact formula Eq. (3.3).

Consider first the two important special cases of Eq. (3.3):

a. If there is a single domain, $D' = D'' = D_0 = D$, $N' = N'' = N_0 = N$, then

$$\langle (N - \langle N \rangle)^2 \rangle = \langle N \rangle + \int_{D \times D} (F_{12} - F_1 F_2) \tag{3.3a}$$

b. If $D_0 = 0$ (D' and D'' are disjoint), then

$$\langle (N' - \langle N' \rangle)(N'' - \langle N'' \rangle) \rangle = \int_{D' \times D''} (F_{12} - F_1 F_2) \tag{3.3b}$$

In case a, the first term on the right $\langle N \rangle$, can be called a *fluctuation* ("noise") and the second term, $\int (F_{12} - F_1 F_2)$ a *correlation* (this is not a standard distinction but it is a useful one). We are considering only cases in which n is very large and the domains D' and D'' are small so that $\langle N \rangle \ll n$; however, $\langle N \rangle$ can be large or small compared to 1. If $\langle N \rangle \ll 1$ (the probability of finding even one molecule is small), the first term (fluctuation) dominates the second (correlation). If $\langle N \rangle \gg 1$, the correlation term dominates the fluctuation (unless $F_{12} - F_1F_2$ is small). More precisely (and more usefully) in the case of large $\langle N \rangle$, either 1 or 2 holds.

1. $\langle (N - \langle N \rangle)^2 \rangle$ and $\int (F_{12} - F_1F_2)$ are large compared to $\langle N \rangle$ and are approximately equal.
2. $\langle (N - \langle N \rangle)^2 \rangle$ and $\int (F_{12} - F_1F_2)$ are both comparable to or smaller than $\langle N \rangle$.

Note that $\int F_1F_2 = \langle N \rangle^2 \gg \langle N \rangle$. In other words, for large $\langle N \rangle$, chaos $F_{12} \sim F_1F_2$ is asymptotically equivalent to the statement that N is peaked about its mean.

The weak chaos condition, $\int (F_{12} - F_1F_2) \ll \langle N \rangle^2$, for a single domain D, is necessary and sufficient for N to have a peaked distribution about its mean, $\langle (N - \langle N \rangle)^2 \rangle \ll \langle N \rangle^2$, and either condition also implies, in the general case Eq. (3.3), that $\langle (N' - \langle N' \rangle)(N'' - \langle N'' \rangle) \rangle$ and $\int (F_{12} - F_1F_2)$ are small compared to $\langle N \rangle' \langle N \rangle''$.

In slightly different words, the chaos condition, $F_{12} \sim F_1F_2$ is equivalent to setting up a "pure" state in which N (in whatever low-dimensional space it is defined) is reproduced with negligible variation. Other properties of molecular chaos are described in Section 5.

An alternative formalism ("Klimontovich") can be used to derive similar identities for a distribution (δ-function) form of F_1. Consider a fluctuating one-particle number density (in this example, fluctuating signifies that F_1 is a random variable):

$$\hat{F}_1 = \sum \delta(x_1 - y_i), \int \hat{F}_1 \, dx_1 = n \tag{3.4}$$

where $y_i(t)$ represents the trajectory of the dynamical system (characteristics of Liouville's equation). As before, we are given an a priori probability distribution $f(Y)$ (as smooth as you like), which is used to define the expectation or average $\langle \ \rangle$. Clearly

$$\langle \hat{F}_1 \rangle = nf_1 = F_1 \tag{3.5}$$

where f_1 and F_1 are given as before in terms of f. Define the fluctuating *pair correlation*,

$$\hat{F}_{12} = \sum_{i \neq j} \delta(x_1 - y_i)\delta(x_2 - y_j), \int \hat{F}_{12} \, dx_1 \, dx_2 = n(n-1)$$

$$= \hat{F}_1\hat{F}_2 - \delta(x_1 - x_2)\hat{F}_1 \tag{3.6}$$

Averaging with respect to f, we observe that

$$\langle \hat{F}_{12} \rangle = F_{12} \tag{3.7}$$

and

$$\langle (\hat{F}_1 - F_1)(\hat{F}_2 - F_2) \rangle = F_{12} - F_1 F_2 + \delta(x_1 - x_2) F_1 \tag{3.8}$$

which is the distribution (in the sense of Schwartz) form of Eq. (3.3). In our terminology, the δ-function in Eq. (3.8) and the term $\int F_1$ in Eq. (3.3) represent *fluctuation* or *noise*, while the remainder corresponds to *correlation*.

For the Boltzmann limit (cf. Section 4), and in any problem in which the imperfection level by which experiments can be repeated exceeds the irreducible thermodynamic noise, the fluctuation terms in Eqs. (3.3) and (3.3a) can be ignored compared to the correlations [this is equivalent to dropping the δ term in Eq. (3.8)]. This situation is our primary concern in this chapter. The reason for our emphasis is that long-range correlations can be handled relatively easily but are largely ignored in the bulk of the statistical mechanics literature. This correlation analysis, when separated from thermal fluctuations, allows a clear separation, in nonequilibrium statistical mechanics, of dense gas correlations (zero in the Boltzmann limit) from long-range correlations (as in turbulence at arbitrarily low density).

4. THE BOLTZMANN AND FLUID DYNAMIC LIMITS

In any reduction of the amount of information that defines a state, for example, from $f^{(n)}$ to $f^{(r)}$ ($r = 2, 3, \ldots$) to $f^{(1)}$ to macroscopic, one might expect each step not only to reduce the amount of information, but also to reduce the accuracy with which the remaining information is described. In some special cases we may hope for more; there is some loss of information, but the lower order description is self-determinate (e.g., there is a loss of information but not of accuracy in a triangular matrix of coupled equations). Finally, in the best of possibilities, a lower order description may contain all the higher order information, in some limit or special case (cf. Refs. 9, 12, 17).

The prime example of the last possibility is classical thermodynamic equilibrium, in which the limit $n \to \infty$ is sufficient to determine all distribution functions, fluctuations, and so on, in terms of a small number of discrete parameters. Because of this miracle, we find the simple limit $n \to \infty$, together with the major tool of equilibrium statistical mechanics, the virial expansion, to be presented with impossible tasks in nonequilibrium. But it is easy to see that with probably only two exceptions, namely the macroscopic fluid and Boltzmann limits, equations involving $f^{(2)}$, $f^{(3)}$, and so on can be expected to give at most incremental improvements (cf. retaining more terms in power series or

Fourier series). For example, the Boltzmann collision term is useful because it is local in x and t; but at the $f^{(3)}$ or higher level, it can be expected to be macroscopically global in both space and time (cf. Grad, Ref. 9; Cohen, Ref. 4). At best, this would not give a very useful kinetic equation (this effect is closely related to the appearance of logarithmic terms in expansions of transport coefficients).

The greater richness of nonequilibrium phenomena can be partly illustrated by the increase in dimensionless parameters. Of the four scale lengths, intermolecular force scale (σ), mean distance between molecules ($\nu^{-1/3}$, where $\nu = n/V$), mean free path (L), and imposed disturbance scale (l), three are independent ($L \sim 1/\nu\sigma^2$), and there are two dimensionless ratios. Both the Knudsen number (L/l) and the occupied volume ($\nu\sigma^3$) can be taken as measures of the dilution of the gas. For simplicity, we let the molecules shrink as n approaches infinity, take $l \sim 1$ to be the dimension of the container, and do not distinguish ν from $n(V = 1)$. The limit $n \to \infty$ requires $\sigma \to 0$, also $m \to 0$ (the molecular mass). The velocity scale is held fixed. Macroscopic quantities such as mass density (in physical or phase space), mass flow velocity, temperature, and pressure remain fixed in this limit. The relative rate at which $n \to \infty$ and $\sigma \to 0$ determines the type of limiting medium, but it is, in all cases, a continuous medium, since the intermolecular spacing shrinks to zero.

In the *fluid* limit $n\sigma^3$ is held fixed. The ratio of potential to kinetic energy remains fixed, so the limit is a real gas (or liquid). Since $n\sigma^2 \to \infty$, the mean free path shrinks to zero, as do the transport coefficients; thus the limit is a macroscopic, ideal fluid that presumably satisfies the Euler equations of motion. There is little rigorous information about this limit except in the spatially homogeneous case, which is exactly the thermodynamic limit. (One serious nonequilibrium attempt is given by Morrey.[25])

The other limit that has been studied is $n\sigma^2$ held fixed, termed the Boltzmann limit by Grad.[9] Since $n\sigma^3 \to 0$, the limit is an ideal gas, $pV = nkT$, with stresses and heat flow given by molecular flow alone (no force contribution). On the other hand, since the mean free path is fixed, collisions remain a finite mechanism for the evolution of the gas. In this limit, binary collisions dominate and become instantaneous and localized in space. The Boltzmann equation is the formal result of this limit if molecular chaos is assumed, otherwise it is the Boltzmann hierarchy (see Section 8). A rigorous proof, for an appreciable fraction of a collision time, is given by Lanford[23,24] in a very deep and difficult piece of work.

Starting from the Boltzmann equation, the small mean-free-path limit has been rigorously shown to give fluid equations (with certain modifications) by Grad,[11] with refinements in a number of other papers, notably Caflisch.[1] There are also certain limiting cases of the Boltzmann equation that give the Navier–Stokes equations directly, rather than as a higher order correction to the Euler equations.[18] It is not clear whether this can be accomplished starting from Liouville's equation.

5. INITIAL CHAOS

Whether taken as a postulate, a theorem, or a conjecture, as self-evident, or as a deep result, the molecular chaos statement is a part of any justification of the Boltzmann equation. For a comprehensive discussion of this subject, in particular of the necessity to split it into two parts, "Initial Chaos" and "Subsequent Chaos," see Ref. 9. Although only a few decades ago the principal use of the Boltzmann equation was to evaluate transport coefficients for use in macroscopic equations, it is now fully established that its chief purpose is to describe phenomena on a space and time scale of the mean free path and mean collision time. Since there is no mechanism to produce chaos in a fraction of a collision time (even for much longer times; see Section 6), it must be initially present. Initial chaos cannot be a question of molecular dynamics. It can only follow from the fact that the number of degrees of freedom is large, or from the fact that the gas is rarefied, or from some conscious or unconscious rules that are followed in preparing experiments.

The first mathematical result of this type was presented by Grad[8] and published in Ref. 10. Following is a brief summary of this material.

For simplicity consider a phase space that is the product of n replicas of the unit interval, $0 < x < 1$, and introduce a normalized probability density $f^{(n)}(x_1 \cdots x_n)$, $0 < x_i < 1$, where $f^{(n)}$ is symmetric under permutation of its arguments. To make explicit the symmetry, the problem is discretized (coarse grained) by subdividing the unit interval into r equal subintervals Δ_i (r is considered to be large but fixed while $n \to \infty$). The discrete probability is represented by the values of $\int f^{(n)}$ over each of the r^n cubes. By the symmetry, the joint probability that x_1 lies in interval Δ_{i_1}, x_2 in Δ_{i_2}, and so on can be replaced by the probability that k_1 of the x_i are in Δ_1, k_2 in Δ_2, and so on, where $\Sigma_1^r k_i = n$. We represent this probability by $F^{(n)}(k_1 \cdots k_r)$, which is now a (discrete-valued) function in a space of *fixed dimension* r as $n \to \infty$. Next we normalize $k_i/n = y_i$, $0 < y_i < 1$, suitably renormalize $F^{(n)}$, and use the same symbol for the normalized probability density $F^{(n)}(y_1 \cdots y_r)$. This distribution contains all the information resident in the original coarse-grained $f^{(n)}$.

An elementary computation (counting) shows that the coarse-grained, one-particle (Boltzmann) distribution in $\Delta_1 \cdots \Delta_r$ is given by the mean occupation numbers $\langle y_1 \rangle \cdots \langle y_r \rangle$ with similar mean values for the two-particle and higher order joint distribution functions.[7]

To obtain a nontrivial limit as $n \to \infty$, consider $[F^{(n)}]^{1/n}$, define a sequence $F^{(n)}$ to be *convergent* if

$$\lim [F^{(n)}]^{1/n} = \psi(y_1 \cdots y_r) \tag{5.1}$$

and call ψ the generator of $F^{(n)}$. By the normalization property, $\max \psi = 1$. If the original $f^{(n)}$ represents *independent* random variables,

$$f^{(n)} = f_1 f_2 \cdots f_n \tag{5.2}$$

then an easy calculation (based on Sterling's formula) yields a well-known result of combinatorial statistical mechanics (in a slightly unfamiliar setting),

$$\psi(y) = \psi^*(y) \equiv \left(\frac{p_1}{y_1}\right)^{y_1}\left(\frac{p_2}{y_2}\right)^{y_2} \cdots \left(\frac{p_r}{y_r}\right)^{y_r} \tag{5.3}$$

where $(p_1,\ldots,p_r)$ is the (discrete) one-particle distribution; for an independent $f^{(n)}$ [Eq. (5.2)], $\log \psi^*$ is closely related to Boltzmann's H-function. Of course, ψ^* is uniquely determined by the one-particle distribution.

In complete generality, we denote by $F_1^{(n)}$, $F_{12}^{(n)},\ldots$ the one-coordinate, two-coordinate, and so on discrete distributions.

Evidently *convergence* of $F^{(n)}$ as in Eq. (5.1) [with $\psi(y)$ essentially arbitrary] is a much more general concept than independence (for which ψ is unique). In between in generality is *condition C*: ψ reaches its maximum at an isolated point, $y_i = y_i^0$. Since ψ^n will be peaked near this maximum, $\langle y_i \rangle \to y_i^0$ for large n; thus y_i^0 represents the one-particle distribution. A trivial consequence of the peaking property is molecular chaos (and somewhat more); that is, for any fixed integer m,

$$F_{1\cdots m}^{(n)} \to F_1 \cdots F_m \tag{5.4}$$

where $F_{12}^{(n)}$ represents the joint two-point distribution derived from $F^{(n)}$, and so on.

It is evident that a continuous function ψ, chosen "at random," will reach its maximum at a single point "with probability one" (e.g., this is a theorem, using Wiener measure). Therefore, among a class of functions ψ of class C that is much larger than *independence*, (represented by $\psi = \psi^*$) the *chaos* condition (5.4) (asymptotic independence for the first few marginal distributions) is very likely.

This result follows from symmetry and large n alone. A reinterpretation and simplification of this result (presented in Ref. 15 and published in Ref. 19) that follows almost immediately from the identity Eq. (3.3) is that chaos is equivalent to a generator ψ (or consequent sequence $F^{(n)}$) that represents a *pure state* in the limit $n \to \infty$. The peaking property C states exactly not only that the expected one-particle distribution is the average, but that it is almost always observed. The virtue of this interpretation is that it gives comparable standing to pure states (chaotic) and mixed states (nonchaotic); both are physically relevant. That the chaotic nature of an initial state depends on the preparation of the experiment is shown by the example in Section 2.

There is evident in the literature a strong psychological bias to always justify chaos and derive the Boltzmann equation. Since this has been believed to be the only correct result, it has been frequently (incorrectly) "proved." It is interesting to note that exactly the same chaos argument holds in the physical rather than a phase space of a molecule. In other words, the derivation of macroscopic fluid dynamics from particle dynamics requires the same distinction between pure and mixed states (and a similar argument must be made in velocity space for a master equation).

There is a close connection between the initial chaos theorem and the Hewitt–Savage theorem,[20] which states, approximately, that for an infinite-dimensional, symmetric, infinitely additive probability, every finite-dimensional distribution is an average, with respect to some measure, of chaotic (product) distributions. In our terminology, the symmetric infinite-dimensional measure induces a *mixed state* for all finite-dimensional distributions. This theorem has to be applied with caution in statistical mechanics. For example, it is not applicable to the two-space scale Boltzmann limit, and it is not applicable to the conventional thermodynamic limit. There is another difficulty with $n = \infty$ rather than $n \to \infty$, in that it blurs the distinction between chaos and independence; and the Hewitt–Savage theorem does not distinguish between pure and mixed states (as being very likely and not so likely). Nevertheless, there seems to be a strong likelihood of strengthening kinetic theory chaos results by use of such mathematical probability structures.

6. SUBSEQUENT CHAOS

From the discussion in Section 5 on initial chaos and the explicit shock tube example in Section 2 (also see Section 8), it is clear that subsequent chaos (which must be verified for solutions of the Liouville's equation if the Boltzmann equation is to be derived) should be generalized to show not only that an initially chaotic state is preserved, but that an initially correlated state will also be preserved (both situations are physically relevant). The only mathematically rigorous result showing preservation of chaos in the Boltzmann limit is the very deep and powerful one of Lanford.[23,24] Although the proof is valid for a time on the order of one-tenth of a mean collision time, this is a very substantial result because, in the Boltzmann limit, the total number of collisions in this "short" time approaches infinity, the H-function decreases monotonically, and it becomes totally impossible to predict the precise path of any molecule. In other words, dynamical (n-particle) determinism is transferred to determinism of the one-particle distribution.

The question of preservation of initial correlations is also covered by Lanford's analysis, but much more can be said on a simpler though less rigorous basis. From the theory of the Boltzmann equation (e.g., Refs. 11, 13, and 14) it is known rigorously (under suitable restrictions) that the nonfluid part of an initial distribution will decay in a time on the order of the mean collision time, whereas the fluid component (suitably defined) will last for a macroscopically long time and eventually will decay on a viscous time scale (in ordinary air, the ratio of these two time scales can be on the order of 10^{10}). The distinction can be traced directly to the fact that the Boltzmann collision operator has a certain null space, namely, corresponding to conservation of mass, momentum, and energy; this null space leads directly to the macroscopic equations of conservation for these quantities. Moreover, it has been shown for a general fluid, not only a gas, that corresponding to any classically conserved quantities, $\varepsilon_1 \cdots \varepsilon_r$ (e.g. mass, momentum, energy, angular momentum, or any other), there is a macroscopic conservation equation.[7] In addition, for any pair of conserved quantities, ε_i and ε_j, the two-point correlation $\langle \varepsilon_i(x_1)\varepsilon_j(x_2)\rangle$ also satisfies a macroscopic conservation equation.[28] In the Boltzmann limit, the null space of the corresponding two-point collision operator will include all products of simple summational invariants.[28]

It follows that under appropriate circumstances (essentially long-range disturbances compared to the mean free path), all correlations between macroscopically conserved quantities will be maintained for macroscopically long times and eventually will decay to thermodynamic equilibrium together with the single-coordinate conserved quantities. In other words, an initially correlated state will preserve the correlations until an ultimate approach to thermodynamic equilibrium of the mean state.[14]

7. ASCENDING HIERARCHY FOR LINEAR SYSTEMS

Let $u(x, t)$ satisfy the linear evolution equation

$$\frac{\partial u}{\partial t} = Lu \tag{7.1}$$

Here x is a space of any finite dimension, L is linear, operating on $u(x)$ with t as a parameter, and, since we perform only elementary operations, the precise norms and spaces applying to u and L are not specified. For $u(x_1, x_2)$ defined in the product space of x with itself, we denote by L_1 the operator L on u as a function of x_1 with x_2 held fixed as a parameter, and similarly for L_2.

Let x_1 and x_2 be two symbols for the argument of u in Eq. (7.1) and set $u_i = u(x_i)$; an elementary computation yields

$$\frac{\partial}{\partial t}(u_1 u_2) = (L_1 + L_2)(u_1 u_2) \tag{7.2}$$

and so on for higher order products. Next, consider $u(x)$ to be a stochastic variable with an expectation $\langle\ \rangle$; it is sufficient to consider $\langle\ \rangle$ to be any linear, idempotent operator that commutes with $\partial/\partial t$ and with L. Then

$$\frac{\partial}{\partial t}\langle u_1 u_2\rangle = (L_1 + L_2)\langle u_1 u_2\rangle \tag{7.3}$$

and similarly for higher order correlations. If Eq. (7.1) is a determinate equation for the evolution of $u(x, t)$, each correlation $\langle u_1 u_2\rangle$, $\langle u_1 u_2 u_3\rangle$, $\langle (u_1 - \langle u_1\rangle)(u_2 - \langle u_2\rangle)\rangle$ and so on satisfies a determinate evolution equation, uncoupled with any other. Incidentally, both $\langle u\rangle$ and $u - \langle u\rangle$ satisfy the same equation, Eq. (7.1), as does u.

Mathematically, the operator $L_1 + L_2$ is very similar to the original operator L; in other words, there is not much more information available from higher order correlations than there is in the original determinate one-point equation. On the other hand, a vast amount of physical information can be obtained from what may be an elementary mathematical extension. The fact that for a linear (or linearized) system, determinate equations for all correlations follow without approximation, truncation, or cutoff (beyond what approximations may be needed for practical solution of the original system), explains the relative ease with which a large class of physical problems concerning fluctuations is solved, compared to the difficulty in handling nonlinear problems of this sort.

Next consider a vector $u^i(x, t)$, $i = 1 \cdots n$, which satisfies the linear evolution equation (sum on r)

$$\frac{\partial u^i}{\partial t} = L^{ir} u^r \tag{7.4}$$

As before, setting $u_1^i = u^i(x_1)$, $u_2^j = u^j(x_2)$, we observe that

$$\frac{\partial}{\partial t}\langle u_1^i u_2^j \rangle = \left(L_1^{ir} \delta_{js} + L_2^{js} \delta_{ir} \right) \langle u_1^r u_2^s \rangle \tag{7.5}$$

(Note that $\langle u_1^i u_2^j \rangle \neq \langle u_1^j u_2^i \rangle$.) In other words, the set of n^2 two-point correlations (and, of course, each set of higher order correlations) satisfies a determinate evolution system, uncoupled to any others.

We give a few elementary examples to illustrate the *linear ascending hierarchy*. If $L = \partial^2/\partial x^2$, then Eq. (7.1) is a heat equation; Eq. (7.2) is simply the two-dimensional heat equation

$$\frac{\partial u_{12}}{\partial t} = \left(\frac{\partial^2}{\partial x_1^2} + \frac{\partial^2}{\partial x_2^2} \right) u_{12} \tag{7.6}$$

If the original linear, determinate equation is a wave equation

$$\frac{\partial^2 u}{\partial t^2} = \frac{\partial^2 u}{\partial x^2} \tag{7.7}$$

it is first put into the form [Eq. (7.4)]

$$\frac{\partial u^1}{\partial t} = \frac{\partial u^2}{\partial x}$$

$$\frac{\partial u^2}{\partial t} = \frac{\partial u^1}{\partial x} \tag{7.8}$$

and becomes a slightly unusual fourth-order hyperbolic system for the four correlations $\langle u_1^i u_2^j \rangle$. Eliminating to obtain a single fourth-order equation (satisfied by any one of the four correlations), yields

$$\left(\frac{\partial^2}{\partial t^2} - \left(\frac{\partial}{\partial x_1} + \frac{\partial}{\partial x_2}\right)^2\right)\left(\frac{\partial^2}{\partial t^2} - \left(\frac{\partial}{\partial x_1} - \frac{\partial}{\partial x_2}\right)^2\right)\phi(x_1, x_2, t) = 0 \quad (7.9)$$

The characteristic cone is the square with corners $x_1 = \pm 1$, $x_2 = \pm 1$ [This is the convex hull of two degenerate cones, the diagonals of the square, which belong to the two second-order factors in Eq. (7.9).] From this example it is evident that although the operator in Eq. (7.5) is obtained directly from that of Eq. (7.4), its properties are not instantly visible.

As a third example take the linearized Boltzmann equation, written in the form

$$Df = Lf, \qquad D = \frac{\partial}{\partial t} + \xi_r \frac{\partial}{\partial x_r} = \frac{\partial}{\partial t} + \xi \cdot \frac{\partial}{\partial x} \quad (7.10)$$

[The convection operator, $\xi \cdot \partial/\partial x$ can be transposed and combined with the collision operator, L, thereby reducing the problem exactly to Eq. (7.1), but the results are more transparent in the present notation.] As before, writing $f_1 = f(x_1, \xi_1)$ and D_1 and L_1 for the operators on x_1 and ξ_1, respectively, also defining

$$D_{12} = \frac{\partial}{\partial t} + \xi_1 \cdot \frac{\partial}{\partial x_1} + \xi_2 \cdot \frac{\partial}{\partial x_2} \quad (7.11)$$

$$f_{12} = \langle f_1 f_2 \rangle, \qquad \psi_{12} = \langle (f_1 - \langle f_1 \rangle)(f_2 - \langle f_2 \rangle) \rangle \quad (7.12)$$

we obtain

$$D_{12} f_{12} = (L_1 + L_2) f_{12} \quad (7.13)$$

$$D_{12} \psi_{12} = (L_1 + L_2) \psi_{12} \quad (7.14)$$

and similarly for higher order correlation expressions. The linear ascending Boltzmann hierarchy is at each level totally decoupled and determinate just like any other linear system.

8. THE NONLINEAR BOLTZMANN HIERARCHY

The standard (descending) hierarchy in statistical mechanics starts from the determinate Liouville equation for $f^{(n)}$ and integrates down to obtain coupled equations for the lower order distributions (and, continuing the descent, for macroscopic state variables). None of these equations, singly or in combination, is determinate except for the Liouville equation itself. Formally letting $n \to \infty$ produces minor changes and yields the infinite Bogolubov–Born–Greene–Kirkwood–Yvon (BBGKY) hierarchy in which each equation is coupled to the next higher, and integration of any equation gives the entire lower order hierarchy. For reasons discussed in Ref. 9 this limit, $n \to \infty$, is not enough to give any significant simplification (except in thermodynamic equilibrium). In other words, there is no reason to expect, from a simple limit $n \to \infty$, that there is any decoupling or closure of any part of the system.

In the Boltzmann limit of a rarefied gas, one obtains a formally similar, infinite, coupled hierarchy, but with instantaneous, binary collisions as in the actual Boltzmann equation (this argument is indicated by Grad,[9] is elaborated on by Sastri[27] and Cercignani[2] and is rigorously derived by Lanford[23]). In the same abbreviated notation as in Section 7, this (descending) Boltzmann hierarchy is

$$\begin{aligned} D_1 f_1 &= [f_{1'\alpha'}] \\ D_{12} f_{12} &= [f_{12'\alpha'}] + [f_{1'2\alpha'}] \\ D_{123} f_{123} &= [f_{123'\alpha'}] + [f_{12'3\alpha'}] + [f_{1'23\alpha'}] \\ &\text{etc.} \end{aligned} \tag{8.1}$$

We note that only binary collisions enter into the Boltzmann hierarchy (although if one were to carelessly take too literally setting $x_3 = x_\alpha$ in $f_{123'\alpha'}$, since x_3 and x_α are only $\sigma \to 0$ apart, one would seem to be led to multiple collisions). For comparison, note the Boltzmann equation (determinate)

$$D_1 f_1 = [f_{1'} f_{\alpha'}] \tag{8.2}$$

Next we follow the ascending procedure of Section 7 and, starting from Eq. (8.2), construct an ascending sequence of correlations $\langle f_1 f_2 \rangle, \langle f_1 f_2 f_3 \rangle$, assuming implicitly that the determinate (but *nonlinear*) Eq. (8.2) is precisely obeyed in the Boltzmann limit (already attained) and that $\langle \ \rangle$ represents an expectation over some a priori probability density on a suitably defined space of random elements f_1. This is a second probability structure, logically independent of that inherent in the Liouville equation.

To be more precise in our notation, we write $\hat{f}_1$ for the argument in Eq. (8.2) to indicate that it is a random (fluctuating) distribution, and set

$$\langle \hat{f}_1 \rangle = f_1 \tag{8.3}$$

Instead of Eq. (8.1), we have

$$D_1 \hat{f}_1 = [\hat{f}_{1'} \hat{f}_{\alpha'}] \tag{8.4}$$

$$D_1 f_1 = [\langle \hat{f}_{1'} \hat{f}_{\alpha'} \rangle] \tag{8.5}$$

From Section 3, we are led to make the identifications (in the limit $n \to \infty$)

$$f_{1\cdots r} = \langle \hat{f}_1 \hat{f}_2 \cdots \hat{f}_r \rangle \tag{8.6}$$

A trivial calculation, as in Section 7, leads to the same *ascending* Boltzmann hierarchy as the descending hierarchy Eq. (8.1).

In other words, suppose macroscopic experimental imperfections are included in the a priori Liouville distribution f^n; although the expectation $f_1 = \int f^{(n)}$ may be unique (as $n \to \infty$), $\hat{f}_1$ retains a significant fluctuation about f_1. This lack of uniqueness of $\hat{f}_1$ leads to the Boltzmann hierarchy, Eq. (8.1), as originally described. However, we may consider a restriction of $f^{(n)}$ that converges with negligible dispersion to a unique $\hat{f}_1$ that satisfies the determinate Boltzmann equation (8.4) in the limit. In the shock tube problem, $f^{(n)}$ represents a mixed state in experiment 3. The restrictions of $f^{(n)}$ represent one of the pure states of experiments 1 and 2. The second probability structure implicit in the mean value $\langle\ \rangle$ in Eqs. (8.5) and (8.6) refers to applying the coin toss to the Navier–Stokes or Boltzmann equations instead of to Liouville's equation. Superposing the second a priori probability *after* taking the Boltzmann limit gives the same hierarchy, Eq. (8.1). This answers (positively) one of the questions raised in Ref. 19 but only for the infinite Boltzmann hierarchy, which, unfortunately, has very little mathematical theory, and only a slowly growing empirical theory,[3,27,31,33] none of which applies to a turbulent regime.

The fact that a linear system gives rise to a completely determinate and decoupled ascending hierarchy, as compared to the infinite, coupled, nonlinear hierarchy, is very important for applications. It explains, for example, why so many linear physical problems (e.g., scattering of small amplitude light or sound waves from a medium or from rough walls) can be solved using standard mathematical techniques (roughly speaking, the same techniques that are used for the original, determinate equation without fluctuations).

The almost universal procedure for obtaining information from the infinite hierarchy Eq. (8.1) is to introduce some truncation or closure recipe. For example, one can set

$$\bar{f}_{1\cdots r} = f_1 f_2 \cdots f_r \tag{8.7}$$

The bar over f signifies a replacement of $f_{1\cdots r}$ by some expression involving $f_1, f_{12}, \ldots, f_{1\cdots r-1}$. The choice of Eq. (8.7) is a particularly poor choice of truncation, since it offers essentially no improvement for increasing r; for example, f_1 satisfies exactly the same Boltzmann equation for any level of truncation. Nevertheless, this type of truncation is frequently used.[6]

It is conventional to replace the infinite sequence $\{f_1, f_{12}, f_{123}, \ldots\}$ by a formally equivalent sequence $\{f_1, \psi_{12}, \psi_{123}, \ldots\}$, which is more amenable to truncations. For example,

$$f_{12} = f_1 f_2 + \psi_{12} \tag{8.8}$$

and

$$f_{123} = f_1 f_2 + \{f_1 \psi_{23}\}_3 + \psi_{123} \tag{8.9}$$

serve to define ψ_{12} and ψ_{123} in turn; the notation $\{\ \}_s$ signifies a symmetric sum, not over all permutations of the subscripts $1 \cdots r$, but of those which yield distinct terms, and the index s is a check, indicating how many terms (not $r!$) occur in the sum. The continuation "etc" has been intentionally omitted after Eq. (8.9) because there are a number of possible sequels—for example, depending on whether terms like $\{\psi_{12}\psi_{34}\}_3$ are included in f_{1234}. Sastri[27] compares a number of these possibilities.

In the original sequence $\{f_1, f_{12}, \ldots\}$ we note that each successive term includes all the previous information; for example, $\int f_{123}\, dx_3 = f_{12}$, $\int f_{123}\, dx_2\, dx_3 = f_1$. These identities can be considered to be compatibility conditions on the sequence. Similarly, each equation in the Boltzmann hierarchy contains all the preceding equations. The equivalent compatibility for ψ_{12} and ψ_{123} is local (confined to one level), that is, $\int \psi_{12}\, dx_2 = 0$, $\int \psi_{123}\, dx_3 = 0$, and "etc." is now added as a natural requirement for the continuation (so far undefined) of the series (8.8) and (8.9),

$$\int \psi_{1 \cdots r}\, dx_r = 0 \tag{8.10}$$

Sastri[27] has introduced further compatibility criteria for judging the adequacy of a truncation scheme, using the guiding principle that it should at least have the opportunity of improving with increasing order. He considers a class of $\bar{f}^{(r)}$ that are polynomials in $f^{(1)} \cdots f^{(r-1)}$ (this excludes, e.g., Kirkwood's superposition formula, but includes the Mayer cluster series, among others).

Sastri's first criterion is that $\int \bar{f}_{1 \cdots r}\, dx_r$ should equal $f_{1 \cdots r-1}$, not a truncation or approximation to the latter. Second, if $\bar{f}_{1 \cdots r, \alpha}$ is used in the collision term of the hierarchy equation for $Df_{1 \cdots r}$, integrating down should give the exact lower order hierarchy system (in particular, *not* the Boltzmann equation for f_1). Thus in the truncated, coupled system, only the final equation is not exact. Sastri then shows that for a certain class of polynomial truncations, the second desideratum just formulated follows from the first.

In the conventional virial (or density) expansion, the Mayer cluster series enters naturally to account successively for two-body, three-body, and so on clusters of intermolecular collisions. Since the Boltzmann limit has taken the potential energy density $n\sigma^3$ to be zero, only two-body collisions are relevant; thus the Mayer (or cumulant) argument is no longer persuasive. We are led to a simpler choice[27] by considering the ascending hierarchy starting from the fluctuating determinate Boltzmann equation Eq. (8.4) and introducing the fluctuation relative to the mean,

$$\hat{f}_1 = f_1 + \hat{\delta}_1, \qquad \langle \hat{\delta}_1 \rangle = 0 \tag{8.11}$$

into Eq. (8.6). First we observe

$$f_{12} = \langle (f_1 + \hat{\delta}_1)(f_2 + \hat{\delta}_2) \rangle = f_1 f_2 + \langle \hat{\delta}_1 \hat{\delta}_2 \rangle$$

$$f_{123} = f_1 f_2 f_3 + \{ f_1 \langle \hat{\delta}_2 \hat{\delta}_3 \rangle \}_3 + \langle \hat{\delta}_1 \hat{\delta}_2 \hat{\delta}_3 \rangle$$

and conclude that

$$\psi_{12} = \langle \hat{\delta}_1 \hat{\delta}_2 \rangle \qquad \psi_{123} = \langle \hat{\delta}_1 \hat{\delta}_2 \hat{\delta}_3 \rangle \tag{8.12}$$

If we now say "etc." by *defining*

$$\psi_{1 \cdots r} = \langle \hat{\delta}_1 \cdots \hat{\delta}_r \rangle \tag{8.13}$$

we obtain the unique resolution

$$f_{1 \cdots r} = f_1 \cdots f_r + \{ f_1 f_2 \psi_{3 \cdots r} \} \binom{r}{2} + \cdots + \psi_{1 \cdots r} \tag{8.14}$$

in which every term (except the first) contains a single ψ-factor. This is *not* the standard cumulant representation.

The basis for the fluctuating ascending hierarchy of ψ-correlations is the first-order fluctuation equation

$$D_1 \hat{\delta}_1 = L_1 \hat{\delta}_1 + [\hat{\delta}_{1'} \hat{\delta}_{\alpha'} - \psi_{1'\alpha'}], \qquad \psi_{1'\alpha'} = \langle \hat{\delta}_{1'} \hat{\delta}_{\alpha'} \rangle \tag{8.15}$$

where

$$L_1 \hat{\delta}_1 = [f_{1'} \hat{\delta}_{\alpha'} + f_{\alpha'} \hat{\delta}_{1'}] \tag{8.16}$$

is the linearized Boltzmann collision operator (linearized about f_1, which happens not to be a solution of the Boltzmann equation).

Constructing the ascending hierarchy of products of $\hat{\delta}$, and using the definition (8.13), yields

$$D_{12} \psi_{12} = (L_1 + L_2) \psi_{12} + [\psi_{1'2\alpha'}] + [\psi_{12'\alpha'}] \tag{8.17}$$

and, in general

$$D_{1 \cdots r} \psi_{1 \cdots r} = (L_1 + \cdots + L_r) \psi_{1 \cdots r} + [\psi_{1 \cdots r'\alpha'} - \psi_{1 \cdots r-1} \psi_{r'\alpha'}]_r \tag{8.18}$$

$$L_1 \psi_{1 \cdots r} = [f_{1'} \psi_{\alpha' 2 \cdots r} + f_{\alpha'} \psi_{1' 2 \cdots r}] \tag{8.16$'$}$$

where the permutation index r is used together with the Boltzmann collision symbol [ ] exactly as with the permutation symbol { }, and permutations are understood to operate on the indices $1 \cdots r$ but not α (physically, none of the molecules $1 \cdots r$ collide with one another). Note that the factor $\psi_{1\cdots r-1}$ in the last term of Eq. (8.18) is not operated on by collisions. Each $\psi^{(r)}$-equation is coupled to f_1 (through L), also to $\psi^{(r+1)}$, $\psi^{(r-1)}$, and $\psi^{(2)}$. In the present notation, the Boltzmann hierarchy, Eq. (8.1), is simply

$$D_{1\cdots r} f_{1\cdots r} = \left[f_{1\cdots r'\alpha'} \right]_r \tag{8.19}$$

Until now the algebraic manipulations with the hierarchy have been exact. It is common practice to assume that fluctuations are small, that is, $\hat{\delta}_1$ is, in some sense, small compared to f_1. If $\hat{\delta}$ is ordered by ε, then $\psi_{12} = O(\varepsilon^2)$, $\psi_{1\cdots r} = O(\varepsilon^r)$ and keeping only the dominant terms (assumed not to vanish) in each equation, the relevant equations become

$$D_1 f_1 = [f_{1'} f_{\alpha'}] \tag{8.20}$$

$$D_{1\cdots r}\psi_{1\cdots r} = (L_1 + \cdots + L_r)\psi_{1\cdots r} \tag{8.21}$$

and

$$D_1 \hat{\delta}_1 = L_1 \hat{\delta}_1 \tag{8.22}$$

Except for f_1, which satisfies the conventional nonlinear Boltzmann equation, the fluctuations and correlations satisfy linear, uncoupled equations (once f_1 is determined, hence L).

For sufficiently small $\hat{\delta}$, it seems likely that the formal decoupling, Eqs. (8.20)–(8.22), can be proved rigorously using conventional Boltzmann equation theory.[11,13] This would give an existence theorem for the infinite Boltzmann hierarchy for weak nonlinearity and weak fluctuations.

The complications found in statistical mechanics arise from the fact that the Boltzmann limit is not taken. There is an additional parameter, $n\sigma^3 \sim \sigma \sim n^{-1/2}$. The general case of small fluctuations should be expressed in terms of two small parameters, ε (as above, a measure of experimental imperfection or, possibly, weak instability) and $\sigma \sim n^{-1/2}$ (a measure of thermal fluctuations, contained, e.g., in the difference between f_1 and f_1^σ). It is evident that in almost all experiments, the experimental imperfection is orders of magnitude larger than the thermal noise. However, this estimate is not uniformly valid and scale length also enters (through the fact that thermal fluctuations are relatively larger for small domains containing few molecules). For example, in Ref. 5 the two small parameters ε and σ are implicitly ordered relative to one another as they would be in thermodynamic equilibrium (where there are no long-range correlations in a gas). It is not clear that this calculation is correct until a more systematic boundary layer matching procedure is carried out on the two relevant length scales.

9. STABILITY, LONG-TERM PREDICTION, AND TURBULENCE

Any derivation of a macroscopic or low order description from a higher order description is a very strong stability statement. For example, consider an initial value problem for the Boltzmann equation in the limit of small mean free path, ε. An essentially arbitrary initial distribution will, within a short time, roughly $O(\varepsilon)$, converge to an approximately local Maxwellian [discrepancy also $O(\varepsilon)$], and will remain locally Maxwellian, with macroscopic fluid behavior, thereafter.[11] This contraction of an essentially arbitrary distribution function to a much smaller subspace is a strong stability result. The greater detail of the initial state is not entirely forgotten; it remains as an $O(\varepsilon)$ perturbation in the subsequent fluid state [or an $O(\varepsilon^2)$ alteration if the initial state happens to be exactly locally Maxwellian—thereby distinguishing the Boltzmann from the fluid solution even with precisely fluid initial values].[11]

The reduction from Liouville to Boltzmann is also preceded by a transient [for a time $O(\sigma) \sim O(n^{-1/2}) \sim$ duration of a collision] until most of the molecules initially colliding complete their collisions, alternatively until the initial two-sided chaos becomes one sided.[9] It is clear (by estimating the number of incomplete collisions with small relative velocity) that this transient will initially decay algebraically rather than exponentially (and will at some higher order in the expansion parameter σ, eventually give a logarithmic correction—cf. "The Ubiquitous Logarithm" in Ref. 13). Again, the reduction from an n-particle to a 1-particle description is a very strong stability statement.

Now consider the Navier–Stokes equations, without regard to kinetic justification, and assume a strong existence, uniqueness, and continuous dependence on initial data theorem (not yet available in three dimensions, in the large in time). Also consider an unstable initial value problem, for example, an unstable steady channel flow or flow around an object. A small initial disturbance or "error," ε, will give rise to a growing error that may look like εt^{α} or $\varepsilon e^{t/\tau}$. Technically, there is continuous dependence, since both errors approach zero with ε for any $0 < t < t_1$. But, particularly in the second case, $\varepsilon e^{t/\tau}$, it is clear that "continuous dependence" is a mathematical artifact, especially if one is interested in $t \sim 20\tau$ or $t \sim 50\tau$. Even in the first case, εt^{α}, since fluid dynamic data are rarely controllable to within better than relative accuracies like 10^{-2} or 10^{-3}, the same conclusions can usually be drawn.

Exactly the same conclusions can be drawn for solutions of the Boltzmann and Liouville equations of the same unstable physical problem; namely, there is no effective continuous dependence on initial data and initial chaos will be lost after some macroscopic time. Furthermore, there is a built-in discrepancy between solutions of the Boltzmann and Liouville equations for the "same" fluid problem: in the case of the Boltzmann equation depending on the value of the small mean free path, in the case of the Liouville equation depending on both the mean free path and the Boltzmann limit parameter, $n\sigma^3$. Also the "initial error" of magnitude ε depends on the entire one-particle distribution (Boltzmann) or $f^{(n)}$ (Liouville), which may be assumed to be not under conscious experimental control at all.

Before a perturbation of small magnitude ε has a chance to grow too much, the small $\hat{\delta}$, ψ_{12} and so on analysis of Section 8 can be assumed to hold. The determinate linear correlation equation for ψ_{12} is almost exactly (for $\hat{\delta}_1$ it is *exactly*) the linearization of the Boltzmann equation that would be used conventionally to define linear stability of a given steady flow. In other words, correlations will start to grow at the same rate as the main flow (for $\hat{\delta}_1$; for ψ_{12} at twice the rate).

If we are dealing with a macroscopic situation in which perturbations eventually become large and disturb the main flow (this includes the development of steady cellular structures as well as turbulence), two-point and higher correlations will also grow until they are not small compared to the original flow. In other words, chaos will be destroyed. The presence of macroscopic instability therefore offers the possibility (more likely, the certainty) that the validity of the Boltzmann equation (and also the Navier–Stokes equation) is lost as the instability becomes important.

We can adopt the strict mathematical view that $\varepsilon e^{t/\tau}$ can be made arbitrarily small by taking ε sufficiently small (in principle if not in fact). We can then salvage the applicability of chaos and the Boltzmann or Navier–Stokes equations by superimposing a second probability structure as discussed in Section 8, on the (presumably known, in some sense) exact solutions of the determinate equations. This application, however, is more dubious than the case described in Section 8. It is not only abstract duplication of initial conditions that differentiates the two formulations, but a difference in actual physical parameters (mean free path and $n\sigma^2$) that differentiate the equations. The solutions of Navier–Stokes and Boltzmann and Liouville are different; and these differences will grow in some way related to the instability.

It is at present pure hypothesis that, despite the pointwise discrepancy, the *statistical* behavior of turbulent solutions of the Navier–Stokes equations may duplicate those of the Boltzmann equation (i.e., that the two correlation hierarchies are in some sense equivalent).

There is a much more practical reason *not* to make use of this hypothesis, and to treat fluid turbulence directly from the Boltzmann hierarchy (non-chaotic), rather than first take a fluid limit, then work with an independent stochastic structure in fluid dynamics. To justify this, we take a slight detour and examine the different behavior of fluid and Boltzmann equation toward nonlinearity.

First we remark that any fluid system of partial differential equations exhibits nonlinearity essentially everywhere. The differential (convection) part of the Boltzmann equation is strictly linear, and nonlinearity appears only in velocity space through the collision term. In the usual derivation of macroscopic conservation equations, the only nonlinearity (viz., the collision term) drops out identically. The nonlinearity of fluid dynamics appears only when the terms that arise from integrating the linear convection term are interrelated through the very nonlinear Maxwellian distribution. This very different way in which nonlinearity enters is observed in a comparison of solutions to representative macroscopic and kinetic problems.[18]

The point is basically trivial. The mathematical nature and optimal tools for analyzing a physical problem can be completely different when approached by a fluid model or a kinetic model. In turbulent flow, both models offer infinite hierarchies, in one case nonlinear differential systems, in the other case integrodifferential systems (where the differential part is linear). A natural approximation or expansion or truncation of one may be very deeply hidden in the other. The transformation of one to the other (by a generalization of Hilbert or Chapman–Enskog) is a very complex procedure. What the Boltzmann hierarchy offers is entirely new avenues of approach to an intractable problem.

We give just one example. Consider a problem of stationary turbulence and assume (without justification) that $\psi_{123} = 0$. The equation for ψ_{12} is exactly that of linearized Boltzmann stability of the mean flow f_1. In other words, if there is to be a stationary turbulent state, the mean flow must be linearly stable (for large wavelengths this implies fluid stability of the mean flow). The key assumption, $\psi_{123} = 0$, does not seem to have any simple translation into fluid terms.

The literature in this subject is quite sparse. Tsugé[31] has considered Tollmein–Schlichting waves using a generalization of the 13-moment approach. Unfortunately, the problem is linearized, so that the inclusion of two-point correlations does not offer very much additional information. Zhigulev[33] and Chen[3] have expanded the Boltzmann hierarchy using Chapman–Enskog techniques; this basically nonlinear approach is restricted by the assumption that fluctuations are small compared to the basic flow, which is only slightly better than linearization (cf. Section 8).

ACKNOWLEDGMENT

This work was supported by the U.S. Department of Energy, contract number DE-ACO2-76ER-03077 and the Air Foorce Office of Scientific Research, contract number AFOSR-81-0020.

REFERENCES

1. R. Caflisch, *Commun. Pure Appl. Math.* **33**, 651 (1980).
2. C. Cercignani, *Theory and Application of the Boltzmann Equation* (Elsevier, New York, 1975).
3. T. Q. Chen, to be published (1982) (private communication).
4. E. G. D. Cohen, *Acta Phys. Austriaca, Suppl. X* 157 (1973).
5. M. H. Ernst and E. G. D. Cohen, *J. Stat. Phys.* **25**, 153 (1981).

6. R. E. Fox and G. E. Uhlenbeck, *Phys. Fluids* **13**, 2881 (1970).
7. H. Grad, *J. Phys. Chem.* **56**, 1039–1048 (1952).
8. H. Grad, *Phys. Rev.* **91**, 1031 (1953).
9. H. Grad, "Principles of the Kinetic Theory of Gases," in *Handbuch der Physik*, Vol. 12, S. Flugge, Ed. (Springer-Verlag, Berlin, 1958), pp. 205–294.
10. H. Grad, *J. Chem. Phys.* **33**, 1342–1348 (1960).
11. H. Grad, *Phys. Fluids* **6**, 147–181 (1963).
12. H. Grad, "Microscopic vs. Macroscopic Models," in *Mathematical Models in Physical Sciences*, S. Drobott, Ed. (Prentice-Hall, Englewood Cliffs, NJ, 1963); pp. 3–16.
13. H. Grad, "Asymptotic Equivalence of the Navier–Stokes and Nonlinear Boltzmann Equations," *Symposia in Applied Mathematics*, Vol. 17 (American Mathematical Society, Providence, RI, 1965), pp. 154–183.
14. H. Grad, *J. SIAM* **13**, 259 (1965).
15. H. Grad, "Long-Range Correlation in a Gas," Yeshiva University, Statistical Mechanics Conference, November 30, 1966.
16. H. Grad, *J. SIAM* **14**, 935–955 (1966).
17. H. Grad, "Levels of Description in Statistical Mechanics and Thermodynamics," in *Delaware Seminar on the Foundation of Physics* (Springer-Verlag, New York, 1967), pp. 49–76.
18. H. Grad, "Singular and Nonuniform Limits of Solutions of the Boltzmann Equation," in *Transport Theory*, Vol. 1, *Proceedings of Symposia in Applied Mathematics*, (AMS, SIAM, Providence, R.I., 1969).
19. H. Grad, "Singular Limits of Solutions of Boltzmann's Equation," in *Rarefied Gas Dynamics* K. Karamcheti, Ed. (Academic Press, New York, 1974), pp. 37–53.
20. E. Hewitt and L. J. Savage, *Trans. Amer. Math Soc.*, **80**, 470-501 (1955).
21. E. Hopf, *J. Rational Mech. Anal.* **1**, 87 (1952).
22. L. D. Landau and E. M. Lifshitz, *Fluid Mechanics* (Addison-Wesley, New York, 1959), Chap. 17.
23. O. E. Lanford, "Time Evolution of Large Classical Systems," in *Dynamical Systems and Applications*, J. Moser, Ed., Lecture Notes in Physics, Vol. 38 (Springer-Verlag, Berlin, 1975).
24. O. E. Lanford, *Soc. Math. Fr., Astérisque* **40**, 117 (1976).
25. C. B. Morrey, *Commun. Pure Appl. Math.* **8**, 279 (1955).
26. G. V. Ramanathan, *Il Nuovo Cim.* **49B**, 31 (1979).
27. C. C. A. Sastri, "Long-Range Correlations in Kinetic Theory," Report MF-72, Courant Institute of Mathematical Sciences, New York University, New York (1973) (unpublished).
28. C. C. A. Sastri, *J. Stat. Phys.* **13**, 43 (1975).
29. H. Spohn, "Fluctuation Theory for the Boltzmann Equation," in *Studies in Statistical Mechanics*, E. W. Montroll and J. L. Lebowitz, Eds. (North-Holland, New York, 1981).
30. H. Spohn, "Boltzmann Hierarchy and Boltzmann Equations," to be published, (private communication).
31. S. Tsugé, *Phys. Lett.* **33A**, 145 (1970).
32. S. Tsugé and K. Sagara, *J. Stat. Phys.* **12**, 403 (1975).
33. V. N. Zhigulev, *Sov. Phys.-Doklady* **10**, 1003 (1966).

SUFFICIENT CONDITIONS TO SINGLE OUT THE GIBBS MEASURE FROM OTHER TIME-INVARIANT MEASURES

SHELDON GOLDSTEIN
Department of Mathematics
Rutgers University
New Brunswick, New Jersey

Abstract. *Conditions are established that single out particular stationary probability measures on the phase space of dynamical systems. The microcanonical and canonical ensembles and infinite volume Gibbs states are discussed.*

The proposal of J. W. Gibbs[1] that the appropriate description of a macroscopic system in thermodynamic equilibrium is provided by certain probability measures on the phase space of the system—what we now call Gibbs measures—has for quite some time been universally accepted. Yet a *complete* derivation, from first principles, of the validity of Gibbs's proposal has never been provided. Acceptance has been based more on the fruitfulness of the system, on the fact that it leads to correct predictions, than on any a priori justifications.

Here I discuss some attempts at a partial justification of Gibbs's proposal, emphasizing conditions that single out the Gibbs measure from other measures on phase space. The conditions should be such that any probability measure describing thermodynamic equilibrium must, or at least plausibly should, satisfy them. Even if such conditions are found, however, a complete justification of Gibbs's proposal will not have been provided until it is shown that

Work supported in part by National Science Foundation grant PHY-7803816.

thermodynamic equilibrium should be described by *probability measures*. I do not discuss this problem here.

This chapter treats several related problems:

1. To find conditions that single out the canonical ensemble

$$\mu_\beta \sim e^{-\beta H}\, dq\, dp$$

where $\beta = 1/kT$, $k =$ Boltzmann's constant, $T =$ the temperature of the system, and H is the Hamiltonian, from the class of all probability measures on the phase space of a system consisting of a finite number of particles confined to a box.

2. To find conditions that single out the microcanonical ensemble μ_E, the "projection" of the Liouville measure $dq\, dp$ onto the energy surface $H = E$ of a finite system of particles in a box, from the class of all probability measures on the energy surface.

3. To find conditions that single out measures of the form

$$\rho(H)\, dq\, dp$$

for a finite system in a box.

4. To find conditions that single out infinite volume Gibbs states[11] from the class of all probability measures on the phase space of a system consisting of an infinite number of particles moving in all of space.

For most of the discussion that follows, I take the point of view that a solution for problem 2 or 3 obviates the need for a solution of problem 1, since measures given by a density $\rho(H)$ with expected energy $\langle H \rangle = E$ are for the most part physically equivalent to μ_E; the choice $\rho(H) \sim e^{-\beta H}$ may be regarded as a matter of convenience. In this discussion I deviate from standard usage and regard μ_β and μ_E as well as measures of the form $\rho(H)\, dq\, dp$ as Gibbs measures.

1. ENTROPY

A condition that I do not regard as appropriate is (for problem 2) that the measure $d\mu = f d\mu_E$ on the energy surface should have maximal entropy $-\int(f(x) \ln f(x) \mu_E(dx)$. The maximum of course occurs only for $f = 1$, and thus μ_E is the probability measure of maximal entropy.

I believe that this condition begs the question. The fact that the formula for the entropy involves crucially the density f of μ *with respect to* μ_E already singles out μ_E as the appropriate *a priori measure*. That the (information theoretic) entropy (relative to μ_E) is maximized by μ_E itself merely restates the same thing. For example, if noncanonical coordinates x_i were used to define the phase space volume element $dx = dx_1\, dx_2 \cdots$, a different measure μ'_E—the projection of dx on the energy surface $H = E$—would be obtained, and it, of course, would maximize the entropy (relative to *it*).

2. STATIONARITY

Any probability measure that represents thermodynamic equilibrium must be stationary—it must be the same at all times. More precisely, a measure μ is stationary if

$$\mu(T^t A) = \mu(A)$$

where T^t is the flow on phase space induced by Hamilton's equations and A is any (Borel) set. The measures [e.g., $\rho(H)\, dq\, dp$ or μ_E] we wish to single out are all stationary. This is a consequence of the conservation of energy and the preservation of volume (Liouville's theorem) for Hamiltonian flows.[2] And, in fact, Gibbs's primary consideration in his selection of ensemble was stationarity: the obvious candidates for stationary measures are given by densities $\rho = \rho(H)$, which depend only on the energy. However, in many systems, if not in most, there will be other stationary measures; the existence of such measures is clear for systems in which the motion is very regular, which have (many) constants of the motion in addition to the energy. However, they can even occur in systems in which the motion, the flow, is highly irregular. For example, for a system in a cube consisting of a finite number n of hard spheres, which undergo elastic collisions but are otherwise noninteracting, there exist (provided n is not too large) stationary measures supported by phase points for which all the spheres are moving parallel to one another and perpendicular to one of the sides.

Thus stationarity alone is (usually) not sufficient, and the problem is then, as stated in the title, to single out Gibbs measures from other stationary measures.

3. ABSOLUTE CONTINUITY

The stationary measures for hard spheres described above are measures in which the motions of the spheres are very highly correlated; for example, all the velocities are parallel. In particular, the measures are *singular*—they are supported by a set of volume zero in the phase space. [If we are considering measures on the energy surface $H = E$, these stationary hard sphere measures will be supported on a set of (μ_E) measure zero on the energy surface.] Such measures are excluded by the condition of absolute continuity.

A measure μ on the phase space (energy surface) is *absolutely continuous* if it is given by a density, $d\mu = \rho\, dq\, dp$ ($d\mu = \rho\, d\mu_E$).

The microcanonical ensemble μ_E is the only absolutely continuous stationary probability measure on the energy surface $H = E$ if and only if the flow T^t is *ergodic* on the energy surface [with respect to the stationary measure μ_E–ergodic theory studies flows (or mappings) that preserve a given measure]. Ergodicity is more commonly formulated in other ways (e.g., the equality of time averages and phase averages). For a discussion of the mathematical aspects of ergodicity, see Arnol'd and Avez,[3] and for a further treatment of the relevance of ergodicity to the problem considered here, see the review article by Penrose.[4]

How common is the ergodicity of the Hamiltonian flow on the energy surface? Ergodicity is now regarded as much less common than was believed 30 years ago. At that time it was of course known that integrable systems, which have many independent constants of the motion, are not ergodic. It was believed, however, that perturbations of integrable systems would be ergodic unless the perturbation possessed some symmetry. This is now known to be false. As a consequence of Kolmogorov–Arnol'd–Moser (KAM) theory (see Ref. 3), it is known that systems sufficiently near-integrable are not ergodic. The extent to which systems far from integrable are ergodic is by no means clear. However, Sinai[5] has shown that the hard sphere gas is ergodic, at least for two or three hard spheres.

4. STABILITY

Since we do not and cannot know the Hamiltonian H of our system precisely, to be a physically reasonable description of thermodynamic equilibrium a measure must not only be stationary for the flow T^t generated by H, it should also be approximately stationary for the flows $T^t_{H'}$ generated by Hamiltonians H' close to H, in the sense that $\mu \circ T^t_{H'}$ stays close to μ for all $t > 0$. Here $\mu \circ T^t_{H'}$ is the measure obtained by evolving μ under the flow $T^t_{H'}$. A basically equivalent formulation of this condition on a stationary measure μ is the following: for each Hamiltonian $H' = H_\varepsilon = H + \varepsilon V$ sufficiently close to H, there should exist a probability measure μ_ε near μ and stationary under $T^t_\varepsilon \equiv T^t_{H'}$

$$\mu_\varepsilon \circ T^t_\varepsilon = \mu_\varepsilon$$

$$\mu_\varepsilon \to \mu, \qquad \varepsilon \to 0$$

This condition, a germ of which already appears in Gibbs,[1] was precisely formulated by Haag et al.[6] in their investigation of the stationary states of quantum systems. Probability measures (or quantum states) satisfying this condition are called *stable*.

It is easy to see that measures on phase space of the form $\rho(H)\,dq\,dp$ are stable, since we may set

$$d\mu_\varepsilon = \text{const}\,\rho(H_\varepsilon)\,dq\,dp \approx \rho(H)\,dq\,dp$$

On the other hand, consider a system composed of two subsystems between which there are no interactions. Suppose the first subsystem has Hamiltonian H_1 and the second has Hamiltonian H_2. Then $c_1 e^{-\beta_1 H_1}\, dq_1\, dp_1$ is stationary for the first subsystem, $c_2 e^{-\beta_2 H_2}\, dq_2\, dp_2$ is stationary for the second subsystem, and the product $ce^{-(\beta_1 H_1 + \beta_2 H_2)} = \mu$ is stationary for the composite system. If $\beta_1 = \beta_2$, then μ is stable. If $\beta_1 \neq \beta_2$, μ is presumably not stable, since it should converge under the time evolution to a state in which both subsystems are at the same temperature, if a small interaction between the subsystems is introduced.

It is therefore reasonable to suppose that stability should single out Gibbs states. The only case in which this can be established in a straightforward manner, and without additional assumptions, is for finite quantum systems. The argument goes as follows. Suppose the density matrix ρ is stationary, $[\rho, H] = 0$. If H had a nondegenerate spectrum we could conclude that $\rho = \rho(H)$. If ρ is also stable, then by considering perturbations that split the degeneracy in various ways, we may conclude that $\rho = \text{const} \times \text{identity}$ on energy levels [i.e., $\rho = \rho(H)$].

To obtain a similar theorem for finite classical systems, a stronger form of stability has been used. Roughly speaking, μ is strongly stable if it is stable with μ_ε (see definition of stability) given by a smooth density $\rho_\varepsilon(x)$, $d\mu_\varepsilon = \rho_\varepsilon\, dq\, dp$, which is differentiable in ε at $\varepsilon = 0$ with smooth derivative $g = (d/d\varepsilon)\rho_\varepsilon|_{\varepsilon=0}$. For a finite classical system in which the totality of periodic orbits forms a dense set and the energy surfaces are connected, all strongly stable stationary measures have densities of the form $\rho = \rho(H)$.[7]

The density condition on periodic orbits is not as strong as it may seem. In fact, this condition is generic for C^1 Hamiltonian systems.[8]

The strong stability used here is perhaps stronger than necessary, but stability alone is not sufficient: the persistence of invariant tori, as provided by KAM theory, easily gives the stability of stationary measures that are not Gibbs measures.[7] (Perhaps for systems with more than two degrees of freedom, stability, strengthened by requiring that ρ_ε be smooth, will suffice.)

Stability results have also been obtained for infinite classical systems. It is shown by Aizenmann et al.[9] that just as for infinite quantum systems,[6] the only stable stationary measures that also satisfy several other technical conditions are Gibbs measures (or mixtures of Gibbs measures).

5. STATIONARITY WITH STOCHASTIC BOUNDARIES

The introduction of the stability condition was motivated by the observation that we do not precisely know the true Hamiltonian for our system. For realistic systems the situation is actually much worse: the system is never completely closed, so that its motion is not given precisely by any Hamiltonian at all that does not include the interactions of elements outside the system. If our Hamiltonian is not to include contributions from all particles in the universe, the actual motion of the system is perhaps best given by the Hamiltonian motion supplemented by random effects representing the interactions with particles outside the system. The simplest way to introduce such effects is to alter the rules for reflection of particles from the walls of our systems. Consider the dynamics, for n interacting particles moving in a box, given by Hamilton's equations (with Hamiltonian H) between collisions with the walls. Assume the rule that no matter what the velocity of a particle before collision with a wall, it comes off the wall with the velocity distribution corresponding to an inverse temperature β; that is, the post collisional velocity is a random variable with distribution $\sim \mathbf{n} \cdot \mathbf{v} e^{-(1/2)\beta m v^2}$, where $\mathbf{n}$ is the inward unit normal to the wall. This dynamics models the evolution of a system inside a box whose walls are held at inverse temperature β. The motion obtained is neither conservative—energy is not conserved in "reflections"—nor deterministic, but rather defines a Markov process. It is easy to see that the canonical ensemble $\mu_\beta \sim e^{-\beta H}$ is stationary for this process. Moreover,[10] it is the only stationary measure. Any initial measure converges to μ_β (in variation norm).

The result just described asserts that μ_β is singled out by the condition of stationarity under an evolution that embodies the effects of walls held at inverse temperature β. But it does more than this. It suggests that to find measures describing the steady state (nonequilibrium) heat flow that arises when the temperature varies along the walls, we should look for stationary measures of the dynamics that differs from the one described above only in that the parameter β describing the velocity distribution of particles leaving the wall now depends on the point x of collision: $\beta = \beta(x)$, the inverse temperature of the wall at x. It can be shown[10] that there exists a unique stationary measure $\mu = \mu_{\{\beta(x)\}}$ for this process. The measure μ is equivalent to Lebesgue measure (in the sense that it is given by a nowhere vanishing density). Moreover, as t approaches infinity every initial measure (even a δ-measure) converges (in variation norm) to μ, which depends continuously on the temperature function $\beta(x)$ and, in particular, approaches μ_β, the canonical ensemble, as $\beta(x) \to \beta$ (constant).

6. LONG-RANGE MARKOV PROPERTY FOR INFINITE SYSTEMS

A natural condition on measures on the phase space for an infinite system is that they satisfy, if not a local Markov property, at least a long-range Markov property: the conditional probability for what will be observed locally should not be sensitive to what is happening very far away. A useful formulation of this condition is that the measure be a Gibbs measure for at least some potential ϕ.[11] Gurevich and Sukhov[15] have shown that if such a measure μ satisfies the stationary Bogolubov–Born–Green–Kirkwood–Yvon (BBKGY) hierarchy for the potential h, then μ is a Gibbs measure for h. Note that this result does not require that the infinite system time evolution (given by h) be well defined on the "support" of μ, which in fact has been established only if μ

is a Gibbs state for h.[12] Satisfaction of the stationary BBKGY hierarchy is an expression of "stationarity" that does not require the existence of a well-defined evolution.

7. BEHAVIOR UNDER COMPOSITION

Suppose we say that a measure μ for a system S has a property *completely* if for any integer $n \geqslant 2$ the n-fold product measure, $\mu \times \cdots \times \mu$ (n factors), has the property for the system $S \times \cdots \times S$ (n factors) composed of n independent copies of S (each copy having the same Hamiltonian). It is easy to see that if a measure has the property of being given by a density $\rho = \rho(H)$ completely, then $\rho \sim e^{-\beta H}$ (only $n = 2$ is needed for this). In particular, if the measure $d\mu = \rho\, dq\, dp$ on the finite system S (satisfying the conditions in the strong stability result of Section 4) is completely strongly stable, then $\rho \sim e^{-\beta H}$.

Pusz and Woronowicz[13] and Lenard[14] have investigated a property of a state (or density matrix) ρ that they call passivity. A form of the second law of thermodynamics, passivity requires that no perturbation that acts for only a finite amount of time can decrease the energy of the system (on the average with respect to ρ). Lenard[14] shows for quantum systems described by a finite-dimensional Hilbert space that a completely passive state is of the form $\rho \sim e^{-\beta H}$ (or is a ground state).

A similar result for infinite quantum systems appears in Ref. 15, where it is also shown that passive states possessing good clustering properties are quantum Gibbs states (i.e., satisfy the KMS condition). It is in Ref. 15 that the notions of passivity and complete passivity were introduced.

REFERENCES

1. J. W. Gibbs, *Elementary Principles of Statistical Mechanics* (Yale University Press, New Haven, CT, 1902).
2. H. Goldstein, *Classical Mechanics* (Addison-Wesley, Reading, MA, 1950).
3. V. I. Arnol'd and A. Avez, *Ergodic Problems of Classical Mechanics* (Benjamin, New York, 1968).
4. O. Penrose, *Rep. Prog. Phys.* **42** 1937 (1979).
5. Ya. G. Sinai, *Russ. Math. Surv.* **25**:2, 137 (1970).
6. R. Haag, D. Kastler, and E. B. Trych-Pohlmeyer, *Commun. Math. Phys.* **38**, 173 (1974).
7. J. L. Lebowitz, M. Aizenman, and S. Goldstein, *J. Math. Phys.* **16**, 1284 (1975).
8. S. Newhouse, *Am. J. Math.* **99** (1977).
9. M. Aizenman, G. Gallavotti, S. Goldstein, and J. L. Lebowitz, *Commun. Math. Phys.* **48**, 1 (1976).
10. S. Goldstein, J. L. Lebowitz, and E. Presutti, "Mechanical Systems with Stochastic Boundaries," in *Colloquia Mathematica Societatis Janos Bolyai*: 27. Random Fields, Esztergom (Hungary), 1979 (published in 1982).

11. R. L. Dobrushin, *Funct. Anal. Appl.* **3**, 27 (1969); O. E. Lanford III and D. Ruelle, *Commun. Math. Phys.* **13**, 194 (1969).
12. O. E. Lanford III, *Commun. Math. Phys.* **9**, 179 (1968); Ya. G. Sinai, *Vest. Moscow Univ.* **1**, 152 (1974); C. Marchiori, A. Pellegrinotti, and E. Presutti, *Commun. Math. Phys.* **40**, 175 (1975).
13. W. Pusz and S. L. Woronowicz, *Commun. Math. Phys.* **58**, 273 (1978).
14. A. Lenard, *J. Stat. Phys.* **19**, 575 (1978).
15. B. M. Gurevich and Yu. M. Sukhov, *Commun. Math. Phys.* **49**, 63 (1976); B. M. Gurevich, Ya. G. Sinai, and Yu. M. Sukhov, *Usp. Math.-Nauk* **28**, 5, 49 (1973).

HOW RANDOM IS A COIN TOSS?

JOSEPH FORD
School of Physics
Georgia Institute of Technology
Atlanta, Georgia U.S.A.

Abstract. *This chapter discusses the paradox of random versus deterministic approaches in nonlinear dynamical systems using a combination of ergodic theory and algorithmic complexity theory. Integrability, Bernoulli systems, hierarchy of chaos, mixing systems, C-systems, and K-systems are treated. Future research in nonlinear dynamical systems receives considerable attention in the second part of the chapter. The contradiction between the mathematical requirements of infinite precision concerning observations and the obviously finite precision offered by physical systems is investigated, and the consequences are pointed out.*

1. INTRODUCTION

The coin in the chapter title symbolizes a quite ancient and profound paradox whose resolution perhaps contains the seeds of startling future developments in science. Specifically, an honest coin was among the first of Newtonian systems to be regarded as strictly deterministic or completely random, depending on the point of view adopted. Strictly speaking of course, this paradox can be placed into a much more global setting. Contemporary nonlinear dynamics is quite ambivalent in its view of the random/determinate character of trajectories for a whole class of Hamiltonian systems that yield wild orbital behavior; just as for the coin, these Hamiltonian systems can be regarded as both "deterministic" and "random," (the quotation marks are to emphasize both the ambivalence and the paradox).

As a consequence, my first major goal here is to resolve the random/determinate paradox using concepts developed in algorithmic complexity theory.[1] This resolution then permits a meaningful discussion of the second major theme, which addresses the question: how significant is nonlinear dynamics? In particular, is it possibly only a passing fad as Goldstein[2] hints? Or is it only an interesting amusement, irrelevant for serious science, as Balescu[3] maintains? Or, taking a more generous view, are its results analogous to the discovery of the laser or of the Mossbauer effect, highly significant but only a part of the evolutionary mainstream of science? Indeed, one may reasonably argue that the soliton with its accompanying inverse scattering theory as well as much of contemporary ergodic theory do, in fact, fall precisely in this evolutionary category. However, there are a few holding the view, which is maintained here, that nonlinear dynamics is the rapidly growing embryo of a future science differing dramatically from that of the present.

For 300 years, science has used a linear theory to describe a nonlinear world. Nonlinear dynamics is at last permitting us to grapple with reality on something resembling its own terms. Aside from Poincaré and a few other nonlinear pioneers, science is at most only 30 years into this nonlinear epoch, but the field is now rapidly accelerating as the contributions in this volume reveal. As a consequence, the future not only looks bright but it is also assuming an increasingly exotic form. This chapter presents my view of this exotic future toward which nonlinear dynamics appears to advance. It poses the serious problem of believably linking the future to the present. I have tried to steer a reasonable course between the capriciousness of cataloguing wild guesses and the tedium of lengthy supporting arguments or detailed proofs, where they exist. In the end, I shall please neither those who wish a finished picture of the future nor those who require that all statements be rigorously proved. Nonetheless, I hope to convince the moderately tolerant reader that nonlinear dynamics has a bright and perhaps profound future, independent of the correctness of specific details given here.

In the past, the seeds of significant theoretical change have frequently been found in the paradox of an existing theory. With this in mind, let us begin to examine the random/determinate paradox inherent in nonlinear, Newtonian dynamics.

2. A DETERMINISTIC AND/OR RANDOM MODEL

Let us consider the forward iterates of the following deterministic, measure preserving,[4] first-order difference equation, or mapping of the unit interval upon itself,

$$X_{n+1} = 2X_n \pmod 1 \tag{1}$$

where (mod 1) means subtract off the integer part until X_{n+1} is returned to the unit interval. Equation (1) has the immediate solution

$$X_n = 2^n X_0 \pmod 1 \tag{2}$$

where X_0 is an initial point in the unit interval. An even more revealing form of the solution may be obtained by writing X_0 as the binary decimal

$$X_0 = 0.1100010 \cdots \tag{3}$$

and noting that forward iterates are obtained merely by moving the decimal to the right. It is difficult to imagine a simpler example of a fully solvable, deterministic system than Eq. (1).

Let us now shift gears and regard the digit sequence of X_0 in Eq. (3) as specifying the results of a random infinite coin-toss sequence. This is surely permissible, since the set of all X_0 on the unit interval [0, 1] is one to one with the set of all possible infinite coin-toss sequences. The solution for X_n given in Eq. (2) may then be regarded as merely a tossing of the coin n times given X_0. In this view, Eq. (1) tosses the coin once given X_n. In short, Eq. (1) describes the simplest possible random process. Indeed, mathematicians call it a Bernoulli shift.[4] Moreover, physical scientists sometimes use Eq. (1) to generate sequences of random numbers on a computer—strictly, pseudorandom numbers due to finite computer arithmetic. Here again, it is difficult to imagine a simpler example of a random process than Eq. (1). Thus, we have now fully arrived at the paradox: How can the deterministic Eq. (1) be unpredictable and how can the random Eq. (1) possess an underlying determinism?

The preceding discussion of Eq. (1) is intended only to illustrate the random/determinate paradox at the intuitive level; rigor is added in a later section. Nonetheless even at this point, one gains the feeling that determinism and randomness may possibly coexist in any system whose orbits depend sensitively on precise initial state; in Eq. (1) for example, two initially close points separate exponentially upon sequential iterations, which of course yield a highly sensitive dependence of orbit on initial state. Trajectories for such systems are determinate because each trajectory is uniquely defined once the initial state is specified. However, the same trajectories are also random because they can be placed in a 1 : 1 correspondence with some random set; once this correspondence has been established, each single orbit is then a realization of the particular random process. Thus to move toward a final resolution of the random/determinate paradox, let us investigate the general character of orbits for classical Newtonian or Hamiltonian systems.

3. THE HIERARCHY OF CHAOS

3.1. Liouville–Arnol'd (LA) Integrability[5]

A Hamiltonian system $H_N(q_k, p_k)$ having N degrees of freedom is LA-integrable provided there exists a well-behaved canonical transformation[6] bringing the system Hamiltonian to the form depending on canonical momenta alone. System orbits in (q_k, p_k) phase space, in general, lie upon N-dimensional manifolds specified by the N, constant of the motion, new momenta variables $P_n(q_k, p_k)$. Specifically, in the transformed coordinates we have $P_n = P_{n0}$ and $Q_n = \omega_n(P_k)t + Q_{n0}$, where $\omega_n = \partial \overline{H}(P_k)/\partial P_n$; thus when the system motion is bounded, the orbits not only lie on tori but, in general, the orbits densely cover these tori.[5] Many authors argue that randomness is totally absent from

LA-integrable systems, but in so doing they ignore the chaotic behavior[5] that characterizes bounded integrable systems because of orbits densely covering the tori. Indeed, this chaotic behavior is the ultimate source of randomness observed in harmonic oscillator systems.[7] In particular, the motion of a heavy particle immersed in an infinite chain of light harmonic oscillators may be rigorously shown to obey the Langevin equation; indeed, Shuler and co-workers[8] have established that this result remains valid even for relatively short, finite chains.

3.2. Ergodicity

A Hamiltonian system $H_N(q_k, p_k)$ of N degrees of freedom is said to be ergodic provided the energy surface $H_N = E$ cannot, under the phase space flow generated by $H_N(q_k, p_k)$, be decomposed into two (or more) invariant, disjoint sets of positive measure. Loosely speaking, ergodicity means that almost every orbit densely covers the energy surface, spending equal times in equal areas. Randomness in an ergodic system is manifest in the equality of time and phase space averages:

$$\lim_{T\to\infty} T^{-1}\int_0^T G[q_k(t), p_k(t)]\, dt = \int G(q_k, p_k)\delta(H - E)\prod_k dq_k\, dp_k \quad (4)$$

As Sinai points out,[9] Eq. (4) may also be regarded as the law of large numbers, where the left-hand side denotes the repeated trials and the right-hand side denotes the a priori probability being approached. A Hamiltonian flow is also said to be ergodic on any surface provided the flow is indecomposible on this surface; thus, integrable Hamiltonian flow is ergodic on tori.

3.3. Mixing[5]

For a system ergodic on the energy surface, an initially close bundle of initial states travel everywhere on the energy surface, but they do so in lock step, as a unit. To disperse this coherently moving initial bundle, we must introduce mixing into the system. Physically speaking, a Hamiltonian system $H_N(q_k, p_k)$ is mixing, provided each small energy surface area element at time zero evolves under the flow into a thin filament that densely covers the entire energy surface. Randomness now becomes the type anticipated by the founders of statistical mechanics.[10] Specifically, ensemble averages of velocity correlations decay to equilibrium and the system asymptotically forgets its initial state.

3.4. A Conjecture

Despite the increasing complexity and the pseudorandom character of ergodic and mixing system orbits, there is a sense in which integrable, ergodic, and mixing systems remain strictly deterministic and should not in any way be regarded as truly random, at least not in the same sense as Eq. (1). For integrable, ergodic, and mixing systems, if one partitions the energy surface into small cells and, for any specified orbit, determines at one-second intervals the cell through which the orbit passes from $t = -\infty$ to $t = 0$, the subsequent sequence of future cells is thereby completely determined. In mathematical jargon,[5] these systems all have zero Kolmogorov–Sinai (KS) entropy.

Alternatively stated, as indicated earlier the orbits of LA-integrable systems are ergodic, albeit only on N-dimensional tori embedded in the full $2N$-dimensional phase space. Nonetheless, of itself, ergodicity alone does not necessarily imply pathological orbit behavior beyond the pale of analysis; indeed, the ergodic orbits on a torus have an extremely simple analytical representation. Now ergodic orbits on a torus are not mixing, but conceptually at least only a trivial modification is required to make them so. Recall that ergodic orbits get almost everywhere but that they travel always together in lock step. To convert an ergodic set of orbits into a mixing set, we leave unmodified the spacial structure of the ergodic orbits, but we require that each orbit be traveled under time evolution at a different rate. Clearly, this modification requires no serious increase in analytical complexity. Thus as with ergodicity alone, mixing alone does not necessarily connote any analytical pathology.

In view of these facts, I propose the following definition and conjecture:

Definition. *A system is said to be A-integrable (after the late V. M. Alekseév) provided its chaotic behavior is no stronger than mixing (KS entropy = 0).*

Conjecture. *A Hamiltonian system is A-integrable if, when embedded into a Hamiltonian system having greater degrees of freedom, the latter Hamiltonian is integrable in the usual sense.*

The basic notion underlying this conjecture is that the toroidally ergodic orbits of LA-integrable systems, when "projected" onto the energy surfaces of systems of lesser degrees of freedom, can become fully ergodic or mixing orbits. We leave further discussion of these matters to another place.

3.5. C-Systems and K-Systems

A Hamiltonian system $H_N(q_k, p_k)$ is, crudely speaking, a C-system or a K-system, provided most initially closed system orbits separate exponentially with time. Although suitable for the purposes of this chapter, this definition is not precise and the rigorous minded should consult higher authority.[5]

The randomness of C-systems or K-systems may be observed by making sequential measurements of observables $G(q_k, p_k)$. Suppose that we measure $G(q_k, p_k)$ using a specified, finite precision at one-second intervals from $t = -\infty$ to $t = n$; then knowledge of the measured set $\{G_k\}_{-\infty}^{n}$ does not precisely determine G_{n+1}. Indeed, only statistical-type estimates can be made for the allowed values of G_{n+1}. Although the sequential measurements of G_n are determining the precise system orbit at an exponential rate, the orbits themselves are diverging at an even greater exponential rate. Thus it is with C-systems and K-systems that we first observe the beginning of that true randomness characteristic of a coin-toss sequence.

3.6. Bernoulli Systems

Adequate for our purposes, we may define a Bernoulli or B-system as a C-system or K-system for which sequential measurements of at least one observable G are statistically completely independent. Specifically, knowledge of the sequence $\{G_k\}^n_{-\infty}$ provides absolutely no information about which allowed value of G_{n+1} will be observed, any allowed value being equally probable. Clearly, the sequential values of $G(q_k, p_k)$ would appear to be as random as a sequence generated by a roulette wheel.

But now let us inquire whether there is any true and meaningful distinction to be made between the random sequences generated by deterministic Bernoulli system orbits and those generated by purely random coins or roulette wheels assumed to have no underlying determinism. To answer this question, we turn to algorithmic complexity theory.

4. ALGORITHMIC COMPLEXITY THEORY[1]

As discussed in Sections 3.5 and 3.6, with each orbit of a C-, K-, or B-system we may associate a doubly infinite sequence of integers (or symbols of some alphabet). Indeed frequently, as with Eq. (1), this correspondence is exhaustive and 1 : 1. Here we call a single, specified orbit "random" if its associated sequence of integers is random. Algorithmic complexity theory, as we now show, establishes the meaning and conditions for randomness of such infinite integer sequences.

Let us begin by discussing finite integer sequences. Specifically, we lose no essential generality by considering finite sequences $\{G_k\}^n_{k=1}$ of binary integers for which $G_k = 0$ or 1. Define now the complexity $K^{(n)}_M$ of the n-digit binary sequence $\{G_k\}^n_{k=1}$ as the bit length of the minimum computer program (on machine M) required to compute or print out the sequence $\{G_k\}^n_{k=1}$. Kolmogorov has shown (see Ref. 1) that there exists a universal computer having the minimum $K^{(n)}_M$; hence the subscript M can be dropped. Consider now the complexity $K^{(n)}$ of the simple n-sequence consisting of all 1's. A minimal program might read, "PRINT 1, n times." The bit length of this program is approximately $\log_2(n)$ for large n. Indeed, for any sequence computable by a relatively short, finite algorithm, its complexity $K^{(n)}$ is approximately $\log_2(n)$ for large n. On the other hand, the minimum program for any binary n-sequence need contain no more than approximately n bits, since any n-sequence can be printed from the program "PRINT

$[G_1, G_2, \ldots, G_n]$," where the G_k are explicitly written into the program. Moreover, a sequence has this maximum complexity when the simplest way to compute or specify the sequence is, in essence, to provide a copy of the sequence; alternatively, for a sequence of maximum complexity, there is no algorithm that computes the sequence whose bit length is smaller than the bit length of the sequence itself. With this prologue, it is now quite reasonable to assert that a sequence is certainly deterministic when its $K^{(n)}$ is on the order of $\log_2(n)$ and, with Kolmogorov and others,[1] to define a random sequence as one having $K^{(n)}$ on the order of n. It can be shown that maximum complexity sequences do exist; indeed, they are generic.

As the length of the finite sequences tends to infinity, one might be tempted, as Kolmogorov was, to define a random infinite sequence as one for which $K^{(n)} \cong n$ for all n. Unfortunately, Martin-Lof[1] has proved this to be an attractive but empty definition. In even, truly random infinite sequences, there can be long N-segments of the sequence with segment complexity $K^{(N)} \cong N - \log_2(N)$. To circumvent this problem, let us follow the members of the Kolmogorov school and define the Kolmogorov complexity of an infinite sequence as

$$K = \lim_{n \to \infty} \left[\frac{K^{(n)}}{n} \right], \tag{5}$$

where the limit exists except for a measure-zero set of sequences. An infinite sequence is said to be random provided its Kolmogorov complexity is positive (i.e., nonzero). Abusing the language slightly, we shall continue to refer to a random (infinite) sequence as one having maximum Kolmogorov complexity. The appeal of this complexity definition for a random infinite sequence rests with the fact that maximum complexity ensures that the sequence cannot be determined or computed more simply than presenting a copy of the sequence; alternatively stated, the information contained in the sequence cannot be compressed. Nonetheless, this definition suffers the defect of not explicitly stating that a sequence of maximum Kolmogorov complexity will possess all the properties expected of a truly random sequence. It is to this point that we now turn.

Consider computable tests (i.e., tests expressible by algorithms as large as we please but nonetheless finite) for randomness that are necessary but not sufficient. Then define the universal test for randomness as that composite test which includes all possible computable tests, past, present, or future. Martin-Lof has proved the remarkable theorem: except for a sequence set of measure zero, a sequence passes the universal test if and only if it has maximum Kolmogorov complexity. We may now quite justifiably define a random sequence to be one having maximum complexity, since no human will ever be able to distinguish this definition from earlier definitions of randomness.

This theorem immediately leads us to inquire about the abundance of random sequences, but we first must define a measure for sets of sequences. Recall that we may put the set of all possible infinite sequences, which use the symbols of some finite alphabet as sequence elements, into 1 : 1 correspondence with the points of some interval of the line, the plane, and so on. For example, the set of all binary decimals having no integer part form a sequence set that is 1 : 1 with the unit interval of a line. Thus, we may take the measure of a subset of sequences to be the measure of the corresponding real point set. Finally then, Martin-Lof has proved the theorem: almost all sequences are random. This theorem immediately implies that most of the individual orbits for Eq. (1) are truly random even though Eq. (1) is obviously strictly deterministic. We return to this point in the next section.

As additional consequences of algorithmic complexity theory, we note the mildly startling fact that the digit sequences in transcendental numbers such as π and e are not random, since they can be computed from quite short algorithms. Moreover, if we define an uncomputable number (or infinite digit sequence) as one having positive Kolmogorov complexity, almost all real numbers are uncomputable. This definition is somewhat unconventional but quite useful for our purposes. (An uncomputable number is a random digit string that cannot be computed by any finite algorithm, no matter how large.) Thus one arrives at the incredible fact that although the set of all real numbers is logically and mathematically well defined, almost none of them can individually be written down or even very well comprehended by a (finite) human. On the other hand, although the comprehensible set of computable numbers having null Kolmogorov complexity has measure zero, paradoxically this set is not countable and has therefore the cardinality of the continuum.

But the notion of uncomputability extends far beyond number theory alone. Recall that the symbolic dynamic[1] or infinite sequence representation of an orbit for C-, K-, or B-systems is random, in essence, by definition. As a consequence, almost all orbits for systems having positive KS entropy[5] are uncomputable. This line of reasoning leads us to the somewhat startling theorem: for Newtonian systems having positive KS entropy, Newtonian dynamics must, in effect, compute random numbers; therefore, for such systems, Newtonian dynamics is an uncomputable algorithm. On the other hand, for chaotic systems no worse than mixing that have null KS entropy, Newtonian dynamics is computable, providing further support for the conjecture of Section 3.4. We further expand on these notions next by directly confronting the random/determinate paradox.

5. RESOLUTION OF THE RANDOM/DETERMINATE PARADOX

In the popularly accepted view, all properties of a deterministic system can be uniquely defined and precisely computed, thereby eliminating all randomness or uncertainty. On the other hand, though all properties of a random system are uniquely defined, they cannot a priori be precisely computed; indeed, it is frequently stated that randomness precludes any underlying determinism. Nonetheless, despite this polarized view, contemporary nonlinear dynamics presents strong evidence that the determinate systems of classical mechanics can exhibit random behavior.[11] Perhaps the most commonly accepted resolution of this paradox involves some type of "coarse graining" argument that permits randomness in deterministic systems because of imprecise knowledge of initial and/or final system state.[10] However, deterministic Newtonian dynamics can exhibit a much more profound randomness than envisioned by these standard "coarse graining" arguments, as we now indicate.

If we consider the earlier Eq. (1), $X_{n+1} = 2X_n \pmod 1$, as a deterministic Newtonian model, it certainly does provide a deterministic, finite algorithm for computing any X_n given X_0. But why in Newtonian dynamics is the specification or determination of X_0 regarded as such a trivial matter? According to algorithmic complexity theory, X_0 is in general an uncomputable, random digit string beyond finite, human capability to compute, write down, or measure. Moreover, in this example, the full burden of chaos or its lack is determined precisely by the random properties of the initial state. Indeed, this burden is similarly placed in most chaotic Newtonian systems. Quite generally for chaotic Newtonian systems, the Newtonian or Runge–Kutta algorithm for computing an orbit is quite short; consequently the positive Kolmogorov complexity of the orbit must lie mostly with the random character of the initial state. Almost all orbits of a deterministic, chaotic system are individually random; no "coarse graining" is required. In Eq. (3), this is especially transparent, since X_0 is simultaneously an initial condition and an orbit described as an explicit realization of a random coin-toss sequence. In summary, traditional classical mechanics lays claim to determinism on the grounds that a complete orbit is uniquely defined once the initial conditions are specified; it thereby ignores not only the uncomputability but also the orbital randomness hidden in the initial conditions.

But now wary readers will surely object; noting that integrable systems also have short Runge–Kutta programs as well as uncomputable initial conditions, they will inquire why integrable systems are not also random. To meet this quite reasonable objection, let us first note that the usual formulation of Newton's equations generally provides a finite algorithmic link between uncomputable initial state and uncomputable final state. This usual formulation thus ignores any fundamental distinction that might exist between integrable and chaotic systems, perhaps because of a firm, historically grounded belief in the deterministic character of both types of systems. However, we can regain this distinction without emphasizing the randomness of uncomputable initial and/or final state by transforming to a Newtonian representation in which all states are specified by computable numbers. For this, we need only recall that despite having measure zero, the set of computable numbers is 1 : 1 with the continuum.

Thus let us now label all states using computable numbers. Then in this representation, all systems (integrable, ergodic, and mixing) having null K–S entropy continue to possess orbits with null Kolmogorov complexity; in short, Newtonian dynamics is a computable algorithm for these systems. However, the Newtonian algorithm in this representation for random systems (C-, K-, and B-) has become a bloated, uncomputable monster possessing positive Kolmogorov complexity. Specifically, the Runge–Kutta program is now of infinite length. Finally then, it is perhaps now clear that orbital complexity is the real issue, not uncomputability of initial conditions. For a chaotic system, not only are most orbits wildly erratic but all possible orbits occur. In Eq. (1), for example, an orbit having any desired sequence of zeros and ones in its decimal representation, Eq. (3), is an actual solution. Let us now return to the resolution of the random/determinate paradox.

Section 1 presented intuitive arguments indicating that the difference equation Eq. (1) is both random and deterministic. If we now use the algorithmic definition of random, we perceive that Eq. (1) is quite rigorously both random and determinate. But how can this be, given the traditional views of these notions? Quite clearly, the traditional notion of determinism tacitly requires, but does not count as significant, the notion of infinite computational and/or observational precision. Equally clearly, the traditional notion of randomness precludes underlying determinism only by tacitly disavowing the possibility of infinite computational and/or observational precision. Algorithmic complexity theory resolves the paradox by, in essence, defining "random" as the determinism of infinite algorithms. In short, if one admits infinite computational and/or observational precision, algorithmic complexity theory makes it clear that chaotic systems may be regarded as deterministic and/or random, as one pleases. Thus there is, at last, no contradiction and no paradox.

Nonetheless, mathematicians[1] have resolved the random/determinate paradox only by paying a steep and heavy price, which physical scientists may not wish to accept. Tacitly or explicitly, mathematical theory has for at least 300 years assumed infinite observational and/or computational precision. Over the same period, physical theory has painfully and sometimes traumatically been made aware of limitations and of finitudes in both man and the physical world. In Section 6 we peer into the future of this line of thought.

6. NONLINEAR DYNAMICS: QUO VADIS?

Let us begin our attempt to predict significant developments in the future of nonlinear dynamics by asking how willing physical scientists should be to accept the notion of infinite observational and/or computational precision, which pervades all mathematics. If we accept this notion, all physical science can be regarded as strictly deterministic, with probability being reduced to a convenient tool for describing dice or card games and next week's weather. Indeed, if we do accept this notion, wouldn't the $đQ$ term in the first law of thermodynamics disappear, as Bridgman suggests,[12] leaving us with $dU + dW = 0$? Moreover, wouldn't the second law of thermodynamics then merely be a statement regarding our fumble-fisted, inaccurate handling of theory and laboratory equipment, and couldn't we then circumvent the second law of thermodynamics once our computational and/or observational precision improved? Would not all of statistical mechanics become an unnecessary luxury, contrary to the view of Krylov?[13] Finally in this view, how could we reconcile a world of infinite precision in which probability is only a convenience with the theory of quantum mechanics in which probability is a necessity? In this regard, let us note that the present version of quantum mechanics is a hybrid in which irreducible uncertainty and probability coexist with infinite observational precision for a set of commuting observables. In any event, although these remarks may not be particularly significant to some, they may be sufficiently disturbing to others to warrant attention to the following predictions.

For there to exist a true randomness without underlying determinism in the physical as opposed to the mathematical world, there must exist a natural upper limit to observational and/or computational precision. Should this limit be found to exist, it would join a long list of earlier limits imposed by nature but initially unrecognized by man. Not until the Greeks did man discard the matter continuum for discrete atoms. Avogadro counted these atoms (in a box) and found their number to be finite. In this century, the charge, mass, and intrinsic spin of these atoms were found to be both discrete and finite. On another level, Einstein denied man his Newtonian dream of infinite speed. Planck denied man his energy continuum. Heisenberg called man's attention to the limits of observational precision for certain complementary variables. What we here propose, motivated by the random/determinate paradox and algorithmic complexity theory, is to deny man his Newtonian dream of infinite computational and/or observational precision for all observables. That is, we are proposing to extend the Heisenberg uncertainty principle.

By now it may be already clear that the source of almost all truly fundamental physical problems is the continuum of the real number system that underlies all physical theory, for, quite obviously, the continuum is physically meaningful only under the prior assumption of infinite precision. If as physical scientists we part company with the mathematicians and disallow infinite precision, the first casualties in the continuum are the uncomputable numbers having infinite, random digit strings. These numbers are the worst offenders, since they have positive Kolmogorov complexity and their associated computational algorithms are of infinite size. Next to go are the decimally unending rationals like $\frac{1}{3}$; although their basic algorithms are finite, each basic algorithm must be iterated an infinite number of times, yielding a computer program whose bit length goes to infinity with n like $\log_2(n)$. For the same reason, we must eliminate the infinitely large and the infinitely small, since $\infty = (1 + 1 + 1 + \cdots)$ and $0 = (1 + 1 + 1 + \cdots)^{-1}$ are both infinite algorithms. We are not eliminating the label "zero" for the point at the origin, we are only discarding $\lim_{x \to y}(x - y) = 0$.

The original continuum has now been reduced to a denumerable point set for which the lower bound, the upper bound, and the minimum distance between points have yet to be fixed. To establish these bounds, let us consider the measurement process. All measurements are made with an appropriate meter (e.g., voltmeter, ammeter) that as far as possible places its internal states into a 1 : 1 correspondence with those of the observed system. The meter then announces its observation or internal state through a dial pointer or, if modern, a digital computer readout or printout. Finally, the measurement process is completed by a human who reads the dial pointer or the computer output. Now what is the ultimate limit on the accuracy of this ideal meter plus computer plus human measuring device? Quite clearly, the greater the number of internal states of this ideal device, the greater the accuracy and the broader the range. In principle, one could enlarge the size of the meter and/or computer to near the size of the universe or, more cogently, to a size possessing internal states whose number is far in excess of those in the human mind. But disregarding future evolution, natural or induced, we have no ability to change the informational bit capacity of the human brain. Thus, it is this limited bit capacity of the brain that provides the ultimate natural bounds on the point set replacing the continuum. This is not to assert that other, coarser, natural bounds unknown to us may not preempt human frailty, but if no earlier limit exists, human bit capacity may prove to be the definitive final limit. But how large is this bit capacity? At the moment surely, no one knows, since not even the information storage mechanism is known; however, most authorities agree that it is finite and not arbitrarily large.

At last, the original continuum has been reduced to a bounded, finite point set that is 1 : 1 with the integer point set $(1, 2, 3, \ldots, N)$, where N is the largest integer whose digit string can be stored in the brain. Although we are here emphasizing the brain as the limiting factor providing us with a finite point set, our real assertion is that the future will provide some limiting factor that prevents man, starting from the God-given integer 1, from counting all the way up to aleph null. In any event, if we now measure an observable A, then the minimum value or quantum of A, call it A_m, is proportional to unity while the maximum value A_M is proportional to N. As an immediate consequence, we then obtain the dimensionless Kolmogorov–Alekseév uncertainty relation

$$\frac{\mathrm{NA}_m}{\mathrm{A}_M} \geqslant 1 \tag{6}$$

which is valid for all observables. One notes here that all observables are quantized and have a finite spectrum, including space and time. This result suggests that all observables at this submicroscopic level of observation should be represented by spin operators, as has been previously suggested by Finkelstein and others.[14] If one recalls that the Heisenberg uncertainty relationship precludes classical orbits in the (q, p) plane and then notes that the Kolmogorov–Alekseev uncertainty relationship precludes orbits in any plane [e.g., the (q_1, q_2) plane], one anticipates that all operators are noncommuting, again indicating that the appropriate algebra at the submicroscopic level is that of noncommuting, spinlike operators. Finally, Eq. (6) suggests itself as being the uncertainty relationship associated with an as yet undeveloped integer dynamics that may prove to be the submicroscopic, ultimate theory of the universe. If such is the case, why have we not seen some macroscopic manifestation of this underlying theory analogous to the macroscopic observation of the pure quantum state of superconductivity or superfluidity? The answer is that perhaps we have. Generally, at the macroscopic level these submicroscopic spin variables combine in such a way as to mimic a continuum, as with q and p, say; however, the intrinsic spin variables remain spinlike even at the macroscopic level. If we recall that the spinlike qualities of submicroscopic operators implied by the inequality Eq. (6) were obtained by limiting the precision of measurement, it is especially interesting to note that Dirac theory, which combines quantum mechanics and relativity, obtains the intrinsic spin of the electron and the positron precisely because of the limits on angular momentum imposed by h in the small and by c in the large.

7. CONCLUDING REMARKS

The random/determinate paradox has been illustrated and resolved, using a combination of ergodic theory and algorithmic complexity theory. Along the way, it was indicated that ergodic and mixing systems may be no less integrable than the systems usually given that designation. On the other side of the coin, it was shown for chaotic systems (C-, K-, and B-) that Newtonian dynamics represents an uncomputable and perhaps undefinable algorithm. Finally, using the random/determinate paradox as a signpost, a few significant future pathways of nonlinear dynamics have been exposed. The presentation has been such a blend of unfamiliar fact with motivated fantasy that many, if not most, may find it amusing but not very convincing. Briefly, let me indicate why these proposals are not as radical as they may appear to be.

First, there is now a host of nonlinear dynamicists whose work convinces them that chaotic systems are both strictly deterministic and purely random, although they may not have developed the rigorous viewpoint expressed herein. Indeed, there are mathematicians[15] who can rigorously prove this point starting from a quite different standpoint. Integrability has always been a troublesome notion,[16] and many workers recognize that LA-integrability is too narrow a base to be all-inclusive. Moreover, Newtonian dynamics never in its wildest dreams contemplated the possibility of what we now know as chaotic systems; thus, there is no reason to expect Newtonian dynamics to properly treat such systems. Finally, the arguments leading to Eq. (6), as well as the equation itself and its implications, have been criticized by several of my colleagues, not because they are wild fantasy, but because they are such standard results as to be stale and old hat. Recently, I chanced to overhear Otto Rössler[17] talking to a friend during a conference break and, to my amazement, he was describing an independent development of my comments in Section 6, without having heard or read it. I also recently learned of

"Conference on the Physics of Computation" at MIT, which gathered a whole lecture hall full of people to whom my remarks would be regarded as quite mundane. In short, these ideas may be strange, but you'll likely hear a great deal more of them.

ACKNOWLEDGMENTS

To Prof. Boris Chirikov, Novosibirsk, I am profoundly indebted for airborne tutorials on algorithmic complexity theory that covered not only the subject but several continents as well. Without his teachings, I would never have come to write this chapter. However, as with all teachers, Prof. Chirikov should not be held accountable for the writings of his pupils.

REFERENCES

1. G. J. Chaitin, *Sci. Am.*, May 1975; A. K. Zvonkin and L. A. Levin, *Usp. Mat. Nauk* **25**:6, 85 (1970); A. A. Brudno, *Usp. Mat. Nauk* **33**:1, 207 (1978); P. Martin-Lof, *J. Inf. Control* **9**, 602 (1966); V. M. Alekseév and M. V. Yakobson, *Phys. Rep. 75*, 287 (1981).
2. See the introduction to H. Goldstein, *Classical Mechanics*, 2nd ed. (Addison-Wesley, Reading, MA, 1980).
3. See the appendix on the ergodic problem in R. Balescu, *Equilibrium and Nonequilibrium Statistical Mechanics* (Wiley, New York, 1975).
4. See P. Billingsley, *Ergodic Theory and Information* (Wiley, New York, 1965); Example 1.6, p. 7.
5. V. I. Arnol'd and A. Avez, *Ergodic Problems of Classical Mechanics* (Benjamin, New York, 1968). This is now a standard reference on ergodic theory; integrability is discussed in Appendix 26.
6. See J. Moser, *Stable and Random Motions in Dynamical Systems* (Princeton University Press, Princeton, NJ, 1973), p. 41.
7. See, for example, P. Mazur and E. Montroll, *J. Math. Phys.* **1**, 70 (1960), and the references listed therein.
8. R. I. Cukier, K. E. Shuler, and J. D. Weeks, *J. Stat. Phys.* **5**, 99 (1972).
9. Ya. G. Sinai, *Introduction to Ergodic Theory* (Princeton University Press, Princeton, NJ, 1976).
10. See D. ter Haar, *Elements of Statistical Mechanics* (Rinehart, New York, 1954), Appendix I.
11. See American Institute of Physics, *Conference Proceedings*, Vols. 46 and 57 (AIP, New York, 1978, 1979). Also consult G. Laval and D. Grésillon, Eds., *Intrinsic Stochasticity in Plasmas* (Editions de Physique Courtaboeuf, Orsay, France, 1979).
12. P. W. Bridgman, *Science* **75**, 420 (1947); *Proc. Am. Acad. Arts Sci.* **82**, 301 (1953).
13. N. S. Krylov, *Works on the Foundations of Statistical Physics* (Princeton University Press, Princeton, NJ, 1979).
14. See D. Finkelstein, *Phys. Rev.*, **184**, 1261 (1969), and *Int. J. Theor. Phys.* (to appear, 1981), as well as the references to earlier work contained therein.
15. M. Kac, private communication.
16. See A. Wintner, *The Analytical Foundations of Celestial Mechanics* (Princeton University Press, Princeton, NJ, 1941), p. 144.
17. O. E. Rössler, private communication.

Part 2

DYNAMICS

ONE MECHANISM FOR THE ONSETS OF LARGE-SCALE CHAOS IN CONSERVATIVE AND DISSIPATIVE SYSTEMS

ROBERT H. G. HELLEMAN
Theoretical Physics Group
Twente University of Technology
Enschede
The Netherlands

Abstract. *An exciting new development in nonlinear dynamics, "Period doubling to chaos,"[1,58,59] is derived here, approximately but analytically, for conservative as well as dissipative systems.*

A transition from regular (quasi)-periodic behavior to chaotic behavior occurs in most mechanical systems as μ,[1] *which may be the energy (whence "ergodic behavior"), the Reynolds number (whence "turbulent behavior"), or some other parameter, is varied. Local, small-scale, chaotic behavior is known to arise about (linearly) unstable periodic orbits.*[1-9] *Hence the transition above can take place if more and more periodic orbits turn unstable as we increase* μ.

At the μ*-value where one periodic orbit turns unstable, another stable periodic orbit is usually created with twice the period of the original orbit, and so on. Thus, at* μ_k, *a period-*2^k *(in some units) orbit "bifurcates" from a period-*2^{k-1} *orbit. Surprisingly, the "Feigenbaum sequence"* $\{\mu_k\}$ *converges geometrically to some finite critical value* μ_∞. *At that point all periodic orbits created have turned unstable and large-scale chaotic behavior results.*

This critical value μ_∞, *as well as the "universal" rates of convergence* δ *and* α, *respectively for the* μ_k *and the orbits, are calculated, analytically and explicitly,*

Appendix C is written by **ROBERT S. MACKAY**, Plasma Physics Laboratory, Princeton University, Princeton, New Jersey.

using one "renormalization" scheme valid for both conservative and dissipative systems. The derivation also shows why the δ- and α-values are "universal," but different for conservative and dissipative systems.

1. INTRODUCTION

Over the past decades it has become apparent that *most* mechanical systems with more than one degree of freedom exhibit a transition from the regular (quasi)-periodic behavior we know from graduate mechanics textbooks to chaotic behavior, now described in a number of reviews.[1-9, 46-48] The latter might be called "ergodic behavior" in a conservative system or "turbulent behavior" in a dissipative system. The transition from regular to chaotic behavior becomes noticeable as we increase some parameter μ, which may be the energy, the Reynolds number, or an external driving amplitude. Even at very low nonlinearities (i.e., small values of μ) some chaotic behavior can still be present, although the motion *looks* very regular to the naked eye. It usually is this small-scale chaotic behavior that prevents us from obtaining exact solutions for nonlinear dynamical systems or even useful asymptotic approximations, valid over a long time.

The existence of small-scale chaotic—and even random—behavior has been established mathematically, and numerically, about linearly unstable periodic orbits.[1-9,24] When more and more periodic orbits turn unstable, as μ is increased, *large-scale* chaotic behavior sets in. One exciting new development of the past 6 years is the discovery that there exist *finite* critical values of μ beyond which infinite numbers of periodic orbits are unstable. This agrees well with the experimental observation that turbulence sets in beyond a finite critical value of the Reynolds number. Some of the recent new theories have indeed been confirmed by experiments on the onset of turbulence.[1,2,7,8,58-60] In conservative systems the problem is more pathological, or subtle.[1-6,18-29,57] Here there are many critical values of μ predicted, and observed, at which different regions of phase space become (predominantly) chaotic. A remaining problem is to find out when, if ever, the combined chaotic regions occupy most of phase space. Yet, these theories advance our understanding here also, at least from the local to the regional level.

In recent theory it is noted that when one periodic orbit turns unstable, as μ is increased, another stable orbit is usually created, with *twice* the period of the first one. The new orbit itself turns unstable at a higher value of μ and another stable orbit is created, with *four* times the period of the first one, and so on. Thus a period-2^k (in some units) orbit "bifurcates," at some μ_k, from a period-2^{k-1} orbit. Surprisingly, the resulting "Feigenbaum sequence" $\{\mu_k\}$ converges to some finite critical value μ_∞, with asymptotic behavior:

$$\mu_k \underset{k\to\infty}{\sim} \mu_\infty - a\delta^{-k} \tag{1.1}$$

It turns out that the constant δ has one "universal" value for dissipative systems, Eq. (5.7), and another "universal" value for conservative systems, Eq. (4.7). The numerical discovery of Eq. (1.1), in dissipative systems, was made by three different groups of researchers,[10–12] independently and virtually simultaneously, a not infrequent occurrence in science. The period-2^k orbits converge geometrically to some nonperiodic limit orbit, as $k \to \infty$, at an asymptotic rate α that is again a universal constant of the same type [cf. Eqs. (4.9) and (5.9)]. Although more and more periodic orbits branch off as $\mu \to \mu_\infty$, it is only for $\mu > \mu_\infty$ that we have an infinite number of unstable "branches" in a finite region, about the limit orbit. Here we encounter large-scale chaotic behavior. Therefore these Feigenbaum-"trees" of periodic orbits, or *Feigenbäume*, are likely to play an important role in the investigation of globally chaotic behavior.

The foregoing self-similar behavior, over infinitely many scales, in μ and in phase space, naturally gave rise to a number of "renormalization theories" for these phenomena.[10,11,13,14,7,18,25,1,60,61] Here I present a renormalization theory[1] that is valid for both conservative and dissipative systems. This enables us to find the common properties of conservative and dissipative Feigenbäume as well as their essential differences. Although Derrida and Pomeau's (independently developed) renormalization theory[14,13] is similar, in principle, to the present one,[1] both differ from Feigenbaum's original renormalization theory for dissipative systems.[10,7] *Exact* results are available at the moment for the existence of a "universal" dynamics near the critical point μ_∞.[54,56] Most of these theories use systems with only one or two dynamical variables (but cf. Ref. 56) at discrete moments in time. This has the advantage that only one or two nonlinear difference equations need to be studied, a familiar device in nonlinear dynamics[1–9]: when one erects a two-dimensional surface (of "section") in the larger phase space of a system of coupled nonlinear differential equations, the intersection points of the orbits with this plane do have a dynamics, described by a mapping of the plane into itself, of the above-mentioned type.

Section 2 combines those two nonlinear first-difference equations into one second-difference equation and discusses the conditions under which the lowest order (i.e., quadratic) approximation to such nonlinear equations can always be reduced to the same standard form. In Section 3 we calculate exactly, and explicitly, *one* period doubling bifurcation for a mapping in this standard form. Doubling the time unit and scaling the coordinates about the new orbit, we find that the lowest order approximation can again be reduced to the same standard form. Starting the period doubling all over we obtain our renormalization scheme. It is shown in Section 4 that an "area preserving" mapping (the surface of section mapping for a conservative system[1–6,9] remains area preserving under this renormalization. In Section 5 it is shown that "area contracting" maps (the surface of section mappings for dissipative systems) *increase* their rate of area contraction and reduce to a one-dimensional map in the limit of infinite renormalization. This is why the asymptotic rates δ [Eq. (1.1)] and α are universal to all "dissipative maps" and have the same values, calculated in Section 5, as for the one-dimensional ("logistic") equation. Some of the chaotic behavior arising beyond μ_∞ is discussed in Section 5.1, mainly for this one-dimensional case. The conservative rates δ [Eq. (1.1)] and α, are calculated in Section 4. The second rate of contraction β, in the second direction of the (conservative) phase plane (cf. Ref. 18), is derived by MacKay in Appendix C, using the present renormalization theory. Since we work with the lowest order, quadratic approximation to the local, nonlinear (analytic) mapping about any periodic orbit,[40,31], these phenomena are believed to be "universal" to most such mappings and thus to most nonlinear systems.

2. LOCAL STANDARD FORM FOR MAPPINGS OF CONSTANT JACOBIAN

Here we discuss the conditions under which the lowest order, quadratic approximation to a nonlinear (analytic) mapping of the plane into itself can be reduced to one second-difference equation of the standard form [Eq. (2.6)], which we employ throughout this chapter.

Consider the most general mapping of a two-dimensional surface (of section) into itself:

$$T:\ X_{t+1} = f(X_t, Y_t) \quad \text{and} \quad Y_{t+1} = g(X_t, Y_t), \quad t = 0, 1, 2, \ldots \tag{2.1}$$

For our present purposes we take analytic nonlinear functions f and g. The main restriction in this whole chapter is a confinement to mappings with a constant Jacobian, that is, with

$$\det\left(\frac{\partial(X_{t+1}, Y_{t+1})}{\partial(X_t, Y_t)}\right) = B \quad \text{everywhere} \tag{2.2}$$

(i.e., at every point of the X, Y-plane). This is exact in several important cases: for example, for intersecting (conservative and dissipative) storage rings[1] (cf. Appendix B), for Hénon's (conservative and dissipative) mappings,[1,16,17] and for the Lorenz attractor.[1,2] If the Jacobian [Eq. (2.2)] is not a constant everywhere, it does become "more and more constant" locally when we work in a smaller and smaller region (e.g., about a periodic orbit[40]), as we shall do in the renormalization procedure of this chapter. Hence the existence and ordering of various local phenomena (but not their precise location) is much the same as when the Jacobian is a constant [Eq. (2.2)].[9,16,17]

Conservative and dissipative mappings are distinguished by their B-value:

$$B = \pm 1, \quad \text{conservative mappings} \tag{2.3}$$

$$-1 < B < 1, \quad \text{dissipative mappings} \tag{2.4}$$

The value of B determines the rate of "dissipation" since the Jacobian of the mapping is also the rate of area "contraction" per single mapping as we see in Note 30 (when $|B| > 1$ we take $t' \equiv -t$ and the mapping becomes dissipative in t', with Jacobian $1/B$.[30]) An exact relation between B and the (constant) damping coefficient in the complete phase space, reproducing Eqs. (2.3) and (2.4), is derived for the dynamics of intersecting storage rings [cf. Eq. (B10) and Appendix B]. For $B = 1$ there is an obvious relation between Eq. (2.2) and Liouville's theorem or Poincaré invariants in general.[9,1]

The mapping T of Eq. (2.1) can always be transformed into another mapping T' such that a periodic orbit of the mapping T becomes a fixed point of T'.[40] Here we investigate the local behavior about a periodic orbit of T by expanding the corresponding T' about its fixed point, up to and including second-order Taylor terms[31] in the deviations x_t, y_t from the periodic orbit $\hat{X}_t$, $\hat{Y}_t$, where

$$X_t \equiv \hat{X}_t + x_t \qquad \text{and} \qquad Y_t \equiv \hat{Y}_t + y_t \tag{2.5}$$

Thanks to the condition given in Eq. (2.2) the resulting quadratic expressions for x_{t+1} and y_{t+1} (in terms of x_t, y_t) can always be combined into one quadratic second-difference equation in one variable, as we see in Appendix A. This expression can always be brought into *our standard form*:

$$y_{t+1} + By_{t-1} = 2Cy_t + 2y_t^2 \qquad t = 0, 1, 2, \ldots \tag{2.6}$$

as we see in Note 33. Our phase plane now is the y_t, y_{t+1} plane.[1] We obtain orbits in it by choosing, and plotting, the initial point y_0, y_1, by computing y_2 from Eq. (2.6) at $t = 1$, and plotting y_1, y_2, and so on. For $B = 0$ Eq. (2.6) reduces to a mapping of the y_t line into itself. At $B = 0$ Eq. (2.6) is equivalent to the "logistic equation"

$$x_{t+1} = ax_t(1 - x_t) \tag{2.7}$$

as we see in Note 38, or to the form used in Refs. 7 and 8:

$$x_{t+1} = 1 - \mu x_t^2 \tag{2.8}$$

as we see in Note 36. At $B = 1$ Eq. (2.6) is obviously equivalent to Hénon's conservative mapping,[16] as we see in Ref. 37. At $B = -b$, Eq. (2.6) is equivalent to Hénon's dissipative mapping,[17]

$$x_{t+1} - bx_{t-1} = 1 - ax_t^2 \tag{2.9}$$

as we see in Note 35.

Hence the only free parameter, at fixed B [Eq. (2.2)], in our standard form [Eq. (2.6)] is C. Note also that for each orbit (or phenomenon) at one value of C there exists an identical (but shifted) mirror-orbit (or mirror-phenomenon) at another C value, $\overline{C}$, with

$$\overline{C} = 1 + B - C \tag{2.10}$$

as we see in Note 34. Thus all phenomena appear twice, at C-values symmetric about C^*, with

$$C^* \equiv \frac{1+B}{2} \tag{2.11}$$

For historical reasons, we study C-values below C^*. In the study of the logistic equation, Eq. (2.7), one favors $a > a^*$. The different notations are easily translated into each other via Note 34 and Eq. (2.10) (cf. Ref. 1).

3. ONE PERIOD-DOUBLING BIFURCATION AND RENORMALIZATION

In this section we calculate exactly a stable period-1 orbit for our standard equation (2.6) and a stable period-2 orbit that bifurcates from that period-1 orbit when the latter turns unstable with reflection (w.r.). It is shown that after some scaling and counting t modulo 2, the quadratic (i.e., "regional") part of the mapping about this period-2 orbit becomes identical to our original standard equation (2.6). Starting the same calculation all over for this new mapping, we thus obtain the renormalization procedure employed in the next sections.

An obvious period-1 orbit of Eq. (2.6)[40] is $y_t = 0$. Its variational equation and solutions are

$$\begin{gathered} y_{t+1} + By_{t-1} = 2Cy_t \\ \text{whence} \quad y_t = a\lambda_1^t + b\lambda_2^t, \quad \text{with} \quad \lambda^2 - 2C\lambda + B =: 0. \end{gathered} \tag{3.1}$$

These solutions remain bounded if and only if $|\lambda_{1,2}| \leqslant 1$, whence we find:

$$\text{the origin } y_t = 0, \quad \text{is stable if and only if} \quad |C| \leqslant \frac{1+B}{2} \tag{3.2}$$

One easily checks, by substitution in Eq. (2.6), that a period-2 orbit is:

$$\begin{gathered} \hat{y}_t = a + |b|(-1)^t, \qquad \text{with } 2a \equiv -\frac{1+B}{2} - C \\ \text{and} \quad 4b^2 \equiv \left(C + \frac{1+B}{2}\right)\left(C - 3\frac{1+B}{2}\right) \end{gathered} \tag{3.3}$$

It exists, real, when $C \leqslant -(1+B)/2$ (or above $3(1+B)/2$[34]), that is, exactly the same C-value where the origin turns unstable (w.r.) [cf. Eq. (3.2)]. Denoting the C-value at which such a stable period-2^k orbit is first[34] "created" by C_k, we just obtained:

$$C_0 = \frac{1+B}{2} \quad \text{and} \quad C_1 = -\frac{(1+B)}{2} \tag{3.4}$$

At C_k a period-2^k orbit bifurcates off a period-2^{k-1} orbit ($k \geqslant 1$). Substituting $y \equiv \hat{y} + \Delta y$, the original mapping [Eq. (2.6)] is transformed into:

$$\Delta y_{t+1} + B\,\Delta y_{t-1} = (2C + 4\hat{y}_t)\,\Delta y_t + 2\Delta y_t^2 \tag{3.5}$$

with the periodic $\hat{y}$ [Eq. (3.3)]. Finally, we add Eq. (3.5) at $t = 2\tau + 1$ to B times Eq. (3.5) at $t = 2\tau - 1$. Into this we substitute Eq. (3.5) at $t = 2\tau$ and obtain *exactly*:

$$\Delta y_{2\tau+2} + B'\,\Delta y_{2\tau-2} = 2C'\,\Delta y_{2\tau} + 2e\,\Delta y_{2\tau}^2 + 2\big[\Delta y_{2\tau+1}^2 + B\,\Delta y_{2\tau-1}^2\big] \tag{3.6}$$

with
$$B' \equiv B^2 \qquad \text{N. B.} \tag{3.7}$$

and
$$C' \equiv \frac{de}{2} - B = -2C^2 + 2(1+B)C + 2B^2 + 3B + 2 \;\text{N. B.} \tag{3.8}$$

where
$$d \equiv 2C + 4\hat{y}_0 \quad \text{and} \quad e \equiv 2C + 4\hat{y}_1 \tag{3.9}$$

In Note 39 we find that the square brackets of Eq. (3.6) also contain a term proportional to $\Delta y_{2\tau}^2$ (plus higher orders). Rescaling those two quadratic terms, and counting t modulo 2, we arrive at *our renormalized mapping*:

$$y'_{\tau+1} + B' y'_{\tau-1} = 2C' y'_\tau + 2y'^2_\tau + \text{higher orders,} \tag{3.10}$$

with

$$y'_\tau \equiv \alpha\,\Delta y_{2\tau} \tag{3.11}$$

$$\text{where} \qquad \alpha \equiv e + \frac{d^2}{(1+B)}$$

(cf. Note 41), as we see in Note 39. To second order Eq. (3.10) is the same as our original mapping [Eq. (2.6)], with "renormalized" coefficients. Hence there is again a period doubling bifurcation at $C' = -(1 + B')/2$ [cf. Eqs. (3.2)–(3.4)] and substituting Eqs. (3.7) and (3.8) in it we derive

$$C_2 = \frac{1 + B - \sqrt{6B^2 + 8B + 6}}{2} \tag{3.12}$$

exactly, since only the linear part of Eq. (3.10) is used to calculate Eq. (3.12) from the exact results, Eqs. (3.1)–(3.9).

4. CONSERVATIVE FEIGENBAUM SEQUENCES

We continue the renormalization procedure of Section 3 for conservative mappings, that is, $|B| = 1$, [Eq. (2.3)], and calculate various rates of convergence and contraction that are "universal" for conservative systems.

When $B = \pm 1$ we see that $B'' = B' = 1$, and so on, since $B' \equiv B^2$ [Eq. (3.7)]. The C-renormalization [Eq. (3.8)] now reduces to

$$C' = -2C^2 + 4C + 7 \tag{4.1}$$

(at $B = -1$ we start with period 2, i.e., with $B' = 1$ [39]). The C' in the left-hand side of Eq. (4.1) is the coefficient in the new scaled equation (3.10), about the new, bifurcated, orbit, Eq. (3.3). It might be interesting to know at which C'-value a bifurcation takes place, but we prefer knowing the C-value in the original equation (2.6) to which this corresponds. This C-value can here be solved from the right-hand side of Eq. (4.1). For instance, when we substitute $C' = C_1 = -1$ in Eq. (4.1) and solve for C from its right-hand side, we obtain $C = C_2 = 1 - \sqrt{5}$, the same as Eq. (3.12) at $B = 1$. In general, when k bifurcations have taken place off the new orbit (i.e., at $C' = C_k$), exactly $k + 1$ bifurcations have taken place off the original orbit (i.e., $C = C_{k+1}$ then).

The bifurcation points therefore satisfy

$$C_k \underset{r}{=} -2C_{k+1}^2 + 4C_{k+1} + 7 \tag{4.2}$$

in the renormalization approximation [i.e., using Eq. (3.10) to second order only]. Iterating Eq. (4.2) we could obtain the subsequent bifurcation points. The parabola on the right-hand side of Eq. (4.2) and the straight line on the left-hand side do have two intersection points, that is, the recursion relation Eq. (4.2) has fixed points, $C_{k+1} = C_k$. Here we consider, at $B = 1$, the fixed point C_∞,[34] with

$$C_\infty \underset{r}{=} \frac{3 - \sqrt{65}}{4} \underset{r}{\approx} -1.2656\ldots \tag{4.3}$$

while numerically[27, 18]

$$C_\infty = -1.266311276922099\ldots \tag{4.4}$$

a relative error of 5.5×10^{-4}; that is, the numerically observed bifurcation points C_k have an accumulation point at Eq. (4.4). For C_∞ at $B = -1$, see Eq.

(5.13). About C_∞ we try an asymptotic expansion:

$$C_k \underset{k\to\infty}{\sim} C_\infty + a\delta^{-k} \tag{4.5}$$

[cf. Eq. (1.1)]. Substituting this in Eq. (4.2) and comparing the asymptotic terms, we find for the *conservative Feigenbaum constant*

$$\delta \underset{r}{=} -4C_\infty + 4 \underset{r}{=} 1 + \sqrt{65} \underset{r}{\approx} 9.06\ldots \tag{4.6}$$

while numerically[18, 13, 14, 25–29]

$$\delta = 8.721097200\ldots \tag{4.7}$$

a relative error of 3.9×10^{-2} [employing the exact period-4 solution of Eq. (2.6), cf. footnote 155 of Ref. 1, we can renormalize over two bifurcations at once and obtain $\delta = 8.87\ldots$]. Hence, the C_k converge geometrically (asymptotically). We can now evaluate the rescaling constant α (3.11)

$$\alpha \underset{r}{=} e + \frac{d^2}{2} \qquad \underset{r}{\approx} -4.0955\ldots \tag{4.8}$$

while numerically[41, 18, 25–29, 13, 14]

$$\alpha = -4.018076704\ldots \tag{4.9}$$

with the e and d from (3.9), a relative error of 1.9×10^{-2}. Hence, the orbits converge geometrically (asymptotically). The relative error for the limit value Eq. (4.3) is better, of course, than for the rates δ and α at which the limit is approached.

This conservative Feigenbaum was found numerically by several groups.[18, 25–29, 13, 14] A doubly logarithmic plot of it by van Zeyts[27, 1] is reproduced in Fig. 1, and several of its phase plots in Fig. 2. These numerical studies appear to confirm that the δ and α above are universal constants. Thus, even if we know only two consecutive bifurcation points, called μ_0 and μ_1, we "universally" expect an infinite number of unstable Feigenbaum branches at about 13% above this interval, since

$$\mu_\infty \underset{r}{\approx} \mu_0 + (\mu_1 - \mu_0)\frac{\delta}{\delta - 1} \underset{r}{\approx} \mu_0 + 1.1295(\mu_1 - \mu_0) \tag{4.10}$$

[cf. Eqs. (4.5)–(4.7)]. Note that conservative Feigenbäume "produce" chaotic behavior that differs visibly from that of a dissipative Feigenbaum: already before the C_k reach C_∞ there is small-scale chaotic behavior about each unstable orbit of a conservative system.[1] On the other hand, the stable orbits of

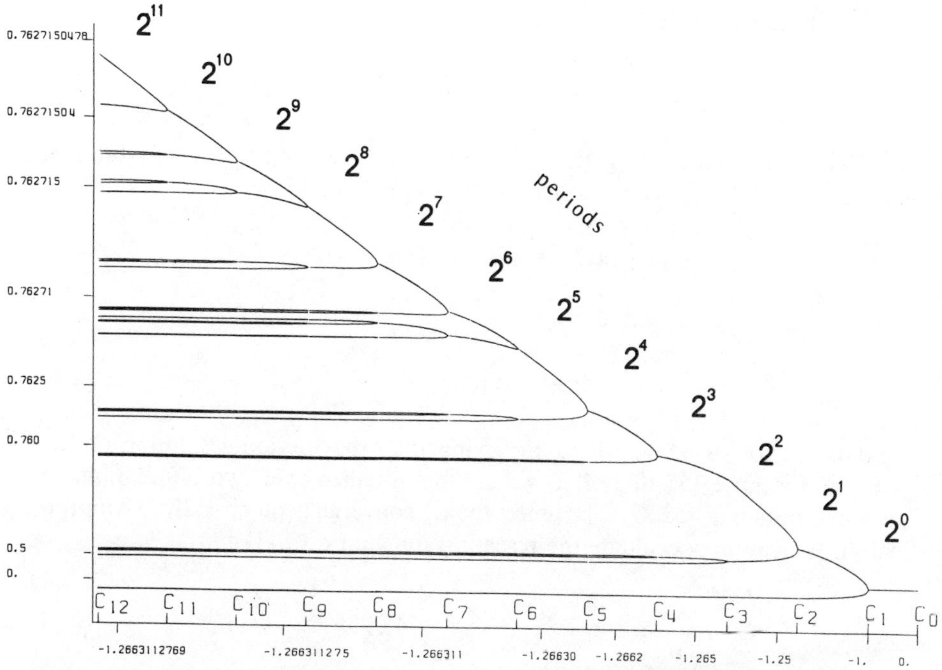

Figure 1. Period-Doubling Feigenbaum sequence $\{C_k\}$ for a conservative system, Eq. (2.6) with $B = 1$.[27] Vertically plotted are the y_t of the elliptic orbit of period 2^k, splitting off at C_k from the elliptic orbit of period 2^{k-1}, which continues as a hyperbolic with reflection orbit, not plotted here. Note the constant rates δ [Eq. (4.7)]—at which the C_k converge—and α [Eq. (4.9)] at which the orbits converge, in this doubly logarithmic plot [because of a symmetry in the orbits there are only $\frac{1}{2}2^k + 1)$ different y_t-values at period 2^k, $k \geqslant 1$]. To the left of C_∞, Eq. (4.4), an infinite number of hyperbolic (w.r.) orbits remain. Note the dissipative Feigenbaum in Figs. 3 and 4.

a dissipative Feigenbaum are "attractors."[1] In that case most nearby orbits move away and approach a stable periodic orbit.[1] Hence, no chaotic behavior can be produced by a dissipative Feigenbaum until after C_∞. In the conservative Feigenbaum, however, small chaotic regions link up to form larger chaotic regions after each bifurcation[42,27,1] to produce large-scale chaos even before C_∞. This large-scale chaos can only be seen if the region it is produced in is enclosed by one or more Kolmogorov–Arnol'd–Moser (KAM) tori.[1] In the dissipative case it is more easily found, since the region it is produced in can be attracting at many values of C [cf. Section 5.1].

Another difference is that in a conservative system, *infinite* numbers of periodic orbits bifurcate from any stable periodic orbit,[1-6] for example, from the origin in Eq. (2.6) with $B = 1$, each of which can produce its own Feigenbaum. We therefore expect an infinite number of Feigenbäume, and so on, intertwined over all scales of length.[1] In a dissipative system there may be an infinite number of Feigenbäume (at $B = 0$, embedded in the chaotic

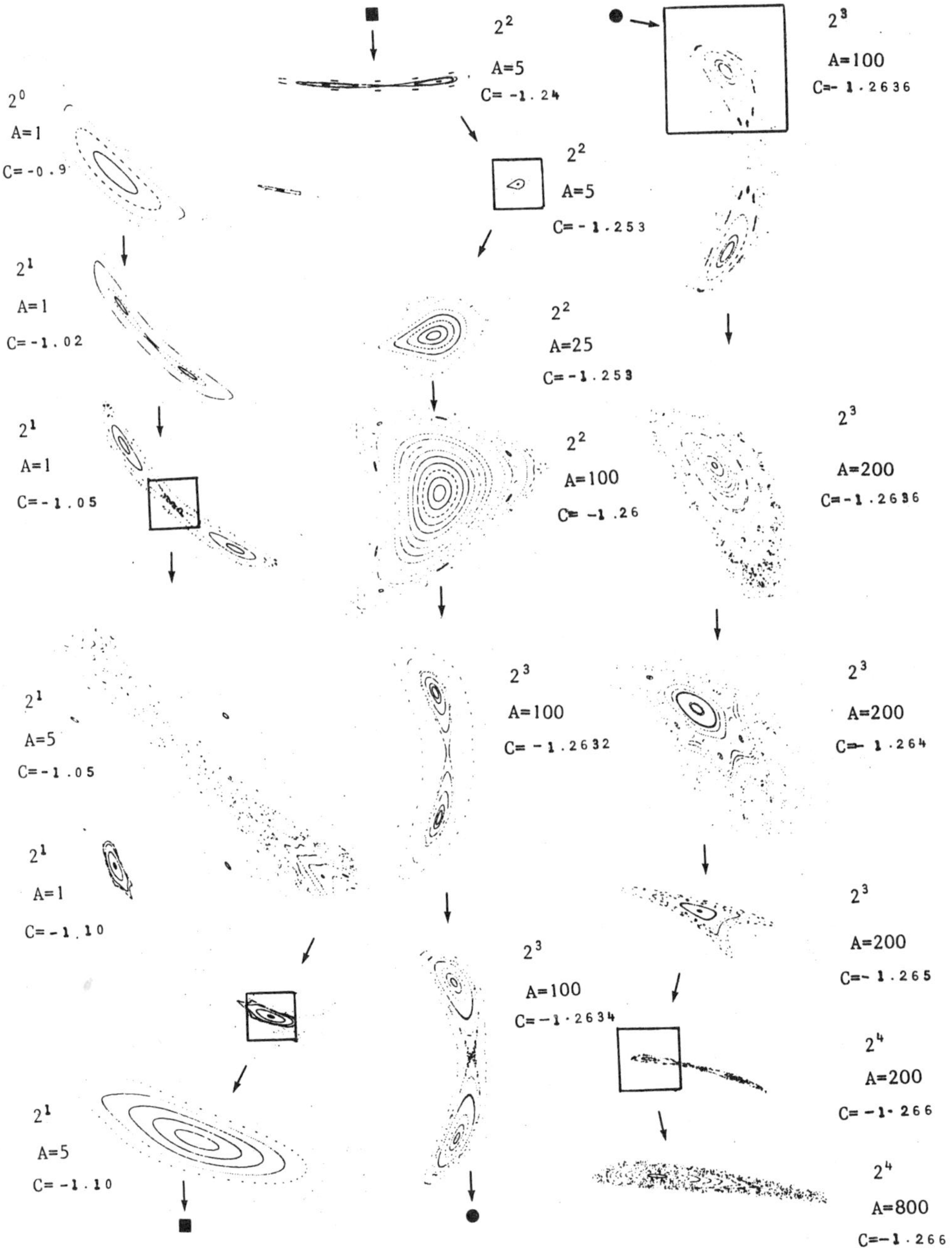

Figure 2. Phase plots of the period-doubling conservative Feigenbaum [Eqs. (4.1)–(4.9)],[27] that is, y_t horizontally versus y_{t+1} vertically. A sequence of magnifications A of the regions about the orbits of period 2^k, starts at the top, on the left, going downward. The C is the parameter in (2.6).

bands[7]) but not over all scales of length, since the stable orbits have finite "regions of attraction."[1]

Finally, T. Bountis and I checked several "integrable" conservative mappings[1] (of Ref. 57) and never obtained a full-fledged Feigenbaum: either the parabola on the right-hand side of the analogue of Eq. (4.2) did not have any intersection points with the straight line on the left-hand side, or the intersection points were unstable fixed points (i.e., the few existing C_k move away from the fixed points). It would be very gratifying indeed if one could prove that *infinities* of intertwined Feigenbäume do not exist in integrable systems, since such systems do not have any chaotic behavior at all.[1,57]

5. DISSIPATIVE FEIGENBAUM SEQUENCES

We continue the renormalization procedure of Section 3 for dissipative mappings [i.e, $|B| < 1$, Eq. (2.6)], and calculate the rates of convergence, and contraction, that are "universal" to dissipative systems. The approach is analogous to the one of Section 4.

In the previous section on conservative systems, a stable periodic orbit has $|\lambda_1| = |\lambda_2| = 1$ [cf. Eq. (3.1)]; that is, nearby-orbits "circle" the stable periodic orbit forever.[1-6] However, in this section on dissipative systems, a stable periodic orbit has eigenvalues with magnitudes less than one [cf. Eq. (3.1) with $|B| < 1$]. Hence stable periodic orbits in this section have a region of "attraction" about them within which all other orbits approach the periodic orbit closer and closer as $t \to \infty$.[1] In this chapter a periodic orbit is called a "repeller" when $|\lambda_1| > 1$, for at least one of its eigenvalues.

When $|B| < 1$ we see that $|B''| < |B'| < |B| < 1$, and so on, since $B' = B^2$, Eq. (3.7). Hence, $B^{(k)}$, the kth renormalization of B has the property

$$B^{(k)} = B^{2^k} \underset{k \to \infty}{\longrightarrow} 0 \tag{5.1}$$

Apparently, *every dissipative mapping* [Eq. (2.6)] *becomes a first-difference mapping, regionally, in the renormalization limit*. This also explains why all such dynamical mappings have the same asymptotic ($k \to \infty$) rates δ and α, irrespective of the value of $|B|$ (< 1), as observed.[7,8,10-14,54-56] Therefore, the C-renormalization relation Eq. (3.8) becomes, asymptotically,

$$C' = -2C^2 + 2C + 2 \tag{5.2}$$

[cf. Eq. (4.1)]; that is, for $B = 0$ its fixed point, $C = C'$, of interest[34] is:

$$2C_\infty \underset{r}{=} \frac{1 - \sqrt{17}}{2} \underset{r}{\simeq} -1.56155\ldots \tag{5.3}$$

while numerically

$$2C_\infty = -1.569945671870945\ldots \tag{5.4}$$

at $B = 0$,[27,52,7,8,10–14] a relative error of 5.3×10^{-3}. For the logistic equation (2.7) this implies[34,38] $a_\infty = 2 - 2C_\infty = 3.5699\ldots$, as observed.[52] The bifurcation points C_k now satisfy:

$$C_k \underset{r}{=} -2C_{k+1}^2 + 2C_{k+1} + 2 \tag{5.5}$$

[cf. Eqs. (5.2), (4.1), and (4.2)]. Again we try the asymptotic expansion Eq. (4.5), this time in Eq. (5.5), see that the C_k converge geometrically, and find for the dissipative Feigenbaum constant:

$$\delta \underset{r}{=} -4C_\infty + 2 = 1 + \sqrt{17} \underset{r}{\approx} 5.12\ldots \tag{5.6}$$

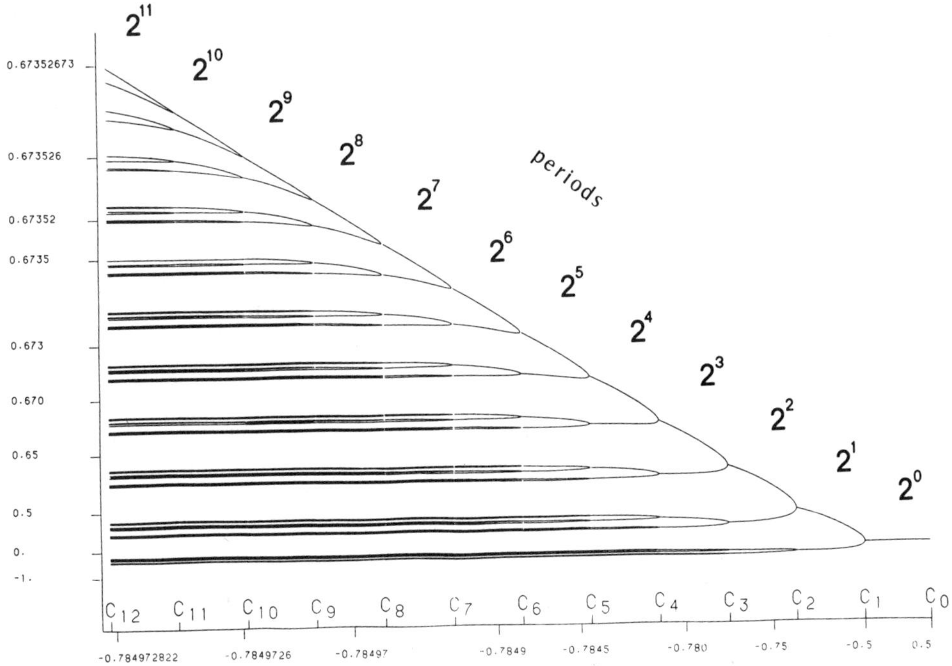

Figure 3. Period doubling Feigenbaum sequence $\{C_k\}$[27] for a dissipative system Eq. (2.6) $B = 0$, equivalent to the logistic equation Eq. (2.7) with $a = 2 - 2C$.[34,38] Vertically plotted are the y_t of the "attractor" of period 2^k, splitting off—at C_k—from an attractor of period 2^{k-1}, which continues as a "repeller,"[1] not plotted here. Note the constant rates δ [Eq. (5.7)]—at which the C_k converge—and α [Eq. (5.9)] at which the orbits converge, in this doubly logarithmic plot. To the left of C_∞ an infinite number of repellers remain (cf. Fig. 4). Also note the conservative Feigenbaum in Fig. 1.

while numerically[10,27,11,14]

$$\delta = 4.6692016091029909\ldots \tag{5.7}$$

a relative error of 9.7×10^{-2} [cf. Eqs. (4.6) and (4.7); a renormalization over two bifurcations yields[51]: $\delta \underset{r}{\approx} 2 + 2\sqrt{2} \simeq 4.83\ldots$. Evaluating the rescaling constant α, Eqs. (3.11-9) we now find:

$$\alpha \underset{r}{=} e + d^2 \qquad \underset{r}{\approx} -2.2399\ldots \tag{5.8}$$

while numerically[10,27,11,12]

$$\alpha = -2.50290787509589284\ldots \tag{5.9}$$

with the e, d of (3.9-3), a relative error of 1.05×10^{-1}. The dissipative Feigenbaum at $B = 0$ has been found numerically and theoretically by several groups.[10-14,7,8,51,52,54-56] A doubly logarithmic plot of it, by van Zeyts, is reproduced in Fig. 3.[27,1] These studies confirm that the δ and α above are universal constants. Thus, even if we know only two consecutive bifurcation points, called μ_0 and μ_1, we "universally" expect an infinite number of unstable Feigenbaum branches at about 27% above this interval, since:

$$\mu_\infty \underset{r}{\approx} \mu_0 + \frac{(\mu_1 - \mu_0)\delta}{\delta - 1} \underset{r}{\approx} \mu_0 + 1.2725(\mu_1 - \mu_0) \tag{5.10}$$

[cf. Eqs. (5.5)–(5.7)]. The asymptotic rates δ, α only require that the asymptotic value of $B^{(k)}$ in Eq. (5.1) be used in the renormalization, Eq. (3.8). Hence they are the same for all values of B, with $|B| < 1$. The critical value C_∞ however does depend on B. So far in this section, we have calculated it at $B = 0$ [Eq. (5.3)]. Thus, instead of the asymptotic renormalization Eqs. (5.2) and (5.5), we should now use the full renormalization [Eqs. (3.7), (3.8), and B^k]. Note that the C', in the left-hand side of Eq. (3.8) appears in the equation of motion, Eq. (3.10), with a damping B' $(= B^2)$, whereas the C, in the right-hand side, appears in the equation of motion, Eq. (2.6), with damping B $(= \pm\sqrt{B'})$. Before Eq. (4.2) I explained that as we iterate from the left to the right (i.e., from C' to C), we get higher and higher bifurcation points [cf. Eqs. (4.2) and (5.5)]. Denoting the kth bifurcation point of the parameter C for an equation of motion Eq. (2.6) with some damping b by $C_k(b)$ the full renormalization, Eq. (3.8), therefore gives:

$$C_k(b^2) \underset{r}{=} -2C_{k+1}^2(b) + 2(1 + b)C_{k+1}(b) + 2b^2 + 3b + 2 \tag{5.11a}$$

Each time we iterate this, from left to right, we obtain a higher order bifurcation point, for a different damping in Eq. (2.6). When we take the critical C-value (i.e., $k = \infty$) in Eq. (5.11a) and iterate, we do remain at a critical point, but one with a different damping in Eq. (2.6); that is, we have a B-renormalization for the C_∞. Hence, denoting the critical value at B^{2^j}, cf. Eq. (5.1), by $C(j)$, we find from Eqs. (5.11a) and (5.1):

$$C(j+1) \underset{r}{=} -2C^2(j) + 2(1 + B^{2^j})C(j) + 2B^{2^{j+1}} + 3B^{2^j} + 2 \quad (5.11b)$$

For $0 < B < 1$ we know two boundary conditions on Eq. (5.11b), namely, $C(+\infty) = C_\infty$ [Eq. (5.3)] and $C(-\infty) = C_\infty$ [Eq. (4.3)]. Since we want to remain at a critical point, we take this $C(+\infty)$ in the left-hand side of Eq. (5.11b), iterate Eq. (5.11b) from $j = +\infty$ back to $j = 0$ and obtain $C_\infty(B)$.

In practice we would start from a relatively small j_m (e.g., one with $B^{2^{j_m}} < 10^{-3}|B|$) because of the relative error in $C(\infty)$, Eq. (5.3). On the other hand for $B = -1$, one iteration suffices, of course. Thus, taking the C_∞ of Eq. (4.3) in the left-hand side of Eq. (5.11b) at $j = 0$, we obtain at $B = -1$:

$$C_\infty \underset{r}{=} -\frac{\left(\sqrt{2 + 2\sqrt{65}}\right)}{4} \underset{r}{\simeq} -1.0643\ldots \quad (5.12)$$

while numerically[27,34]

$$C_\infty = -1.074063\ldots \quad (5.13)$$

a relative error of 9.1×10^{-3}. For $B = -0.3$, Hénon's value[17], Eq. (5.11b), predicts $C_\infty(-0.3) = -0.7301\ldots$; that is, for the parameter a in Eq. (2.9) it yields[34,35] $a_\infty = 1.044\ldots$, while numerically[1] $a_\infty = 1.058\ldots$, a relative error of 1.3×10^{-2}.

5.1. Beyond the Critical Value, Onset of Chaos

Here, I briefly consider the case $B = 0$ in Eq. (2.6), that is,[36,38]

$$y_{t+1} = 2Cy_t + 2y_t^2 \quad (5.14)$$

A number of mathematical and numerical results are available in this case.[7,1,2,10–12,27,50,52,54–61] We have seen that for $C < C_\infty \simeq -0.78497\ldots$, cf. Eq. (5.4), the (period-2^k) Feigenbaum branches are unstable; that is, their variational equations have unbounded solutions, and nearby-orbits move away rapidly. Also an infinite number of unstable Feigenbaum orbits accumulate about its nonperiodic limit orbit. Nevertheless, *all* Feigenbaum orbits are contained within an *interval of attraction*

$$-\tfrac{1}{2} \leqslant y_t \leqslant \tfrac{1}{2} - C \qquad \text{for} \quad -1 < C \leqslant \tfrac{1}{3} \quad (5.15)$$

since *all* points of Eq. (5.15) are mapped *in*to a smaller part of Eq. (5.15): $-\frac{1}{2}C^2 \leqslant y_{t+1} \leqslant \frac{1}{2} - C$.[43–45,50] *All* points outside the interval of Eq. (5.15) become unbounded under repeated mapping with Eq. (5.14), as we see in Notes 44 and 50. So, orbits near the unstable Feigenbaum branches are repelled but cannot move out of the interval (5.15) and away to infinity. Actually, contained in the interval of attraction is a smaller interval that is forever mapped *onto* itself, an "*invariant interval*":

$$-\tfrac{1}{2}C^2 \leqslant y_t \leqslant \tfrac{1}{2}C^4 - C^3, \qquad \text{for} \quad -1 \leqslant C < -0.62\ldots \tag{5.16}$$

as pointed out to me by Mulders.[43,50] The left-hand side is the minimum of the y_t-parabola in Eq. (5.14). The right-hand side is the image of that minimum under Eq. (5.14)[45]; that is, this is the minimum invariant interval (in one piece...).[43,44,50] Beyond $C = -1$ the boundaries of Eqs. (5.16) and (5.15) cross and there are always orbits that become unbounded, whence the C-restrictions in Eqs. (5.15) and (5.16).[45,50]

Beyond C_∞ of Eq. (5.4) none of the (period-2^k) Feigenbaum branches are stable; that is, none are "attractors." Thus, one remaining question is whether

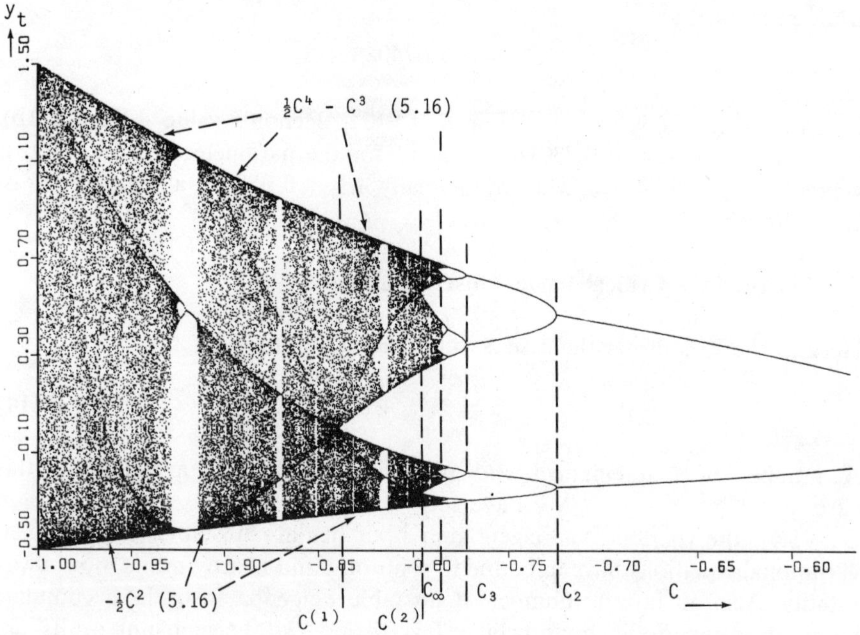

Figure 4. Dissipative Feigenbaum for the "logistic" one-dimensional dynamical mapping Eq. (5.14), that is, Eq. (2.6) at $B = 0$. Vertically plotted are the y_t of Eq. (5.14), after the transients have decayed[27] (after Collet & Eckmann[7]), in this doubly linear graph. A reverse bifurction of chaotic bands takes place at $C^{(1)}$, $C^{(2)}$ etc. Within the chaotic bands new Feigenbäume of different basic (i.e., "trunk") periods are visible—for example, the ones of trunk periods 3, 5, and 6, near the "holes" in the chaotic bands. The part to the right of C_∞ is replotted logarithmically in Fig. 3. Also note the bifurcations in Fig. 5 and the reverse bifurcations in Fig. 6.

the orbits repelled by the infinite number of unstable Feigenbaum branches will chaotically fill up the invariant interval, Eq. (5.16), for $-1 \leqslant C \leqslant C_\infty$, Eq. (5.4). Numerical evidence for the y_t behavior in this C range is presented in Fig. 4, which plots the asymptotic behavior of the y_t orbits after their transients have died down; the y_t are only plotted for $t > 1000$ (and: < 4000). Chaotic behavior, as well as new Feigenbaum branches, with different basic—"trunk" —periods, can indeed be seen in the interval of Eq. (5.16).[7,52]

As an example of chaotic behavior I mention the exact solution of Eq. (5.14) at $C = -1$:

$$y_t = \tfrac{1}{2} + \cos(2^{t+1}\phi) \qquad C = -1 \tag{5.17}$$

which is checked by substitution. This solution is due to Ulam and von Neumann [cf. Refs. 7 and 52]. The ϕ is determined by the initial y_0. Although Eq. (5.17) contains an infinite number of periodic orbits (at certain rational fractions of 2π for ϕ, mod 2π), including the period-2^k Feigenbaum orbits, *all* are unstable. As a matter of fact, two orbits starting near each other ($\phi \simeq \phi'$) separate at an exponential rate [cf. small values of ϕ (mod 2π) in Eq. (5.17)[6,3–5]]; that is, we have a "sensitive dependence on initial conditions.[46,7,1] Actually the general solution Eq. (5.17) is "mixing," ergodic,[46,47] and "δ-correlated"[50] for most choices of ϕ.[12]

A striking feature of Fig. 4 is the "*reverse bifurcation of the chaotic bands.*" At $C^{(1)}$ ($= -0.83928675521416113\ldots$) one chaotic band splits into two bands; at $C^{(2)}$ ($= -0.79628609205348932\ldots$) the two bands split into four, and so on.[7,27,49] Moreover, the $C^{(k)}$ converge to the same critical value C_∞, at the same geometric rates δ [Eqs. (5.7), (4.5)] and α [Eq. (5.9)] as the earlier C_k. This is made particularly clear by the logarithmic plots and theory of Ref. 7.

Reverse bifurcation can be obtained from our C-renormalization as well, at least for the C-values at which $\{y_t\}$ is truly ergodic. In that case the global behavior of most orbits is fully determined by the local behavior in any small (finite) interval.[47] Since we did analyze the local, or regional, behavior in the renormalization analysis of Sections 3–5, the renormalization of Eqs. (5.5)–(5.9) applies here as well: changing the sign of the a in Eq. (4.5) we still converge geometrically, at the same rate δ [Eq. (5.6)], albeit from the other side of C_∞, as do the $C^{(k)}$. In fact, the (k)th bifurcation of the bands, at $C^{(k)}$, takes place at the y-value of the unstable period-2^k orbit (not plotted in Fig. 4, but cf. $C^{(1)}$, $y = 0$).[7] It is at those $C^{(k)}$ that one can prove ergodicity[7] and show the numerical correlation functions to be (Kronecker) δ-functions.[12,50] Exactly for which other C-values we have ergodicity, or for which fraction of $[-1, C_\infty]$, is an open problem.[7,46]

Normal and Reverse bifurcations also exist in systems of rate equations, for example in *the Rössler attractor*:

$$\dot{x} = -y - z \tag{5.18}$$

$$\dot{y} = x + g y \tag{5.19}$$

$$\dot{z} = f + xz - \mu z \tag{5.20}$$

with parameter g, f, μ.[48] The divergence of this flow in the three-dimensional phase space is $g + x(t) - \mu$. Hence for $x < \mu - g$ any volume element shrinks continually (by the Gauss theorem) as it flows through space (i.e., we have a dissipative system). Such rate equations are widely used throughout physics, chemistry, hydrodynamics, meteorology, population dynamics and biology. The Rössler equations have a limit cycle for some choice of the parameters, as shown in Fig. 5*A*. Its first few period doubling bifurcations are shown in Figs. 5*B* and 5*C*. Its μ_∞ is observed near $\mu \simeq 4.20$.[48] Below the x, y-projections of the orbits the corresponding power spectra of z are plotted [i.e., the squares of the amplitudes of the Fourier components of $z(t)$]. Note that each period doubling introduces a half-frequency subharmonic (and many higher harmonics). Reverse bifurcations can be seen in Fig. 6, beyond μ_∞. The δ-value for this system is observed to be the same as in Eq. (5.7). Similar Feigenbaum sequences have also been found in systems of three, five, and seven rate equations for the (spatial) Fourier modes of the Navier–Stokes equation in hydrodynamics, a periodically driven nonlinear oscillator in the conservative and dissipative cases, in other systems,[1,7,8] and in experimental turbulence.[1]

Finally, embedded in the chaotic bands of Fig. 4 is an infinity of small Feigenbäume with different trunk periods. The one with trunk period 3 is visible near $C \simeq -0.93\ldots$, born at $C = \frac{1}{2} - \sqrt{2}$ (its ordinary bifurcations, into attractors of periods 3×2^k, move leftward in Fig. 4 as did our original trunk-1 Feigenbaum). Therefore there must be a trunk-6 Feigenbaum embedded in the two chaotic bands between $C^{(1)}$ and $C^{(2)}$, as can be seen in Fig. 4 near $C \simeq -0.81\ldots$, and so on. Bountis, van Zeyts,[27] and I have analytically

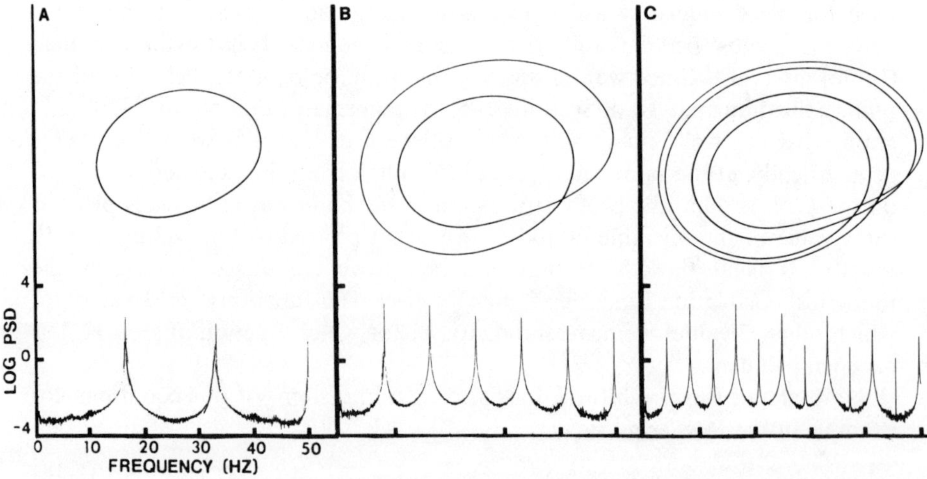

Figure 5. Period doubling bifurcations of the Rössler attractor [Eqs. (5.18)–(5.20)] projected on the x, y-plane: $g = f = 0.2$; $\mu = 2.6$ (a), $= 3.5$ (b), 4.1 (c). Note the subharmonics in the power spectra of $z(t)$. For $\mu > \mu_\infty$ see Fig. 6. (From Crutchfield et al.[48])

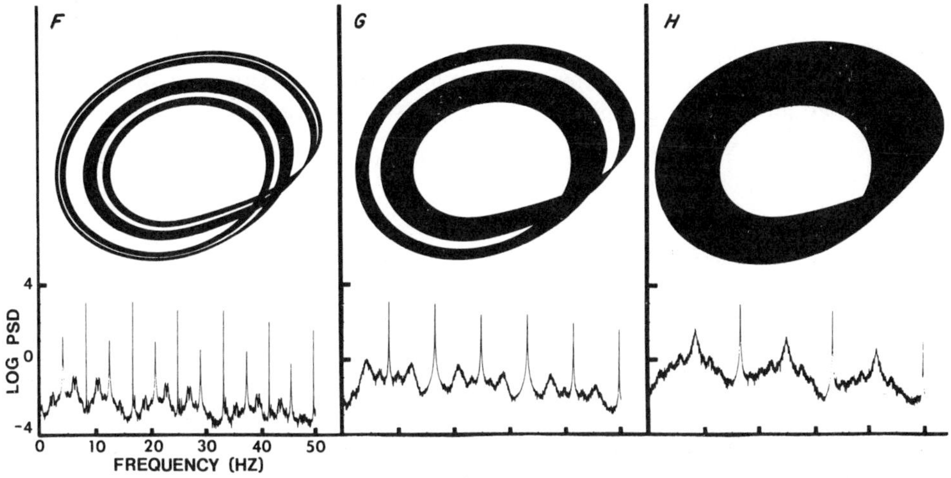

Figure 6. Chaotic bands appear in the Rössler attractor [Eqs. (5.18)–(5.20)] above the critical point, $\mu_\infty \simeq 4.20$, of the Feigenbaum sequence in Fig. 5. Note the reverse bifurcations; that is, at $\mu \simeq 4.23$ (F) we have four chaotic bands, at $\mu \simeq 4.30$ (G) Two bands and at $\mu \simeq 4.60$ (H) one band. Note Fig. 5. Reverse bifurcations are also visible in Fig. 4. (From Crutchfield et al.[48])

(and numerically) continued such trunk-m Feigenbäume from $B = 0$, via finite B,[11] to $B = 1$, a conservative system. We suggest the existence of an analogue of the Birkhoff theorem[1, 3–6, 47] on the existence of pairs of stable and unstable periodic orbits, for any B with $-1 \leqslant B \leqslant 1$ in Eq. (2.6). It seems to us as if the many conservative Feigenbäume for $C > C_\infty$ [Eq. (4.4)] at $B = 1$ have, at $B = 0$, been condensed into the chaotic band with $-1 < C < C^{(1)}$.... A gradual transition from the first into the second situation takes place as we let B decrease from 1 to 0.

Although much remains to be done, in particular on the analogous phenomena for continuous differential equations, it is exciting to detect any progress at all on these venerable problems of chaotic behavior in deterministic systems.[60]

6. ACKNOWLEDGMENTS

I thank Robert MacKay for correcting some of my earlier algebra[41, 1] and for extending the present renormalization schema to also yield the second rate of contraction β [18] in the second direction of the phase plane (Appendix C). I am very grateful to Johan van Zeyts,[27] Jan Mulders,[50] and Tassos Bountis for allowing me to mention in Sections 4, 5, and 5.1 some of their numerical and analytical work, most of which has not yet appeared in print. The results on intersecting storage rings (Appendix B) were stimulated by the encouragement, support, and enthusiasm of Mel Month and David Sutter. Several suggestions by Jouke Heringa, for improvements in the text, were appreciated and followed.

I am grateful to Elly Reimerink for typing (the draft) beyond the call of duty (let alone business hours). This study was supported in part under Contract DE-AC03-77ER01538.

APPENDIX A: DERIVATION OF THE STANDARD FORM

We have seen in Note 40 that any mapping T [Eq. (2.1)] can be transformed into another one T' such that a given periodic orbit of T becomes a fixed point of T', at the origin of the phase plane. Expanding the analytic mapping T' about this origin, in a Taylor series of terms up to and including order p, the x_{t+1} and y_{t+1} each becomes a polynomial of order p in x_t, y_t. Under our constraint of a constant Jacobian, equal to B [Eq. (2.2)], we can now employ one of Engel's results,[15,32] which holds when p is a prime number. It allows us to recombine those two first-difference equations into one second-difference equation of only one variable and of the same degree p,

$$y_{t+1} + By_{t-1} = 2Cy_t + \sum_{k=2}^{p} a_k y_t^k \tag{A1}$$

as we see in Note 32, when p is a prime number. Hence any analytic mapping T [Eq. (2.1)] or T' (Note 40), with constant Jacobian can be approximated arbitrarily well[1] by our standard form, Eq. (A1), when we take arbitrarily high prime-order polynomials. This mapping about the origin (of the y_t, y_{t+1} phase plane now) contains only p parameters (one of the a_k can be scaled out). Compared to the number of parameters of the *two* original pth-order polynomials in two variables we have a great simplification, produced by the constraint Eq. (2.2). For instance when $p = 2$ we have only two parameters in Eq. (A1) (after scaling y), whereas the original equations without constraint would have at least six parameters (after scaling x and y) [cf. Eqs. (A2), A3)]. Engel's results were pointed out earlier by Moser[15] and by Hénon,[16,17] who also gave an explicit derivation for the case with $p = 2$, a stable origin [Eq. (3.2)] and $B = 1$.[16] In Ref. 17 Hénon notes that it can trivially be extended to other B-values. As another explicit example I derive here our standard equation (2.6), that is, Eq. (A1) with $p = 2$, for an *unstable* origin and general B. This is the case that is actually required for this chapter, where the origin turns unstable after a period doubling bifurcation.

First consider the four linear terms in the foregoing Taylor expansion of T', or T [Eq. (2.1)], about the origin. Its four coefficients constitute a real 2×2 matrix $\mathbf{L}$ whose determinant must equal B if the constraint of Eq. (2.2) is to hold at the origin. This matrix $\mathbf{L}$ has a simple quadratic eigenvalue equation and one easily checks that if its eigenvalues λ_1, λ_2 were complex we would have $|\lambda_1| = |\lambda_2| = \sqrt{B}$ ($B > 0$, always, in that case). Since we can confine ourselves to $|B| \leqslant 1$ [cf. Eqs. (2.3), (2.4) and beyond] we see that when the origin is unstable, that is, when $|\lambda_1| > 1$, both eigenvalues must be real. Hence, transforming to real variables on which **L** is diagonal, we find for the most general quadratic mapping ($p = 2$), of an x, y-plane with an *unstable* origin, into itself:

$$x_{t+1} = \lambda_1 x_t + u x_t^2 + w x_t y_t + v y_t^2 \tag{A2}$$

$$y_{t+1} = \lambda_2 y_t + u' x_t^2 + w' x_t y_t + v' y_t^2 \tag{A3}$$

with

$$|\lambda_1| > 1 \qquad \text{and} \qquad \lambda_1 \lambda_2 = B, \qquad |B| \leqslant 1 \tag{A4}$$

the Jacobian Eq. (2.2) at the origin, with λ_1, λ_2 and the other parameters real numbers. If w happens to be negative we can turn it positive by transforming to $y' \equiv -y$. Similarly, if u happens to be negative we turn it positive by changing to $x' \equiv -x$. Whence we always have

$$u \geqslant 0 \qquad \text{and} \qquad w \geqslant 0 \tag{A5}$$

We now impose the constraint Eq. (2.2), of a constant Jacobian equal to B, for all values of x, y. Substituting our equations (A2) and (A3) into this constraint and comparing equal powers of x and y we obtain:

$$\frac{u}{u'} = \frac{w}{w'} = \frac{v}{v'} \equiv \frac{1}{z} \tag{A6}$$

and

$$-\frac{\lambda_1}{\lambda_2} = \frac{2u}{w'} = \frac{w}{2v'}, \tag{A7}$$

that is, only a slight variation of Hénon's equations for the case of a stable origin.[16] Combining Eqs. (A6) and (A7) we find

$$w^2 = 4uv \tag{A8}$$

$$\text{whence } v \geqslant 0 \text{ also,} \qquad \text{and} \qquad w'^2 = 4u'v' = w^2/z^2$$

[cf. Eq. (A5)]. According to analytic geometry this implies that both conic sections in our equations (A2) and (A3) are parabolae [as a result of the constraint Eq. (2.2)]. Thus, transforming the x, y variables into

$$\chi_t \equiv x_t\sqrt{u} \qquad \text{and} \qquad \eta_t \equiv y_t\sqrt{v} \tag{A9}$$

our equations (A2) and (A3) become

$$\chi_{t+1} = \lambda_1\chi_t + \sqrt{u}\,(\chi_t + \eta_t)^2 \tag{A10}$$

$$\eta_{t+1} = \lambda_2\eta_t + z\sqrt{v}\,(\chi_t + \eta_t)^2 \tag{A11}$$

where Eqs. (A8) and (A5) are used. Eliminating the quadratic term between Eqs. (A10) and (A11) yields:

$$\frac{\chi_{t+1}}{\lambda_1} + \frac{\eta_{t+1}}{\lambda_2} = \chi_t + \eta_t \tag{A12}$$

after repeated use of Eqs. (A6)–(A8). Further transforming with

$$s_t \equiv \chi_t + \eta_t \qquad \text{and} \qquad d_t \equiv \chi_t - \eta_t \tag{A13}$$

changes Eq. (A12) into

$$d_{t+1}\left(\frac{1}{\lambda_1} - \frac{1}{\lambda_2}\right) = 2s_t - s_{t+1}\left(\frac{1}{\lambda_1} + \frac{1}{\lambda_2}\right) \tag{A14}$$

an explicit linear equation for one of the variables in terms of the other.[32] Transforming the nonlinear form (A10) to the same variables, Eq. (A13), we get

$$s_{t+1} + d_{t+1} = \lambda_1(s_t + d_t) + 2\sqrt{u}\,s_t^2 \tag{A15}$$

Substituting the d expression (A14) twice in Eq. (A15) we arrive at:

$$s_{t+1} + Bs_{t-1} = 2Cs_t + Ds_t^2 \tag{A16}$$

with

$$2C \equiv \lambda_1 + \lambda_2 \qquad \text{and} \qquad D \equiv \frac{\sqrt{u}\,(\lambda_1 - \lambda_2)}{\lambda_1} \tag{A17}$$

A simple scaling with

$$q_t \equiv \frac{Ds_t}{2} \tag{A18}$$

finally produces *our standard equation*:

$$q_{t+1} + Bq_{t-1} = 2Cq_t + 2q_t^2 \tag{A19}$$

[cf. Eq. (2.6)]. I thank Charles Karney and Mike Lieberman for inquiring about the generality of Eq. (A19) and discussing with me some of the arguments appearing in this appendix and in Section 2.

APPENDIX B: THE STANDARD FORM IS EXACT FOR INTERSECTING STORAGE RINGS

Here we study the motion of protons and electrons in the intersecting storage rings of high energy physics. I show that a "surface of section" mapping of this motion is exactly described by a second-difference equation of the standard form (A1) [cf. Eq. (2.6) and Ref. 1].

The particles above are supposed to travel along an ideal "circular" ring, whose circumference can be as long as 30 km.[20] In practice the particles exhibit unavoidable ("betatron") oscillations about this ideal orbit. The main dynamics problem in that field is to keep those oscillations small enough, over up to 10^{11} revolutions..., to ensure that the particles do not exceed the width of the vacuum chamber and be lost.[20–22]

The oscillations of the particles in one ring alone are very well described[20,22] by the simple harmonic oscillator

$$\frac{d^2y}{d\phi^2} = -Q^2y - Rp \qquad \text{with} \quad p \equiv \frac{dy}{d\phi} \tag{B1}$$

where y is the particle's displacement perpendicular to the plane of the ring and ϕ is its azimuthal angle, along the ring, which is proportional to the time. The "tune" Q is determined by the gradient of the containing magnetic field[20–22] and R is the damping due to radiation. This "resistance" R may be complex but vanishes in the ("conservative") case of protons. The effect of the second, colliding, beam is to add a nonlinear term:

$$\frac{d^2y}{d\phi^2} = -Q^2y - Rp + \mathfrak{B}\mathfrak{F}(y)\left[\sum_{t=0,1,2,..}^{\infty} \delta(\phi - t2\pi)\right] \tag{B2}$$

(cf. Refs. 21 and 1). The second beam intersects the first ring over a very small ϕ interval only, whence the δ-functions. Because they are small, few electrons (or protons) actually collide. Yet, as they pass briefly through the second beam they feel a strong nonlinear electromagnetic force $\mathfrak{B}\mathfrak{F}(y)$.[20–22] To get enough true collisions the particles must undergo $\approx 10^{10}$–10^{11} such near-collisions, without y becoming extremely large and exceeding the width of the vacuum chamber. This number of revolutions is about the same as the earth has ever made about the sun.... The integer t in Eq. (B2) counts this number of revolutions and beam–beam crossings.

In between two subsequent δ-pulses of Eq. (B2) the equation is linear, that is, the same as Eq. (B1), which has the solution

$$y(\phi) = a\exp(q_+\phi) + b\exp(q_-\phi) \tag{B3}$$

with

$$q_\pm^2 + Rq_\pm + Q^2 \equiv 0 \tag{B4}$$

$$\text{whence} \quad q_+ + q_- = -R \quad \text{and} \quad q_+ q_- = Q^2,$$

a and b are determined from $y(0)$ and $p(0)$. We introduce as new variables the (right-limits) of the values at multiples of 2π:

$$y_t \equiv y(t2\pi+) \quad \text{and} \quad p_t \equiv p(t2\pi+) \tag{B5}$$

Given these values as initial conditions one can trivially calculate a and b in Eq. (B3) and find the later y_{t+1} and p_{t+1} expressed in the initial y_t, p_t as:

$$\begin{pmatrix} y_{t+1} \\ p_{t+1} \end{pmatrix} = \mathbf{M}\begin{pmatrix} y_t \\ p_t \end{pmatrix}$$

where

$$\mathbf{M} \equiv \frac{1}{(q_+ - q_-)}\begin{pmatrix} q_+e_- - q_-e_+ & e_+ - e_- \\ q_+q_-(e_- - e_+) & q_+e_+ - q_-e_- \end{pmatrix} \tag{B6}$$

with

$$e_\pm \equiv \exp(q_\pm 2\pi) \tag{B7}$$

That is, $\mathbf{M}$ is the usual propagator matrix for Eq. (B1) [i.e., also for Eq. (B2) with $\mathcal{B} = 0$). For $\mathcal{B} \neq 0$ we integrate out Eq. (B2), from $\phi = (t+1)2\pi - \varepsilon$ to $\phi = (t+1)2\pi + \varepsilon$ and take $\varepsilon \downarrow 0$, to obtain

$$p((t+1)2\pi+) - p((t+1)2\pi-) = 0 + 0 + \mathcal{B}\mathcal{F}(y_{t+1}) \tag{B8}$$

since $y(\phi)$ remains a continuous function [only in the $\mathcal{B} = 0$ case is the $p(\phi)$ continuous as well]. Equation (B6) propagates p from $p(t2\pi+)$ to $p((t+1)2\pi-)$ for any $\mathcal{B}$ in Eq. (B2), since no δ-pulses are encountered in this interval. Further propagation to $p((t+1)2\pi+)$ can now be obtained from Eq. (B8). Hence the total propagator for Eq. (B2) becomes:

$$\begin{pmatrix} y_{t+1} \\ p_{t+1} \end{pmatrix} = \mathbf{M}\begin{pmatrix} y_t \\ p_t \end{pmatrix} + \begin{pmatrix} 0 \\ \mathfrak{B}\mathfrak{F}(y_{t+1}) \end{pmatrix} \tag{B9}$$

with the matrix **M** from Eq. (B6). These two first-difference equations are already much simpler to work with than the differential equation, Eq. (B2), but we prefer to reduce the system (B9) even further. Note that the expression for y_{t+1} in Eq. (B9) remains linear. Hence we can reduce the two first-difference equations in Eq. (B9) to one second-difference equation, as mentioned in Appendix A and in Note 32. We do it by first solving for p_t from the y_{t+1} expression in Eq. (B9):

$$p_t = \frac{(q_+ - q_-)y_{t+1} - (q_+ e_- - q_- e_+)y_t}{e_+ - e_-} \tag{B10}$$

a linear combination of y_t and y_{t+1}. So p_{t+1} is the same linear combination, but of y_{t+1} and y_{t+2}. Substituting this and Eq. (B10) into the second component of Eq. (B9) we find, after some collecting and canceling,

$$y_{t+2}\left\{\frac{q_+ - q_-}{e_+ - e_-}\right\} + y_t\left\{\frac{e_+ e_- (q_+ - q_-)^2}{(e_+ - e_-)(q_+ - q_-)}\right\} =$$

$$y_{t+1}\left\{\frac{(q_+ - q_-)(e_+ + e_-)}{e_+ - e_-}\right\} + \mathfrak{B}\mathfrak{F}(y_{t+1}) \tag{B11}$$

Finally we multiply Eq. (B11) by $(e_+ - e_-)/(q_+ - q_-)$, and take $\tau \equiv t + 1$, to obtain:

$$y_{\tau+1} + By_{\tau-1} = 2Cy_\tau + E\mathfrak{F}(y_\tau) \tag{B12}$$

with

$$B \equiv \exp(-2\pi R) = e_+ e_- \tag{B13}$$

where

$$2C \equiv \exp(2\pi q_+) + \exp(2\pi q_-) = e_+ + e_- \tag{B14}$$

and

$$E \equiv \mathfrak{B}\frac{e_+ - e_-}{q_+ - q_-} \tag{B15}$$

[cf. Eqs. (B4)–(B7)]. Note that in the conservative case we have $R = 0$, that is,

$B = 1$, $C = \cos(2\pi Q)$ and $E = (\mathcal{B}/Q)\sin(2\pi Q)$. Thus the conservative version of Eq. (B12) is:

$$y_{t+1} + y_{t-1} = 2Cy_t + \frac{\mathcal{B}S}{Q}\mathcal{F}(y_t) \tag{B16}$$

with

$$S \equiv \sin(2\pi Q) \qquad \text{and} \qquad C = \cos(2\pi Q) \tag{B17}$$

where we changed notation ($\tau \to t$) [cf. Eqs. (2.13)–(2.16) of Ref. 1]. Choosing $\mathcal{B} = (q_+ - q_-)/(e_+ - e_-)$ in Eqs. (B15) and (B12), we finally obtain Eq. (B9) in our *standard form*:

$$y_{t+1} + By_{t-1} = 2Cy_t + \mathcal{F}(y_t) \tag{B18}$$

This choice of $\mathcal{B}$ is not a restriction, since $\mathcal{F}(y)$ is still arbitrary in Eqs. (B2) and (B15). Choosing $\mathcal{F}(y) = 2y^2$ we obtain our standard equation (2.6), exactly. Choosing other functions $\mathcal{F}(y)$, of greater importance for storage rings,[20–22] we still recover Eq. (2.6) in a neighborhood of a periodic orbit, as discussed in Appendix A. Note that we have in Eq. (B18) the *exact* "surface of section" mapping for the differential system Eq. (B2) and that in Eq. (B13) we have the *exact* relation between the rate of "area contraction" B of the mapping and the damping coefficient R of the differential system.

APPENDIX C: RENORMALIZATION CALCULATION OF THE SECOND RESCALING PARAMETER β, by *Robert S. MacKay*

Our standard equation (2.6) becomes, in the conservative case,

$$y_{t+1} + y_{t-1} = 2Cy_t + 2y_t^2 \tag{C1}$$

In Appendices A and B Eq. (C1) was obtained from two coupled first-difference equations [e.g., Eqs. (A10) and (A11) or (B9)]. There are many other ways in which Eq. (C1) can be realized as two coupled first-difference equations. Note that in Eq. (3.11) the rescaling coefficient α was obtained for only one of the two variables in such a system of difference equations. It turns out that the rescaling coefficient β of the second variable is different from the α in Eq. (3.11).[18] Here I derive an expression for β [viz. Eq. (C13)] from the present renormalization scheme, which is a good approximation of its numerical value Eq. (C14).[18]

The system of first-difference equations I chose is:

$$y_{t+1} = -x_t + Cy_t + y_t^2 \tag{C2}$$

$$y_t = +x_{t+1} + Cy_{t+1} + y_{t+1}^2 \tag{C3}$$

One easily checks that this system is still equivalent to Eq. (C1) by letting $t \to t-1$ in Eq. (C3) and adding it to Eq. (C2). Hence the y in Eqs. (C2) and (C3) is the same as in Eq. (C1). This choice of Eqs. (C2) and (C3) clearly brings out the time reversibility of the conservative mapping (C1).[1,18] Proceeding in a manner analogous to the renormalization of Section 3, we see that Eqs. (C2) and (C3) have period-1 orbits at $x_t = 0$ and $y_t = 0$, respectively [Eq. (3.2)]. A period-2 orbit bifurcates from it, with $\hat{x}_0 = \hat{x}_1 = 0$ and $\hat{y}_0$, $\hat{y}_1$ which are the same as before in Eq. (3.3), of course. Substituting $x \equiv \hat{x} + \Delta x$ and $y \equiv \hat{y} + \Delta y$ the system of Eqs. (C2) and (C3) becomes

$$\Delta y_{t+1} = -\Delta x_t + (C + 2\hat{y}_t)\,\Delta y_t + \Delta y_t^2 \tag{C4}$$

$$\Delta y_t = +\Delta x_{t+1} + (C + 2\hat{y}_{t+1})\,\Delta y_{t+1} + \Delta y_{t+1}^2 \tag{C5}$$

Taking Eq. (C4) at $t = 2\tau + 1$ and adding it to Eq. (C5) at $t = 2\tau$ we obtain

$$\Delta y_{2\tau+2} = -\Delta y_{2\tau} + e\,\Delta y_{2\tau+1} + 2\,\Delta y_{2\tau+1}^2 \tag{C6}$$

where e was defined in Eq. (3.9). This is the same as Eq. (3.5) at $B = 1$. Substituting for the $\Delta y_{2\tau+1}$ the expression Eq. (C4) at $t = 2\tau$ we find, after some algebra,

$$\Delta y_{2\tau+2} = -\frac{\beta}{\alpha}\Delta x_{2\tau} + C'\,\Delta y_{2\tau} + \alpha\,\Delta y_{2\tau}^2 + \text{higher orders} \tag{C7}$$

where

$$C' \equiv \frac{de}{2} - 1 \tag{C8}$$

with d, e as in Eq. (3.9), where

$$\alpha \equiv e + \frac{d^2}{2} \tag{C9}$$

as in Eq. (3.11) and where

$$\beta \equiv e\alpha \qquad \text{N.B. (C10)}$$

The expressions (C8) and (C9) are the same[41,1] as the earlier ones [Eqs. (3.8) and (3.11)] at $B = 1$. However Eq. (C10) is new. Finally, we scale the y as before, in Eq. (3.11), and the x as

$$x'_\tau \equiv \beta \Delta x_{2\tau} \tag{C11}$$

Under the combined rescalings Eq. (C7) changes into

$$y'_{\tau+1} = -x'_\tau + C' y'_\tau + y'^2_\tau + \text{higher orders} \tag{C12}$$

To second order Eq. (C12) is the same as our original Eq. (C2), with "renormalized" coefficients. A similar renormalized version of Eq. (C3) is easily obtained from the above as well. Hence we could start all over and obtain another period doubling bifurcation, and so on. Each time x will be scaled by β [cf. Eqs. (C10) and (C11)]. Evaluating Eq. (C10), with the aid of Eqs. (3.9) and (3.3), the conservative C_∞ [Eq. (4.3)], and the conservative α [Eq. (4.8)], we arrive at

$$\beta \underset{r}{=} e^2 + \frac{ed^2}{2} \qquad \underset{r}{\simeq} 16.909\ldots \tag{C13}$$

while numerically

$$\beta = 16.363896879\ldots \tag{C14}$$

(cf. Ref. 18), a relative error of 3.3×10^{-2}. In Ref. 18 slightly different values were obtained from a slightly different renormalization scheme. The conservative α and δ were calculated in Refs. 18 and 1, from these two schemes. Note that $\beta > \alpha^2$, both numerically and in our renormalization.

REFERENCES AND NOTES

1. R. H. G. Helleman, "Self-Generated Chaotic Behavior in Nonlinear Mechanics," in *Fundamental Problems in Statistical Mechanics*, Vol. 5, E. G. D. Cohen, Ed. (North Holland, Amsterdam and New York, 1980), pp. 165–233, and references.
2. *Nonlinear Dynamics*, R. H. G. Helleman, Ed., *Ann. N.Y. Acad. Sci.* **357**, 1–507 (1980): Proceedings of the International Conference, New York, December 1979.
3. M. V. Berry, "Regular and Irregular Motion," in *Topics in Nonlinear Dynamics*, S. Jorna, Ed., American Institute of Physics Conference Proceedings Series, Vol. 46 (AIP, New York, 1978), pp. 16–120.
4. M. V. Berry, "Regularity and Chaos in Classical Mechanics—Illustrated by Three Deformations of a Circular Billiard," *Eur. J. Phys. A*, to appear (Physics, U., Bristol, U.K.) (1982)
5. A. J. Lichtenberg and M. A. Lieberman, *Regular and 'Stochastic' Motion*, (Springer Verlag, Berlin, 1982).
6. J. Ford, "The Statistical Mechanics of Classical Analytic Dynamics," in *Fundamental Problems in Statistical Mechanics*, Vol. 3, E. G. D. Cohen, Ed. (North Holland, Amsterdam and New York, 1975), pp. 215–255.
7. P. Collet and J.-P. Eckmann, *Iterated Maps on the Interval as Dynamical Systems*, Progress in Physics, Vol. 1 (Birkhäuser-Verlag, Basel & Boston, 1980).

8. J.-P. Eckmann, "Roads to Turbulence in Dissipative Dynamical Systems," *Rev. Mod. Phys.* **53**, 643–654 (1981).

9. Y. M. Trève, "Theory of Chaotic Motion with Application to Controlled Fusion," in *Topics in Nonlinear Dynamics*, S. Jorna, Ed., American Institute of Physics Conference Proceedings Series, Vol. 46 (AIP, New York, 1978), pp. 147–220.

10. M. J. Feigenbaum, "The Universal Metric Properties of Nonlinear Transformations," *J. Stat. Phys.* **21**, 669–706 (1979); **19**, 25, (1978); *Phys. Lett. A* **74**, 375 (1979); *Commun. Math. Phys.* **77**, 65–86 (1980); Springer Lecture Notes, Physics, Vol. 93, p. 163–166 (1979); Ref. 2, pp. 330–336 (1980); and lecture at the Gordon Conference on Dynamical Instabilities (June 1976).

11. P. Coullet and C. Tresser, "Critical Transition to 'Stochasticity' for some Dynamical Systems," *J. Phys. Lett.* **41**, L255 (1980); P. Coullet, C. Tresser, and A. Arneodo, *Phys. Lett.* **72A**, 268 (1979); **79A**, 259 (1980); *Eur. J. Phys.* **13A**, L123 (1980); *Phys. Lett. A* **81**, 197 (1981); "On the Relevance of Period Doubling Cascades at the Onset of Turbulence," *Phys. Lett.*, to appear (Mec. Statist., U. de, Nice, France) (1982); *C. R. Acad. Sci. (Paris)* **287**, 577 (1978); *J. Phys. C (Paris)* **5**, 25 (1978).

12. S. Grossmann and S. Thomae, "Invariant Distributions and Stationary Correlation Functions," *Z. Naturforsch.* **32a**, 1353–1363 (1977) (I was unaware of this article, at the time I wrote Ref. 1); "Correlations and Spectra of Periodic Chaos Generated by the Logistic Parabola," *J. Stat. Phys.* **26**, 485–504 (1981); "A Scaling Property in Critical Spectra of Discrete Systems," *Phys. Lett. A*, **83**, 181 (1981).

13. B. Derrida and Y. Pomeau, "Feigenbaum's Ratio of Two-Dimensional Area Preserving Maps," *Phys. Lett. A*, **80**, 217–219 (1980); B. Derrida, Y. Pomeau, and A. Gervois, *Eur. J. Phys.* **12A**, 269 (1979); *Ann. Inst. H. Poincaré* **29**, 305 (1978).

14. B. Derrida, Critical Properties of One Dimensional Mappings, in *Bifurcation Phenomena in Mathematical Physics and Related Topics*, Bardos and Bessis, Eds. (Reidel, Dordrecht, Netherlands, 1980), pp. 137–154; *J. Phys. C (Paris)* **5**, 49 (1978).

15. W. Engel, "Ganze Cremona-Transformationen von Primzahlgrad in der Ebene," *Math. Ann.* **136**, 319–325 [especially result (21)] (1958); **130**, 11–19 (1955); J. Moser, *Bol. Soc. Mat. Mex.* 176–180 (1960); But note: L. A. Campbell, *Math. Ann.* **205**, 243–248 (1973) [especially the note added in proof]; S. S. Abhyankar and T. Moh, *J. Für Mathematik* **276**, 148–166 (1976); D. Wright, *J. Pure Appl. Algebra* **12**, 235–251 (1978) (I thank Robert MacKay for pointing out the last three references).

16. M. Hénon, "Numerical Study of Quadratic Area-Preserving Mappings," *Q. Appl. Math.* **27**, 291–312 (1969).

17. M. Hénon, "A Two-dimensional Mapping with a Strange Attractor," *Commun. Math. Phys.* **50**, 69–77 (1976).

18. J. M. Greene, R. S. MacKay, F. Vivaldi, and M. J. Feigenbaum, "Universal Behavior in Families of Area Preserving Maps," *Physica* **3D**, 468–486 (1981); J. Greene, "Some Order in the Chaotic Regimes of Two-Dimensional Maps," in this Volume.

19. G. Schmidt and J. Bialek, "Fractal Diagrams for Hamiltonian 'Stochasticity,'" preprint (1981); G. Schmidt, "'Stochasticity' and Fixed Point Transitions," *Phys. Rev. A* **22**, 2849–2854 (1980).

20. M. Month and J. C. Herrera, Eds., *Nonlinear Dynamics and the Beam–Beam Interaction*, American Institute of Physics Conference Proceedings Series, Vol. 57 (AIP, New York, 1979).

21. R. H. G. Helleman, "Exact Results for Some Linear and Nonlinear Beam–Beam Effects," in Ref. 20, pp. 236–256, and its references.

22. M. Month, Ed. Proceedings of the Workshop on the Beam-Beam Interaction, SLAC-Pub. 2624 (Stanford Linear Accelerator Center, Stanford, CA, 1980); "A Review of Beam–Beam Phenomena," BNL25703 (Brookhaven National Laboratory, Upton, NY, 1979).

23. R. H. G. Helleman, "On the Iterative Solution of a 'Stochastic' Mapping," in *Statistical Mechanics and Statistical Methods*, (Plenum Press, New York, 1977), pp. 343–370.

24. T. C. Bountis and R. H. G. Helleman, "On the Stability of Periodic Orbits of Two-Dimensional Mappings," *J. Math. Phys.* **22**, 1867–1877 (1981).

25. D. F. Escande and F. Doveil, "Renormalization Method for Computing a 'Stochastic' Threshold," several preprints (1982) (Laboratoire de Physique des Milieux Ionisés, Ecole Polytechnique, Palaiseau, France); lectures at the 1980–1982 "Dynamics Days Twente" spring conferences, Twente University of Technology, Enschede, Netherlands; *Phys. Lett. A* **83**, 307 (1981); *J. Stat. Phys.* **26**, 257 (1981).

26. G. Benettin, C. Cercignani, L. Galgani, and A. Giorgilli, "Universal Properties in Conservative Dynamical Systems," *Lett. Nuovo Cimento* **28**, 1–4 (1980); **29**, 163–166 (1980).

27. J. B. J. van Zeyts, Internal D-1 Report, Theoretical Physics Group, Twente University of Technology, Enschede, Netherlands, 1980–1981.

28. T. C. Bountis, "Period Doubling and Universality in Conservative Systems," *Physica* **3D**, 577–589 (1981).

29. G. Contopoulos and M. Zikides, "Periodic Orbits and Ergodic Components of a Resonant System," *Astron. Astrophys.* (1980?); Contopoulos, lecture at the 1981 "Dynamics Days Twente" spring conference, Twente University of Technology, Enschede, Netherlands.[65]

30. Upon mapping the points of any closed loop in the X, Y-plane with Eq. (2.1), we see that the area of the loop changes as:

$$\oiint_{\text{mapped loop}} dX_{t+1}\, dY_{t+1} = \oiint_{\text{loop}} J_t\, dX_t\, dY_t$$

where

$$J_t \equiv \det\left(\frac{\partial(X_{t+1}, Y_{t+1})}{\partial(X_t, Y_t)}\right)$$

That is, whenever this Jacobian is a constant B Eq. (2.2), the areas of all loops contract by the same factor B, upon each mapping.

31. If accidentally all second-order Taylor terms in x_t, y_t about the fixed point happen to vanish, we should consider another periodic orbit instead. Good candidates are the periodic orbits that bifurcate from the fixed point (cf. conservation of Poincaré index) when they turn unstable, with reflection.[1] In that case we can still find a Feigenbaum (cf. Sections 3 and 4) that is "rooted" in the original fixed point. If our special fixed point also happens not to turn "unstable with reflection" at any parameter value, we have a very special periodic orbit in which no Feigenbaum is rooted.

32. In fact, for any Taylor expansion up to and including terms of order p, where p is a prime number, the two resulting first-difference equations can be transformed into one second-difference equation of one variable z_t and a polynomial nonlinearity of order p. All this can be inferred from Eq. (21) in Ref. 15 because whenever the variables can be transformed so that one of the first-difference equations becomes linear,[15] we can explicitly solve for one of its variables and substitute it in the second, nonlinear, equation to obtain a second-difference equation of the above-mentioned type. When $p = 2$ we obtain an expression of the form: $Az_{t+1} + Bz_{t-1} = 2Cz_t + Dz_t^2$. There is no constant term, since the fixed point was put at the origin,[40] $z_t = 0$. See Appendix A.

33. Consider the quadratic expression in Note 32. No generality is lost by setting $A = 1$ (or dividing all terms by A). The transformation $z_t \equiv 2y_t/D$ then puts the expression into our standard form Eq. (2.6). See Appendix A.

34. Transforming Eq. (2.6) with $y_t \equiv u_t - C + (1 + B)/2$ and $\bar{C} \equiv 1 + B - C$ [Eq. (2.10)], we find:

$$u_{t+1} + Bu_{t-1} = 2\bar{C}u_t + 2u_t^2$$

This is the same as Eq. (2.6), apart from $C \to \overline{C}$. Thus all phenomena exist twice, at C-values symmetric about C^* [Eq. (2.11)]. Note that the coordinates of the period-1 attractor and the period-1 repeller are interchanged by the $y \to u$ transformation; that is, the repeller becomes an attractor at C^* [Eq. (2.11)], and vice versa. In Sections 2–5 we might take $\overline{C}$ as the parameter μ of Section 1.

35. Transforming Eq. (2.9) with $y_t \equiv a(e - x_t)/2$, with e solved from $-a^2e^2 + (b-1)ae + a = 0$, we obtain the standard form, Eq. (2.6), with $C = -ae$ real, if and only if $(b-1)^2 > -4a$ (cf. Note 34).

36. Transforming Eq. (2.8) with $x \equiv -2y_t/\mu + g$, where g is solved from $-\mu^2 g^2 - \mu g + \mu = 0$, we obtain the standard form, Eq. (2.6), with $C = -g\mu$ real, if and only if $4\mu > -1$ (cf. also Note 34).

37. Transforming Hénon's conservative mapping,[16] equivalent with[23]: $x_{t+1} + x_{t-1} = 2Cx_t + x_t^2$, with $x_t \equiv 2y_t$ we obtain the standard form, Eq. (2.6).

38. Transforming the "logistic equation" (2.7) with $y_t \equiv -ax_t/2$, we obtain the standard form, Eq. (2.6) with $C = a/2$ (cf. also Note 34).

39. With $r(\tau) \equiv \Delta y_{2\tau+1}/\Delta y_{2\tau-1}$, the square bracket term of Eq. (3.6) becomes $(r^2 + B)\Delta y_{2\tau-1}^2$, and Eq. (3.5) yields $(r+B)\Delta y_{2\tau-1} = d\,\Delta y_{2\tau} + \mathcal{O}(\Delta^2)$. Combining these we find the square bracket term of Eq. (3.6) to be $\Delta y_{2\tau}^2 d^2(r^2+B)/(r+B)^2 +$ *higher orders*. As a function of r only, this expression has a minimum $(B > 0)$,[41] at $r = 1$. Thus the constant second-order contribution from the square bracket term is $\Delta y_{2\tau}^2 d^2/(1+B)$ (when $B < 0$, we start with the second bifurcation, that is, with T^2 since $B^9 > 0$, cf. Eq. (3.7)[40]; whence the last term in the α-expression, Eq. (3.11).

40. If a mapping T has a periodic solution (e.g., of period m), then T^m has a fixed point (period 1). In that case we continue with the latter mapping and translate the variables by constants (i.e., $T^m \to T'$), so that the fixed point of T' lies at the origin.

41. The α-expressions, Eqs. (3.11), (4.8) and (5.8)[39] differ slightly from the one in Eq. (3.24) of Ref. 1. I am grateful to R. S. MacKay (Plasma Physics Laboratory, Princeton University, Princeton, NJ) for correcting my analysis in Note 155 of Ref. 1 (the term $\beta\Delta y_{2\tau}^2$ there is of second order, away from the bifurcation point).

42. Similar observations were also reported by M. Lieberman and D. Escande during the 1981 Workshop on Long-Time Prediction in Nonlinear Conservative Dynamical Systems (Austin).

43. If we consider separately the part of the interval to the left of the minimum of Eq. (5.14) at $y = -\frac{1}{2}C$, and the other part to the right, the parabola on the left-hand side of Eq. (5.14) becomes *monotonic*, and the results Eqs. (5.15) and (5.16) are easily proven on each of these two parts.[44,50]

44. For $y_0 > \frac{1}{2} - C$, the subsequent y_t's become unbounded, since $y_{t+1}/y_t = 2(y_t + C) > 1$ in that case,[50] because of the mapping, Eq. (5.14). For $y_0 < -\frac{1}{2}$, we have[43,44]: $y_1 > \frac{1}{2} - C$ $(C < 1)$. Thus Eq. (5.15) is the maximum interval of attraction.

45. For $y_0 = -\frac{1}{2}C$ we obtain from Eq. (5.14): $y_1 = -\frac{1}{2}C^2$ [the minimum of the y_t-parabola in Eq. (5.14)], $y_2 = \frac{1}{2}C^4 - C^3$. The latter exceeds $\frac{1}{2} - C$ for $C < -1$, whence[44] the restriction to $C \geqslant -1$ in Eqs. (5.15) and (5.16).[50] Under the C-restrictions on Eq. (5.15), $y = -\frac{1}{2}C$ lies in the interval Eq. (5.15). Under the $C < -0.62$ restriction on Eq. (5.16) it also lies in the interval Eq. (5.16).

46. D. Ruelle, "Dynamical Systems with Turbulent Behavior," in *Mathematical Problems in Theoretical Physics*, G. Dell'Antonio, S. Dopplicher, and G. Jona-Lasinio, Eds., Lecture Notes in Physics, Vol. 80 (Springer-Verlag, Berlin, 1978), pp. 341–360; M. V. Jakobson, *Commun. Math. Phys.* **81**, 39–88 (1981).

47. V. Arnol'd and A. Avez, *Ergodic Problems of Classical Mechanics* (Benjamin, New York and Amsterdam, 1968).

48. O. E. Rössler "Continuous Chaos—Four Prototype Equations," *Ann. N.Y. Acad. Sci.* **316**, 376–392 (1979); in: *Synergetics: A Workshop*, H. Haken, Ed. (Springer-Verlag, Berlin, 1977)

pp. 184–199; *Phys. Lett. A* **57**, 397–398 (1976); **71**, 155 (1979); and J. Crutchfield, D. Farmer, N. Packard, R. Shaw, G. Jones, and R. J. Donnely, "Power Spectral Analysis of a Dynamical System," *Phys. Lett. A* **76**, 1–4 (1980); same authors with H. Froehling, in Ref. 2, pp. 453–472; O. Gurel, Z. Naturforschung, **61A**, 219 (1977).

49. G. Mayer-Kress and H. Haken, "The Influence of Noise on the Logistic Model," *J. Stat. Phys.*, **26**, 149–171 (1981); *Phys. Lett. A* **82**, 151–155 (1981); E. N. Lorenz, "Noisy Periodicity and Reverse Bifurcation," in Ref. 2, pp. 282–291; S. Chang and J. Wright, *Phys. Rev.* **23A**, 1419 (1981).

50. J. Mulders, Internal *D*-2 Report, Theoretical Physics Group, Twente University of Technology, Enschede, Netherlands, 1981.

51. R. M. May and G. F. Oster, "Period-Doubling and the Onset of Turbulence, an Analytic Estimate of the Feigenbaum Ratio," *Phys. Lett. A* **78**, 1–3 (1980).

52. R. M. May, "Simple Mathematical Models with Very Complicated Dynamics," *Nature* (*London*) **261**, 459–467 (1976).

53. L. J. Laslett, "Some Illustrations of 'Stochasticity,'" in American Institute of Physics Conference Proceedings Series, Vol. 46 (AIP, New York, 1978), pp. 221–247.

54. M. Campanino, H. Epstein, and D. Ruelle, "On Feigenbaum's Functional Equation,"(Institut des Hautes Etudes Scientifiques, Bures sur Yvette, France), to appear in: *Topology* (81); "On the Existence of Feigenbaum's Fixed Point," *Commun. Math. Phys.* **79**, 261–302 (1981); H. Epstein and J. Lascoux, "Analyticity Properties of the Feigenbaum Function," preprint IHES/P/81/27; O. E. Lanford III, "Remarks on the Accumulation of Period-Doubling Bifurcations," in *Mathematical Problems in Theoretical Physics*, Lecture Notes in Physics, Vol. 116; (Springer-Verlag, Berlin, 1980), pp. 340–342; "Smooth Transformations of Intervals," Seminaire Bourbaki 80/81, no. 563; "A Computer-Assisted Proof of the Feigenbaum Conjectures," IHES preprint P/81/17.

55. P. Collet, J.-P. Eckmann, and O. E. Lanford III, "Universal Properties of Maps on an Interval," *Commun. Math. Phys.* **76**, 211–254 (1980).

56. P. Collet, J.-P. Eckmann, and H. Koch, "On Universality for Area-Preserving Maps of the Plane," *Physica* **3D**, 457–467 (1981); "On Period Doubling Bifurcations for Families of Maps in $\mathbb{R}^n$," *J. Stat. Phys.*, (1981).

57. T. C. Bountis, H. Segur, and F. Vivaldi, "Integrable Hamiltonian Systems and the Painlevé Property," *Phys. Rev. A* **25**, 1257–1264 (1982).

58. M. Feshbach, "Physics and the A.P.S. in 1980," *Phys. Today* **34**, esp. 44–45 (April 1981).

59. G. B. Lubkin, "Period-Doubling Route to Chaos Shows Universality," *Phys. Today* **34**, 17–19 (March 1981) "Feigenbaum's Number," *Scientific American*, p. 116 (Sept. 1981); R. Landauer, "Reviewing the Reviewers," *Amer. J. Biophys.* (1981).

60. *Chaotic Behavior of Deterministic Systems*, G. Iooss, R. H. G. Helleman, and R. Stora Eds., *Proceedings of the July 1981 "Les Houches" Summer School* (North Holland, Amsterdam and New York, 1982).

61. R. H. G. Helleman, "Feigenbaum Sequences in Conservative and Dissipative Systems," in *Chaos and Order in Nature*, H. Haken, Ed. (Springer-Verlag, Berlin and New York, 1981), pp. 232–248.

PERIOD DOUBLING AS A UNIVERSAL ROUTE TO STOCHASTICITY

ROBERT S. MACKAY
Plasma Physics Laboratory
Princeton University, Princeton, New Jersey

Abstract. *In dissipative systems, period doubling is known to provide a route, with universal properties, from coherent to chaotic behavior.*[1] *This chapter describes an analogous transition in area preserving maps, which represent the simplest conservative systems. This transition almost completely eliminates closed invariant curves from the vicinity of an originally stable periodic orbit.*

1. THE UNIVERSAL PERIOD DOUBLING SEQUENCE

As an example, consider the DeVogelaere form of the area preserving quadratic map:

$$T: x' = -y + f(x) \qquad (1)$$

$$y' = x - f(x')$$

with

$$f(x) = px - (1-p)x^2 \qquad (2)$$

where p is a parameter. For $|p| < 1$ the fixed point at the origin is elliptic. As p decreases through -1 it loses its stability and gives birth to a stable two-cycle (Fig. 1). As p decreases further this two-cycle itself goes unstable and gives birth to a stable four-cycle (Fig. 2).

This period doubling process repeats infinitely often, but in a finite range of parameter, accumulating at some value $p^* = -1.26631127692$. The parameter values p_n, at which the nth period doubling occurs, are found to converge

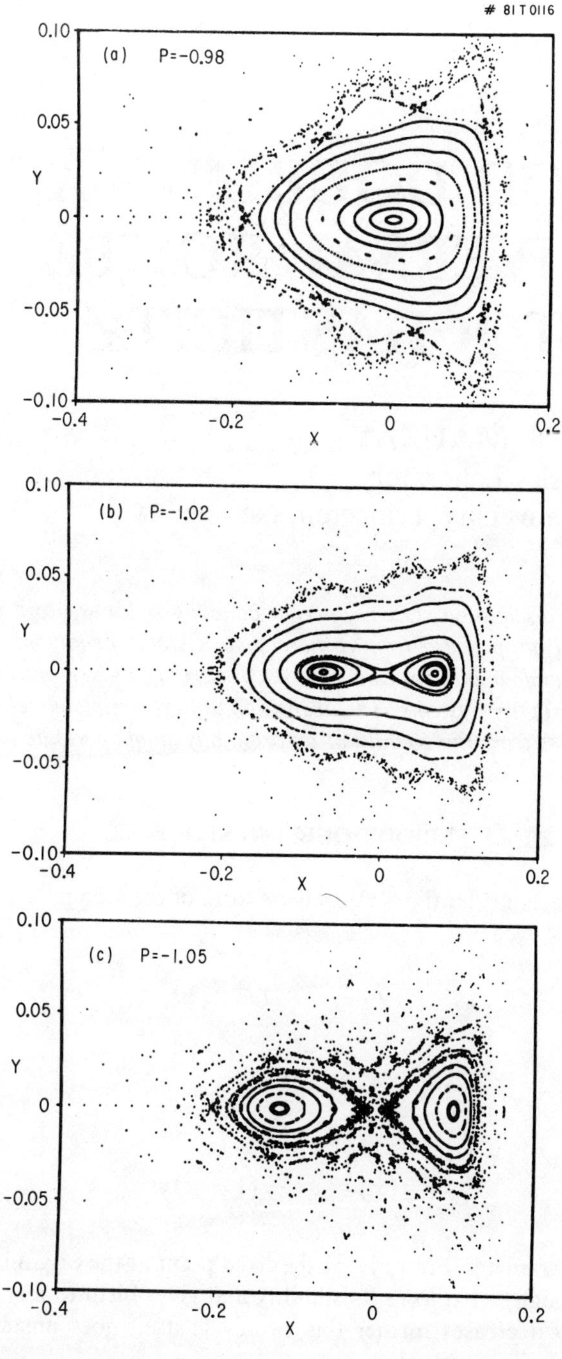

Figure 1. Some orbits of the map (1, 2) for three parameter values, showing the first-period doubling.

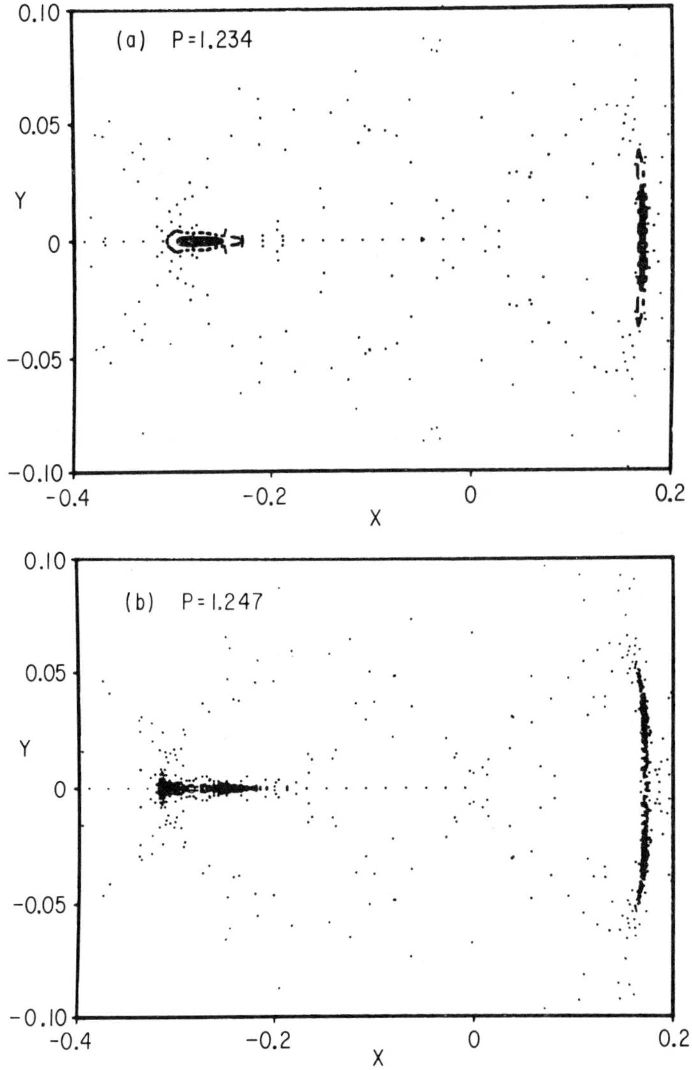

Figure 2. Some orbits of the map (1, 2) for two parameter values, showing the second-period doubling.

asymptotically geometrically to p^*, with ratio $1/\delta$, where:

$$\delta = 8.721097200\ldots \tag{3}$$

The period doubling process possesses further self-similarity, easiest revealed by exploiting the symmetry evident in Figs. 1 and 2. A map S is said to be a *symmetry* of a map T if:

$$S^2 = \text{identity} \tag{4}$$

S reverses orientation, and

$$S^{-1}TS = T^{-1}$$

The last condition implies that *S* transforms *T* into its inverse. Thus maps with symmetry are said to be *reversible*.[2] For example, DeVogelaere maps [Eq. (1)] are reversible with the following symmetry:

$$S: x' = x \qquad (5)$$

$$y' = -y$$

The fixed points of a symmetry often form lines, called its *symmetry lines*. It can be shown[3] that given a fixed point on a symmetry line, there is a *dominant symmetry* such that precisely two points of each periodic orbit of its period doubling sequence lie on the fixed line of the dominant symmetry. Thus the period doubling sequence can be followed by looking for periodic points on this *dominant line* only, rather than in the whole plane. Incidentally, the DeVogelaere form of the quadratic map is used to give the dominant symmetry the simple form, Eq. (5).

Plotting the positions on the dominant line of such periodic points against parameter yields a slice of the *period doubling tree* (Fig. 3). It can be seen to repeat itself on smaller and smaller scales, accumulating at a point (x^*, p^*),

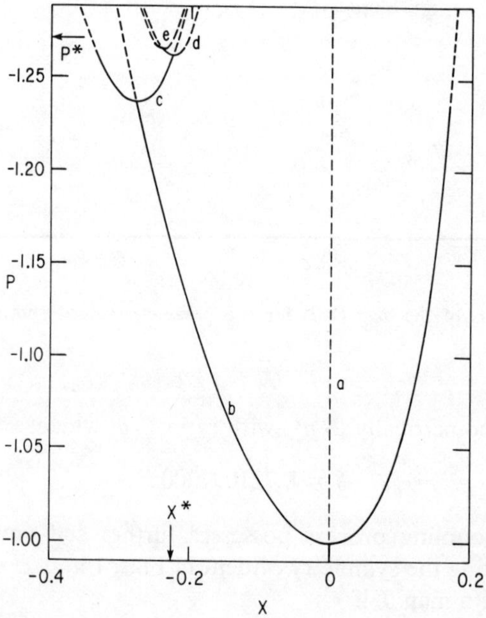

Figure 3. The section of the period doubling tree lying on the dominant symmetry.

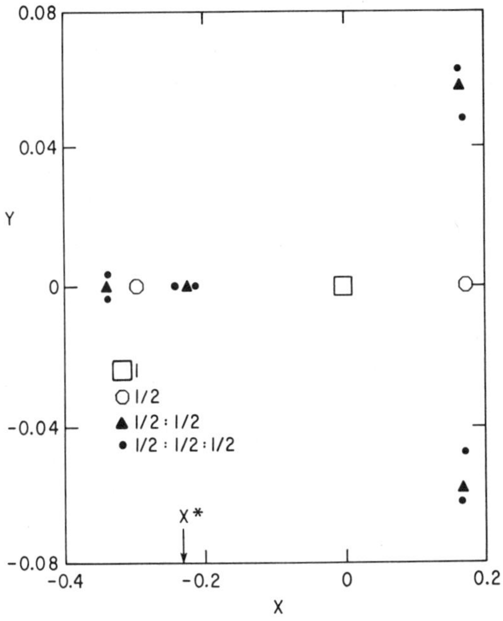

Figure 4. The first few periodic orbits of the period doubling sequence at the accumulation parameter value p^*.

with rescaling factors δ in parameter, and α along the dominant line:

$$\alpha = -4.018076704\ldots \tag{6}$$

The rescaling factors δ and α were already known.[4–6] We found[7] that this self-similarity extends off the dominant line too (Fig. 4), with a rescaling factor β across the line:

$$\beta = 16.363896879\ldots \tag{7}$$

In fact, the whole map shares properties similar to those of its period doubling sequence, as follows. At the accumulation parameter value p^*, T converges under the operation A of squaring (i.e., composing with itself) and rescaling by α, β in x, y about $x^*, 0$. The limit T^* is a map invariant under A. Furthermore the whole one-parameter family converges under the operation D of squaring and rescaling in parameter as well as in x, y, to a one-parameter family T_p^* (T^* is the special case of T_p^* when $p = p^*$). For more details see Ref. 7.

This behavior is all the more remarkable in that it appears to be universal; that is, almost all one-parameter families of reversible area preserving maps with period doubling sequences have the same δ, α, β, T^*, and T_p^* (up to simple changes of coordinates and parametrization).[4–9]

2. COMPLETE STOCHASTICITY

One of the most striking consequences of period doubling, as Figs. 1 and 2 indicate, is the destruction of closed invariant curves, and the corresponding reduction in the area they trap. Figure 5 shows, for a range of parameter values, the points of the dominant line whose orbits are bounded. Also for some parameter values beyond p^* we found that all points of a large grid in the plane escaped. Thus, as conjectured by J. B. McLaughlin[10] and R. H. G. Helleman,[1] for example, it might appear that beyond p^* there are no closed invariant curves in the vicinity. We would call this a transition to *complete stochasticity*, but the stochasticity turns out to be not quite complete. There are small universal islets of stability beyond p^*.[20]

In general, of course, there can also be large, nonuniversal regions of stability, since T_p^* exerts its influence only in a neighborhood of the original periodic orbit. Thus, for instance, the period doubling sequence of the stable five-cycle of the island chain born at $p = \cos(2\pi/5) = 0.30901699\ldots$, accumulates at $p = 0.119353761903\ldots$, leaving plenty of invariant curves around the fixed point, but destroying completely the fifth-order island chain. As another example, the period doubling sequence of the fixed point of the standard map accumulates at $k = 2\pi \times 1.05597806\ldots$,[4] but at $k = 2\pi$ an islet of stability pops out by tangent bifurcation, whose fixed point does not lose stability until $k^2 = (2\pi)^2 + 16$,[11] that is, $k = 2\pi \times 1.1854471\ldots$.

This transition should be distinguished from that to "connected stochasticity,"[12] when the last of the original invariant curves of an originally integrable map is destroyed, permitting orbits unconfined in action. That occurs at $k = 0.971635\ldots$ in the standard map, well before the fixed point has doubled at all.

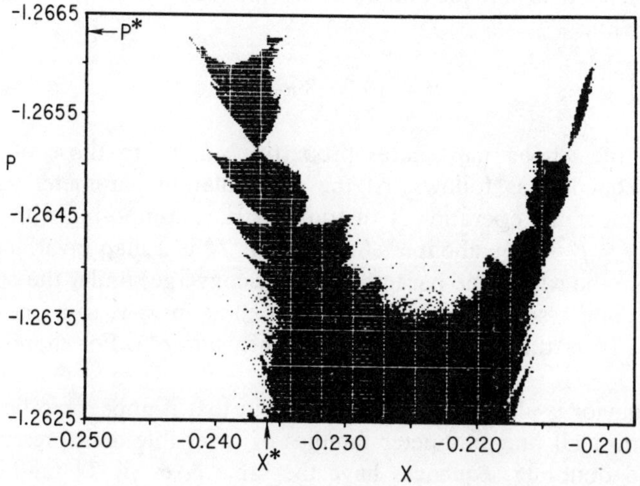

Figure 5. Points of the dominant line, for a range of parameter values, that remain bounded under T_p^*.

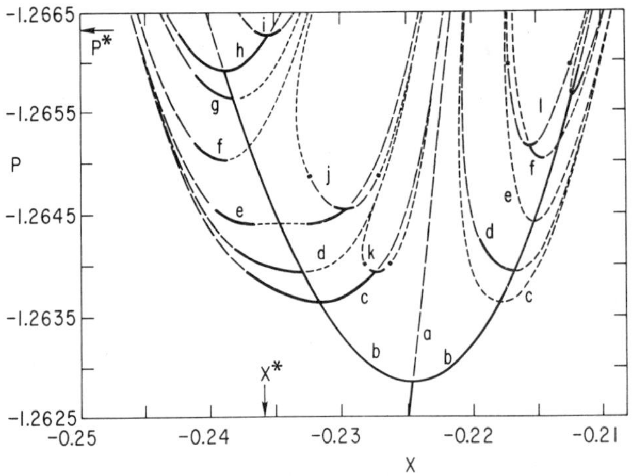

Figure 6. Some of the branches of the bifurcation tree of T_p^* with periods as follows: a, 1; b, 2; c, 12; d, 10; e, 8; f, 6; g, 10; h, 4; i, 8; j, 16; k, 24; l, 12. Thick line indicates elliptic; long dashes, inversion hyperbolic; short dashes, ordinary hyperbolic.

3. CONCLUDING REMARKS

We conclude with three questions:

1. What about maps with no symmetry? These are rare in practice. For example, all maps derived from a Hamiltonian even in the momenta are reversible.[13] We know only one example that can be shown to have no symmetry,[14] although numerical evidence indicates that Rannou's map[15] has no symmetry either.[16] Reversibility is not generic, however, in one-parameter families of area preserving maps. This follows from the work of Rimmer.[17] Nevertheless, we expect that nonreversible maps will exhibit the same universal behavior. This is supported by finding that Rannou's map has period doubling with the same δ,[16] and by analogy with the one-dimensional case where the universal one-parameter family of maps has a symmetry property. It is even, and yet it attracts both even and noneven maps.

2. What about bifurcations other than period doubling? A period n-tupling bifurcation of a periodic orbit requires a multiplier* to pass through an nth root of unity. The (only) multiplier λ of a periodic orbit of a one-dimensional map is always real, so the only bifurcations are tangent ($\lambda = +1$) and period doubling ($\lambda = -1$). In two-dimensional area preserving maps the two multipliers are reciprocal reals or conjugate points on the unit circle. Thus in the latter case all bifurcations are possible (Fig. 6). Note the connection between Figs. 5 and 6, in particular the temporary destruction of closed invariant curves associated with the period tripling bifurcation. One can follow

*The multipliers of a periodic orbit are the eigenvalues of the linearization of the map around the orbit.

infinite bifurcation sequences other than period doubling, such as period tripling, and find that they have self-similarity too.[7] Period doubling, however, is the most significant, since it is the last bifurcation before a stable periodic orbit loses its stability, and it leads to complete stochasticity.

3. What about higher dimensional systems? The results for one-dimensional maps extend to higher dimensional dissipative systems.[18] In contrast, it appears that the two-dimensional area preserving results do not extend to higher dimensional symplectic maps.[19] It is an open question, however, whether they might have their own universal period doubling sequences.

ACKNOWLEDGMENTS

I thank John Greene for his inspiration, guidance, and encouragement, Jon Schonfeld for arranging for me to speak, and H. Greenwood of Keele University, England, for the use of computing facilities. This work was supported by the U.S. Department of Energy contract DE-AC02-76-CHO-3073, and the U.K. Science Research Council grant B/80/3001.

REFERENCES

1. M. J. Feigenbaum, this volume; R. H. G. Helleman, and R. S. MacKay, "One Mechanism for the Onset of Large-Scale Chaos in Conservative and Dissipative Systems," this volume.
2. R. Devaney, *Trans. Am. Math. Soc.*, **218**, 89 (1976).
3. R. S. MacKay, "The Dominant Symmetry of Reversible Maps," notes (1981).
4. G. Benettin, C. Cercignani, L. Galgani, and A. Giorgilli, *Lett. Nuovo Cimento* **28**, 1 (1980).
5. T. C. Bountis, *Physica 3D*, 577 (1981).
6. F. Vivaldi, and J. Ford, private communication on period doubling in a forced Duffing oscillator.
7. J. M. Greene, R. S. MacKay, F. Vivaldi, and M. J. Feigenbaum, *Physica 3D*, 468 (1981).
8. G. Benettin, L. Galgani, and A. Giorgilli, Lett. Nuovo Cimento *29*, 163 (1980).
9. P. Collet, J.-P. Eckmann, and H. Koch, *Physica 3D*, 457 (1981).
10. J. B. McLaughlin, *J. Stat. Phys.* **24**, 375 (1981).
11. B. V. Chirikov, *Phys. Rep.* **52**, 263 (1979).
12. J. M. Greene, *J. Math. Phys.* **20**, 1183 (1979).
13. J. M. Greene, in *Nonlinear Orbit Dynamics and the Beam–Beam Interaction*, M. Month and J. C. Herrera, Eds., American Institute of Physics Conference Proceedings Series, Vol. 57 (AIP, New York, 1979), p. 257.
14. J. N. Mather, private communication.
15. F. Rannou, *Astron. Astrophys.* **31**, 289 (1974).
16. J. M. Greene, private communication.
17. R. Rimmer, *J. Differ. Equations* **29**, 329 (1978); "Generic Bifurcations from Fixed Points of Involutory Area Preserving Maps," P. Math. Res. Paper 79-9, La Trobe University, Melbourne, Australia (1979).
18. P. Collet, J.-P. Eckmann, and H. Koch, *J. Stat. Phys.* **25**, 1 (1981).
19. R. S. MacKay and T. C. Bountis, "Period Doubling in 4-D Symplectic Maps," notes (1981).
20. R. S. MacKay, "Islets of Stability Beyond Period Doubling," Physics Letters *87A*, 321 (1982).

SOME ORDER IN THE CHAOTIC REGIMES OF TWO-DIMENSIONAL MAPS

JOHN M. GREENE
Center for Studies of Nonlinear Dynamics
La Jolla Institute
La Jolla, California

Abstract. *The first part of this chapter is concerned with a compact and useful method of dealing with real 2×2 matrices as two complex numbers. Next, a symmetric matrix is given whose eigenvalues yield an approximation to the Lyapunov numbers of a map. Finally, it is shown that in general the corresponding eigenvector directions at a given point converge in the limit of long orbit segments, yielding an invariant direction field for the map. Essentially all loss of information associated with positive Lyapunov number is along this direction field. Thus the exponential separation of orbits is not in random directions. These concepts have all been known previously in the abstract, but is interesting to see how they are illustrated by numerical computations.*

1. INTRODUCTION: NOTATION

There are a number of rather separate ideas that I want to bring together coherently. I have a new shovel, a place to dig, and some thoughts about what might be found. Each of these is a subject in itself, but I hope to show that they all go together.

First, a notation for 2×2 matrices as two complex numbers is clearly feasible, since a 2×2 matrix has four independent parameters. In fact it is a remarkable and useful idea. It is known as a Cayley algebra,[1] but seems not to have been widely used in our generation.

---

*Permanent address: Princeton Plasma Physics Laboratory
Princeton University
Princeton, New Jersey

A 2 × 2 matrix can be denoted by a single letter M or by its four components M_{ij}. A third representation, which I have been using for some time, is to expand in spin matrices.[2,3]

$$M \equiv a\begin{pmatrix} 1 & 0 \\ 0 & 1 \end{pmatrix} + ib\begin{pmatrix} 0 & -i \\ i & 0 \end{pmatrix} + c\begin{pmatrix} 0 & 1 \\ 1 & 0 \end{pmatrix} + d\begin{pmatrix} 1 & 0 \\ 0 & -1 \end{pmatrix} \tag{1}$$

The fourth method, developed here, is to combine these four coefficients into two complex numbers,

$$\begin{aligned} M &\equiv [a - ib, d + ic] \\ &= [\xi, \mu] \\ &\equiv \begin{pmatrix} \xi_R + \mu_R & \mu_I - \xi_I \\ \mu_I + \xi_I & \xi_R - \mu_R \end{pmatrix} \end{aligned} \tag{2}$$

where R and I subscripts denote real and imaginary parts, respectively. It is convenient to introduce a complex notation for vectors at the same time,

$$\begin{aligned} \mathbf{X} &\equiv \begin{pmatrix} x \\ y \end{pmatrix} \equiv [x + iy] \\ &\equiv \chi \end{aligned} \tag{3}$$

Some combination of Greek letters and square brackets will be used to distinguish this notation from the other possibilities.

The linear operations of addition and multiplication by a real constant are straightforward in this notation. The determinant takes the neat form

$$\det[\xi, \mu] = |\xi|^2 - |\mu|^2 \tag{4}$$

The remarkable thing about this formulation lies in its rules for multiplication. The simplest of these is the product of two vectors,

$$\chi_1\chi_0 = \tfrac{1}{2}(\chi_1\chi_0^* + \chi_1^*\chi_0) \tag{5}$$

where the asterisk denotes complex conjugate. Multiplying a vector by a matrix yields

$$[\xi, \mu]\chi = \xi\chi + \mu\chi^* \tag{6}$$

More interesting is the scalar product of two matrices,

$$[\xi_1, \mu_1] \cdot [\xi_0, \mu_0] = [\xi_1\xi_0 + \mu_1\mu_0^*, \xi_1\mu_0 + \mu_1\xi_0^*] \tag{7}$$

Note that the position of the asterisks makes the operation noncommutative. Clearly, the unit matrix is [1, 0].

The utility of this formulation should become clearer after its use later, but a few comments can be made now. The complex notation expresses a 2×2 matrix in terms of two angles and two lengths, that is, in terms of the phases and magnitudes of two complex numbers. The two lengths are related by the determinant, Eq. (4). Years of experience in multiplying such matrices yields the insight that generally the numbers in the product are bigger than those of the factors, as if two magnitudes were being multiplied. On occasion, however, the numbers in the product are smaller, as if there were some sort of phase relation, specifying the way the matrices fit together, that is needed to completely determine the product. That kind of intuition is reflected perfectly in Eq. (7). For instance, the magnitude $|\xi_1\xi_0 + \mu_1\mu_0^*|$ depends on the difference between the angles of $\xi_1\xi_0$ and $\mu_1\mu_0^*$. The complex notation reveals the essence of 2×2 matrices in a way that is inaccessible to other notations.

It is instructive to see how these angles change under rotation of the coordinate system. The rotation matrix is given by

$$R \equiv \begin{pmatrix} \cos\Theta & -\sin\Theta \\ \sin\Theta & \cos\Theta \end{pmatrix}$$

$$= [\exp i\Theta, 0] \tag{8}$$

Then

$$RMR^{-1} = [\exp i\Theta, 0][\xi, \mu][\exp -i\Theta, 0]$$

$$= [\xi, \mu \exp 2i\Theta] \tag{9}$$

Thus ξ is stationary and μ rotates twice as fast as the coordinates. This can be traced back to the spin representation of Eq. (1).

This section is completed with the presentation of some other common matrix operations. The inverse of a matrix is given by

$$[\xi, \mu]^{-1} = \frac{[\xi^*, -\mu]}{|\xi|^2 - |\mu|^2} \tag{10}$$

and the transpose by

$$[\xi, \mu]^\dagger = [\xi^*, \mu] \tag{11}$$

The eigenvalues of a matrix λ can be given in terms of reduced eigenvalues, $\hat{\lambda}$,

$$\hat{\lambda} \equiv \frac{\lambda}{\left(|\xi|^2 - |\mu|^2\right)^{1/2}} \tag{12}$$

$$\hat{\lambda} + \hat{\lambda}^{-1} = \frac{\xi + \xi^*}{\left(|\xi|^2 - |\mu|^2\right)^{1/2}} \tag{13}$$

Thus they depend only on the determinant and the real part of ξ. This is a convenient way to express things because the rest of the chapter deals with matrices of unit determinant and there is no distinction between eigenvalues and reduced eigenvalues. The reduced eigenvalues are real if the right-hand side of Eq. (13) is greater than two in absolute magnitude; otherwise they are complex and lie on the unit circle. One must be wary in this case, because this kind of complex number does not fit with complex notation for matrices.

In terms of these quantities the eigenvectors for real positive eigenvalues are given by

$$\chi_\lambda = \left((\hat\lambda\xi - \hat\lambda^{-1}\xi^*)\mu\right)^{1/2} \tag{14}$$

Showing that this is correct is somewhat roundabout. First note from Eq. (6) that

$$[\xi, \mu]\chi_\lambda = \left[\xi + |\mu|\left(\frac{\hat\lambda\xi^* - \hat\lambda^{-1}\xi}{\hat\lambda\xi - \hat\lambda^{-1}\xi^*}\right)^{1/2}\right]\chi_\lambda \tag{15}$$

To show the term in the square brackets is the eigenvalue λ, first multiply the numerator and denominator of the second term by the complex conjugate of the denominator. This removes the square root in the numerator and permits the reduction of the denominator,

$$\begin{aligned}(\hat\lambda\xi^* - \hat\lambda^{-1}\xi)(\hat\lambda\xi - \hat\lambda^{-1}\xi^*) &= (\hat\lambda^2 + \hat\lambda^{-2})|\xi|^2 - \xi^2 - \xi^{*2}\\ &= (\hat\lambda + \hat\lambda^{-1})^2|\xi|^2 - (\xi + \xi^*)^2\\ &= (\hat\lambda + \hat\lambda^{-1})^2|\mu|^2\end{aligned} \tag{16}$$

with the aid of Eq. (13). The remainder of the demonstrations follows directly from Eqs. (12) and (13).

Finally, the powers of a matrix can be expressed as

$$\begin{aligned}[\xi, \mu]^n = {} & \left(|\xi|^2 - |\mu|^2\right)^{(n-1)/2}\\ & \times\left[\frac{(\hat\lambda^{n+1} - \hat\lambda^{-n-1})\xi - (\hat\lambda^{n-1} - \hat\lambda^{-n+1})\xi^*}{\hat\lambda^2 - \hat\lambda^{-2}}, \frac{\hat\lambda^n - \hat\lambda^{-n}}{\hat\lambda - \hat\lambda^{-1}}\mu\right]\end{aligned} \tag{17}$$

Note that each of the coefficients is real even when $\hat\lambda$ is complex. In the special case that $\hat\lambda = 1$, this becomes

$$[\xi, \mu]^n = \left(|\xi|^2 - |\mu|^2\right)^{(n-1)/2}\left[\tfrac{1}{2}(n+1)\xi - \tfrac{1}{2}(n-1)\xi^*, n\mu\right] \tag{18}$$

2. MAPS AND LYAPUNOV NUMBERS

Now that we have the potential for a better understanding of the product of matrices, we need a place to apply the method. A good candidate is the calculation of Lyapunov numbers for two-dimensional maps, since they are based on the infinite product of 2×2 matrices. The next few paragraphs introduce the ideas necessary for this calculation. This is not the place for a full review, and the interested and unsatisfied reader should investigate other sources (e.g., Refs. 4 and 5).

A two-dimensional map is a pair of equations

$$x_1 = f(x_0, y_0)$$
$$y_1 = g(x_0, y_0) \tag{19}$$

that takes a point in the plane (x_0, y_0) and produces a new point (x_1, y_1). Iterating this many times produces a sequence of points (x_n, y_n) that is called an orbit. The rest of this chapter deals exclusively with area preserving maps that transform a region with a given area into another region with the same area. Such maps capture the essential nature of Hamiltonian systems. Anything learned from them can be applied immediately to an interesting class of dynamical systems.

Given an orbit, it is instructive to consider the behavior of nearby orbits. This is best done by linearizing the map in the neighborhood of the given orbit and constructing the equation satisfied by the difference between two orbits, (x'_n, y'_n) and (x_n, y_n). Thus

$$\begin{pmatrix} \delta x_1 \\ \delta y_1 \end{pmatrix} = J_0 \begin{pmatrix} \delta x_0 \\ \delta y_0 \end{pmatrix} \tag{20}$$

where

$$J_n = \begin{pmatrix} \dfrac{\partial f(x_n, y_n)}{\partial x_n} & \dfrac{\partial f(x_n, y_n)}{\partial y_n} \\ \dfrac{\partial g(x_n, y_n)}{\partial x_n} & \dfrac{\partial g(x_n, y_n)}{\partial y_n} \end{pmatrix} \tag{21}$$

and

$$\delta x_n = x'_n - x_n$$
$$\delta y_n = y'_n - y_n \tag{22}$$

The coefficients of the matrix J_n are evaluated on the given orbit, (x_n, y_n).

Over a longer orbit segment the displacement satisfies

$$\begin{pmatrix} \delta x_n \\ \delta y_n \end{pmatrix} = M_n \begin{pmatrix} \delta x_0 \\ \delta y_0 \end{pmatrix} \tag{23}$$

where M_n is the ordered product of J matrices,

$$M_n = J_{n-1} \cdots J_0 \tag{24}$$

Area conservation yields

$$\det J_i = \det M_n = 1 \tag{25}$$

In general, these nearby orbits pull away from each other. More specifically, consider a cluster of points on a circle surrounding the initial point on an orbit segment. Further down the orbit, after a number of iterations of the map, the circle will be stretched into an ellipse. The aspect ratios of such ellipses, for different initial conditions and orbit lengths, is an interesting set of numbers. When you put cream in your coffee and give it a stir, these numbers measure the degree to which globules are stretched into wisps. It is the first step toward mixing. In particular, the aspect ratios measure the degree of ill conditioning of the operation of recovering initial conditions from a final state. Thus they are related to questions of predictability.

It has been shown[6] that in general the major radius of each ellipse grows exponentially with orbit length, on the average, in the limit of very long orbits. The rate of exponential increase is known as the Lyapunov number of that orbit. Thus the study of the ellipses can be called the calculation and approximation of Lyapunov numbers.

Each ellipse can be calculated directly from M_n, the displacement matrix for the corresponding orbit segment defined in Eq. (24). Inverting Eq. (23) yields

$$\begin{pmatrix} \delta x_0 \\ \delta y_0 \end{pmatrix} = M_n^{-1} \begin{pmatrix} \delta x_n \\ \delta y_n \end{pmatrix} \tag{26}$$

The condition that the initial points lie on a circle then produces the equation for the ellipse at the end of the orbit segment,

$$\begin{aligned} \delta x_0^2 + \delta y_0^2 &= 1 \\ &= (\delta x_n, \delta y_n)\left(M_n^\dagger\right)^{-1} M_n^{-1} \begin{pmatrix} \delta x_n \\ \delta y_n \end{pmatrix} \end{aligned} \tag{27}$$

Thus the eigenvalues of the matrix $(M_n^\dagger)^{-1} M_n^{-1}$ are the inverse of the square of the major and minor axes of the ellipse. Inverting again, and the eigenvalues of

$$N_n \equiv M_n M_n^\dagger \tag{28}$$

are directly the square of the major and minor axes. Furthermore, the eigenvectors are in the direction of those axes. The matrix N_n is symmetric, its eigenvalues are real and positive, and its eigenvectors are orthogonal. It is trivially generalized to higher dimensional systems. It is thus a compact representation of a method of approximating Lyapunov numbers that has been given previously, by Bennetin et al.[7]

We now have matrices to multiply, and a notation for multiplying them. Denoting

$$M_n \equiv [\xi_n, \mu_n] \tag{29}$$

then

$$\begin{aligned} N_n &= [\xi_n, \mu_n][\xi_n^*, \mu_n] \\ &= [|\xi_n|^2 + |\mu_n|^2, 2\xi_n\mu_n] \end{aligned} \tag{30}$$

The eigenvalues of N_n are

$$\lambda_n^\pm = (|\xi_n| \pm |\mu_n|)^2 \tag{31}$$

and the eigenvectors can be written

$$\begin{aligned} \chi_n^+ &= \frac{(\xi_n\mu_n)^{1/2}}{(|\xi_n||\mu_n|)^{1/2}} \\ \chi_n^- &= i\chi_n^+ \end{aligned} \tag{32}$$

The Lyapunov number l is defined in terms of limiting eigenvalues of N_n for very long segments,

$$l \equiv \lim \frac{1}{n} \ln(|\xi_n| + |\mu_n|) = \lim \frac{1}{n} \ln|\xi_n| \tag{33}$$

When people look for a measure of the "size" of a matrix, they frequently use its eigenvalue. From Eq. (13) this means that they effectively use the real part of ξ as this measure. The results above, however, suggest that $|\xi|$ is probably a more useful measure of the "size" of a matrix.

3. APPLICATION

What can be learned from all this? I do not have the full answer to that question, but I hope to intrigue some readers into exploring the following ideas. Much of this has been discussed by Ruelle,[8] but further work might prove instructive.

The eigenvectors of N_n, $(\xi_n\mu_n)^{1/2}$, assign a direction to every point in the plane. If we know that an orbit is initially somewhere in a little circle around (x_0, y_0), near the final point (x_n, y_n), we only know that it is somewhere along a needle that points in the direction $(\xi_n\mu_n)^{1/2}$. Thus this is the direction along which information is stretched. Now, fix this final point and vary the length of the orbit. This yields a series of directions at each point, each associated with a different initial point and its matrix, N_n, and parameterized by the orbit length. The remarkable thing is that these directions are all close together and approach a limit as the orbit length becomes very long. This result can be studied most effectively in the complex notation introduced above.

The first step in this demonstration is to consider the product of two displacement matrices,

$$\begin{aligned} M_{i+j} &= M_i M_j \\ &= [\xi_i, \mu_i][\xi_j, \mu_j] \\ &= [\xi_{i+j}, \mu_{i+j}] \end{aligned} \tag{34}$$

These two matrices arc associated with adjacent segments of an orbit, each segment of arbitrary length, as given in Eq. (24). It is straightforward to show that

$$\begin{aligned} |\xi_{i+j}|^2 + |\mu_{i+j}|^2 &= \left(|\xi_i|^2 + |\mu_i|^2\right)\left(|\xi_j|^2 + |\mu_j|^2\right) \\ &\quad + 2\left(\xi_i\xi_j\mu_i^*\mu_j + \xi_i^*\xi_j^*\mu_i\mu_j^*\right) \\ &= S\left(|\xi_i|^2 + |\mu_i|^2\right) \end{aligned} \tag{35}$$

where the real factor S

$$S \equiv |\xi_j|^2 + |\mu_j|^2 + 2\frac{\xi_i\xi_j\mu_i^*\mu_j + \xi_i^*\xi_j^*\mu_i\mu_j^*}{|\xi_i|^2 + |\mu_i|^2} \tag{36}$$

is a measure of the ratio of the magnitudes of the matrices M_{i+j} and M_i. In terms of S

$$\xi_{i+j}\mu_{i+j} = S\xi_i\mu_i + \frac{\xi_i^2\xi_j\mu_j - \mu_i^2\xi_j^*\mu_j^*}{|\xi_i|^2 + |\mu_i|^2} \tag{37}$$

after using the determinant condition, $|\xi|^2 - |\mu|^2 = 1$. If $|\xi_i|$ is large and S is not very small, the first term on the right is of order $|\xi_i|^2|\xi_j|^2$ and the second term is of order $|\xi_j|^2$. Thus the second term is smaller than the first by a factor of $|\xi_i|^{-2}$. This is independent of the size of $|\xi_j|$. That is, wherever $|\xi_i|$ becomes infinite in the limit of very long orbits, the direction Θ_i,

$$\exp i\Theta_i \equiv \frac{(\xi_i \mu_i)^{1/2}}{(|\xi_i||\mu_i|)^{1/2}} \tag{38}$$

approaches a limit.

Preliminary calculations seem to show that the phases of all the other numbers, ξ, μ, and their products, are essentially random, but the phase of $\xi\mu$ at a given final point is steady.

The question of convergence of the stretching direction needs to be examined in more detail. To the extent the limit in Eq. (33) converges smoothly to a finite number, $|\xi_i|$ is large for large i, and there is convergence of the stretching direction. There are, however, exceptional cases that need to be considered.

Along a stable periodic orbit the displacement matrices M_n are powers of the matrix for a single period. Since the orbit is assumed to be stable, the eigenvalues of the single-period matrix have unit magnitude. From Eq. (17) high powers of such matrices do not have large magnitudes, $|\xi|$. Thus there is no convergence of direction at points on stable periodic orbits. However, there can be weak convergence of direction for the periodic and almost periodic orbits of integrable systems, according to Eq. (18). This is in agreement with the results of Casati et al.[9]

The problem of convergence in general has to do with bounding the quantity S of Eq. (36). The significance of the second term on the right of Eq. (37) depends on S being finite or large when $|\xi_j|$ is large. The minimum value of S, for the most unfavorable phase of $\xi_i\xi_j\mu_i^*\mu_j$, is

$$S_{\min} = (|\xi_j| - |\mu_j|)^2 + 2|\xi_j||\mu_j|\frac{(|\xi_i| - |\mu_i|)^2}{|\xi_i|^2 + |\mu_i|^2}$$

$$= \frac{1}{(|\xi_j| + |\mu_j|)^2} + \frac{2|\xi_j||\mu_j|}{(|\xi_i| + |\mu_i|)^2(|\xi_i|^2 + |\mu_i|^2)} \tag{39}$$

where the determinant condition, Eqs. (25) and (4), has been used. This can be quite small if $|\xi_j|$ is large. However, this small value is obtained only for phases of $\xi_j\xi_j\mu_i^*\mu_j$ very close to the most unfavorable phase. To the extent that this phase is random, small values of S are improbable, but they do happen. A possible consequence of this is that the convergence may not be uniform. Thus, the field of directions $(\xi_n\mu_n)^{1/2}$ is smooth everywhere, expect where $|\mu_n| = 0$, for all finite n, but it may not be smooth in the limit.

This result indicates that there is a bit of order in the regions of chaotic orbits. Information is not shredded wildly, but is smeared out in given directions at each point. Limitations to the utility of this are discussed later.

Looked at another way, at each point a stretching direction is defined in the limit. This direction is invariant under the map, since an additional iteration does not make any difference. This has some similarities with a C-system, as discussed for example in Arnol'd and Avez.[10] The chief difference seems to be that in a C-system the invariant direction does not depend on taking a limit and the same direction is obtained for all orbit lengths. It might be possible to generalize the notion of C-system.

There is another set of invariant directions defined for a map. These are associated with the unstable manifolds of all the unstable periodic orbits. That is, they are derived from the invariant curves that are the infinite extensions of the unstable eigenvectors of each unstable periodic point. None of these curves can cross each other, since iterating backward each has a unique limit point. Period doubling bifurcations[11,12] produce vast numbers of unstable periodic orbits in the regions of chaotic behavior! The invariant curves from each writhe and squirm together across the map. Each is then an integral curve for a set of directions.

It is tempting to believe that the invariant stretching direction and the invariant unstable manifold direction are coincident. Numerical work is supportive of this conjecture. It is not clear how to prove it.

Finally, some numerical results are presented graphically (Figs. 1 and 2). These show integral curves for the direction field, $(\xi_n \mu_n)^{1/2}$, for two different cases. They were prepared using an integration routine for solving differential equations. For this purpose, the standard map[13] was chosen

$$y_1 = y_0 - \frac{k}{2\pi} \sin 2\pi x_0$$

$$x_1 = x_0 + y_1 \tag{40}$$

The parameter in this map was fixed at $k = 1.064563616002$, and a small region around $x = 0.5$, $y = 0.102777$ was examined in two approximations. First, 20 iterations of the map were used to define the stretching direction, $(\xi_n \mu_n)^{1/2}$, then 60 iterations were used. There is general agreement between the two approximations to the invariant stretching direction, showing a trend toward convergence, but there are differences.

The overall structure of loops in Figs. 1 and 2 can perhaps be understood as follows. Pictures of unstable manifolds show folding as they approach another unstable fixed point. Presumably the loops are associated with an approach to the unstable fixed point at $x = 0.5$, $y = 0.0$. Thus the results are consistent with the invariant stretching direction being coincident with the direction of the invariant unstable manifolds.

The particular values of the mapping parameters were chosen because they define the point of a tangent bifurcation of two periodic orbits that close after

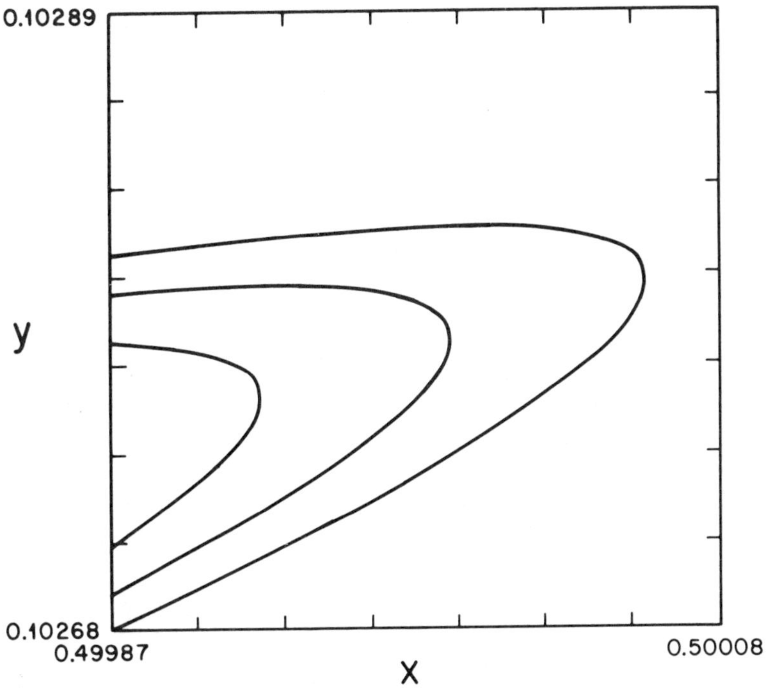

Figure 1. Integral curves of the expanding eigenvector direction field calculated from 20 iterations of the standard map with $k = 1.064563616002$.

29 iterations of the map. That is, for k values slightly less than the value chosen there are no periodic orbits of that length in the vicinity, but for slightly larger values of k there is both a stable and an unstable orbit nearby. The stable orbit goes through it complete cycle of period doubling bifurcations well before k reaches 1.06457, but within the interval there are stable orbits with vanishing Lyapunov numbers and no limiting stretching direction. Presumably, this is related to a place where S is close to $S_{\min}$ of Eq. (39) and the conditions for convergence of Eq. (37) are not satisfied. Tangent bifurcations happen randomly for all values of k and for periodic orbits of all lengths. I have never found any organizing principle for this phenomenon. Thus, it made an interesting subject for this numerical study.

Figure 1, using 20 iterations to define the direction field, does not show any indication of the period-29 orbits. It only shows the looping described above in connection with the overall loop structure. Along these curves, the values of $|\xi_{20}|$ ranged from 30 to 350.

Figure 2 does show the influence of the periodic orbits. Along the outer curves, away from the periodic orbits, the values of $|\xi_{60}|$ range from 3.10^4 to 7.10^6, but near the point of the central orbit, values of $|\xi_{60}|$ as small as 50 were found. It is quite clear that there is a nearby critical point in the direction field where $|\mu_{60}| = 0$. As a result, the direction fields in Figs. 1 and 2 are quite similar, but their connection into integral curves is quite different.

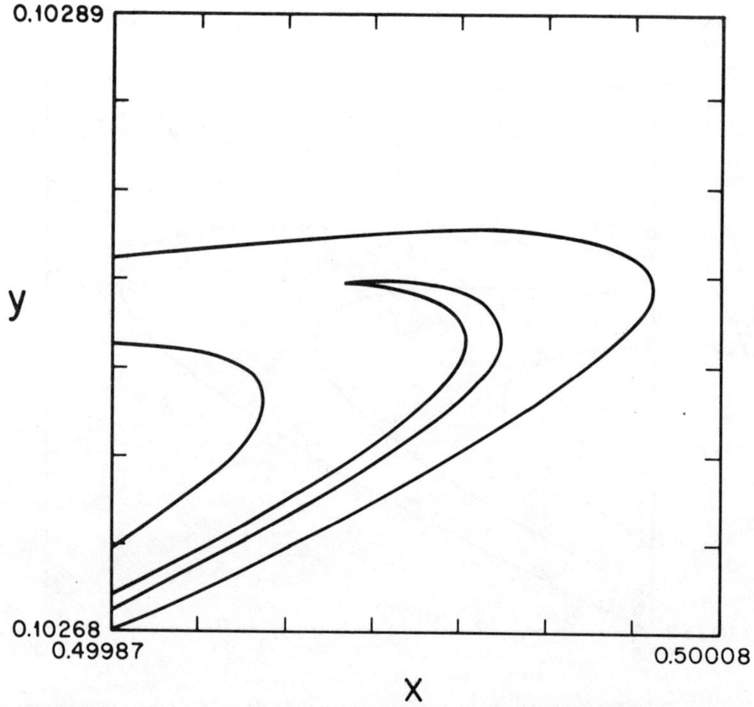

Figure 2. Integral curves of the expanding eigenvector direction field calculated from 60 iterations of the standard map with $k = 1.064563616002$.

A possible generalization of this is that there is convergence of the direction field of $(\xi_n \mu_n)^{12}$ in the limit of large n, for almost all points, but there is no convergence of the integral curves of this field. Thus, information is stretched in a nearly invariant direction at every point, but it is mixed by the random critical points of the various approximations to this direction field.

There are a number of ideas here, not all of them consistent. I would be most happy to be shown wrong on any of these points.

ACKNOWLEDGMENTS

I am much indebted to Dr. Weiss for sharing his ideas, and particularly for introducing me to Cayley algebras. Conversations with Dr. Littlejohn have been most helpful, and thanks are due to Dr. Goroff for pointing out Ruelle's paper (Ref. 8).

REFERENCES

1. *Encyclopedia Dictionary of Mathematics*, S. Iyanaga and Y. Kawada, Eds. (MIT Press, Cambridge, MA, 1980), p. 193.

2. J. M. Greene, *J. Math. Phys.* **9**, 760 (1968).
3. J. M. Greene, *J. Math. Phys.* **20**, 1183 (1979).
4. *Topics in Nonlinear Dynamics*, S. Jorna, Ed., American Institute of Physics Conference Proceedings Series, Vol. 46 (AIP, New York, 1978).
5. R. H. G. Helleman, in *Fundamental Problems in Statistical Mechanics*, Vol. 5, E. G. D. Cohen, Ed. (North Holland, Amsterdam, 1980), p. 165.
6. V. I. Osledec, *Tr. Mosk. Mat. O.* **19**, 179 (1968) (*Trans. Moscow Math. Soc.* **19**, 197).
7. G. Bennetin, L. Galgani, A. Giorgilli, and J. M. Strelcyn, *Meccanica* **15**, 9, 21 (1980).
8. D. Ruelle, Institut des Hautes Études Scientifiques, Publications Mathematiques, No. 50 (IHES, Bures sur Yvette, France, 1979).
9. G. Casati, B. V. Chirikov, and J. Ford, *Phys. Lett. A* **77**, 91 (1980).
10. V. I. Arnol'd and A. Avez, *Ergodic Problems of Classical Mechanics*, (Benjamin, New York, 1968), pp. 55–79.
11. P. Collet, J.-P. Eckmann, and H. Koch, *Physica* **3***D*, 457 (1981).
12. J. M. Greene, R. S. MacKay, F. Vivaldi, and M. J. Feigenbaum, *Physica* **3***D*, 468 (1981).
13. B. V. Chirikov, *Phys. Rep.* **52**, 265 (1979).

RENORMALIZATION APPROACH TO NONINTEGRABLE HAMILTONIANS

D. F. ESCANDE
Laboratoire de Physique des Milieux Ionisés
Groupe de Recherche N° 29 du Centre National
de la Recherche Scientifique
École Polytechnique
Palaiseau, France

Abstract. *This chapter reviews the results obtained with F. Doveil and A. Mehr by an approximate renormalization method applied to the Hamiltonian* $H(v, x, t) = v^2/2 - M \cos x - P \cos k(x - t)$. *This method is in the spirit of Kolmogorov–Arnol'd–Moser (KAM) theory and allows the fairly accurate computation of universal quantities, the threshold* $s(v)$ *of breakdown of KAM tori or of destabilization of cycles with average velocity* v, *and the threshold of large-scale stochasticity. The link between tori and nearby cycles is demonstrated, the mean residue of cycles is analytically computed, and Greene's main assertions on the standard mapping are proved. The graph of* $s(v)$ *is shown to be a fractal. Expressions are given for the width of stochastic layers that correct previous estimates. It is shown how to apply the results for* H *to a large class of two-degree-of-freedom (one-degree time-dependent) Hamiltonian systems.*

1. INTRODUCTION

Consider the Hamiltonian

$$\mathcal{H}(p, z, \tau) = \frac{p^2}{2m_0} - e \sum_{i=1}^{2} V_i \cos(k_i z - \omega_i \tau + \varphi_i) \tag{1}$$

which describes the motion of a particle of charge $-e$, mass m_0 in the potential of two longitudinal waves with potential V_i, wave number k_i and frequency ω_i; z is the position, p the momentum, and τ the time. A shift in the origin of time and position, allows us to make $\varphi_i = 0$ and to replace the V_i's by the $|V_i|$'s. Let $x = k_1 z - \omega_1 \tau$, $k = k_2/k_1$, $\Delta v = \omega_2/k_2 - \omega_1/k_1$, $M = eV_1/m_0 \Delta v^2$, $P = eV_2/m_0 \Delta v^2$, and $t = k_1 \Delta v \tau$. Then, the evolution of x as a function of t is given by the Hamiltonian

$$H(v, x, t) = \frac{v^2}{2} - M\cos x - P\cos k(x - t) \tag{2}$$

which is defined by the parameters (M, P, k). If the roles of the waves labeled 1 and 2 are reversed, a similar Hamiltonian H_e is obtained, which is nothing but Eq. (2) with the parameters $(P, M, 1/k)$.

H is the simplest time-dependent Hamiltonian with one degree of freedom having two resonant terms. Figure 1 sketches the stroboscopic plot of the orbits of this Hamiltonian for $k = 2$. The stroboscopic flashes are at times $t = nT$, where $T = 2\pi/k$ is the time period of H. The trapping domains of resonance M centered at $v = 0$ and of resonance P centered at $v = 1$ are shown. In fact there exist stochastic layers[1] at the border of these domains. Two higher order trapping domains and a passing orbit with average velocity v_0 also appear. This figure corresponds to a small value of the stochasticity parameter $s = 2\sqrt{M} + 2\sqrt{P}$, which is the sum of unperturbed widths of the primary resonances M and P. For a value of s of the order of 1, a large-scale stochastic instability appears in between the velocities 0 and 1, and a typical trajectory within this domain appears as a set of points erratically distributed on a surface. This transition has been experimentally proved by the heating of

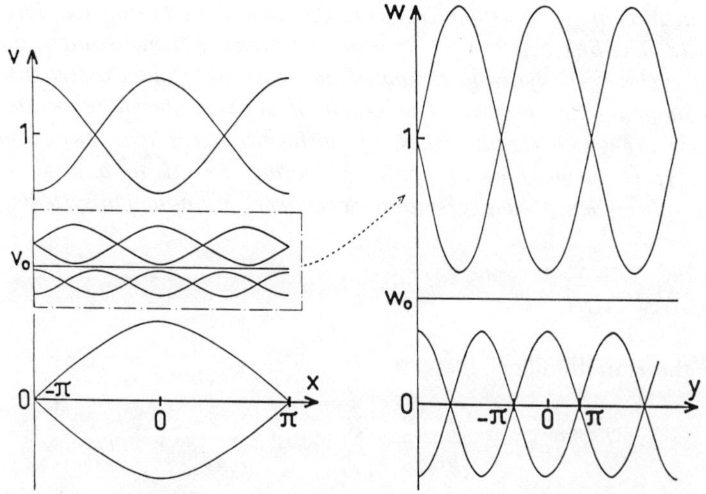

Figure 1. Graphical description of the renormalization method for $k = 2$.

a plasma column by a large amplitude standing wave,[2] and has been seen on a synchronous dipole motor,[3] which also can be modeled by the Hamiltonian Eq. (1), where m_0 is the inertia momentum of the rotor and $eV_1 = eV_2$ is the amplitude of the two oppositely rotating torques.

The transition to the large-scale stochastic instability of Hamiltonian systems with two degrees of freedom has long been a problem in many fields of physics: astronomy, statistical mechanics, accelerator physics, and plasma physics. Much work has been done in that respect (see, e.g., Ref. 1 and references therein), and many approximate methods have been developed for determining the threshold of this transition (see, e.g., the review by Lichtenberg[4]). The simplest one is due to Chirikov[1] and yields $s = 1$ as an approximate criterion. Comparison with numerical calculations shows that this yields the right order of magnitude, but the true threshold is generally smaller, even though primary resonances repel each other.[5,6] The most accurate methods are due to Greene[7,8] and to Greene and Percival.[9] They necessitate many computer calculations but are in the spirit of KAM theory[10] and yield accurate predictions. References 7 and 8 show numerically that the stabilty of one torus and of nearby (in some sense) cycles is strongly related.

Here I review a renormalization method[6,11–15] that allows the prediction of the stability of cycles and of the existence of KAM tori and shows the connection between them. This method involves several approximations, but comparison with numerical calculations reveals a good agreement. The Hamiltonian Eq. (2) appears as the paradigm for two-degree-of-freedom (or one-degree time-dependent) Hamiltonians. The theory yields a large (denumerable!) number of universal objects and quantities.

Most of the results presented here have been derived since the beginning of 1980 with the collaboration of F. Doveil or A. Mehr. My purpose is not to give the details of the derivations but rather to present the leading ideas and a global review of the main results. The interested reader is referred to papers already published or to be published in the near future.[6,11–15]

The transition to the large-scale stochasticity is related to the destruction of KAM tori.[1,10] In Fig. 1 there appears a line $v = v_0$, which is the trace of such a preserved torus, called hereafter $\mathfrak{T}_0$. $\mathfrak{T}_0$ is enclosed between two higher order resonances, the trapping domains of which are represented as chains of three and four islands in Fig. 1.

This chapter shows that the stability of $\mathfrak{T}_0$ can be fairly accurately described by a Hamiltonian that takes into account only the effect of these two higher order resonances.

Appropriate rescalings and good approximations allow us to give this Hamiltonian the same structure as H but with new parameters (M', P', k') that are explicit functions of the initial ones and coordinates (w, y) that replace (v, x). In the new coordinates $\mathfrak{T}_0$ corresponds to a new velocity w_0, which can be explicitly expressed as a function of k and v_0 (in fact it is dealt with the rotation number Q_0 of $\mathfrak{T}_0$). In a graphical way, it may be said that the little box in (v, x) space is transformed into a large box in the (w, y) space of

Fig. 1. Mathematically there is a renormalization transformation T_r on the four parameters (Q_0, k, M, P). If as we iterate T_r, the amplitudes of the two resonances decrease, the KAM theorem then applies and $\mathcal{T}_0$, preserved in the iteratively considered subsystems, is preserved in the initial system. The breakdown of $\mathcal{T}_0$ corresponds to an increase of the amplitudes. The border between these two situations yields the threshold of destruction of $\mathcal{T}_0$. Consideration of a denumerable set of tori allows us to give a good estimate for the threshold of the large-scale stochastic instability.

The cycles either correspond to the family Φ of higher order resonances in Fig. 1 or they do not. If there is no correspondence, look at the small box as for a KAM torus. After a finite number L of renormalization transformations, a given cycle $\mathcal{C}$ with rotation number Q_0 appears as corresponding to the family Φ for the Lth subsystem. For instance, it corresponds to the chain of three islands of Fig. 1. The theory shows that the stability of $\mathcal{C}$ is governed by the resonances related to the chains of two, three, and four islands of Fig. 1. A Mathieu equation is obtained, the parameters of which depend on Q_0, M, P, and k. The stability diagram of the Mathieu equation directly indicates the stability of $\mathcal{C}$.

This chapter is organized as follows. Section 2 yields the method for obtaining the renormalization transformation, states the Mathieu equation, and gives a special treatment for the stochastic layers of primary resonances. Section 3 studies the renormalization transformation and defines a first set of universal objects; it yields as a consequence the threshold of the large-scale stochasticity for H and makes the link with KAM theorem. Section 4 shows the connection between tori and nearby cycles, and the fractal nature of the graph of $s(v_0)$, which is the threshold of first destabilization of a cycle with average velocity v_0. The mean residue[7] is analytically calculated, and Greene's main assertions on the standard mapping[7] are proved. Section 5 indicates how the results for H can be directly applied to Hamiltonians of the types

$$H_2(\mathbf{J}, \boldsymbol{\theta}) = H_0(\mathbf{J}) + \varepsilon \sum_{i \in \mathcal{J}} V_i(\mathbf{J}) \cos(\mathbf{p}_i \cdot \boldsymbol{\theta}) \tag{3}$$

and

$$h(J_1, \theta_1, t) = h_0(J_1) + \varepsilon \sum_{i \in \mathcal{J}} V_i(J_1) \cos(p_i \theta_1 + p_i t) \tag{4}$$

where $\mathbf{J}$, $\boldsymbol{\theta}$, and $\mathbf{p}_i$ are two-dimensional vectors and $\mathcal{J}$ is a set with more than one element. The $\mathbf{p}_i$'s are noncollinear so that H_2 and h are nonintegrable. Sections 2.1, 3.1, and 3.2 correspond to Refs. 6, 11, and 12, Section 2.2 to Ref. 13, Sections 3.3, 4.3, and 4.4 to Ref. 14, Sections 4.1 and 4.2 to Ref. 15, and Section 5 to Ref. 6. Section 2.3 is original.

2. BASIC METHODS

When $P = 0$, H reduces to $H_0(v, x) = v^2/2 - M\cos x$. By taking into account the periodicity in x, action-angle variables (I, θ) can be defined[16] for this Hamiltonian (see Appendix). When written in action-angle variables, H_0 becomes $H_0'(I)$. The canonical equations yield $\dot{I} = 0$ and

$$\dot{\theta} = \Omega(I) = \frac{dH_0'}{dI} \tag{5}$$

First consider that k is a rational r/p. Then H has a period $2p\pi$ in x. When written in (I, θ) coordinates H becomes $H'(I, \theta, t)$, which has period $2p\pi$ in θ and can be subjected to Fourier expansion:

$$H'(I, \theta, t) = H_0'(I) - P\sum_{n=i}^{j} V_n(I)\cos\varphi_n \tag{6}$$

where $\varphi_n = (k+n)\theta - kt$, $i = -\infty$, $j = +\infty$, and the V_n's are given in the Appendix.

A given trajectory $\mathfrak{T}_0$ is characterized by its average velocity v_0. When $P = 0$, I is a constant I_0 along $\mathfrak{T}_0$. It results from the definition of (I, θ) that $v_0 = \Omega(I_0)$. Define the rotation number of $\mathfrak{T}_0$ by

$$Q_0 = \frac{k}{v_0} = \frac{k}{\Omega(I_0)} \tag{7}$$

Let $Z = Q_0 - k$ and $n_0 = \text{int}(Z)$. Here is considered a trajectory $\mathfrak{T}_0$ that lies between the resonances M and P. The values of v_0 are between 0 and 1, and $Q_0 > k$. When $k + 1 > Q_0 > k$ it is easy to show that the rotation number Q_0' of $\mathfrak{T}_0'$ (defined with respect to H_e) satisfies $Q_0' > 1/k + 1$. By dealing with either H or H_e, it can be assumed $Q_0 > k + 1$ $(n_0 \geqslant 1)$ for any trajectory $\mathfrak{T}_0$ lying between the resonances M and P.

As long as P is not too large, the action on $\mathfrak{T}_0$ remains close to I_0; the first approximation is to replace H_0' by its second-order expansion about I_0 and $V_n(I)$ by $V_n(I_0)$ $(n = -\infty, +\infty)$.

When both M and P are nonzero, $\dot{\theta}$ is no longer a constant on $\mathfrak{T}_0$, but the foregoing definition of Q_0 implies that for long times the phase of the nth resonant term of Eq. (6) increases on $\mathfrak{T}_0$ like $kt(k + n - Q_0)/Q_0$.

2.1. Renormalization Transformation

Assume first that $Z = Q_0 - k$ is not an integer. As a result $d\varphi_n/dt$ is the smallest for $n = n_0$ and $n = n_0 + 1$. We make the usual assumption that the stability of $\mathfrak{T}_0$ is mainly governed by these two "slow" terms, and all the others are neglected in Eq. (6): $i = n_0$ and $j = n_0 + 1$. This first approximation yields a Hamiltonian similar to Eq. (1) with two resonant terms that correspond to the "small box" of Fig. 1. Let $\delta k = Q_0 - k_0 - n_0$ $(0 < \delta k < 1)$ and λ be 0 or 1 for $\delta k < \frac{1}{2}$ or $> \frac{1}{2}$, respectively (the reason for defining λ is given in Section 3.1). Let $\mu = n_0 + \lambda$ and $\nu = n_0 + 1 - \lambda$, $k_\mu = k + \mu$, $k_\nu = k + \nu$, $y = k_\mu \theta - kt$, and $t' = \gamma t$ with

$$\gamma = \frac{(2\lambda - 1)k}{k_\nu} \tag{8}$$

Transformations similar to those allowing one to go from $\mathcal{H}$ to H define a new Hamiltonian H'', and the evolution of the phase y as a function of the time t' is determined by the Hamiltonian

$$H''(w, y, t') = \frac{w^2}{2} - M'\cos y - P'\cos k'(y - t') \tag{9}$$

where

$$k' = \frac{k_\nu}{k_\mu} \tag{10}$$

$$M' = \alpha P V_\mu(I_0) \tag{11}$$

$$P' = \alpha P V_\nu(I_0) \tag{12}$$

with $\alpha = (k_\mu k_\nu / k)^2 \sigma_0$, with $\sigma_0 = d\Omega(I_0)/dI$. Hamiltonian H'' has the same structure as H. With the new coordinates (w, y, t'), the "small box" has become a "large box." In the limit where $M = P = M' = P' = 0$, it is easy to compute the new constant velocity w_0 on $\mathfrak{T}_0$. According to Eq. (7), this defines the new rotation number

$$Q' = \frac{k'}{w_0} = \frac{Q_0}{k_\mu |k_\mu - Q_0|} \tag{13}$$

Equations (7)–(13) define a renormalization transformation that maps (Q_0, k, M, P) into (Q', k', M', P').

The question of the stability of $\mathfrak{T}_0$ is not yet solved, but arises in the subsystem described by H'': either $Z' = Q' - k'$ is an integer or it is not. If not, it is necessary to iterate the renormalization transformation. This is done in Section 3.1. If Z' is an integer, the solution for Z integer can be applied in the subsystem.

2.2. Mathieu Equation for Describing the Stability of Cycles

Now assume that $Z = Q_0 - k$ is an integer l ($\mathfrak{T}_0$ is a cycle). Then $d\varphi_n/dt$ is the smallest for $n = l - 1$, l, and $l + 1$. Again it is assumed that the stability of $\mathfrak{T}_0$ is mainly governed by "slow" terms. This yields a Hamiltonian $\mathcal{H}'$ of the same type as $\mathcal{H}$ but with one more resonant term. The Poincaré–Birkhoff theorem[10] implies that there only is a finite number of cycles with rotation number Q_0 for $s > 0$. The classical analysis of the nonlinear pendulum shows that they correspond to $\varphi_l = 0 \pmod{\pi}$. Let the cycle be given by $[I(t), \theta(t)]$. A trajectory $[I(t) + u(t), \theta(t) + z(t)]$ close to the cycle is calculated from the tangent flow to the cycle

$$\dot{z} = \sigma_0 u \tag{14}$$

$$\dot{u} = -Pz \sum_{i=l-1}^{l+1} (k+i)^2 V_i(I_0) \cos\left[\varphi_l + \frac{(\varphi_l + kt)(i-l)}{(k+l)}\right] \tag{15}$$

where $I(t)$ has been approximated by I_0 and $\theta(t)$ by $(\varphi_l + kt)/(k+l)$. Eliminating u from Eqs. (14) and (15) yields

$$\frac{d^2 z}{d\tau^2} + (a - 2q\cos 2\tau) z = 0 \tag{16}$$

with

$$\tau = \frac{\varphi_l + kt}{2(k+l)} \tag{17}$$

$$a = \frac{4P\sigma_0\eta_l(k+l)^4 V_l(I_0)}{k^2} \tag{18}$$

$$q = \frac{2P\sigma_0\eta_l\left[(k+l+1)^2 V_{l+1}(I_0) + (k+l-1)^2 V_{l-1}(I_0)\right](k+l)^2}{k^2} \tag{19}$$

and $\eta_l = \cos(\varphi_l)$. Both $|a|$ and $|q|$ are growing functions of s. When $\varphi_l = \pi$ the cycle is always unstable. When $\varphi_l = 0$ the cycle is stable for small values of s in accordance with the Poincaré–Birkhoff theorem. The stability diagram of the Mathieu equation[17] shows that when s grows, the cycle becomes unstable, then again stable, again unstable, and so on. A transition to instability is never permanent. Nevertheless, the width in s of the stability domains strongly diminishes when s increases. This is substantiated by numerical calculations[13, 15] and is consistent with previous mathematical results.[18] This contrasts with the case of the standard mapping where the transition to instability is definitive.[19]

When applied to the Mathieu equation, a standard perturbation method[20] shows that as s overshoots $s_c = s(v)$ for a given cycle, z grows exponentially with a growth rate $\gamma = [2q_c q_c'(s - s_c)]^{1/2} + O[(s - s_c)^{3/2}]$, where $q_c = q(s_c)$ as given by Eq. (19) and $q_c' = dq(s_c)/ds$. The positive Lyapunov coefficient of the cycle, γ, defines a local rate of instability of the system. The exponent $\frac{1}{2}$ is analogous to a critical exponent for γ. For the transition to large-scale stochasticity Chirikov[1] has computed a similar exponent (cf. Section 6).

The threshold of first destabilization of a cycle with average velocity $v_l' = k/(k+l)$ can be computed with Eqs. (18) and (19) and the formula[17] $a = 1 - q - q^2/8 + O(q^3)$. This defines an implicit function $F(M, P, k, l) = 0$. Section 1 demonstrated that the Hamiltonian H_e characterized by the parameters $(s, 1/\rho, 1/k)$ where $\rho = (M/P)^{1/2}$, corresponds to the same dynamical system as H. The cycles with velocities v_n' of Hamiltonian H_e correspond to cycles with velocities $v_n'' = nk/(nk+1)$ for H. This feature leads to a method for describing the stability of the set Φ of cycles with velocities v_n' and v_n''. The estimates for $n = 0$ agree within a few percentage points (4% for $\rho = k = 1$) with the direct numerical computation of the threshold. When n grows, the cycles with velocities v_n' and v_n'' lies closer to resonances M and P, respectively. Such a cycle is therefore destabilized and absorbed in the stochastic layer[1] of M or P, respectively, for smaller and smaller values of s. Since the approximations leading to Eq. (16) are more and more nearly correct as s decreases, this method gives better and better estimates of the threshold of first destabilization of the cycles as n grows. For intermediate values of n, the perturbation of the action along the cycle by the $n = 0$ resonance or resonance P (V_0 is the largest of the V_n's) is sensitive and this method does not give good estimates of $s(v_n')$.

The effect of this perturbing resonance can be taken into account by performing a second action-angle transformation that suppresses resonance V_0. This method[13] describes the set Φ in a global way and involves a Mathieu equation too, with coefficients a and q more intricate than in Eqs. (18) and (19).

Figure 2 shows the threshold $s(v)$ of first destabilization of the orbits with velocities $v = v_n'$ or v_n'' in the case $k = 1$, $\rho = 4$. The triangles correspond to the exact threshold obtained by direct numerical integration of the canonical equations. The agreement of the theory with the numerical points is better than 7% and increases strongly as $s(v)$ decreases. Values of $s(0)$ and $s(1)$ are computed by the first method and correspond to the threshold of first destabilization of the cycles related to resonances M and P. In between, $s(v)$ has a convex shape computed by the second method and goes to zero as v goes to 0 or 1 because of the existence of stochastic layers about the M and P resonances. The same shape of $s(v)$ has already been obtained in Ref. 19 for the standard mapping. As noted in this reference, any of the cycles with velocity v_n' or v_n'', $n > 1$, corresponds to a resonance that has a stochastic layer too. It can therefore be suspected that between two consecutive points of $s(v)$, the same structure as $s(v)$ between $v = 0$ and $v = 1$ should be recovered. Conjecture about this feature appears in Ref. 19, numerical evidence for the standard mapping in Ref. 21, and proof in Section 4.1.

The second method is only good if k/ρ^2 is not too close to 1. In the opposite case, a third method[13] can be used. It amounts to performing the Kolmogorov transformation,[1] which "kills" both resonances M and P, on H. Again a Mathieu equation is obtained. It correctly describes $s(v)$ in the vicinity of its maximum s_m and gives estimates that agree by better than 5% for values of

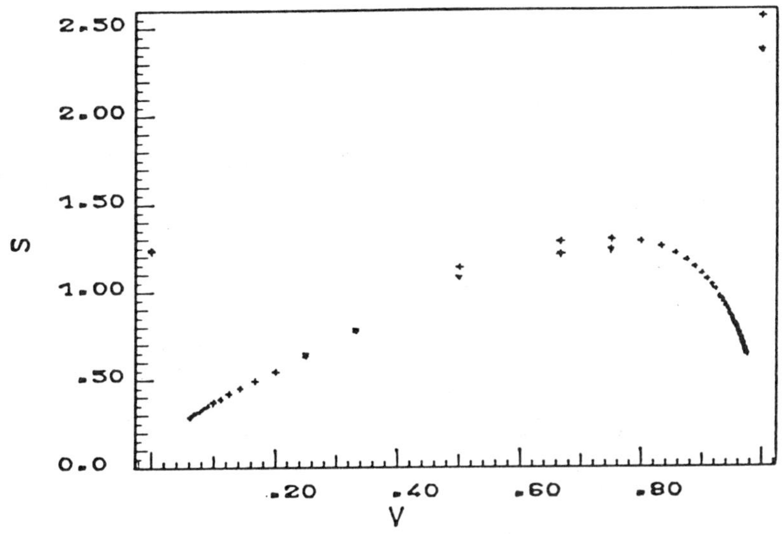

Figure 2. Threshold $s(v)$ for $k = 1$ and $\rho = 4$. Crosses and triangles represent, respectively, theoretical and numerical results.

$s(v)$ larger than $s_m/2$. The remaining part of the curve must be computed by the first method. Since all three methods rely on neglecting many terms, the system appears to be slightly stabler than it actually is, and the estimated $s(v)$ overestimates the actual one.

The residue[7] of the cycle with average velocity v_l' is easy to compute because z is a solution of Eq. (16). Floquet's theory[17] yields that

$$z = P(\tau)(b \cos \nu\tau + c \sin \nu\tau) \tag{20}$$

where P is of period π, b and c are arbitrary constants, and ν is the characteristic exponent. It can be deduced from[17]

$$a = \nu^2 + \frac{q^2}{2(\nu^2 - 1)} + O(q^4) \tag{21}$$

Let τ_0 be the period of the cycle. As is shown later, τ_0/π is an integer. Let $\mathbf{Z}(\tau)$ be the vector of components $[u(\tau), z(\tau)]$. Then

$$\mathbf{Z}(\tau + \tau_0) = \mathcal{M}\mathbf{Z}(\tau), \tag{22}$$

where the matrix $\mathcal{M}$ can be explicitly calculated from Eq. (20). The residue is[7]

$$R = \frac{1}{4}(2 - \operatorname{Tr} \mathcal{M}) = \sin^2\left(\frac{\Psi}{2}\right) \tag{23}$$

where

$$\Psi = \nu \tau_0 \tag{24}$$

In the limit where s goes to zero, it results from Eqs. (18), (19), and (21) that $\nu^2 = a$; the residue and therefore $\mathfrak{M}$ depend only on V_l; that is, the resonance V_l behaves as if isolated.

2.3. Stochastic Layer

The width of stochastic layers has been computed by two different types of approach. One is Chirikov's whiskers mapping,[1] which locally reduces to the standard mapping, and the other[16,22–26] applies the overlap criterion to the Hamiltonian Eq. (6). Up to now no connection has been made between the two approaches. Here I present a simple method that yields the missing link and corrects these previous estimates.

The description of the stability of orbits in Sections 2.1 and 2.2 takes into account only two or three resonant terms in the infinite expansion Eq. (6). This approximation can be released for describing the stochastic layer of resonances M and P. Only resonance M is considered; the case of resonance P is directly obtained through the equivalent Hamiltonian H_e. The stochastic layer has an inner part and an outer part that correspond respectively to the trapped and untrapped orbits of resonance M when $P = 0$. Action-angle variables can be defined[16] too in the trapping domain of resonance M. When written in these variables, H has the expression of Eq. (6) with $\varphi_n = n\theta - kt$ and V_n is given in the appendix. Let β be 1 for untrapped and 2 for trapped trajectories, and $k_n = (2 - \beta)k + n$.

After applying the first approximation of Section 2.1 to Eq. (6), the evolution of $y = \varphi_{n_0}$ is governed by the Hamiltonian

$$H_s(w, y, t) = \frac{1}{2}w^2 - A \sum_{n=-\infty}^{+\infty} \eta_n \cos[\kappa_n y - (n_0 - n)t] \tag{25}$$

where $A = \sigma_0 k_{n_0}^4 P V_{n_0}(I_0)/k^2$, $\kappa_n = k_n/k_{n_0}$, and $\eta_n = V_n(I_0)/V_{n_0}(I_0)$. It can be checked a posteriori that $\kappa_n \simeq 1$ and $\eta_n \simeq 1$ for n in a neighborhood of n_0 that grows with n_0 (for values of s that correspond to destabilization of tori and cycles of interest). The Hamiltonian of Eq. (25) with $\eta_n = \kappa_n = 1$ for every n corresponds[1] to the standard mapping with parameter

$$K = (2\pi)^2 A \tag{26}$$

The stochasticity parameter for Eq. (25), computed for neighboring resonances, is then $s' = 4\sqrt{A} = 2\sqrt{K}/\pi$. For large values of n_0, A has almost the same value for $k_{n_0-1} < Q_0 < k_{n_0+1}$. Therefore all this set of rotation numbers can be described by Eq. (25) computed for $Q_0 = k_{n_0}$. The estimates given in Refs.

16 and 22–26 consisted in retaining only two resonances in Eqs. (2)–(5), and in taking $s' = s'_0 = 1$ for the threshold of the large-scale stochasticity in the vicinity of resonance V_{n_0}, that is, $K = K_0 = (\pi/2)^2 \simeq 2.467$. In fact the right threshold is[7] $K = K_g \simeq 0.9716$.

The appropriate parameter for describing the width of the stochastic layer is $m_1 = 1 - m$, where $m = ([H_0(v, x)/M + 1]/2)^{2\beta-3}$ (see the appendix). For large values of n_0, m_1 is close to zero. This allows us to compute the maximum value m_{10} of m_1 in the stochastic layer with the use of Eqs. (26), (A7), (A10), (A16), and (A18). Let $\varepsilon = 2\sqrt{M}/(\pi k)$; for $\varepsilon \ll 1$, we find

$$m_{10} = \left(\frac{P}{M}\right)\frac{\beta 2^{2\lambda+1}}{K_g \pi^\lambda \Gamma(\lambda)\varepsilon^{\lambda+1}}\exp\left(-\frac{1}{\varepsilon}\right) \tag{27}$$

where $\lambda = 2k$. The corresponding value of n_0 is given by

$$k_{n_0} = \frac{\beta}{\varepsilon\pi^2}\log\left(\frac{16}{m_{10}}\right) \tag{28}$$

This formula shows that for thin stochastic layers (small m_{10}, large n_0), the steplike progression (successive overlap of the resonances V_n) can be approximated by the continuous expression Eq. (27).

Equation (27) is in complete agreement with Eq. (4.60) of Ref. 1 for $\beta = 1$. The discrepancy for $\beta = 2$ (inner layer) can be explained as follows. Consider two orbits symmetrical with respect to the unperturbed separatrix (same value of m_1) that begin near $x = -\pi$. When the values of x are $-\pi, 0, \pi, 2\pi, 3\pi$ on the untrapped trajectory, they are $-\pi, 0, \pi, 0, -\pi$ on the trapped one. For $\varepsilon \ll 1$, the first trajectory has a kick at $x = 0$ and $x = 2\pi$, but the second one gets a kick at only the first value $x = 0$, where the velocity is positive.[1] Therefore, for trapped orbits the factor λ of Eq. (4.55) of Ref. 1 must be replaced by 2λ. This yields a complete agreement with Eq. (27) for $\beta = 2$. The present method thus yields the same standard mapping (same parameter K) as Ref. 1, but the derivation is shorter and includes fewer assumptions. Nevertheless, for $|n_0|$ large it is equivalent to assume $I = I_0$ as here, or $\varphi = \varphi_{sx}$ as in Ref. 1.

As a consequence, Eqs. (10) and (12) of Ref. 23 must be multiplied by a factor $\lambda K_0/(2K_g) \simeq 5\lambda/4$, which enhances the estimate of the domain of magnetic braiding. As noted in that reference, m_{10} has a steep rise when ε grows. In Refs. 25 and 26 the threshold for heating is taken as the value of the parameter ω_p/Ω_i for which m_{10} takes appreciable values. Despite the use of K_0 instead of K_g, the steep rise of $m_{10}(\varepsilon)$ allows good agreement with numerical calculations.

For negative velocities the outer part of the separatrix of resonance M involves resonances V_n for $n < -k$. It results from Eq. (A15) that the factor $\exp(-1/\varepsilon)$ of Eq. (27) must be replaced by $\exp(-3/\varepsilon)$. The same result is obtained from Ref. 1 by taking into account the properties of the Melnikov–Arnol'd integral. This result explains why the outer part of the stochastic layer is not visible in numerical calculations.[16]

The first destabilization of the cycle with average velocity v''_{n_0} corresponds to the destabilization of a primary resonance in the standard mapping, that is, to[19] $K = 4$. Combination of Eqs. (27) and (28) where K_g is replaced by 4 allows an accurate prediction of $s(v'_{n_0})$ for large values of n_0. This confirms, for instance, that resonance $m = 6$, $n = 1$ is stable in Fig. 2 of Ref. 4: it appears as six white islands in the black stochastic layer!

When $\varepsilon \gg 1$, the approximate Eq. (27) is no longer correct. When $\lambda = 2k$ is 1 or 2, Eqs. (A5) and (A15) may be used. For the outer layer the denominator $1 - q^{4k_n}$ is small for all n's, but for the inner layer only one out of two of the denominators $1 - (-1)^{\lambda+n} q^n$ is small. Therefore half the resonant terms of Eq. (25) can be deleted. This Hamiltonian is again equivalent to a standard mapping, but with parameter $K = \pi^2 A$, where A is computed for large V_n. This yields the same expression for m_{10} for both the inner and the outer layers:

$$m_{10} = \frac{\lambda^\lambda P}{2K_g M^{(\lambda+2)/2}} \tag{29}$$

This result agrees with Ref. 1, since for $\varepsilon \gg 1$, a trapped orbit gets a kick at $x = 0$ for negative velocities too. This point is confirmed by numerical calculations of Ref. 24, which yields an Eq. (6) similar to Eq. (29). The factor $128/\pi^2 \simeq 13$ of that equation must be replaced according to Eq. (29) by $2/K_g \simeq 2$. The agreement claimed in this reference between Eq. (6) and numerical calculations is perhaps due to the very indirect way of measuring m_{10} that is used.

Finally the discrepancy between Eqs. (27) and (29) and the estimates of Ref. 22 is due to the incorrect way of calculating the width of the resonances of this reference.

3. STABILITY OF KAM TORI

3.1. Study of the Renormalization Transformation

The knowledge of Q_0 and k is equivalent to that of k, n_0, and δk ($Q_0 = k + n_0 + \delta k$, $0 < \delta k < 1$). Let δk_l, n_l, k_l, M_l, and P_l be the values of the parameters δk, n, k, M, P after l iterations of the renormalization transformation T_r ($l = 0$ corresponds to the initial values of the parameters). The definition of $\lambda(\delta k)$ implies that $n_l \geqslant 1$ for any l. This choice yields the fastest convergence toward smaller scales ("boxes") in the phase space. Because of the approximations made on deriving T_r, this transformation is only approximate and a fast convergence is highly desirable. This is the reason for using H_e when $n_0 = 0$.

The study of T_r shows that δk_{l+1} and n_{l+1} depend only on δk_l. Therefore, the mapping of the unit interval $\delta k \to \delta k'$ plays a dominant role. This mapping has an infinite set of unstable, irrational fixed points

$$\delta k_n^\lambda = \frac{1}{2}\left\{\left[n^2 + (4 - 2\lambda')n + 5\lambda'\right]^{1/2} - n - \lambda'\right\} \tag{30}$$

where $\lambda' = 1 - \lambda$, $\lambda = 0$ or 1, and n is a positive integer. Iterate T_r with the set of initial parameters $(\delta k_n^\lambda, n, k_0, M_0, P_0)$ and call $\mathcal{T}_n^\lambda$ the KAM torus with rotation number $Q_0 = k_0 + n + \delta k_n^\lambda$. Then $n_l = n$ and $\delta k_l = \delta k_n^\lambda$ for all l. It is readily shown that k_l converges toward $k_n^\lambda = 1 - \lambda + \delta k_n^\lambda$. Since n_l and δk_l are fixed, consider the restricted map $\mathcal{M}_n^\lambda$, which acts on (k, M, P) only. Map $\mathcal{M}_n^\lambda$ has a hyperbolic fixed point F_n^λ with a stable manifold $\mathcal{S}_n^\lambda$ of dimension 2 that separates the basins of attraction $\mathcal{D}_i$, $i = \alpha, \omega$, of $\omega = (k_n^\lambda, 0, 0)$ and $\alpha = (k_n^\lambda, \infty, \infty)$. When (k_0, M_0, P_0) lies in $\mathcal{D}_\omega$, then after a finite number L of iterations of $\mathcal{M}_n^\lambda$, M_L and P_L are small enough for the KAM theorem to apply. Therefore $\mathcal{T}_n^\lambda$ is preserved in the Lth subsystem, and consequently in the original system. When (k_0, M_0, P_0) is in $\mathcal{D}_\alpha$ the resonances overlap more and more strongly, and $\mathcal{T}_n^\lambda$ is destroyed.

The renormalization transformation T_r has a denumerable set of hyperbolic fixed points. These points and their attributes (eigenvalues, stable and unstable manifold) are universal objects for Hamiltonians of type 1 but, as shown in Section 5, also for those of types 3 and 4. The surfaces $\mathcal{S}_n^\lambda$ are shown in Sections 3.2 and 5 to play a central role in the determination of the threshold for large-scale stochasticity. The unstable eigenvalue δ_n^λ of $\mathcal{M}_n^\lambda$ plays a role similar to Feigenbaum's[27] δ in Section 4.1.

When iterating T_r with any δk_0, the (n_i, λ_i) constitute a coding of δk_0 equivalent to the one given by the coefficients a_i, $i = 1, 2, \ldots$ of the expansion of δk_0 into a continuous fraction

$$\delta k_0 = [a_1, a_2, \ldots, a_n, \ldots] = \frac{1}{a_1 + 1/(a_2 + 1/\cdots)} \tag{31}$$

Table 1

$(n_i, \lambda_i) \to a_j$		$a_j \to (n_i, \lambda_i)$	
1. Set $j = 0$, $i = 1$		1. Set $j = 1$, $i = 1$	
2. Is λ_i equal to 0?		2. Is a_j equal to 1?	
3. Yes	No	3. Yes	No
Add 1 to j	Add 1 to j	$\lambda_i = 1$	$\lambda_i = 0$
$a_j = n_i + 1$	$a_j = 1$	$n_i = a_{j+1}$	$n_i = a_j - 1$
	Add 1 to j	Add 2 to j	Add 1 to j
	$a_j = n_i$		
4. Add 1 to i.			
5. Go to step 2.			

Table 1 yields two algorithms for computing the a_j's from the (n_i, λ_i)'s, and vice versa. For $\delta k_0 = \delta k_n^0$ the a_j's are $a_j = n + 1$, and for $\delta k_0 = \delta k_n^1$, $a_{2l} = n$ and $a_{2l+1} = 1$; δk_1^1 is the golden mean, $(1 + \sqrt{5})/2$, minus one.

3.2. Threshold of Large-Scale Stochasticity

For easier comparison with previous work,[1,4] H is characterized by k_0, $\rho = (M_0/P_0)^{1/2}$ and $s = 2\sqrt{M_0} + 2\sqrt{P_0}$ instead of (k_0, M_0, P_0). So the well-known overlapping resonance criterion[1] is simply written $s = 1$ for any values of k and ρ.

The KAM torus $\mathfrak{T}_n^\lambda$ corresponds to a certain function $s_n^\lambda(k_0, \rho)$. As s increases for fixed values of k_0 and ρ, we know that all the tori between the resonances M and P are destroyed. Therefore there is a maximum value of $s_n^\lambda(k_0, \rho)$ when n and λ vary, which we call $s_1(k_0, \rho)$. For s less than $s_1(k_0, \rho)$ there is at least one torus $\mathfrak{T}_n^\lambda$ that is preserved. Similar tori can be defined for H_e; this yields a function $s_2(k_0, \rho)$. Let

$$s(k_0, \rho) = \max[s_1(k_0, \rho), s_2(k_0, \rho)]$$

For s less than $s(k_0, \rho)$ there is at least one torus preserved between the resonances M and P. Thus $s(k_0, \rho)$ is a lower bound of the true stochasticity threshold.

Let $X = 2\sqrt{M}$ and $Y = 2\sqrt{P}$. Figure 3 plots the threshold of the large-scale stochasticity $Y(X)$ obtained from $X + Y = s(k_0, X/Y)$ for fixed values of k_0. Because H_e and H are equivalent, this plot also yields the threshold for the various values of $1/k_0$. The dotted line corresponds to the overlap criterion $s = 1$. Comparison with numerical calculations shows[6,12] that $s(k_0, \rho)$ has the right variation with the parameters k_0 and ρ and agrees with the actual threshold by better than 7% for $\frac{1}{5} \leqslant \rho \leqslant 5$, $\frac{1}{2} \leqslant k \leqslant 2$. The $s = 1$ criterion is good only to 30% in this range of parameters. The good agreement between $s(k_0, \rho)$ and the actual threshold shows that the last $\mathfrak{T}_n^\lambda$ is "almost" the last torus to disappear at the threshold. This torus corresponds to $\mathfrak{T}_1^1$ when $\rho = k_0 = 1$. $\mathfrak{T}_1^1$ has a rotation number $Q_1^1 = 1 + g$, where $g = (1 + \sqrt{5})/2$ is the golden mean. The symmetry of H in that case implies that the torus $\mathfrak{T}_g$ with rotation number g disappears for the same value of s as $\mathfrak{T}_1^1$. This result substantiates the numerical evidence of Greene[7] that $\mathfrak{T}_g$ is destroyed very close to the stochasticity threshold in the case of the standard mapping, which can be viewed as a periodization in v of H for $k_0 = \rho = 1$ [this can be seen by comparing Eq. (2) with Eq. (26) for $\kappa_n = \eta_n = 1$].

Since T_r is continuous for k rational, it can be defined for k irrational too. In that case the lack of periodicity of H prevents the existence of KAM tori. Nevertheless there still exists a transition to large-scale stochasticity where a trajectory can diffuse with time from the vicinity of resonance M to the neighborhood of resonance P. The function $s(k_0, \rho)$ yields an estimate of this threshold.

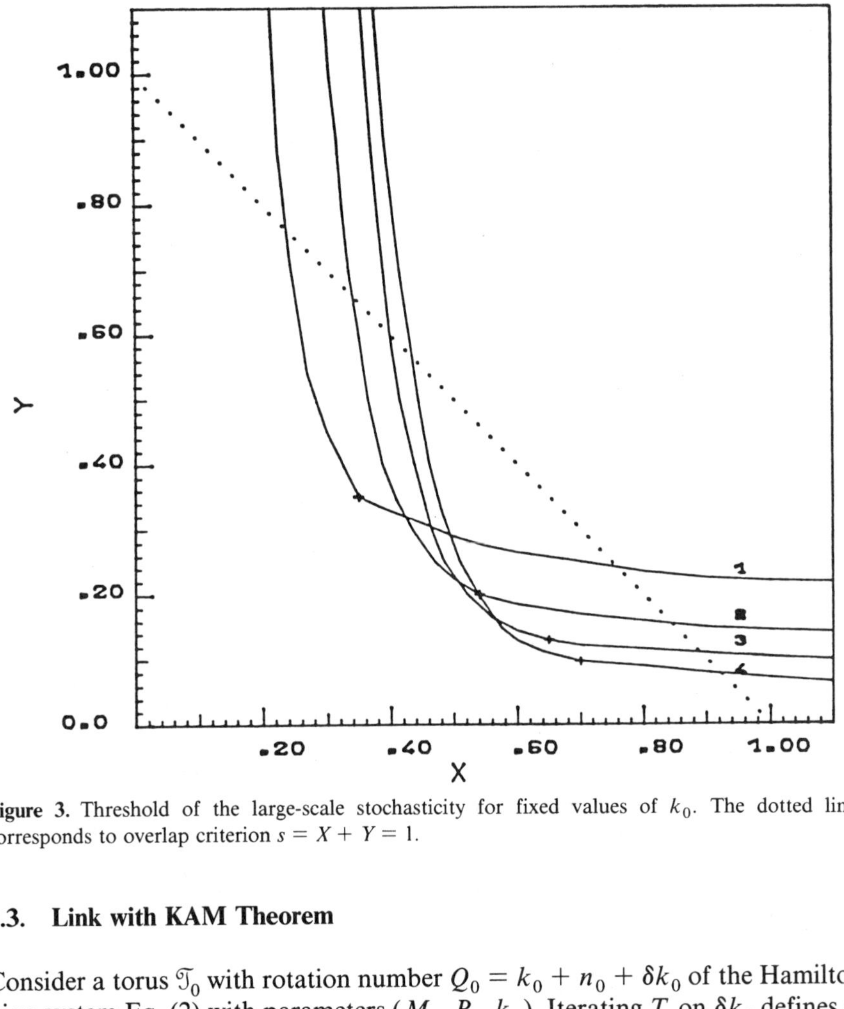

Figure 3. Threshold of the large-scale stochasticity for fixed values of k_0. The dotted line corresponds to overlap criterion $s = X + Y = 1$.

3.3. Link with KAM Theorem

Consider a torus $\mathfrak{T}_0$ with rotation number $Q_0 = k_0 + n_0 + \delta k_0$ of the Hamiltonian system Eq. (2) with parameters (M_0, P_0, k_0). Iterating T_r on δk_0 defines a denumerable set of (n_i, λ_i)'s. Define the matrix

$$\mathfrak{L}_i = \begin{pmatrix} \xi_i & 1 \\ \zeta_i & 1 \end{pmatrix} \qquad i \geqslant 1 \tag{32}$$

where $\xi_i = n_{i-1} + \lambda_i$, $\zeta_i = n_{i-1} + 1 - \lambda_i$, and the matrix

$$\Pi_L = \begin{pmatrix} \delta_L & \gamma_L \\ \beta_L & \alpha_L \end{pmatrix} = \prod_{i=1}^{L} \mathfrak{L}_i \tag{33a}$$

and the vector $\mathbf{X}_i$ with components $(\log M_i, \log P_i)$. With this definition of Π_L it is readily shown that

$$k_L = \frac{\beta_L + \alpha_L k_0}{\delta_L + \gamma_L k_0} \tag{33b}$$

It results from Eqs. (11) and (12) that for small values of M

$$\mathbf{X}_i = \mathbf{C}_i + \mathfrak{L}_i \mathbf{X}_{i-1} \tag{34}$$

where $\mathbf{C}_i$ is a vector with components $(\log d_i, \log b_i)$, where

$$d_i = \beta_i^2 \Sigma_{\xi_i}^{k_{i-1}} z^{\xi_i} \tag{35}$$

$$b_i = \beta_i^2 \Sigma_{\zeta_i}^{k_{i-1}} z^{\zeta_i} \tag{36}$$

with $\beta_i = (k_{i-1} + \xi_i)(k_{i-1} + \zeta_i)/k_{i-1}$, $z = [Q_{i-1}/(2k_{i-1})]^2$ and Σ_l^k is defined in the appendix. Define $\mathbf{C}_i' = \Pi_i^{-1}\mathbf{C}_i$, then, it follows from Eq. (34) that

$$\mathbf{X}_L = \Pi_L(\mathbf{X}_0 + \mathbf{Y}_L) \tag{37}$$

where

$$\mathbf{Y}_L = \sum_{i=1}^{L} \mathbf{C}_i'$$

Now choose δk_0 such that the coefficients of Eq. (31) verify $a_j/\gamma^{j/4} \to 0$ when $j \to \infty$ with $1 < \gamma < \sqrt{2}$. Almost all irrational δk_0's (in the sense of Lebesgue measure) have this property.[28] It is easy to show[14] first that $\|\mathbf{C}_i\|/\gamma^i \to 0$ when $i \to \infty$, and then, by taking into account the fact that $\|\mathfrak{L}_i \mathbf{X}\| > \sqrt{2}\,\|\mathbf{X}\|$ for any vector $\mathbf{X}$ with both components of the same sign, that $\mathbf{Y}_L$ has a limit $\mathbf{Y}_\infty$ when $L \to \infty$. When taking M_0 and P_0 small enough, the sum $\mathbf{X}_0 + \mathbf{Y}_\infty$ has both components negative, then $\|\mathbf{X}_L\|/2^{L/2} \to \infty$, that is, M_L and P_L go to zero faster than $\exp(-C2^{L/2})$ where C is a positive constant.

Therefore for almost all KAM tori, if s is small enough, $T_r^L(H)$ converges toward the integrable Hamiltonian $v^2/2$, which implies that $\mathfrak{T}_0$ is preserved. This is totally consistent with the KAM theorem. Notice that the convergence is very fast: exponential of exponential!

4. LINK BETWEEN TORI AND NEARBY CYCLES

4.1. The Fractal Diagram

Suppose now that δk_0 is a rational. Then the expansion in Eq. (31) is finite; let μ be the maximum value of n. After a finite number L of iterations of T_r on the rotation number $Q_0 = k_0 + n_0 + \delta k_0$, it results from Table 1 that $Q_L = k_L + n_L$ and $\mu/2 \leqslant L \leqslant \mu$. According to Section 2.2, the threshold of first destabilization of the cycle with rotation number Q_0 is given by $F(M_L, P_L, k_L, n_L) = 0$. Combined with T_r^L this corresponds to a function $G_{Q_0}(M_0, P_0, k_0) = 0$. The G is continuous on M_0, P_0, and k_0 but strongly discontinuous on Q_0. Consider the graph $\mathfrak{F}$ of $s_{k_0}(v)$, which can be computed from G with $v = k_0/Q_0$ and

$\rho = (M_0/P_0)^{1/2}$. It is natural to compute this function step by step: first take the points already found in Section 2.2. Those corresponding to $v = 0$ and 1 are said to belong to the zeroth sheet of $\mathfrak{F}$, and the others to the first sheet of $\mathfrak{F}$. Then compute the points that correspond to $N - 1$ iterations of T_r; they define the Nth sheet of $\mathfrak{F}$. Figure 4 shows a part of the three first sheets in the case $\rho = 1$, $k = \frac{1}{2}$. As expected from Section 2.2, it appears that between two points of the Nth sheet, the curve has the same shape as the first sheet between the two points of sheet 0 (Fig. 2). This scale invariance of $\mathfrak{F}$ allows us to call it a fractal.[29]

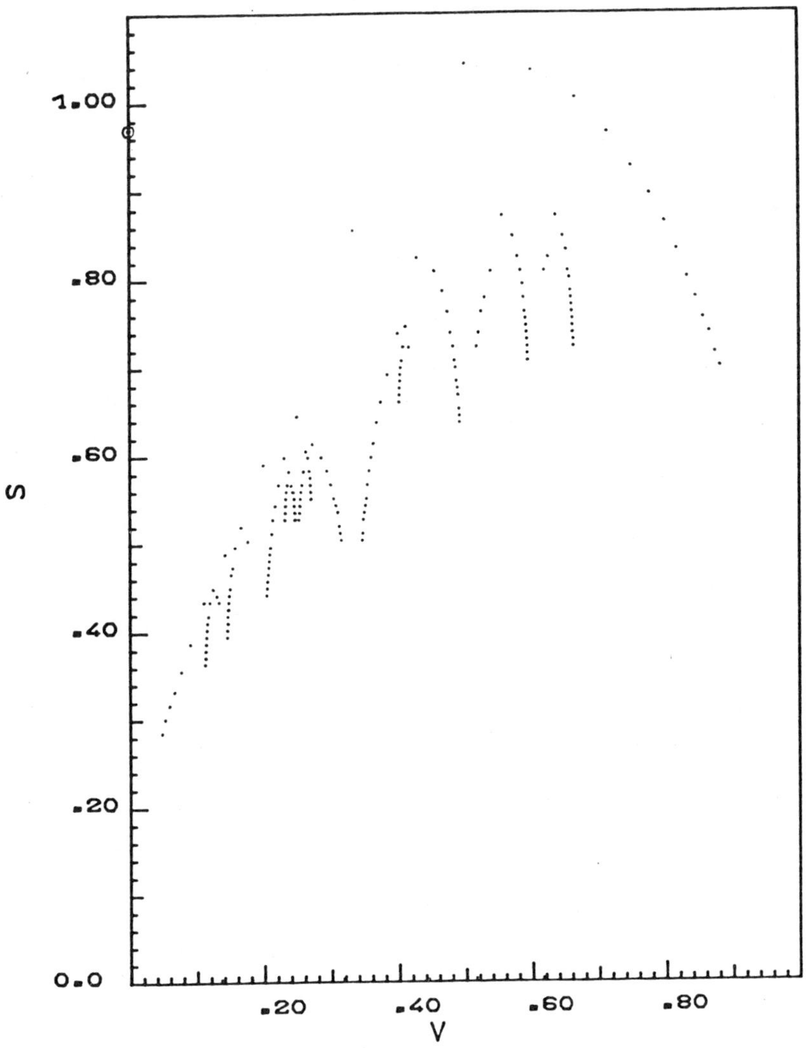

Figure 4. Fractal diagram for $k = \frac{1}{2}$ and $\rho = 1$: $s(1) = 2.9$ is out of scale; $s(0) = 0.97$ is indicated by a circle.

This structure has been conjectured and numerically computed in the case of the standard mapping by G. Schmidt.[19,21] His method involves no approximation, but knowledge of the trajectories is necessary for computing the thresholds. Such knowledge is not needed here, but the threshold is approximate. The errors for the lower and higher order sheets of $\mathcal{F}$ are due mainly to the approximate character of $F(M, P, k, l)$ and T_r, respectively. For computing the central part of the arches of sheets of order higher than 1, k_L/ρ_L^2 is not too close to 1, and the second method of Section 2.2 can be used for obtaining F. Figure 4 was computed this way. Notice that similar graphs can be obtained for the resonances lying inside of the trapping domain of resonances M and P.

4.2. Truncation of Irrational Rotation Numbers

Let δk_0 be an irrational and $\delta k_0^{(\mu)}$ its μth convergent,[28] $\delta k_0^{(\mu)} = [a_1, a_2, \ldots, a_\mu]$. If, according to Table 1, $a_\mu = 1$ combines for the L + first iteration of T_r on δk_0 with $a_{\mu+1}$, delete a_μ and replace $a_{\mu-1}$ by $a'_{\mu-1} = a_{\mu-1} + 1$ (if $\mu = 1$, replace n_0 by $n_0 + 1$). In the following the quantities with index L and no index μ correspond to the iteration of T_r on δk_0.

Consider the stable cycle C_μ with rotation number

$$Q_0^{(\mu)} = k_0 + n_0 + \delta k_0^{(\mu)} \tag{38}$$

After L iterations of T_r, $\mu/2 \leqslant L \leqslant \mu$ then $Q_L^{(\mu)} = k_L + l$, where l is an integer. Table 1 shows that $l = n_L$ if $a_\mu > 1$; if $a_\mu = 1$ either $l = n_L = 1$ or $l = n_L + 1 \geqslant 2$. Therefore the sequence $(n'_i, \lambda'_i)i = 1, \ldots, L$ obtained for $Q_0^{(\mu)}$ is the same as for Q_0 except perhaps for $i = L$.

First let $\delta k_0 = \delta k_n^0$, then $a_j = n + 1$ for every j, $L = \mu$, and $l = n_\mu = n$. For a fixed value of $\rho = (M_0/P_0)^{1/2}$, let s_μ be the threshold of first destabilization of C_μ. For studying the evolution of s_μ for $\mu \to \infty$, a graphical description is very suitable. As k_μ goes to k_n^0 in that limit, assume for simplicity that $k_0 = k_n^0$ for giving a two-dimensional representation of $\mathfrak{M}_n^0$. Figure 5 shows the plane (M, P), the unstable manifold U of F_n^0, the trace S of $\mathcal{T}_n^0$, the curve N defined by $F(M, P, k_n^0, n) = 0$, the line Δ given by $M = \rho^2 P$, and the circle C inside of which $\mathfrak{M}_n^0$ may be approximated by its tangent mapping at F_n^0. Then s_μ corresponds to a point A_μ of Δ such that $T_r^\mu(A_\mu)$ with coordinates $(M_\mu^{(\mu)}, P_\mu^{(\mu)})$ belongs to N. This prevents A_μ from remaining at finite distance from S when $\mu \to \infty$. Therefore A_∞ is on S, and s_∞ is s_n^0, the threshold of destabilization of $\mathcal{T}_n^0$. For large values of μ only a small proportion of the $T_r^i(A_\mu)$'s $1 \leqslant i \leqslant \mu$ is outside C. Thus the tangent mapping of $\mathfrak{M}_n^0$ at F_n^0 rules the evolution of s_μ, $\mu \to \infty$, and it is easy to show[15] that

$$\frac{s_{\mu+1} - s_\mu}{s_\mu - s_{\mu-1}} \to \frac{1}{\delta_n^0} \qquad \mu \to \infty \tag{39}$$

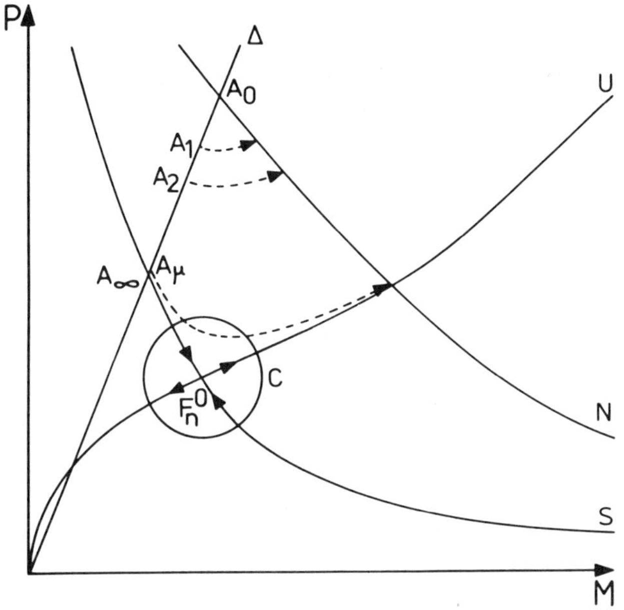

Figure 5. Graphical proof of the importance of F_n^{λ} for describing s_{μ} when $\mu \to \infty$.

The reasoning above is the same as the one that allows the definition of Feigenbaum's δ in Ref. 30.

Now let $\delta k_0 = \delta k_n^1$. Then $a_{2j} = n$ and $a_{2j+1} = 1$ for every j. If $\mu = 2\nu$, then $L = \nu$ and $l = n_{\nu} = n$. If $\mu = 2\nu + 1$, then $L = \nu$ and $l = n_{\nu+1} = n + 1$ (if $\nu = 0$, replace $n_0 = n$ by $n + 1$). A figure similar to Fig. 5 can be drawn in that case. Now a new curve N' defined by $F(M, P, k_n^1, n + 1) = 0$ appears. The same argument as before shows[15] that $s_{\mu} \to s_n^1$ for $\mu \to \infty$ and

$$\frac{s_{\mu+2} - s_{\mu}}{s_{\mu} - s_{\mu-2}} \to \frac{1}{\delta_n^1} \qquad \mu \to \infty \tag{40}$$

The two series $s_{2\nu}$ and $s_{2\nu+1}$ have the same limit and the same universal properties. Numerical evidence for this point has been provided[21] in the case of the standard mapping for the golden mean ($n = 1$). This calculation yields[31] $\delta_1^1 \simeq 2.62$, whereas the present theory gives $\delta_1^1 \simeq 2.63$.

If δk_0 is equivalent[28] to δk_n^{λ} (its a_j's are those of δk_n^{λ}, for j large enough), the same properties as for δk_n^{λ} hold. Nevertheless for any irrational δk_0 no limit [Eqs. (39) and (40)] exists, but s_{μ} still converges[15] toward the threshold related to $Q_0 = k_0 + n_0 + \delta k_0$.

These results prove the basic conjecture of Ref. 7 that a torus is preserved if and only if nearby cycles [the C_{μ}'s defined by Eq. (38)] are stable. Numerical calculations show that Fig. 5 corresponds to the general respective positions of curves S and C. Therefore s_{μ} converges generally towards s_{∞} by upper values.

For computing the position of stochastic domains Schmidt has proposed[19,21] the approximation of $\mathcal{F}$ by its N first sheets plus the continuous curve $\mathcal{F}_N$ built by linear interpolation on the Nth sheet. For a given value $s = s_0$, the points of $\mathcal{F}_N$ above $s = s_0$ correspond to preserved KAM tori. The maximum value s'_N of $\mathcal{F}_N$ corresponds to the threshold to large-scale stochasticity. The results of this section substantiate this method. The mismatch of s'_N with the actual threshold decreases[15] faster than $1/(2.5)^{N-1}$ because all values of δ_n^λ are larger than 2.5.

4.3. Calculation of the Mean Residue

Let us call C_μ the stable or unstable cycle with rotation number $Q_0^{(\mu)} = k_0 + n_0 + \delta k_0^{(\mu)}$. Write $Q_0^{(\mu)} = N_\mu / R_\mu$, where N_μ and R_μ are relatively prime; N_μ is the number of points of C_μ in Fig. 1. Thus the period in the original system is

$$T_0^{(\mu)} = N_\mu \frac{2\pi}{k_0} \tag{41}$$

According to Eq. (8), $t_{i+1} = \gamma'_i t_i$ with $\gamma'_i = (2\lambda_{i+1} - 1)k_i/(k_i + n_i + 1 - \lambda_{i+1})$. Therefore the period of C_μ in the time t_L is

$$T_L^{(\mu)} = \Gamma_L T_0^{(\mu)} \tag{42}$$

with

$$\Gamma_L = \prod_{i=0}^{L-1} |\gamma'_i| = \frac{k_0}{\alpha_L k_0 + \beta_L} \tag{43}$$

where Eqs. (33) have been used. Equations (41)–(43) combined with Eqs. (17) and (49–50) yield the period $\tau_0 = \tau_\mu$, which occurs in Eq. (25):

$$\tau_\mu = \frac{\pi p'_0}{D(p'_0, q_\mu)} \tag{44}$$

where D is the largest common divisor of $p'_0 (k_0 = r'_0/p'_0)$ and q_μ, which is the denominator of $\delta k_0^{(\mu)} = p_\mu / q_\mu$.

For this section of rule of Section 4.2 for $a_\mu = 1$ is slightly changed: if $a_\mu = 1$, always delete a_μ and replace $a_{\mu-1}$ by $a'_{\mu-1} = a_{\mu-1} + 1$. Then either $l = n_L$ but λ_L is replaced by $\lambda'_L = 0$, or $l = n_L + 1$.

First consider the case where s is small. Then (see Section 2.2) $\nu_\mu^2 \simeq a_\mu = a(M_L^{(\mu)}, P_L^{(\mu)}, k_L, l)$ is small, so that the residue of C_μ, $R_\mu \simeq \psi^2/4 \simeq a_\mu \tau_\mu^2/4$. It results from Section 3.3 that for almost all KAM tori the vector $\mathbf{Y}_\infty$ exists. Let (v, w) be its components. It results from Eq. (37) that

$$M_L^{(\mu)} = g_L M_0^{\delta_L} P_0^{\gamma_L} \tag{45}$$

$$P_L^{(\mu)} = h_L M_0^{\beta_L} P_0^{\alpha_L} \tag{46}$$

and that for large L's, g_L is proportional to $\exp(v\delta_L + w\gamma_L)$ and h_L to $\exp(v\beta_L + w\alpha_L)$. Combining Eqs. (45) and (46) with Eqs. (18), (23), (24), and (44) yields

$$R_\mu = B_\mu h_L g_L^l M_0^{\beta_L + l\delta_L} P_0^{\alpha_L + l\gamma_L} \tag{47}$$

with

$$B_\mu = \tau_\mu^2 \frac{(k_L + l)^{4+2l}}{4^l k_L^{2+2l}} \Sigma_l^{k_L} \tag{48}$$

Write $\delta k_0^{(\mu)} = p_\mu / q_\mu$, and let $r_\mu = n_0 q_\mu + p_\mu$. It can be shown[14] by using the properties of convergents[28] that

$$q_\mu = \alpha_L + l\gamma_L \tag{49}$$

$$r_\mu = \beta_L + l\delta_L \tag{50}$$

As a result

$$R_\mu = B_\mu h_L g_L^l M_0^{r_\mu} P_0^{q_\mu} \tag{51}$$

C_μ has an average velocity $v = q_\mu k_0/(q_\mu k_0 + r_\mu)$. Such a velocity corresponds to a resonant term with a phase $(q_\mu k_0 + r_\mu)x - q_\mu k_0 t$, which is obtained in perturbation theory from products of the type $(M_0 \cos x)^{r_\mu}[P_0 \cos k(x-t)]^{q_\mu}$. It is thus natural for R_μ to be proportional to $M_0^{r_\mu} P_0^{q_\mu}$.

With the definition of r_μ and q_μ we can write

$$N_\mu = \frac{r_0' q_\mu + p_0' r_\mu}{D(p_0', q_\mu)} \tag{52}$$

where D is the largest common divisor of p_0' and q_μ. The mean residue[7] of C_μ is $f_\mu = (R_\mu)^{1/N_\mu}$. It comes from the previous results that for M_0 and P_0 of order ε, f_μ has a limit f_∞ if $p_0' = 1$. Then f_∞ is proportional to ε^α where α is a positive number; $\alpha = 1$ if $k_0 = 1$. If p_0' is larger than 1, f_∞ can be defined as the limit of the f_μ's for which $D(p_0', q_\mu) = 1$.

These properties of f_μ are exact in the limit of small values of ε and s; $[\varepsilon = 0(s^2)]$. If s is under s_∞ the threshold of destabilization of the torus with rotation number $Q_0 = k_0 + n_0 + \delta k_0$, a finite number of iterations of T_r is sufficient for applying the preceding results.[14] Thus f_∞ is defined for the same values of δk_0 and is a growing function of s.

For $\delta k_0 = \delta k_n^\lambda$ (or equivalent), and for $s = s_\infty$, it can be shown[14] that $(M_\mu^{(\mu)}, P_\mu^{(\mu)})$ converges toward one point (M_{n0}, P_{n0}) at finite distance from F_n^0 when $\lambda = 0$. For $\lambda = 1$ $(M_\nu^{(2\nu)}, P_\nu^{(2\nu)})$ converges toward (M_{n1}, P_{n1}) and

$(M_\nu^{(2\nu+1)}, P_\nu^{(2\nu+1)})$ toward (M'_{n1}, P'_{n1}). Equations (18), (19), and (21) relate to these points values of $\nu = \nu_{n0\eta}, \nu_{n1\eta}, \nu'_{n1\eta}$, and Eqs. (23) and (44) values of $R = R_{n0\eta}, R_{n1\eta}$, and $R'_{n1\eta}$, where $\eta = \pm 1$ or -1 according to whether the C_μ's are stable or unstable. These are also universal values. As a result, $f_\infty = 1$ for $s = s_\infty$.

Since the approximations leading to T_r are no longer verified when (M, P) goes to infinity, the mean residue cannot be analytically computed for $s > s_\infty$, but must be larger than 1.

4.4. Proof of Greene's Main Assertions on the Standard Mapping

As shown in Section 3.2, the Hamiltonian of Eq. (2) with $k = 1$ and $M = P$ can be used for describing the standard mapping. This is especially good for small values of s because resonances behave as if they were isolated. This fact has been used in Ref. 16 for calculating trajectories of the Hamiltonian in Eq. (1) by superposition of two standard mappings. The results of Section 4.3 (and small additions to be found in Ref. 14) allow us to prove most assertions of Ref. 7. The exceptions are the following:

1. The first part of assertion II (in Ref. 7) does not seem easy to check.
2. In the present renormalization theory, the golden mean torus is "almost" the last to disappear for $\rho = k = 1$. Its destruction occurs at $s \simeq 0.6995$. The torus with rotation number $Q_0 = 2 + \delta k_0$, where $\delta k_0 = [1, 1, 2, 1, \ldots, 1, \ldots]$, disappears at $s \simeq 0.6999$, that is, slightly afterward. This is not the case for the standard mapping.[32] In fact the discrepancy with assertion IV is very small. It is due to the two resonances approximation. Nevertheless in the present theory the candidate to be "almost the last" torus shows up by itself and no numerical calculation is necessary for getting it.
3. For the same reason as before, the prediction for the threshold of the large-scale stochasticity is $s \simeq 0.7$ in the present theory, not $s = 0.627 = 2\sqrt{K_g}/\pi$, where $K_g = 0.9716$ is given in assertion VII.

Assertion V shows that $R_{111} = R'_{111} = \frac{1}{4}$. This implies $\nu_{111} = \nu'_{111} = \frac{1}{3}$. The present theory shows that $Q_0^{(\mu)}$ of Section 4.3 must be taken as $Q_0^{(\mu)} = k_0 + n_0 + \delta k_0^{(\mu)}$, not as the μth convergent of $Q_0 = k_0 + n_0 + \delta k_0$. There is no difference in the case of k_0 an integer, but the discrepancy is very strong for noninteger values of k_0. This could lead to the absence of convergence of f_μ (even in the case of $\delta k_0 = \delta k_n^\lambda$) in a similar way as in Ref. 33.

5. GENERAL CASE

Sections 1–4 deal with the Hamiltonians of Eqs. (1) and (2). This section shows how the previous results can be directly applied to the Hamiltonians of Eqs. (3) and (4). First notice that the evolution of (J_1, θ) given by Eq. (4) is the same as that given by Eq. (3) with $H_0(\mathbf{J}) = h_0(J_1) + J_2$. By applying to Eq. (3) the same type of approximations as in Section 2.1, a Hamiltonian [Eq. (2)] can be defined for each torus or cycle.

Define $\boldsymbol{\omega} = \partial H_0/\partial \mathbf{J}$. In the limit $\varepsilon = 0$, the resonance condition for resonance i is

$$\mathbf{p}_i \cdot \boldsymbol{\omega}(\mathbf{J}) = 0 \tag{53}$$

This defines a value $\mathbf{I}_i$ of $\mathbf{J}$. This value lies on the energy line $H_0(\mathbf{J}) = E$. Therefore the $\mathbf{I}_i$'s are ordered on this line. A given torus or cycle corresponds to a frequency $\boldsymbol{\omega}_0$ that is related in the limit $\varepsilon = 0$ to a value $\mathbf{I}_0$ defined by the condition

$$\boldsymbol{\omega}_0 = \boldsymbol{\omega}(\mathbf{I}_0) \tag{54}$$

Let $\mathbf{q}$ be a vector perpendicular to $\boldsymbol{\omega}_0$, therefore tangent to the energy line at $\mathbf{I}_0$.

The first approximation of Section 2.1 consists here in replacing the $V_i(\mathbf{J})$'s by the constants $V_i(\mathbf{I}_0)$ and $H_0(\mathbf{J})$ by its second-order Taylor expansion in the vicinity of $\mathbf{I}_0$. Now the fact that the energy line is locally a parabola is taken into account. A canonical transformation changes the vectors $\mathbf{J}$ into a new frame directed along $\boldsymbol{\omega}_0$ and $\mathbf{q}$. It is defined by the generating function

$$F(\xi, \eta, \boldsymbol{\theta}) = \boldsymbol{\theta} \cdot (\mathbf{I}_0 + \xi \mathbf{q} + \eta \boldsymbol{\omega}_0) \tag{55}$$

This yields

$$\mathbf{J} = \frac{\partial F}{\partial \boldsymbol{\theta}} = \mathbf{I}_0 + \xi \mathbf{q} + \eta \boldsymbol{\omega}_0 \tag{56}$$

$$\varphi = \frac{\partial F}{\partial \xi} = \mathbf{q} \cdot \boldsymbol{\theta} \tag{57}$$

$$\psi = \frac{\partial F}{\partial \eta} = \boldsymbol{\omega}_0 \cdot \boldsymbol{\theta} \tag{58}$$

Let $(\alpha_i, \beta_i) i = m, p$ be defined by

$$\mathbf{p}_i = \alpha_i \mathbf{q} - \beta_i \boldsymbol{\omega}_0 \tag{60}$$

Then $\mathbf{p}_i \cdot \boldsymbol{\theta} = \alpha_i \varphi - \beta_i \psi$. For not too high values of ε the motion occurs in the vicinity of unperturbed energy line (see, e.g., Fig. 3.1 of Ref. 34). Thus η is of order ξ^2, and the $\xi\eta$ and η^2 terms are neglected. This yields a new Hamiltonian

$$H'(\xi, \eta, \varphi, \psi) = \frac{1}{2} a \xi^2 + \omega_0^2 \eta + \varepsilon \sum_{i \in \mathcal{I}} V_i(\mathbf{I}_0) \cos(\alpha_i \varphi - \beta_i \psi) \tag{61}$$

where $a = \mathbf{q} \cdot (d\boldsymbol{\omega}/d\mathbf{I}) \cdot \mathbf{q}$. The constant term $H_0(\mathbf{I}_0)$ has been deleted because it plays no role in the dynamics. The time evolution of (ξ, φ) can be equivalently described by

$$H''(\xi, \varphi, t) = \frac{1}{2}a\xi^2 + \varepsilon \sum_{i \in \mathcal{J}} V_i(\mathbf{I}_0)\cos(\alpha_i \varphi - \gamma_i t) \tag{62}$$

where $\gamma_i = \omega_0^2 \beta_i$.

The second approximation of Section 2.1 consists here in retaining in $\mathcal{J}$ only the two i's ($i = m, p$) that correspond to the $\mathbf{I}_i$ closest to $\mathbf{I}_0$ on the energy line. Since Eq. (62) has the same structure as Eq. (1), a Hamiltonian of the type in Eq. (2) can be defined for describing the stability of the torus or cycles with frequency ω_0. The parameters are

$$k = \frac{\alpha_p}{\alpha_m} \tag{63}$$

$$M = A_m \tag{64}$$

$$P = A_p \tag{65}$$

with

$$A_i = \frac{\varepsilon |a V_i(\mathbf{I}_0)|}{\Delta v^2} \tag{66}$$

with $\Delta v = \gamma_m/\alpha_m - \gamma_p/\alpha_p$. Frequency ω_0 corresponds to a mean velocity

$$v_0 = \frac{\alpha_p \beta_m}{\alpha_p \beta_m - \alpha_m \beta_p} \tag{67}$$

The results of preceding sections can now be applied to the Hamiltonians of Eqs. (3) and (4). For instance, an estimate of the threshold of the large-scale stochasticity between resonances m and p can be obtained as follows: the tori $\mathcal{T}_n^\lambda$ of H and H_e correspond through Eqs. (53), (60), and (67) to values $\mathbf{I}_{n\lambda}$ and $\mathbf{I}'_{n\lambda}$ of $\mathbf{J}$. The $\mathcal{S}_n^\lambda$'s defined in Section 3.1 allow us to compute the thresholds $\varepsilon_{n\lambda}$ and $\varepsilon'_{n\lambda}$ for these tori. The maximum value of $\varepsilon_{n\lambda}$'s and $\varepsilon'_{n\lambda}$'s yields the threshold of large-scale stochasticity. As for large n's, the $\mathcal{T}_n^\lambda$'s are inside the stochastic layer of resonance M, only a few $\varepsilon_{n\lambda}$'s and $\varepsilon'_{n\lambda}$'s are worth computing. If $V_m(\mathbf{J})$ and $V_p(\mathbf{J})$ do not vary much between $\mathbf{I}_m$ and $\mathbf{I}_p$, Fig. 3 may be directly used for computing the threshold.

6. CONCLUSION

The limited length of this chapter does not allow me to present the results already obtained by a second renormalization method, which is aimed at describing islands in the islands. It uses the Hamiltonian of Eq. (6) for trapped orbits and makes approximations similar to those of Section 2. It estimates Feigenbaum's δ and similar quantities for the bifurcation of periods $3T$, $4T$, and so on. It also proves the following fact, which appears in numerical calculations (see Fig. 4 of Ref. 35 and Fig. 14 of Ref. 36): when a period-T cycle becomes unstable, a figure-eight trajectory shows up in Poincaré maps; it corresponds to the stochastic layer of the unstable cycle, which is separated from the surrounding stochastic sea by KAM tori. These tori disappear before the destabilization of period-$2T$ orbits, which produces two eights, and so on. Therefore, contrary to the dissipative case,[37] chaos "eats" the adiabatic domains from outside and there is no mirror sequence of period doubling bifurcations to chaos.

The present renormalization technique relies mainly on the choice of two principal resonant terms. In fact, what shows up is a definite sequence of canonical transformations for each KAM torus. These transformations consist in completely (not partially, as in Kolmogorov transformation[1]) killing the nearest resonances to the torus. They can be done on Hamiltonians of the type in Eq. (4) without discarding any resonant term. For practical purposes the first approximation of Section 2.2 is nevertheless useful, but the final estimates should be better. This method is presently in progress and is expected, for instance, to improve the analytical estimate of the threshold for the standard mapping. It should also shed some light on the results of Ref. 9 and could contribute to a rigorous mathematical result in the spirit of the KAM theorem.

Though approximate, the present theory already gives good estimates of universal quantities, of various thresholds, and yields the right variation of the thresholds with the parameters k and M/P. Furthermore it is in the spirit of the KAM theory and proves many conjectures that resulted from numerical calculations.[41] It also contributed to giving a global description of the phase space and to explaining a number of plots of Poincaré maps.

The analogy of the present renormalization procedure with that of Wilson[38] reflects the analogy between transition to chaos and phase transition. It can thus be expected that there exist critical exponents for the correlation times and the Kolmogorov entropy. Figure 5.3 of Ref. 1 and the $\frac{1}{2}$ exponent of Section 2.2 substantiate this feeling.

Much work remains before insight into Hamiltonian systems is achieved that will be sufficient for physical applications: for instance, the calculation of diffusion coefficients and correlation times, the study of the effect of dissipation (when do strange attractors replace stable cycles?), and the study of the case of many degrees of freedom.

ACKNOWLEDGEMENTS

I thank the participants in the "Séminaire sur les phénomènes stochastiques," and especially P. Collet, J. Lascoux, F. Ledrappier, G. Schmidt, and

B. Souillard, for fruitful discussions. I am indebted to J. Greene for running again his program on the standard mapping and to G. Gallavotti for his suggestions for making the method more rigorous.

APPENDIX: TRAPPED AND UNTRAPPED ORBITS

Trapped Orbits

Let $m = \frac{1}{2}[H_0(v, x)/M + 1]$. Then

$$I = \frac{8}{\pi}\sqrt{M}\left[E(m) - m_1 K(m)\right] \tag{A1}$$

$$\Omega = \frac{dH_0'}{dI} = \sqrt{M}\,\frac{\pi}{2K(m)} \tag{A2}$$

$$\sigma = \frac{d\Omega}{dI} = -\frac{\pi^2}{16mK^3(m)}\left[\frac{E(m)}{m_1} - K(m)\right] \tag{A3}$$

where $m_1 = 1 - m$ and K and E are the elliptic integrals of first and second kind.[39] The x is related to θ through

$$x = 2\arcsin\left\{\sqrt{m}\, sn\left[(2K(m)\theta/\pi), m\right]\right\} \tag{A4}$$

where sn is the Jacobian elliptic function.[39]

For $2k$ integer the V_n's can be computed exactly[40]:

$$V_n = \left[\frac{\pi}{2K(m)}\right]^{2k} \frac{q^{n/2}}{1-(-1)^{2k+n}q^n}\Sigma_n^k \tag{A5}$$

where

$$q = \exp\left[-\frac{\pi K(m_1)}{K(m)}\right] \tag{A6}$$

and

$$\Sigma_n^{1/2} = 2$$

$$\Sigma_n^1 = 4n$$

$$\Sigma_n^{3/2} = (2n)^2 + 2\left[\frac{2K(m)}{\pi}\right]^2$$

$$\Sigma_n^2 = \frac{8n^3}{3} + \frac{16n[2K(m)/\pi]^2}{3}$$

For small values of q the method given in Ref. 6 yields for any k

$$V_n = \left[\frac{\pi}{2K(m)}\right]^{2k} q^{n/2}\Sigma_n^k \tag{A7}$$

with

$$\Sigma_n^k = \sum_{l=1}^{n} 2^l C_{2k}^l C_{n-1}^{l-1} \tag{A8}$$

where

$$C_z^l = \frac{z(z-1)\cdots(z-l+1)}{l!} \tag{A9}$$

for z real and l integer. For large $n/K(m)$, Ref. 16 yields Eq. (A7) with

$$\Sigma_n^k = \frac{2(2n)^{2k-1}}{\Gamma(2k)} \tag{A10}$$

Untrapped Orbits

Let $m = 2/[H_0(v, x)/M + 1]$, then

$$I = \frac{4}{\pi}\left(\frac{M}{m}\right)^{1/2} E(m) \tag{A11}$$

$$\Omega = \pi\left(\frac{M}{m}\right)^{1/2} \frac{1}{K(m)} \tag{A12}$$

$$\sigma = \frac{\pi^2 E(m)}{4m_1 K(m)^3} \tag{A13}$$

$$x = 2\,\mathrm{am}\left[\pm\frac{K(m)\theta}{\pi}, m\right] \tag{A14}$$

where am is the amplitude of elliptic functions.[39]

For $2k$ integer the V_n's can be computed exactly in a way similar to that used for trapped orbits

$$V_n = \left[\frac{2\pi}{K(m)}\right]^{2k} \left(\frac{q}{m}\right)^k \frac{q^n}{1-q^{4(k+n)}}\Sigma_n^k \tag{A15}$$

where q is given by Eq. (A6), $\Sigma_n^{1/2} = 1$, $\Sigma_n^1 = 1 + n$,

$$\Sigma_n^{3/2} = \frac{(3/2+n)^2}{2} - \frac{(2-m)[2K(m)/\pi]^2}{16}$$

and

$$\Sigma_n^2 = \frac{(2+n)^3}{6} - \frac{(2-m)(2+n)[2K(m)/\pi]^2}{12}$$

For small values of q

$$V_n = \left[\frac{2\pi}{K(m)}\right]^{2k} \left(\frac{q}{m}\right)^k \Sigma_n^k q^n \tag{A16}$$

with

$$\Sigma_n^k = \sum_{l=1}^{n} C_{2k}^l C_{n-1}^{l-1} \tag{A17}$$

where C_z^l is given by Eq. (A9). For large $n/K(m)$ Ref. 16 yields Eq. (A16) with

$$\Sigma_n^k = \frac{(k+n)^{2k-1}}{\Gamma(2k)} \tag{A18}$$

Notice that the definition of (I, θ) for both trapped and untrapped orbits implies that $I = v$ and $\theta = x$ for $M = 0$.

REFERENCES

1. B. V. Chirikov, *Phys. Rep.* **52**, 263 (1979).
2. F. Doveil, *Phys. Rev. Lett.* **46**, 532 (1981).
3. V. Croquette and C. Poitou, *C.R. Acad. Sci. Ser. B* **292**, 1353 (1981); *J. Phys.* (Paris) **42**, L537 (1981).
4. A. J. Lichtenberg, in *Intrinsic Stochasticity in Plasmas*, G. Laval and D. Grésillon, Eds. (Editions de Physique Courtaboeuf, Orsay, France, 1979) pp. 13–40.
5. D. F. Escande, in *Intrinsic Stochasticity in Plasmas*, G. Laval and D. D. Grésillon, Eds. (Editions de Physique Courtaboeuf, Orsay, France, 1979), pp. 41–51.
6. D. F. Escande and F. Doveil, *J. Stat. Phys.* **26**, 257 (1981).
7. J. M. Greene, *J. Math. Phys.* **20**, 1183 (1979).
8. J. M. Greene, in *Nonlinear Dynamics and the Beam–Beam Interaction*, M. Month and J. C. Herrera Eds., American Institute of Physics Conference Proceedings Series, Vol. 57 (AIP, New York, 1979), pp. 257–271.
9. J. M. Greene and I. C. Percival, *Physica* **3D**, 530 (1981).
10. V. I. Arnol'd and A. Avez, *Ergodic Problems of Classical Mechanics* (Benjamin, New York, 1968).
11. D. F. Escande and F. Doveil, Laboratoire de Physique des Milieux Ionisés, Palaiseau, France, unpublished report PMI 1011 (1980).
12. D. F. Escande and F. Doveil, *Phys. Lett. A* **83**, 307 (1981).

13. F. Doveil and D. F. Escande, *Phys. Lett. A* **84**, 399 (1981).
14. D. F. Escande and A. Mehr (to be published).
15. F. Doveil and D. F. Escande, to appear in *Phys. Lett. A*.
16. A. B. Rechester and T. H. Stix, *Phys. Rev. A* **19**, 1656 (1979).
17. G. Blanch, in *Handbook of Mathematical Functions*, M. Abramowitz and I. A. Stegun, Eds. (Dover, New York, 1972), pp. 721–750.
18. R. C. Churchill, G. Pecelli, and D. L. Rod, in *Stochastic Behaviour in Classical and Quantum Hamiltonian Systems*, G. Casati and J. Ford, Eds. (Springer-Verlag, Berlin, 1979), pp. 76–136.
19. G. Schmidt, *Phys. Rev. A* **22**, 2849 (1980).
20. A. H. Nayfeh, *Perturbation Methods* (Wiley, New York, 1973), p. 62.
21. G. Schmidt and J. Bialek, to be published.
22. G. M. Zaslavskii and N. N. Filonenko, *Zh. Eksp. Teor. Fiz.* **54**, 1590 (1968) [*Sov. Phys.: J. Exp. Theor. Phys.* **25**, 851 (1968)].
23. A. B. Rechester and T. H. Stix, *Phys. Rev. Lett.* **36**, 587 (1976).
24. G. R. Smith, *Phys. Rev. Lett.* **38**, 970 (1977).
25. A. Fukuyama, H. Momota, R. Itatani, and T. Takizuka, *Phys. Rev. Lett.* **38**, 701 (1977).
26. A. Fukuyama, in *Intrinsic Stochasticity in Plasmas*, G. Laval and D. Grésillon, Eds. (Editions de Physique Courtaboeuf, Orsay, France, 1979, pp. 207–219.
27. M. J. Feigenbaum, *J. Stat. Phys.* **19**, 25 (1978); **21**, 669 (1979).
28. G. H. Hardy and E. M. Wright, *An Introduction to the Theory of Numbers* (Clarendon Press, Oxford, 1960), pp. 129–168.
29. B. B. Mandelbrot, *Fractals, Form, Chance and Dimensions* (Freeman, San Francisco, 1977).
30. P. Collet, J.-P. Eckmann, and O. E. Lanford III, *Commun. Math. Phys.* **76**, 211 (1980).
31. G. Schmidt and J. Bialek (to be published).
32. J. M. Greene, private communication.
33. J. M. Greene, *J. Math. Phys.*, **9**, 760 (1968).
34. C. R. Menyuk, Ph.D thesis, University of California, Los Angeles (1981).
35. A. J. Lichtenberg, M. A. Lieberman, and R. H. Cohen, *Physica D* **1**, 291 (1980).
36. R. H. G. Helleman, in *Fundamental Problems in Statistical Mechanics*, Vol. 5, E. G. D. Cohen, Ed. (North-Holland, Amsterdam, 1980), pp. 165–234.
37. J. Crutchfield, D. Farmer, N. Packard, R. Shaw, G. Jones, and R. J. Donnelly, *Phys. Lett. A* **76**, 1 (1980).
38. K. Wilson, *Rev. Mod. Phys.* **47**, 773 (1975).
39. L. M. Milne-Thomson, in *Handbook of Mathematical Functions*, M. Abramowitz and I. A. Stegun, Eds. (Dover, New York, 1972), pp. 567–626.
40. G. R. Smith and N. R. Pereira, *Phys. Fluids* **21**, 2253 (1978).
41. Two recent papers [S. J. Shenker and L. P. Kadanoff (1981) and L. P. Kadanoff, *Phys. Rev.* **23**, 1641 (1981)] present assertions on conservative maps, which can be proved from Sections 2.1, 3.1, and 4.3 of this chapter.

CHAOTIC MOTION ALONG RESONANCE LAYERS IN NEAR-INTEGRABLE HAMILTONIAN SYSTEMS WITH THREE OR MORE DEGREES OF FREEDOM

M. A. LIEBERMAN
Department of Electrical Engineering and Computer Sciences
University of California
Berkeley, California

JEFFREY L. TENNYSON
Electronics Research Laboratory
University of California
Berkeley, California

Abstract. *Chaotic motion along resonance layers in phase space appears generically in near-integrable Hamiltonian systems with three or more degrees of freedom. Such motion is forbidden in systems with two degrees of freedom, where only chaotic motion across resonance layers is generic. We review three mechanisms for chaotic motion along resonance layers: Arnol'd diffusion, modulational diffusion, and resonance streaming. The emphasis is on the geometry of the motion in the phase and action spaces, simple physical pictures of the mechanisms, and computational examples.*

Research sponsored by the Office of Naval Research contract N00014-79-C-0674, the National Science Foundation grant ENG-78-26372, and U.S. Department of Energy contract DE-AS03-76F00034-PA# DE-ATOE-76ET53059.

1. INTRODUCTION

It is well known that Hamiltonian systems with one degree of freedom $H(p, q)$ are integrable. For two degrees of freedom $H(p_1, p_2, q_1, q_2)$ integrability is exceptional. In general, resonances between the two degrees of freedom lead to the formation of a dense set of resonance layers in the action space. Within each layer, a chaotic motion appears. Energy conservation prevents large excursions of the motion along the layer. Only motion across the layer is important. For an integrable system with a weak perturbation[2] the chaotic layers are isolated by Kolmogorov–Arnol'd–Moser (KAM) surfaces. Thus motion from one layer to another is forbidden. For strong perturbations, resonance layers can overlap, the intervening KAM surfaces being destroyed. A globally chaotic motion then develops, leading to large excursions in both actions over long times.

For three or more degrees of freedom, strong perturbations also lead to overlap of resonance layers and globally chaotic motion. However, for weak perturbations, two new effects appear:

1. Resonance layers are no longer isolated by KAM surfaces. Generically, the layers intersect, forming a connected web dense in the action space.
2. Conservation of energy no longer prevents large chaotic motions of the actions along the layers over long times. As a result, large, long-time excursions of the actions along resonance layers are generic in systems with three or more degrees of freedom. In contrast, such motions are forbidden in systems with two degrees of freedom. Furthermore, the interconnection of the dense set of layers ensures that the chaotic motion, stepping from layer to layer, can carry the system arbitrarily close to any region of the phase space consistent with energy conservation.

The interconnection of layers and motion along them lies at the heart of the chaotic phenomena reviewed here. We describe using simple physical pictures and computational examples three different mechanisms for motion along resonance layers: Arnol'd diffusion, modulational diffusion, and resonance streaming. These mechanisms have no analogue in systems with one or two degrees of freedom. A quantitative description of the theoretical methods and additional numerical results are found in the monograph by Lichtenberg and Lieberman.[20]

This chapter is in four parts. It is first desirable to consider the geometry of the $2N$-dimensional phase space, the $(2N - 2)$-dimensional surface of section, and various projections, such as the N-dimensional action space. The emphasis is on simple diagrams that illustrate the definition of the resonance layers and give meaning to the notion of motion "across" and "along" layers. Using these diagrams, the different layers that give rise to Arnol'd diffusion, modulational diffusion, and resonance streaming are described qualitatively.

After this, we review in some detail the mechanism of *Arnol'd diffusion*, which gives rise to chaotic motion along resonance layers. Arnol'd diffusion is universal, in that there is no critical perturbation strength. Furthermore, the chaotic motion is intrinsic, that is, generated by the dynamics alone. The diffusion appears as chaotic motion along thin layers of stochasticity surrounding the separatrices associated with nonoverlapping resonances. A model problem, that of a ball bouncing between a flat wall and a periodically rippled wall, is used to illustrate the diffusion. The case of many noninteracting resonances (Nekhoroshev regime) is described qualitatively using a second model problem.

Whereas Arnol'd diffusion appears as chaotic motion along nonoverlapping resonance layers, *modulational diffusion* results when the layers associated with adjacent resonances overlap. The resulting diffusion along the overlapping resonance layer is generally much greater than for Arnol'd diffusion. Again, the chaotic motion is intrinsic to the dynamics; no external noise acts on the system. However, modulational diffusion is not universal. There is a critical perturbation strength below which adjacent resonances do not overlap. Modulational diffusion is generally found in systems that have a slow oscillation in one of the degrees of freedom. We describe here the formation of modulational stochastic layers and present recent numerical results illustrating the chaotic motion along such layers.

The chapter concludes with an examination of the effect of weak, external stochasticity in producing enhanced diffusion along a resonance layer. If the motion in the absence of external noise is oscillation within a resonance layer, the classical transport due to external noise or dissipation can be strongly enhanced along the layer. This effect is called *resonance streaming*, and is illustrated for a model Hamiltonian perturbed by dissipation and by noise.

2. ACTION AND PHASE SPACE

We consider first an integrable Hamiltonian system with N degrees of freedom. In action-angle form

$$H_0 = H_0(\mathbf{I})$$

where $\mathbf{I}$ is the N-tuple of actions. The motion in the $2N$-dimensional phase space $(\mathbf{I}, \boldsymbol{\theta})$ is on an N-dimensional torus defined by the N-tuple of angles $\boldsymbol{\theta}$ conjugate to $\mathbf{I}$:

$$\mathbf{I}(t) = \mathbf{I}_0, \qquad \boldsymbol{\theta}(t) = \boldsymbol{\omega}(\mathbf{I})t + \boldsymbol{\theta}_0 \tag{1}$$

where

$$\omega_j(\mathbf{I}) = \frac{\partial H_0}{\partial I_j} \tag{2}$$

is the N-vector of unperturbed frequencies.

2.1. Action Space

Figure 1 shows the N-dimensional action space. For the unperturbed system the actions are conserved and each trajectory is a stationary point. One may define an $(N-1)$-dimensional energy surface by the condition

$$H_0(\mathbf{I}) = \alpha$$

For example, for free particle motion in N dimensions,

$$H_0 = \sum_{j=1}^{N} I_j^2$$

the energy surfaces are spheres, as shown in Fig. 1.

One may also define an $(N-1)$-dimensional resonance surface by the condition

$$\mathbf{m} \cdot \omega(\mathbf{I}) = 0 \tag{3}$$

where **m** is called the resonance vector and has integer components. Since there is a resonance surface for each resonance vector, these surfaces are dense in the action space. For the free particle, several resonance surfaces are shown as the flat planes in Fig. 1.

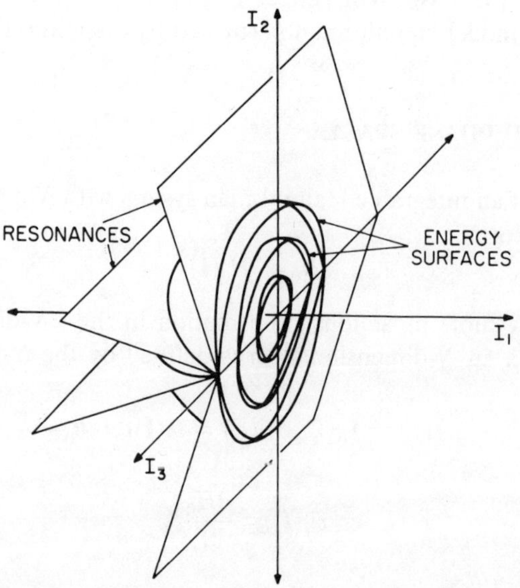

Figure 1. The action space, showing energy surfaces and resonance surfaces for the unperturbed, free particle Hamiltonian.

We consider now the effect of a small perturbation, periodic in $\boldsymbol{\theta}$:

$$H = H_0(\mathbf{I}) + \varepsilon \sum_k V_k(\mathbf{I}) e^{i\mathbf{m}_k \cdot \boldsymbol{\theta}} \tag{4}$$

where k represents the sum over all resonance vectors $\mathbf{m}_k$. The motion in action space is

$$\dot{\mathbf{I}} = -\frac{\partial H}{\partial \boldsymbol{\theta}} = -i\varepsilon \sum_k \mathbf{m}_k V_k e^{i\mathbf{m}_k \cdot \boldsymbol{\theta}} \tag{5}$$

and we see that each component k drives an oscillation in $\mathbf{I}$ in the direction $\mathbf{m}_k$. For most k's the oscillation is nonresonant

$$\mathbf{m}_k \cdot \boldsymbol{\theta}(t) \neq \text{const}$$

and the amplitude of the oscillation in $\mathbf{I}$ is of order ε, as shown in Fig. 2*a*. However, for some value $k = R$ we may find a resonant motion

$$\mathbf{m}_R \cdot \boldsymbol{\theta}(t) = \psi_R = \text{const} \tag{6}$$

where ψ_R is the resonance phase. In the direction of $\mathbf{m}_R$, which we define to be the direction of the resonance action I_R, the amplitude of the oscillation is of order $\varepsilon^{1/2}$. We then have the picture shown in Fig. 2*b*. The direction of $\mathbf{m}_R$ in the action space describes the motion "across" the resonance layer.

As an example, Fig. 3 shows some resonance surfaces and energy surfaces for the two-degree-of-freedom Hamiltonian

$$H_0 = I_1^2 + (6I_2)^2$$

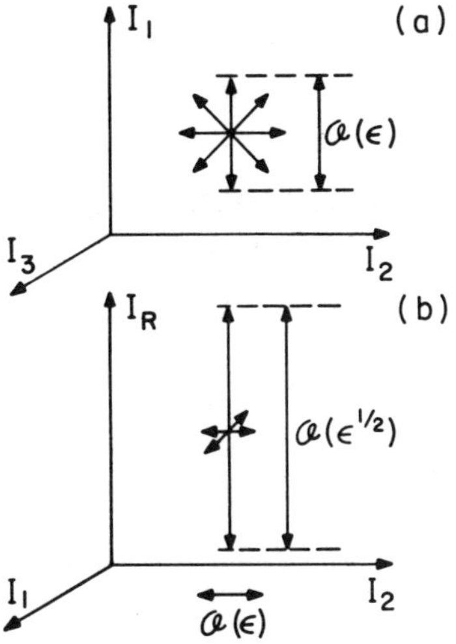

Figure 2. The effect of nonresonant and resonant perturbations. (*a*) Nonresonant perturbations $\mathbf{m}_k$ drive oscillations of order ε. (*b*) A resonant perturbation $\mathbf{m}_R$ drives an oscillation along $\mathbf{m}_R$ of order $\varepsilon^{1/2}$.

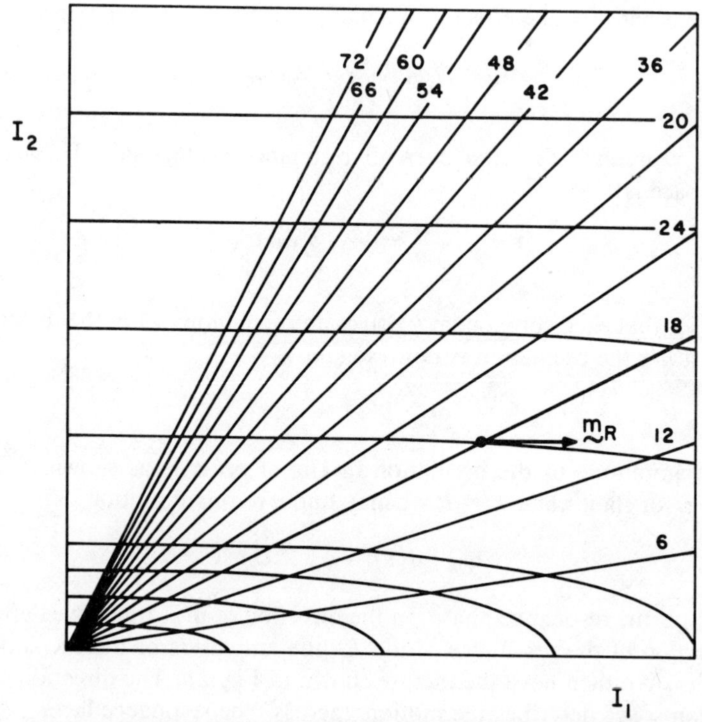

Figure 3. Resonance curves (lines) and energy contours (ellipses) in two-dimensional action space. The Hamiltonian function for this example is $H_0(\mathbf{I}) = I_1^2 + (6I_2)^2$. The resonance labels are the values of m_1 where $\omega_1 m_1 + \omega_2 = 0$.

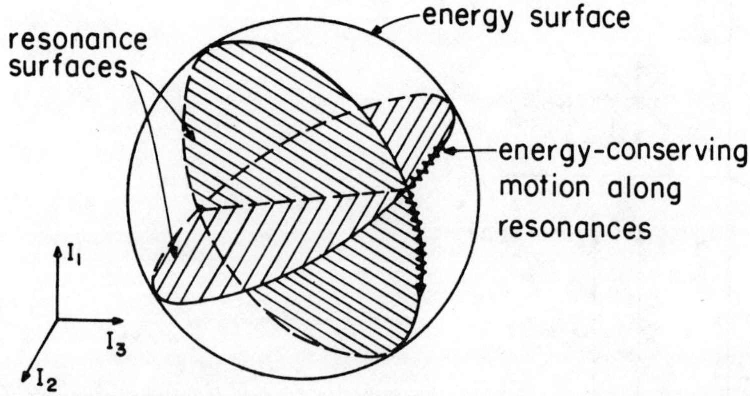

Figure 4. Intersection of two resonance surfaces in an action space having three degrees of freedom. An energy conserving motion (wiggly line) from one resonance surface to another is possible.

The resonances surfaces, from Eq. (3), are lines in the action space given by

$$m_1 I_1 + 36 m_2 I_2 = 0$$

Some of these (for $m_2 = 1$) are plotted in Fig. 3. Note that since

$$\mathbf{m}_R \cdot \frac{\partial H_0}{\partial \mathbf{I}} = 0$$

at resonance, a resonance vector $\mathbf{m}_R$ lies in an energy surface, as shown. In general $\mathbf{m}_R$, the direction of the resonance action excursion, is not perpendicular to the resonance surface. It can be seen from Fig. 3 that even for arbitrary m_1 and m_2, resonance surfaces do not intersect on a constant (nonzero) energy surface. This property is generic for systems with two degrees of freedom.

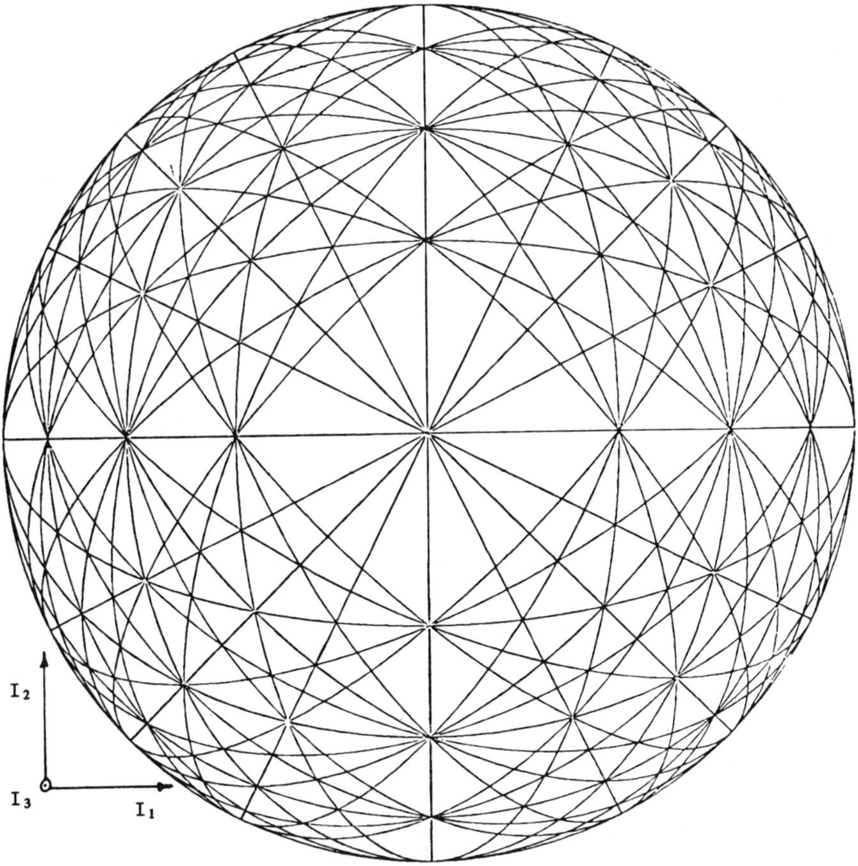

Figure 5. The Arnol'd web for the free particle Hamiltonian. Only some of the intersecting resonances are shown.

For three or more degrees of freedom, resonance surfaces generically intersect, as illustrated in Fig. 4 for a free particle in three dimensions:

$$H_0 = \tfrac{1}{2}\left(I_1^2 + I_2^2 + I_3^2\right)$$

The resonance surfaces are planes that pass through the origin in action space. The two planes intersect at nonzero actions along a line, as shown. The resonance surfaces also intersect a spherical energy surface $H_0(\mathbf{I}) = \alpha$ in great circle meridians (for this example, the resonance vectors $\mathbf{m}_R$ happen to lie perpendicular to the resonance planes). An energy conserving motion from one resonance to another is possible. The motion may proceed along a meridian of one resonance to an intersection, turn sharply, and move along a new meridian. This type of motion is generic to systems with three or more degrees of freedom. The intersection of resonances in the constant energy surface generates a dense interconnected network, the so-called Arnol'd web. The web for this example is illustrated in Fig. 5, with all resonances shown for which $|m_j| \leq 2$.

2.2. Phase Space

Now we consider the geometry of stochastic layers in the $2N$-dimensional phase space. The layers, defined by Eq. (3), are surfaces having dimension $2N - 1$. The KAM surfaces, being perturbed tori defined by the condition

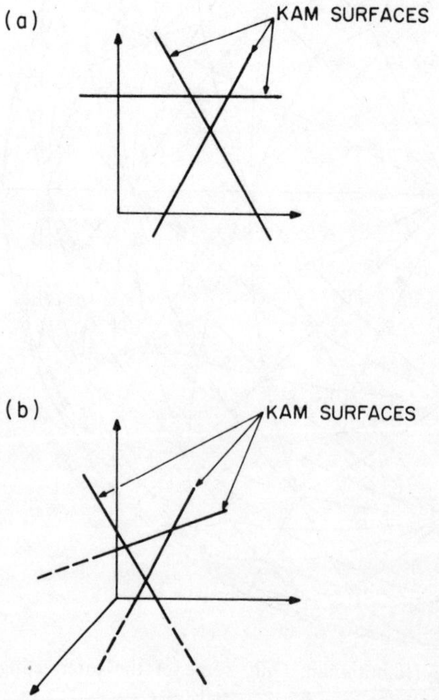

Figure 6. Isolation of regions by KAM surfaces (lines): (*a*) plane divided by lines into a set of closed areas; (*b*) volume not divided by lines into a set of closed volumes.

$\mathbf{I} = \text{const}$, are N-dimensional. The interconnection of resonance layers into the Arnol'd web can then be understood geometrically. For $N \geqslant 3$, the $(2N-1)$-dimensional resonance surfaces cannot be isolated from one another by N-dimensional KAM surfaces. The situation is analogous to that illustrated in Fig. 6, where KAM "lines" isolate regions of a plane, but do not separate a three-dimensional volume into distinct parts.

Now consider a projection of a resonance layer in phase space onto the two-dimensional surface defined by the resonance action I_R and resonance angle ψ_R. Fixing all other actions and angles, we obtain the usual picture (Fig. 7), showing the structure of a stochastic layer in cross section. The stochasticity forms around the separatrix associated with the resonance $\mathbf{m}_R$. The layer thickness is of order $\varepsilon^{1/2}$. Near resonance, the topology of KAM surfaces has changed. Close to resonance, the surfaces perturb to the ellipses shown in the two-dimensional projection in Fig. 7.

If we look at a resonance layer in a three-dimensional projection, adding an additional action variable I_S, we obtain the structure shown in Fig. 8. The resonance layer extends along the unprojected action I_S (although its properties, e.g., thickness, may vary with I_S). The KAM surfaces near exact resonance appear as elliptical tubes within the layer. In Fig. 8, I_S represents one of the $N-1$ action variables (excluding the resonance action I_R) that define motion *along* a layer.

The essential feature of the behavior within resonance layers for three or more degrees of freedom is the existence of long-time chaotic motion along resonance layers (i.e., along I_S). It is easy to see that such motion cannot be driven by the dynamics in near-integrable systems with two degrees of freedom. The change in the Hamiltonian is

$$\Delta H = \Delta H_0 + \varepsilon \Delta H_1 = 0$$

since energy is conserved. Thus

$$\Delta H_0 = \frac{\partial H_0}{\partial I_R} \Delta I_R + \frac{\partial H_0}{\partial I_S} \Delta I_S = \mathcal{O}(\varepsilon) \tag{7}$$

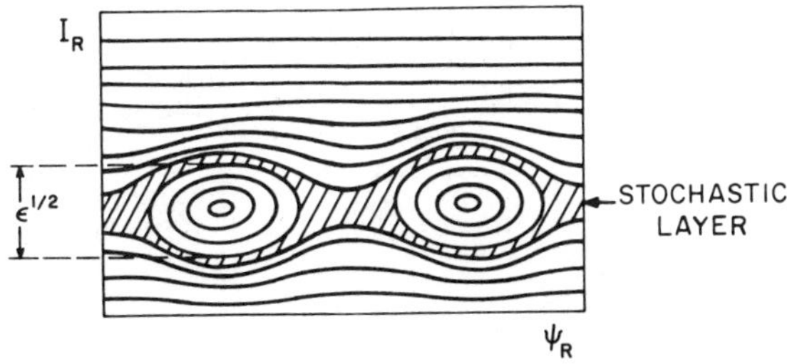

Figure 7. Projection of a resonance layer in phase space onto the two-dimensional surface defined by the resonance action I_R and the resonance phase angle ψ_R.

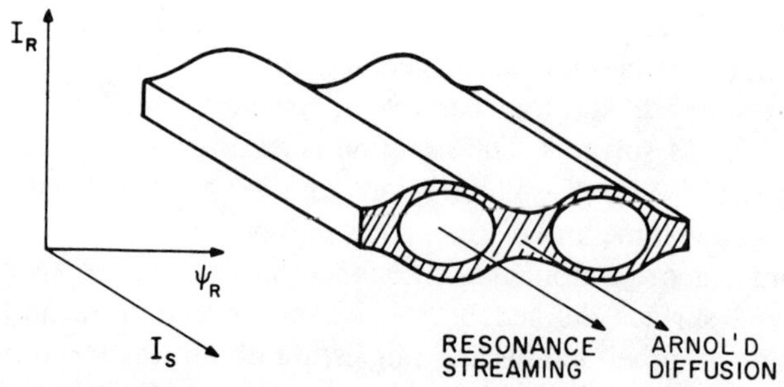

Figure 8. Three-dimensional projection of a resonance layer in phase space. I_s denotes an action variable along the resonance layer. The regions in which Arnol'd diffusion and resonance streaming take place are identified.

But if the resonance action I_R is confined to the resonance layer

$$\Delta I_R = \mathcal{O}(\varepsilon^{1/2}) \tag{8}$$

It follows that

$$\Delta I_S = \mathcal{O}(\varepsilon^{1/2})$$

and large excursions along resonance layers are forbidden.

For three degrees of freedom, Eq. (7) is replaced by

$$\frac{\partial H_0}{\partial I_R}\Delta I_R + \frac{\partial H_0}{\partial I_S}\Delta I_S + \frac{\partial H_0}{\partial I_T}\Delta I_T = \mathcal{O}(\varepsilon) \tag{9}$$

Even if (8) holds, large excursions in the two actions I_S and I_T along the

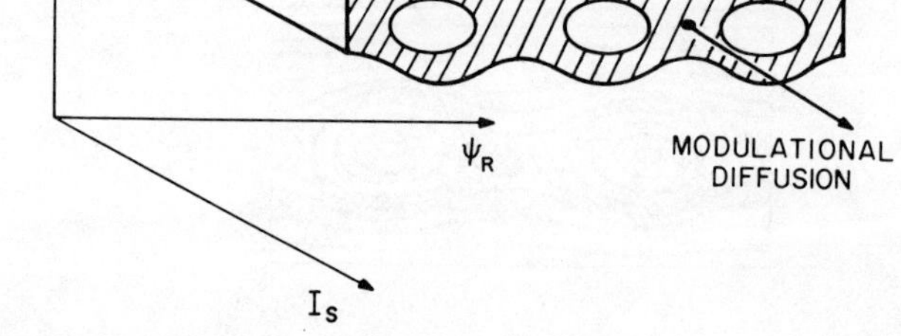

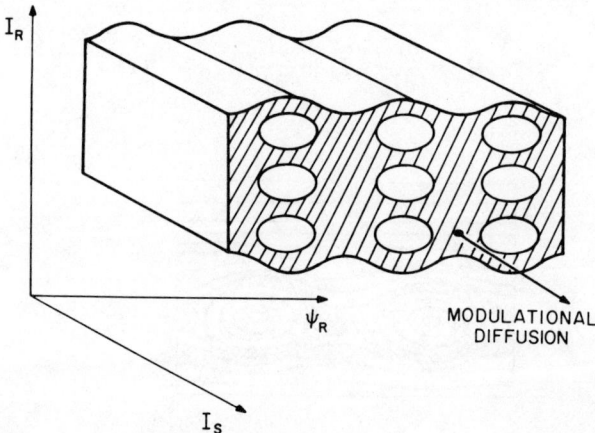

Figure 9. Three-dimensional projection of a set of overlapping resonances in phase space. Modulational diffusion takes place along the layer as shown.

resonance are possible, provided

$$\frac{\partial H_0}{\partial I_S}\Delta I_S + \frac{\partial H_0}{\partial I_T}\Delta I_T = \mathcal{O}(\varepsilon^{1/2})$$

This chapter describes three types of chaotic motion along resonance layers. For Arnol'd diffusion, intrinsic randomness drives a slow diffusion along the stochastic separatrix layer, as illustrated in Fig. 8. In resonance streaming, extrinsic diffusion or dissipation drives a migration through the elliptical KAM tubes within the stochastic separatrix layer. This situation is also shown in Fig. 8. In modulational diffusion (Fig. 9), a slow modulation in one degree of freedom can produce multiplets of sideband resonances. When these overlap, a thick stochastic layer is formed along which diffusive motion appears.

3. ARNOL'D DIFFUSION

As we have seen, for a system with at least three degrees of freedom, all stochastic separatrix layers are connected into a single complex network—the Arnol'd web. The web consists of an intricate system of "freeways, streets, sidewalks, and cracks" that permeates the entire phase space. For an initial condition within the web, the subsequent stochastic motion will eventually intersect every finite region of the phase space, even the predominantly stable regions where the fraction of stochastic intial conditions is small, and even in the limit as the perturbation strength $\varepsilon \to 0$. This motion is the Arnol'd diffusion. The merging of stochastic trajectories into a single web was proved by Arnol'd[3] for a specific nonlinear Hamiltonian. A general proof of the existence of a single web has not been given, but many computational examples are known.

From a practical point of view, there are two major questions concerning Arnol'd diffusion in a particular system:

1. What is the relative measure of stochastic trajectories in the phase space region of interest?
2. For a given initial condition, how fast will the system diffuse along the thin threads of the Arnol'd web?

The extent of the web in phase space can be estimated by means of resonance overlap conditions.[4] Overlap of resonances near the separatrix gives rise to a resonance layer thickness, with stochastic motion occurring *across* the layer as in systems with two degrees of freedom.

Calculation of the diffusion rate along a layer has been given by Chirikov,[4] and Tennyson et al.,[5] and Lieberman[6] for the case of three resonances. For coupling among many resonances, a rigorous upper bound on the diffusion rate has been obtained by Nekhoroshev,[7] but this bound generally overestimates the rate by many orders of magnitude. A statistical treatment of the diffusion regime in which many resonances are important is under development,[4,10,11] and some recent results are described. Extensive numerical simulations of Arnol'd diffusion have been carried out[4,8,11–13] and are summarized in the review article by Chirikov.[4]

3.1. Billiards Problem

A simple example of a system illustrating Arnol'd diffusion is that of a ball bouncing back and forth between a smooth wall at $z = h$ and a fixed wall that is rippled in two dimensions, x and y, at $z = 0$. The surface of section is given in terms of the ball positions in the x_n and y_n directions and the trajectory angles $\alpha_n = \tan^{-1} v_x/v_z$ and $\beta_n = \tan^{-1} v_y/v_z$, just before the nth collision with the rippled wall. The ball motion is shown schematically in Fig. 10, and variables in the x, z-plane are defined in Fig. 11. Assuming that the ripple is small, the rippled wall may be replaced by a flat wall at $z = 0$ whose normal vector is a function of x and y, analogous to the idea of a Fresnel mirror. The simplified difference equations exhibit the general features of the exact equations and may be written in explicit form

$$\begin{aligned}
\alpha_{n+1} &= \alpha_n - 2a_x k_x \sin k_x x + \mu k_x \gamma_c \\
x_{n+1} &= x_n + 2h \tan \alpha_{n+1} \\
\beta_{n+1} &= \beta_n - 2a_y k_y \sin k_y y + \mu k_y \gamma_c \\
y_{n+1} &= y_n + 2h \tan \beta_{n+1}
\end{aligned} \tag{10}$$

where $\gamma_c = \sin(k_x x + k_y y)$, a_x, and a_y are the amplitudes of the ripple in the x- and y-directions, respectively, and μ is the amplitude of the diagonal ripple and represents the coupling between the x- and y-motions. A similar set of equations was examined numerically in pioneering studies by Froeschle[10] and Froeschle and Scheidecker.[11]

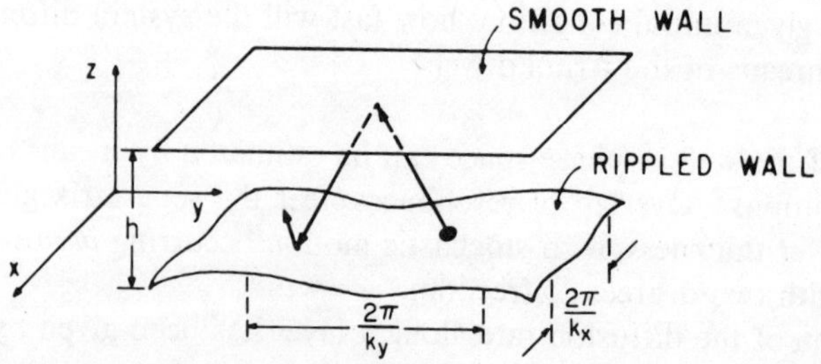

Figure 10. The three-dimensional billiards problem. A point particle bounces back and forth between a smooth wall and a periodically rippled wall.

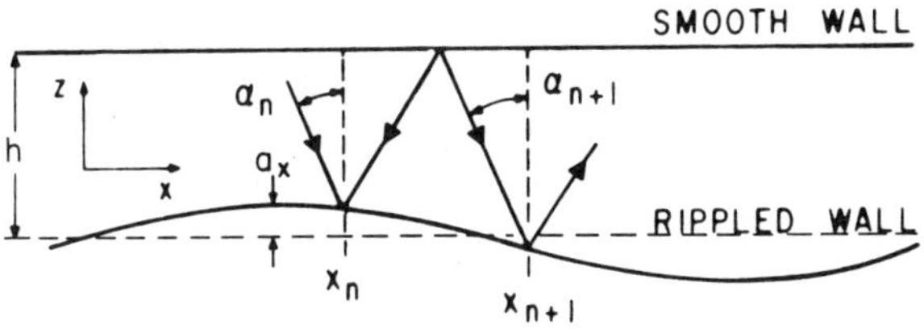

Figure 11. Motion in two degrees of freedom, illustrating the definition of the trajectory angle α_n and the bounce position x_n just before the nth collision with the rippled wall.

If $\mu = 0$, the system breaks into two uncoupled parts describing motion in x–z and y–z separately. Figure 12 shows the motion in the α–x surface of section for the uncoupled case. A number of different orbits are shown, each with different initial conditions. Each particle was run for 1000 iterations. The plot displays the usual features of a system with two degrees of freedom: regular (KAM) orbits, resonance islands, and stochastic orbits. The islands are examples of "higher order" regular (KAM) orbits. The central resonance at $\alpha = 0$, $x = 0$ corresponds to a stable motion for which the ball bounces up and down along z in the valley of the rippled wall. The islands encircling this resonance correspond to "adiabatic" motion in the valley with a small oscillation back and forth in x occurring over many bounce times in z. There are two major stochastic orbits visible in Fig. 12. The thick stochastic layers for α near $\pm\pi/2$ are regions of stochasticity produced by all overlapping resonances with one bounce period in z equal to one or more periods along x. Physically, these motions correspond to grazing angle trajectories (Fig. 12). Isolated from

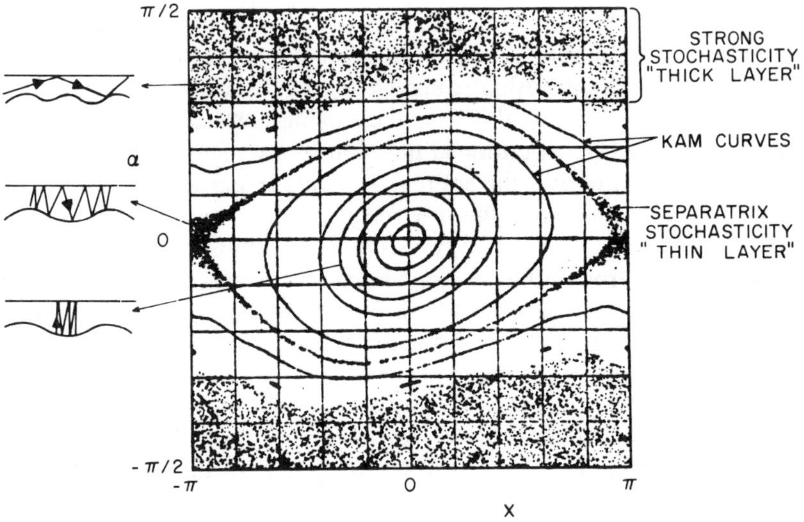

Figure 12. Motion in the α–x surface of section for the uncoupled billiards problem. The parameters are $\mu = 0$; $\lambda_x : h : a_x$ as $100 : 10 : 2$; $\lambda_x = 2\pi/k_x$. Fifteen particles are started at $x = 0$ and allowed to run for 1000 iterations each.

the thick layer by KAM curves spanning the space in x is the thin stochastic layer that has formed near the separatrix associated with the central resonance. Physically, as shown in Fig. 12, the separatrix orbit corresponds to a motion in x for which the ball is either just reflected or just transmitted over a hill. The chaotic motion in this separatrix layer induces an Arnol'd diffusion in the coupled system, which appears as a diffusion in α *along* the separatrix layer of the β–y motion.

3.2. Coupled Motion

A typical numerical calculation showing Arnol'd diffusion in the coupled system is given in Fig. 13. The surface of section for the system is four-dimensional (α, x, β, y), which we represent in the form of a pair of two-dimensional plots (α, x) and (β, y). Thus, two points, one in (α, x) and one in (β, y), are required to specify a point in the four-dimensional section. In Fig. 13 the two plots are superimposed for convenience, and x and y have been normalized to their respective wavelengths, $2\pi/k_x$ and $2\pi/k_y$. The initial condition (Fig. 13*a*) has been chosen on an island encircling the central resonance in x, and within the thin separatrix layer for y. This corresponds to an initial adiabatic motion in x, well confined in the valley, while in y the motion just reaches or passes over a hill. We observe numerically that the y-motion is confined to its separatrix layer until the x-motion reaches its own separatrix layer. The successive stages of the diffusion of the α–x motion are shown in Fig. 13*b*, *c*, and *d*, respectively. In the absence of coupling ($\mu = 0$), the motion in the α–x plane should be confined to a smooth closed curve encircling the central resonance. For a finite coupling, α and x diffuse slowly because of the small randomizing influence of the stochastic β–y motion. The α–x diffusion is the motion along the β–y stochastic layer; that is, it is the Arnol'd diffusion. The diffusion is shown for 1.5×10^5, 3.5×10^6, and 10^7 iterations of the mapping. At this time the α–x motion has diffused out to its own thin separatrix layer. Continued iteration of the mapping shows that the trajectory point diffuses over the entire α–x plane. In particular, the change of direction from diffusion along the β–y separatrix layer to diffusion along the α–x separatrix layer (Fig. 4) has been observed numerically. Similarly, the change of direction from diffusion along a separatrix layer to diffusion along a thick layer (Fig. 12) has been observed. Figure 14 shows these effects in the (α, β) action space (for $x = y = 0$) for the single initial condition of Fig. 13, after 5×10^7 iterations of the mapping. The trajectory has wandered randomly along thin and thick layers in the action space, as shown, spending much of its time in the region of thick layers for both α–x and β–y motion. This corresponds physically to motion with grazing angles of incidence in both the x- and y-directions.

All this, however, is just part of the story. For recall that there exists a dense set of resonance surfaces in the action space. In particular, consider a coupling resonance, where physically the motion is "adiabatically" confined to a valley

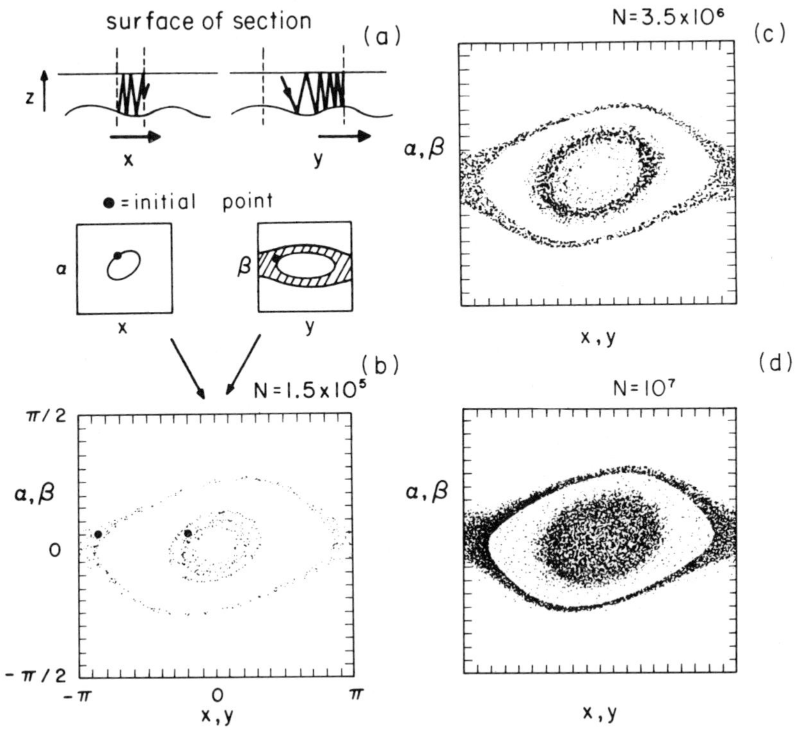

Figure 13. Thin layer diffusion. Initial conditions are close to the central resonance in the α–x space and within the separatrix stochastic layer in the β–y space. Parameters are $\mu/h = 0.004$; $\lambda_x : h : a_x$, and $\lambda_y : h : a_y$ as $100 : 10 : 2$.

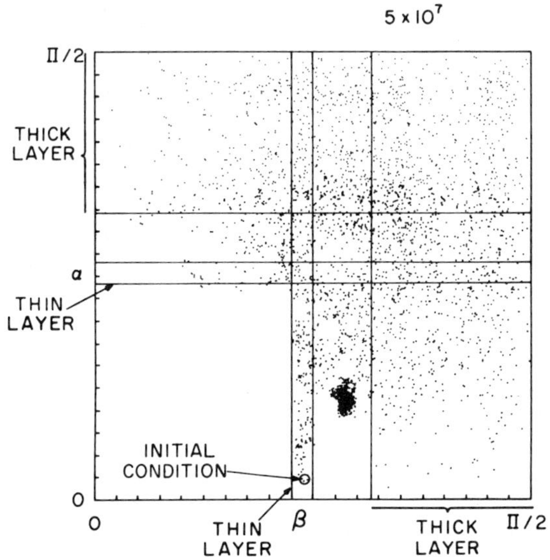

Figure 14. Projection of motion in the α–β action space for $x \approx 0$, $y \approx 0$. The parameters and initial conditions are the same as for Fig. 13. After 5×10^7 iterations. The orbit has wandered in and out of the thick and thin layers of both the α–x and β–y motions.

in both the x- and y-directions as the ball rapidly bounces along z. The ball executes small amplitude oscillations in both the x- and y-directions. If the oscillation frequencies ω_x and ω_y satisfy

$$m_1\omega_x + m_2\omega_y = 0$$

we have a resonance with its stochastic separatrix layer, which is also a part of the Arnol'd web. Including only a single coupling resonance, we have the action space shown in Fig. 15.

We now see a remarkable character of the motion near this coupling resonance in the billiards system. For initial conditions such that the system is placed in the separatrix of the coupling resonance, and thus within the Arnol'd web, the billiard motion initially appears "to be stable," consisting of a fast bounce motion in z and slower, small amplitude oscillations in x and y; in fact, it seems that the motion "is adiabatically confined" to a small neighborhood near $x = y = 0$. However, this is not the case. After a sufficient time, the billiard can be found executing grazing angle motion in both the x- and y-directions. The manner in which the diffusion proceeds is illustrated in Fig. 15. The diffusion typically proceeds first along the coupling resonance, then along the thin layer in x or y, and finally along the corresponding thick layer. With very high probability, the billiard motion rarely "becomes retrapped" in a valley. This follows because the overwhelming fraction of the Arnol'd web is comprised of the "thick stochastic layers," with a negligible (but dense!) fraction of the web in regions such as the coupling resonance, where the motion "appears to be adiabatic." On the other hand, for nearby initial conditions not on the Arnol'd web, the motion is eternally confined to a small neighborhood near $x = y = 0$. Singular behavior indeed!

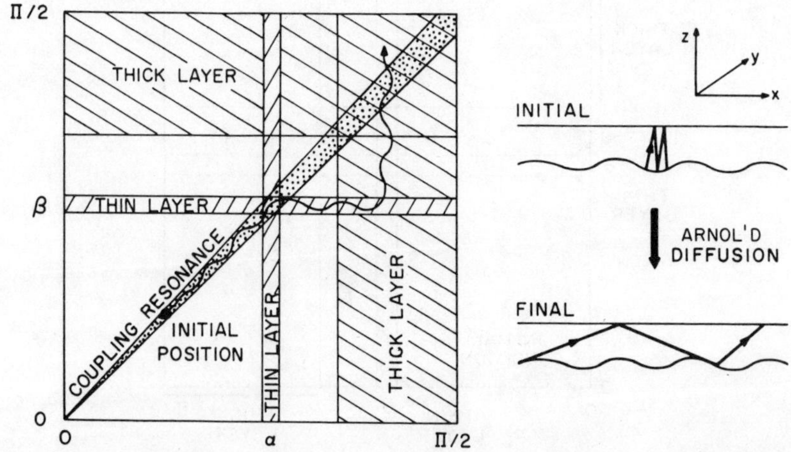

Figure 15. Arnol'd diffusion in the three-dimensional billiards problem, in the angle of incidence space α–β. The initial condition is chosen to be within the separatrix of motion associated with the coupling resonance $\omega_x = \omega_y$. The initial motion with near-normal incidence diffuses toward motion with large angles of incidence. A typical diffusive path is sketched.

3.3. Stochastic Pump Model

The theoretical calculation of Arnol'd diffusion was first performed by Chirikov,[4] and his collaborators. For the billiards problem, the diffusion has been calculated by Tennyson et al.[5] and by Lieberman.[6] The basic theoretical procedure is to break the original three-degree-of-freedom system into two systems, each having two degrees of freedom, which are successively solved. A decomposition is illustrated in Fig. 16, giving the original system and the coupling among the three degrees of freedom due to resonances, as well as the simplest decomposition, the three-resonance, stochastic pump model. In this model, the guiding resonance, along which the Arnol'd diffusion proceeds, is associated with the second degree of freedom. The coupling between the first and second degrees of freedom, described by the Hamiltonian

$$H_{\text{across}}(I_1, I_2\, \theta_1, \theta_2) = \text{const} \tag{11}$$

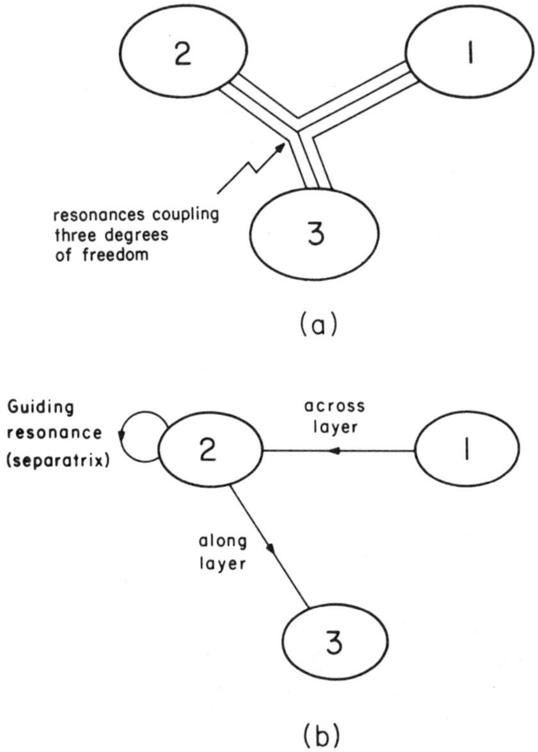

Figure 16. Decomposition of a three-degree-of-freedom system for the theoretical calculation of chaotic motion along a resonance layer. (*a*) Original system with coupling among all degrees of freedom. (*b*) The three-resonance, stochastic pump model for Arnol'd diffusion. (*c*) Model for Arnol'd diffusion in the many-resonance (Nekhoroshev) regime. (*d*) Model for modulational diffusion.

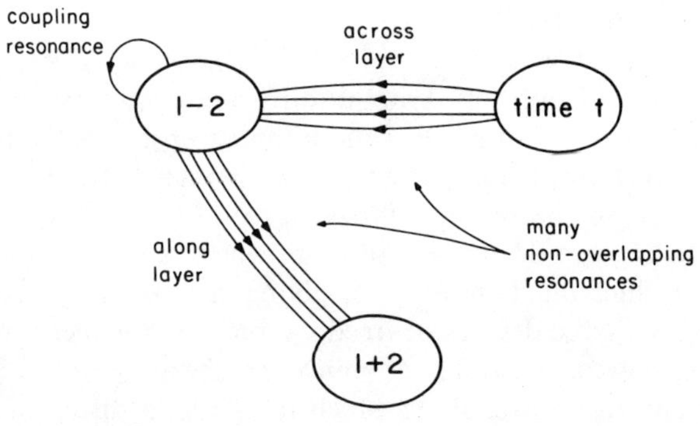

(c)

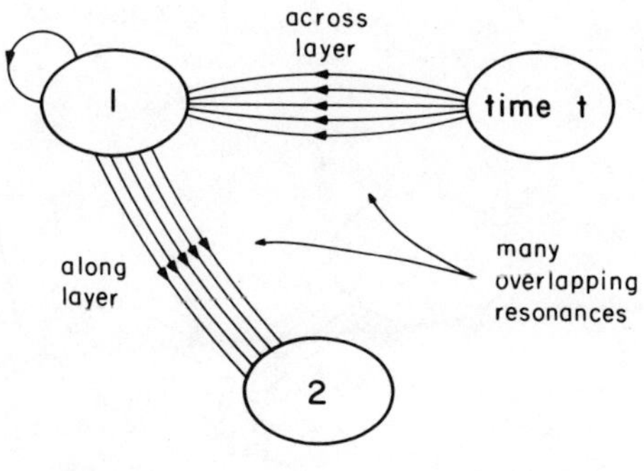

(d)

Figure 16. Continued.

generates the chaotic motion across the separatrix layer. The Arnol'd diffusion is then obtained from

$$H_{\text{along}}(I_2, I_3, \theta_2, \theta_3) = \text{const} \tag{12}$$

which describes the coupling between the second and third degrees of freedom. The motions described by Eqs. (11) and (12) are solved successively, with Eq. (11) first yielding the stochastic variations of $\theta_2(t)$ and $I_2(t)$ in the separatrix layer. These are inserted into Eq. (12), which is solved to obtain the stochastic variation $I_3(t)$ describing the Arnol'd diffusion. For details of these calculations and comparison with numerical simulation for the billiards problem, see Tennyson et al.[5] and Lieberman.[6]

Physically, the model of Fig. 16*b* represents a pumping of stochasticity from the first degree of freedom into the third via the guiding resonance of the second, since as shown by Eq. (9), large chaotic excursions ΔI_1 and ΔI_3 occur while ΔI_2 remains confined to the separatrix layer of width $\varepsilon^{1/2}$. We refer to the motion described by Eq. (11) as the "stochastic pump."

3.4. Many-Resonance Regime

Calculations based on the three-resonance model have been reasonably successful in predicting the Arnol'd diffusion rate that is actually observed numerically.[4,5] However, in the limit of weak coupling among the degrees of freedom, the combined effect of many noninteracting resonances is important, and the three-resonance theory predicts diffusion rates that are much lower than those observed from numerical simulation. The many resonance regime is called the Nekhoroshev region[7-9] after the Soviet mathematician who first derived a rigorous upper bound on the diffusion rate there. However, Nekhoroshev's upper bound is generally many orders of magnitude larger than the actual diffusion rate.

The many-resonance region has been examined numerically,[4,9,12] and some analytic estimates were made[4,8,9] for a simple model of a coupling resonance in a Hamiltonian system. The Hamiltonian studied was[9]:

$$H = \tfrac{1}{2}\left(p_1^2 + p_2^2\right) + \tfrac{1}{4}\left(x_1^4 + x_2^4\right) - \mu x_1 x_2 - \varepsilon x_1 f(t) \tag{13}$$

where the p's are the momenta, the x's are the positions, and

$$f(t) = \frac{\cos \nu t}{1 - A \cos \nu t} = \sum_m \frac{2e^{-\sigma m}}{\sigma} \cos(m\nu t)$$

where $\sigma \simeq (1 - A^2)^{1/2}$ measures the harmonic content of the driving term f. This Hamiltonian describes the motion of two nonlinear oscillators that are coupled quadratically with strength μ. Oscillator 1 is driven by a periodic function of time f with strength ε. Figure 17 shows the resonance surfaces in the $\omega_2(\mathbf{I})$–$\omega_1(\mathbf{I})$ frequency space.

The Arnol'd diffusion was calculated numerically for motion along the coupling resonance $\omega_1 = \omega_2$, shown as the 45° line in Fig. 17. The resonances of the driving term $f(t)$ with oscillator 1,

$$m\nu = \omega_1$$

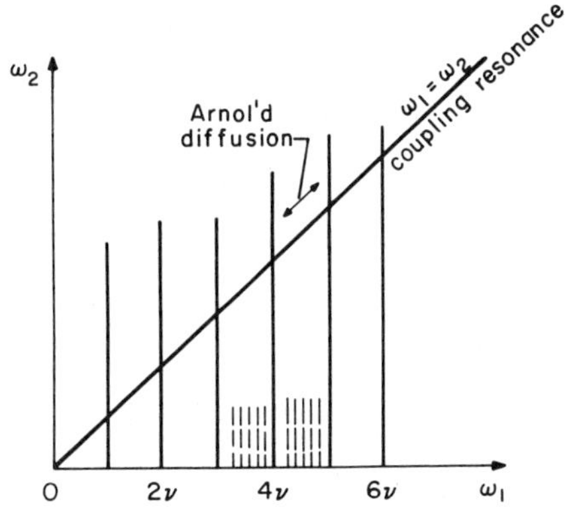

Figure 17. Resonances in the ω_2–ω_1 frequency space for the Hamiltonian of Eq. (13). The diagonal line shows the coupling resonance between oscillators 1 and 2. The solid vertical lines show the resonances of the driving term with oscillator 1. The Arnol'd diffusion has been calculated along the coupling resonance at the point shown. (After Chirikov et al.[9])

are shown as the solid vertical lines. Figure 18 plots numerical calculations of the normalized diffusion rate D at $\sigma = 0.1$, $\omega = 4.5\nu$, versus $\mu^{-1/2}$, where μ is the normalized coupling strength. The theoretical prediction of the three-resonance theory

$$D \propto \exp[-\mu^{1/2}]$$

is shown as the dashed straight line, for the three resonances

$$\omega_1 - \omega_2 = 0$$

$$4\nu - \omega_1 = 0$$

$$5\nu - \omega_1 = 0$$

There is strong deviation from the three-resonance theory for weak coupling strengths $\mu \to 0$. Figure 18 also shows Nekhoroshev's upper bound on the diffusion rate, for which

$$D_{\text{upper}} \propto \exp[-\mu^{-1/9}]$$

The discrepancy at small μ is resolved if we consider the harmonic resonances

$$m\nu - k\omega_1 = 0$$

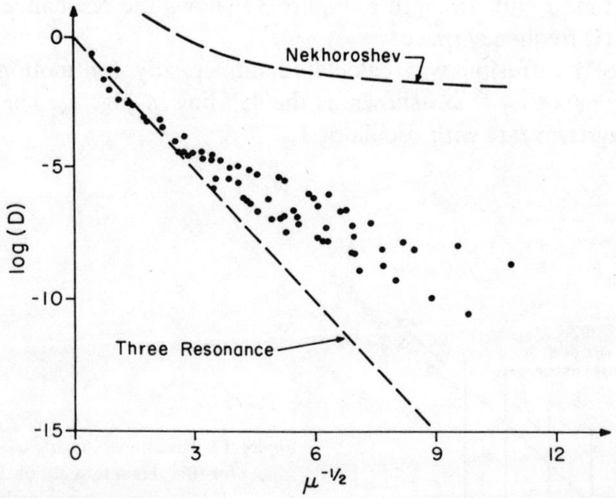

Figure 18. Normalized Arnol'd diffusion rate D versus $\mu^{-1/2}$, where μ is the normalized coupling strength. The dots are numerical calculations, the straight dashed line is the prediction of the three-resonance theory; the scaling of Nekhoroshev's upper bound is also shown. (After Chirikov et al.[9])

between the driving term f and the driven oscillator 1. These resonances, shown as dashed vertical lines in Fig. 17, are excited with small amplitudes, and thus have very thin chaotic separatrix layers. Nevertheless they strongly contribute to the overall Arnol'd diffusion because they lie close to the initial condition $4.5\nu = \omega_1$.

Theoretical calculation of the Arnol'd diffusion in the many-resonance regime has been described by Chirikov[4,8] and Chirikov et al.[9] The theory is based on summing the nonphase-correlated contributions of an infinite set of nonoverlapping resonances, as shown in Fig. 16*c*. The theory is not well developed but yields results that are qualitatively in agreement with numerical calculations.

4. MODULATIONAL DIFFUSION

We turn now to another phenomenon involving chaotic motion along resonance surfaces—the modulational diffusion. This mechanism is similar to Arnol'd diffusion, but as shown in Fig. 9, the chaotic motion is driven along a layer of overlapping resonances in the system. Because of the strong stochasticity in the overlapping layer, modulational diffusion is generally much stronger than Arnol'd diffusion. However, modulational diffusion is not universal. There is generally a resonance overlap condition for formation of the stochastic layer. Below this threshold, one sees only the weaker Arnol'd diffusion.

Modulational diffusion generally appears when the frequency in one degree of freedom is slow compared to the frequencies in the other degrees of freedom. Following Chirikov et al.,[9] we illustrate this for the example of the nonlinear coupled oscillators with the Hamiltonian of Eq. (13), but defining the driving term now as

$$f(t) = \sum_m f_m \cos[m\nu t + \lambda \sin \Omega t] \tag{14}$$

Thus, oscillator 1 is driven at harmonics of the frequency ν and is phase modulated with amplitude λ at the slow modulation frequency Ω.

Expanding Eq. (14) in a Fourier series yields

$$f(t) = \sum_{m,n} f_m J_n(\lambda) \cos[(m\nu + n\Omega)t]$$

The Bessel functions J_n have significant amplitude provided $n \lesssim \lambda$. The result is the formation of a multiplet layer of driving resonances of width approximately $2\Omega\lambda$ centered about each harmonic $m\nu$ of the driving frequency. The multiplets are shown as the vertical sets of lines in the ω_2–ω_1 frequency space in Fig. 19. The modulational diffusion appears as a chaotic motion along a set of overlapping resonances within a multiplet, in the vertical direction in

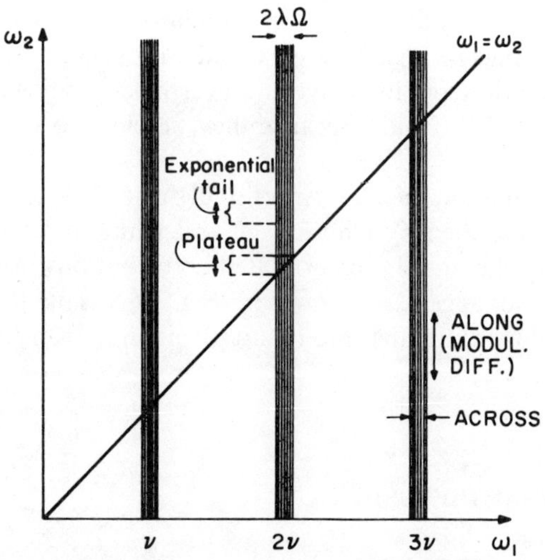

Figure 19. Resonances in the ω_2–ω_1 frequency space for the Hamiltonian of Eq. (13) with modulational driving term [Eq. (14)]. (After Chirikov et al.[13]) The coupling resonance is shown as the diagonal line. The multiplets of resonances that form near harmonics of the driving frequency ν are shown as the sets of vertical lines. The modulational diffusion appears along a multiplet layer. The plateau and exponential tail regimes of the diffusion are indicated.

Fig. 19. The motion along the overlapping multiplet layer is driven by the coupling resonance $\omega_1 = \omega_2$. This picture of modulational diffusion is illustrated in Fig. 16*d*.

4.1. Formation of Multiplet Layer

Near each multiplet

$$\omega_1(I_1) = m\nu$$

the motion can be described by the Hamiltonian

$$H_1 = \frac{(\Delta I_1)^2}{2} + \varepsilon \cos[\theta_1 + \lambda \sin \Omega t]$$

Physically this describes the motion of a ball in a one-dimensional potential well

$$V(\theta_1, t) = \varepsilon \cos[\theta_1 + \lambda \sin \Omega t]$$

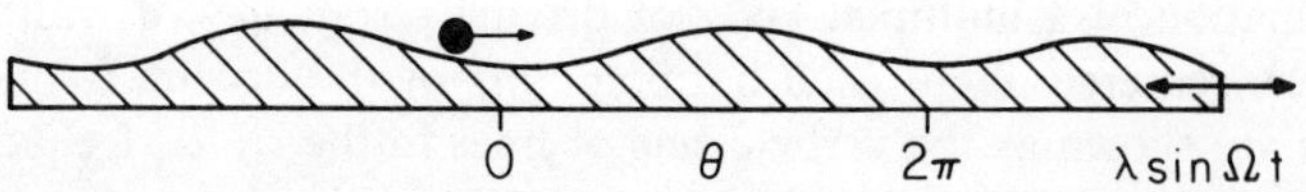

Figure 20. Model for the stochastic pump that generates the chaotic motion across the multiplet layer.

which is slowly shaken back and forth with large amplitude, low frequency oscillations (see Fig. 20).

The conditions for a strong stochastic modulational layer are:

Condition 1. Many resonances in a multiplet:

$$\lambda \gg 1$$

Condition 2. Adjacent multiplet layers do not overlap:

$$\frac{\nu}{\Omega} \gg \lambda$$

Condition 3. Resonances overlap within a multiplet layer:

$$\varepsilon > \frac{\sqrt{\lambda}\,\Omega^2}{23}$$

Condition 3 is derived using the usual resonance overlap criterion[4] that the Chirikov stochasticity parameter K be 1. As the modulation frequency Ω is varied, we find the three regimes illustrated in Fig. 21.

At high frequencies Ω, for which Condition (3) is not satisfied, multiplet resonances do not overlap and we have only the possibility of Arnol'd diffusion. At intermediate frequencies satisfying both Conditions (2) and (3), a thick chaotic layer is formed and we have strong modulational diffusion. At very low frequencies we enter the trapping regime. The overlapping separatrix layers merge and trapping of the ball in a valley of the potential $V(\theta_1)$ can occur (see Fig. 20). The modulational diffusion is weak in this regime.

These three regimes were first described by Tennyson[14] in connection with a simple model of the beam–beam effect for bunched proton beams in the

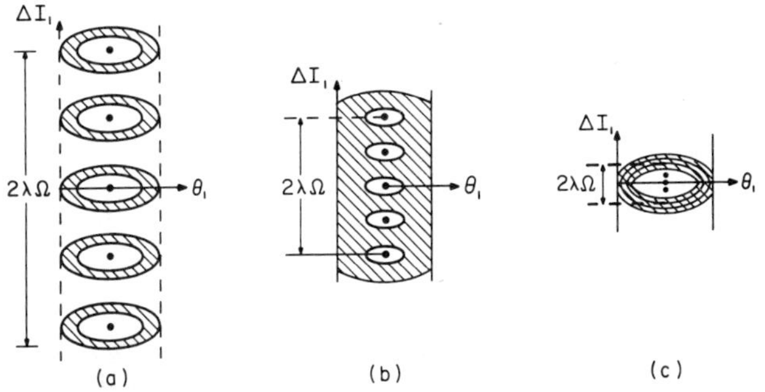

Figure 21. Three regimes in the formation of an overlapping layer of modulational resonances as the modulation frequency Ω is varied. (*a*) High Ω leads to nonoverlapping resonances within the multiplet. There is Arnol'd diffusion but no modulational diffusion. (*b*) Intermediate Ω leads to overlapping of modulational resonances and strong modulational diffusion. (*c*) Low Ω leads to formation of a trapping (regular) regime and weak modulational diffusion.

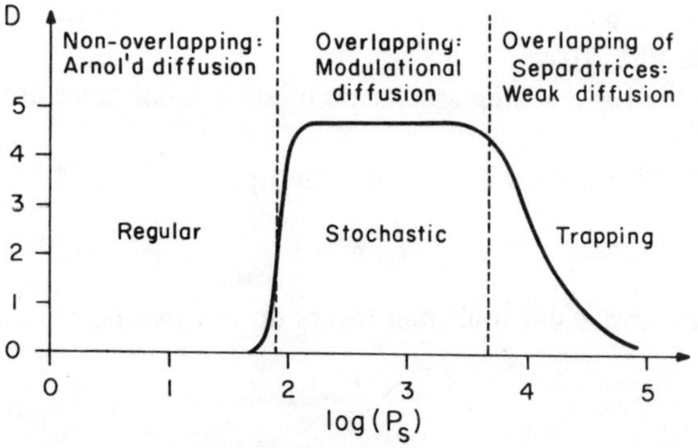

Figure 22. Behavior of the normalized diffusion coefficient describing chaotic motion across a modulational resonance layer. (After Tennyson.[14])

ISABELLE storage ring at the Brookhaven National Laboratory. Figure 22 shows the diffusion coefficient for chaotic motion across the modulational layer as the modulation period $P_s = 2\pi/\Omega$ is increased. Figure 23 shows numerical calculations of the Poincaré section for the vertical motion ($P_Y = V/\omega_0$ vs. Y) of a bunched proton beam as the modulation period is increased. A single modulation resonance becomes visible at $P_s = 200$ and a thick modulation layer has formed at $P_s = 600$.

4.2. Diffusion Along the Layer

We consider now chaotic motion along the set of overlapping resonances, which is the modulational diffusion. For

$$|\omega_2 - \omega_1| \lesssim \lambda\Omega \tag{15}$$

the condition of exact resonance

$$\omega_2 - \omega_1 + n\Omega = 0; \qquad |n| \lesssim \lambda$$

is met within the layer, and we expect a strong diffusion along the layer. The diffusion rate along the resonance can easily be calculated (see Chirikov et al.[13]). A more subtle problem is the calculation of the diffusion rate when the exact resonance condition is not met. This problem has been considered by Chirikov et al.[13,15] for the coupled nonlinear oscillator problem. Figure 24 shows numerical calculations of the normalized modulational diffusion coefficient D_R as the frequency separation $|\omega_2 - \omega_1|$ is varied over a range of parameters λ, Ω, and ε. We note the plateau region when Eq. (15) is satisfied and the exponential drop in the diffusion rate when the exact resonance condition is not met (see also Fig. 19). The shape of the curve (solid line) has been calculated from an analytic theory containing two empirically determined

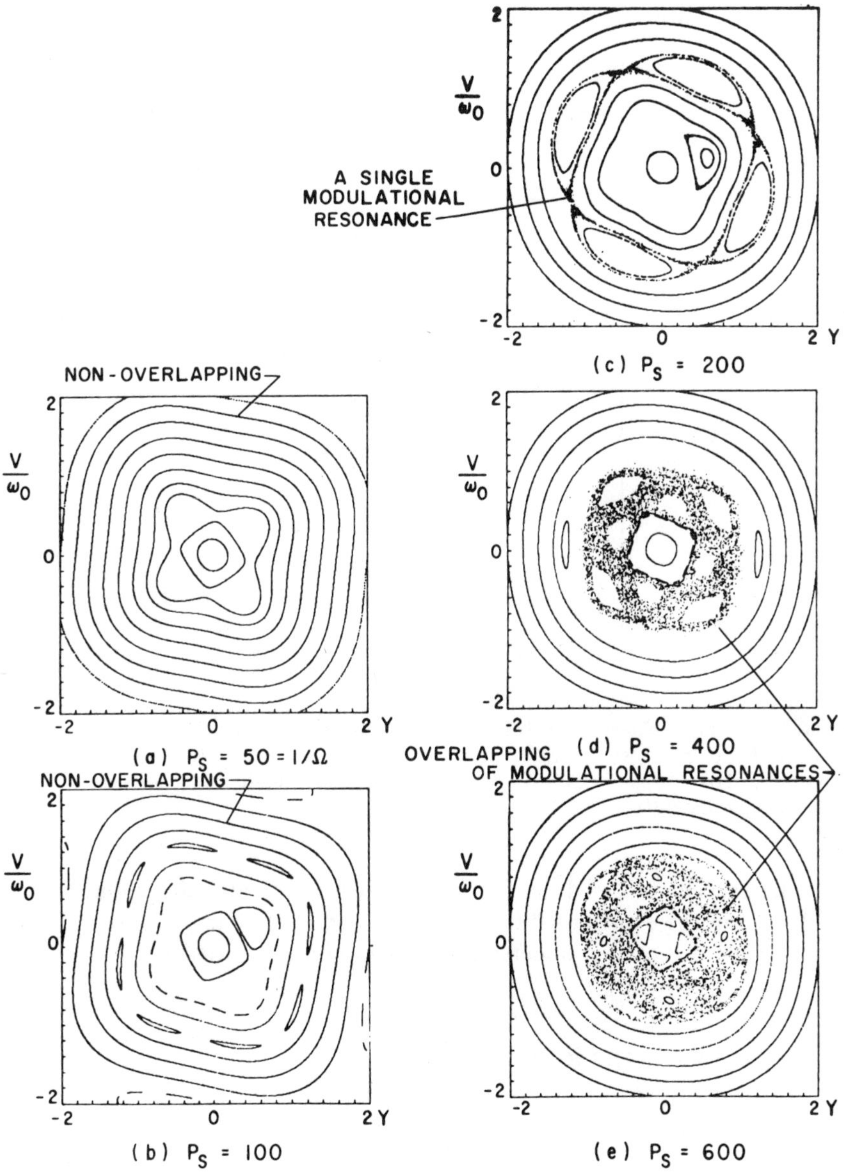

Figure 23. Simulation of bunched proton–proton crossing beams in ISABELLE. The Poincaré section ($P_Y = V/\omega_0$ vs. Y) for the vertical motion is shown as the modulation period $P_s = 2\pi/\Omega$ is increased. A modulational resonance layer is formed. (After Tennyson.[14])

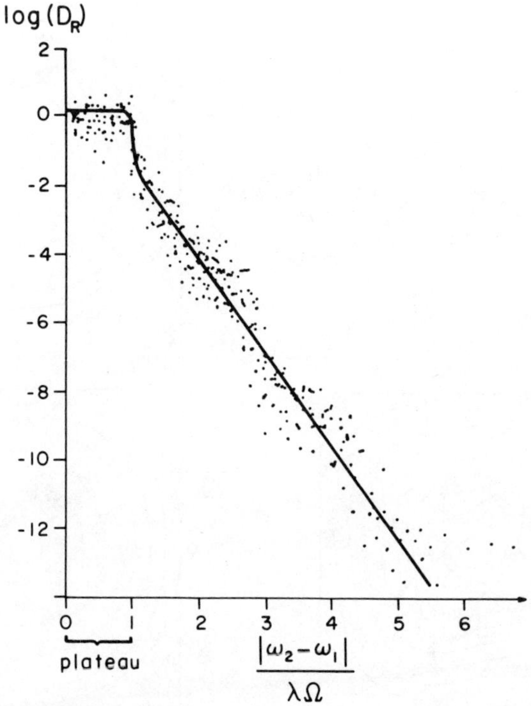

Figure 24. Normalized modulational diffusion coefficient D_R as the frequency separation $|\omega_2 - \omega_1|$ is varied. The points show numerical calculations over a wide range of values of λ, Ω, and ε. The solid curve shows an analytical calculation containing two empirically determined parameters. (After Chirikov et al.[13])

parameters. The sharp drop near the plateau edge and the apparent wavy oscillation in the exponential tail show that there are many interesting features to be further explored in understanding modulational diffusion.

5. RESONANCE STREAMING

5.1. Description

As a last example of chaotic motion along resonance surfaces, we consider the process of resonance streaming.[16,18] Streaming occurs in near-integrable systems that are subject to an externally generated transport process such as a diffusion or dissipation on the action space. In the absence of the external process, we have seen that a perturbed Hamiltonian system undergoes a motion that is a small bounded oscillation in the action space of the unperturbed system (see Fig. 2). Although the oscillation itself may be unimportant, it necessitates an averaging procedure when the long-term behavior is of interest. Specifically, transport phenomena must be described in terms of the motion of the oscillation center, rather than of the instantaneous position. When the phase point is outside nonlinear resonance, the oscillation center transport is almost identical to the unperturbed classical transport. But when the system is resonant, the two can be drastically different in both magnitude and direction. The difference is most pronounced when the resonance vector $\mathbf{m}_k$ is nearly tangent to the resonance surface. In this case the oscillation center can move rapidly along the resonance surface at a rate that is much greater than, but still proportional to, the classical transport rate.

Resonance streaming should not be confused with enhanced classical diffusion due to transport across resonance layers. The transport process is illustrated in Fig. 25*a*, where a large step in the action I_2 results when an initially untrapped phase point becomes trapped at the bottom of the island, swept around to the top of the island, then detrapped. If the external noise is not conservative—that is, if it can change the total energy of the system—it essentially introduces an additional freedom into the original system (see Note 1). Enhanced diffusion across resonance layers may be present in systems with two or more degrees of freedom (including the noise).

Resonance streaming is illustrated in Fig. 25*b*, which shows an orbit moving up and to the left along a resonance surface in the action space. Streaming can occur only in systems with three or more degrees of freedom (including nonconservative noise as a degree of freedom). As shown in Fig. 8, streaming occurs when an orbit becomes trapped inside a resonance tube.

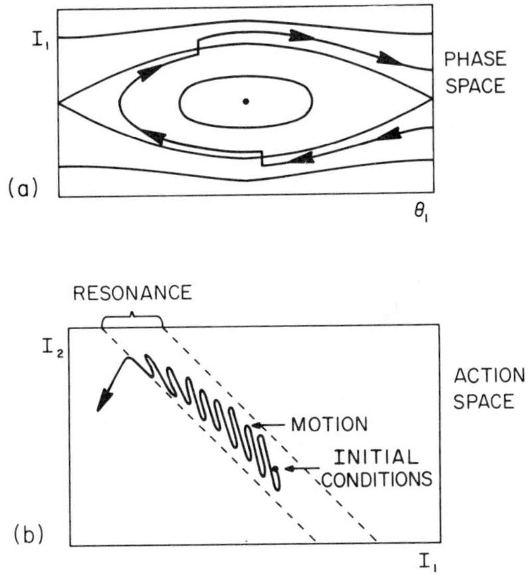

Figure 25. Chaotic motion near resonance layers due to external noise or dissipative processes. (*a*) Enhanced transport across resonances. (*b*) Resonance streaming, which is an enhanced transport along resonances.

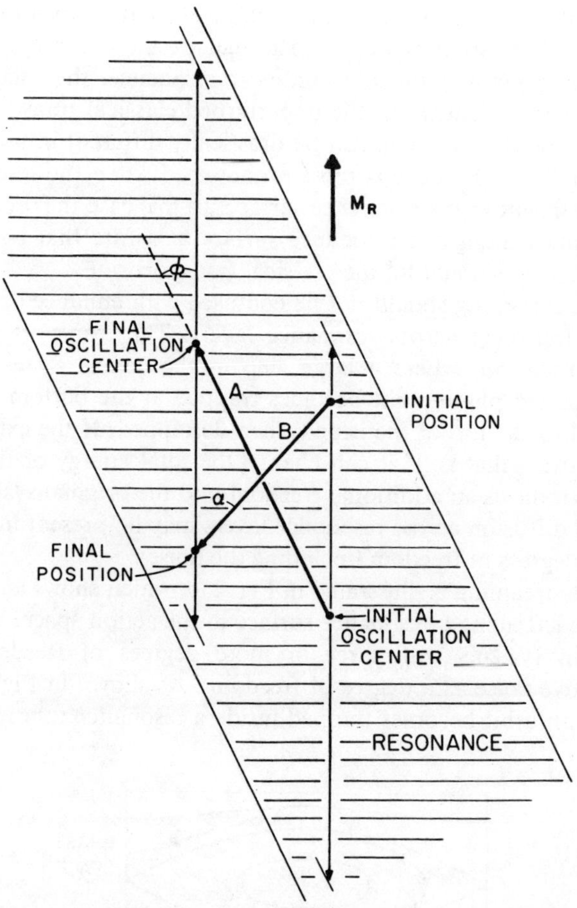

Figure 26. Oscillation center displacement inside a resonance layer. External noise or dissipation causes the phase point, oscillating vertically, to make a jump B. The oscillation center makes a corresponding jump A. Note that A may be greater than B.

To understand the process of resonance streaming, we consider a near-integrable, two-degree-of-freedom system with unperturbed actions I_1 and I_2, having a resonance layer with resonance vector $\mathbf{m}_R$ in the action space (Fig. 26). The initial motion is an oscillation within the resonance layer in the direction $\mathbf{m}_R$, about the initial oscillation center shown Fig. 26. Consider now a sudden nonenergy conserving displacement B to a final position within the layer. The final oscillation center is displaced along the layer a distance A. We note that

$$A = B \sin \Phi \csc \theta \tag{16}$$

and therefore A may be much larger than B if θ is small. Furthermore the oscillation center displacement is always along the resonance. Thus any sequence of displacements (due to noise) that does not take the system out of resonance will result in a net displacement of the oscillation center along the resonance, and this net displacement may be much larger in magnitude than the sum of the classical displacements. For more than three degrees of freedom, the direction of the streaming in the resonance layer is given by the projection of the resonance vector $\mathbf{m}_R$ onto the resonance surface.

A resonance layer may be classified into one of three distinct transport regimes for resonance streaming. These are approximately equivalent to the well-known regimes of neoclassical theory[19]:

1. *The Oscillation-Center (or Weak Transport) Regime.* For the oscillation center concept to be valid, the system must remain in the resonance for a time greater than one oscillation period. This means that the oscillation period must be smaller than the time necessary for the classical transport process to move the phase point from the center of the resonance to its edge. The enhancement ratio Eq. (16) is valid only in the oscillation center regime.

2. *The Classical (or Strong Transport) Regime.* If the external process induces a motion that is faster than that induced by the resonance, the presence of the resonance is inconsequential. The transport is said to be "classical" in this case.

3. *The Plateau Regime.* Since the upper limit of the oscillation center regime does not coincide with the lower limit of the classical regime, there is an intermediate situation. Here the phase point is not in the resonance long enough to complete one oscillation cycle, but the oscillation motion that does occur moves the phase point along the resonance for a distance that is greater than the resonance width.

5.2. Examples of Resonance Streaming

We describe briefly two examples of resonance streaming. We first consider the effect of an external dissipation on a near-integrable Hamiltonian system

$$H = H_0 + \varepsilon H$$

with

$$H_0 = I_1^2 + I_2^2 \tag{17}$$

and

$$\varepsilon H_1 = 10^{-5}\cos(\theta_1 - \theta_2) \tag{18}$$

This system has unperturbed frequencies

$$\omega_1 = 2I_1, \qquad \omega_2 = 2I_2$$

and a resonance surface

$$\omega_1 - \omega_2 = 0$$

or

$$I_1 - I_2 = 0$$

in the action space. For the external process we take a dissipation in I_2 alone, of the form

$$\dot{I}_2 = -10^{-5} I_2 \tag{19}$$

The motion must ultimately be attracted to the line $I_2 = 0$, but may detour along the resonance en route to its ultimate destination.

Figure 27 shows two computer-generated trajectories plotted in the action space. The first trajectory is initially nonresonant. It drifts down to the resonance, makes a small horizontal jump as it crosses, and continues on. The second trajectory becomes trapped inside the resonance as it attempts to cross, and its oscillation center is consequently constrained to move along the resonance curve. The libration angle is about 22° and the ratio of resonant to nonresonant drift speed is 2.55. The resonant libration is slowly damped by the dissipation, and the phase point is drawn toward the center of the resonance (barely perceptible here).

A second example is described by the same Hamiltonian Eqs. (17) and (18), but the external process is a diffusion in the action I_2. The diffusion is defined

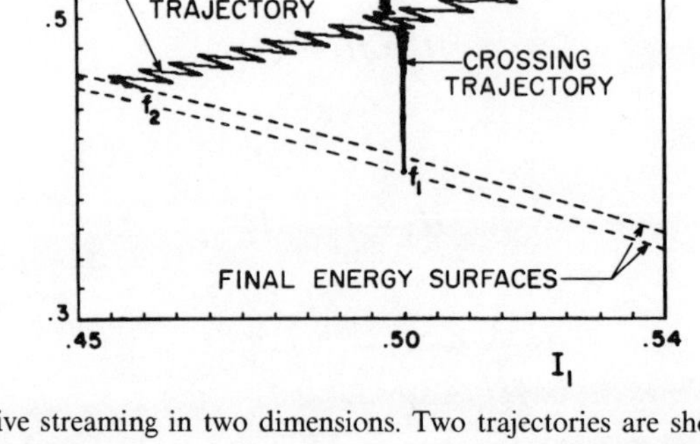

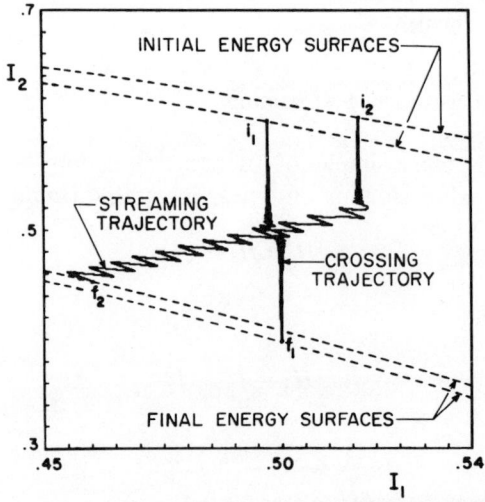

Figure 27. Dissipative streaming in two dimensions. Two trajectories are shown in action space. The first phase point begins at i_1 and descends to f_1, crossing the coupling resonance $\omega_1 = \omega_2$ on its way. The second phase point descends from i_2 but instead of crossing the resonance, becomes trapped and thus streams along the resonance. The time intervals for the two trajectories are the same. The dashed lines show the energy contours at the initial and final positions.

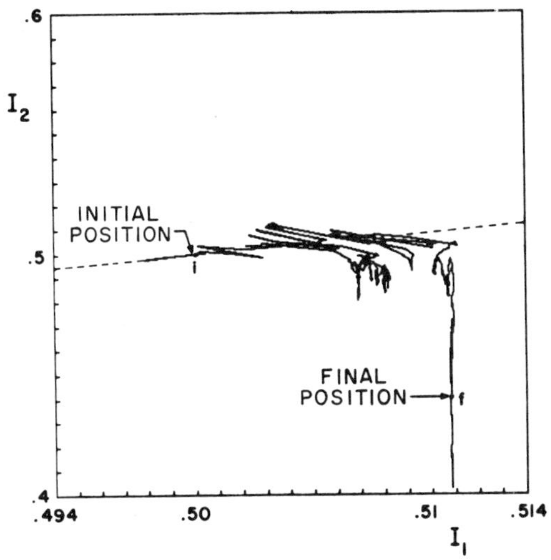

Figure 28. Resonance streaming in two dimensions due to diffusion. The Hamiltonian function is the same as in Fig. 27. The external process is now a vertical diffusion consisting of small steps in ΔI_2, and the time between jumps is $\Delta t = 1$. The trajectory shown is that of the oscillation center, since the position of the phase point is averaged over successive time intervals of length $T = 500$. Successive averaged positions are connected by line segments. The initial position i is at the center of the resonance $\omega_1 = \omega_2$. The particle eventually diffuses out of and away from the resonance, ending the run at f.

by a sequence of small random jumps occurring once each unit time interval;

$$\Delta I_2 = 6 \times 10^{-5} \sin(r_n)$$

where r_n is a random number between 0 and 2π. Figure 28 shows the trajectory of the oscillation center for a phase point that starts at the center of the resonance and diffuses slowly out. Initially the oscillation center diffuses almost horizontally along the resonance curve. Intervals where the trajectory diffuses back and forth across the resonance (see Fig. 25a), can also be seen. Eventually the phase point leaves the resonance, and the oscillation center stops its horizontal motion and begins to diffuse vertically, following the classical displacements.

The theoretical calculation of resonance streaming has been given by Tennyson.[16] Resonance streaming effects may be important in high temperature plasma confinement experiments and in high energy particle storage rings.

6. CONCLUSION

For three or more degrees of freedom, a new phenomenon appears in near-integrable Hamiltonian systems that is not present in systems with one or two degrees of freedom: chaotic motion along resonance layers. Three mechanisms

Table 1. Properties of the Mechanisms of Chaotic Motion Along Resonance Layers

Mechanism	Perturbation Strength	Driving Source for Stochastic Pump	Driving Source Along Resonance Layer(s)
Arnol'd diffusion	$\varepsilon > 0$	One resonance	One resonance
	$\varepsilon > 0$	Many non-overlapping resonances (Nekhoroshev)	Many non-overlapping resonances
Modulational diffusion	$\varepsilon > \varepsilon_{crit}$	Many overlapping resonances	One resonance
Resonance streaming	$\varepsilon > 0$	One resonance	Extrinsic diffusion or dissipation "noise"

for such motion have been described: Arnol'd diffusion, modulational diffusion, and resonance streaming. Table 1 summarizes some important properties of these mechanisms. There are probably many interesting effects to be discovered concerning chaotic motion in multidimensional systems.

REFERENCES AND NOTES

1. A time-varying Hamiltonian $H(\mathbf{p}, \mathbf{q}, t)$ with N **p**'s and **q**'s can be transformed to a time-independent form with $N + 1$ **p**'s and **q**'s by introducing an extended phase space, and is thus equivalent to a system with $N + 1$ degrees of freedom.
2. We generally consider "near-integrable" systems for which $H = H_0 + \varepsilon H_1$, with H_0 integrable, H not integrable, and ε the perturbation strength.
3. V. I. Arnol'd, *Dokl. Akad. Nauk SSSR* **156**, 9 (1964).
4. B. V. Chirikov, *Phys. Rep.* **52**:5, 263–379 (1979).
5. J. L. Tennyson, M. A. Lieberman, and A. J. Lichtenberg, in *Non Linear Dynamics and Beam–Beam Interactions*, M. Month and J. L. Herrera, Eds. American Institute of Physics Conference Proceedings Series, Proc. 57 (1979).
6. M. A. Lieberman, *Ann. N.Y. Acad. Sci.* **357** (1980).
7. N. N. Nekhoroshev, *Usp. Mat. Nauk USSR* **32**, 6 (1977).
8. B. V. Chirikov, *Fiz. Plasmy* **4**:3, 521 (1978).
9. B. V. Chirikov, J. Ford, and F. Vivaldi, in American Institute of Physics Conference Proceedings Series, Vol. 57 (AIP, New York, 1979).
10. C. Froeschle, *Astrophys. Space Sci.* **14**, 110 (1971).
11. C. Froeschle and J. P. Scheidecker, *Astrophys. Space Sci.* **25**, 373 (1973).
12. G. M. Gadiyak, F. M. Izrailev, and B. V. Chirikov, *Proceedings of the Seventh International Conference on Nonlinear Oscillations* (Berlin, 1975), Vol. 2, 1:315.

13. B. V. Chirikov, F. M. Izrailev, and D. L. Shepelyansky, "Dynamical Stochasticity in Classical Mechanics," Preprint 80-209, Institute of Nuclear Physics, Novosibirsk, USSR, (1980).
14. J. L. Tennyson, in *Non Linear Dynamics and Beam–Beam Interactions*, M. Month and J. L. Herrera, Eds. American Institute of Physics Conference Proceedings. Series, Vol. 57 (AIP, New York, 1979).
15. B. V. Chirikov and D. L. Shepelyansky, "Diffusion in the Presence of Many Overlapping Nonlinear Resonances," Preprint 80-211, Institute of Nuclear Physics, Novosibirsk, USSR (1980), (in Russian).
16. J. L. Tennyson, "Enhancement of Classical Transport Processes Along Resonances in Near Integrable Systems with Many Degrees of Freedom," Electronics Research Laboratory Memorandum UCB/ERL M81/7, University of California, Berkeley, January 25, 1981; to appear in Physica **4D**, 1982.
17. J. Tennyson, SLAC PUB 2624 (Stanford Linear Accelerator Center, Stanford, CA, 1980).
18. B. V. Chirikov, *Sov. J. Plasma Phys.* **5**, 492 (1979).
19. A. A. Galeev and R. Z. Sagdeev, *J. Exp. Theor. Phys.* **26**, 233 (1968).
20. A. J. Lichtenberg and M. A. Lieberman, *Regular and Stochastic Motion*, Applied Mathematical Sciences Vol. 38, Springer-Verlag, New York, 1982.

STOCHASTICITY AND ORDER IN A LINEAR QUASI-PERIODIC DIFFERENTIAL EQUATION

A. SALAT
Max-Planck-Institut für Plasmaphysik
Garching, Federal Republic of Germany

J. TATARONIS
Department of Electrical and Computer Engineering,
University of Wisconsin
Madison, Wisconsin

Abstract. *Long-time correlations in the solution of a linear second-order, quasi-periodic differential equation are investigated numerically and graphically. The equation is also equivalent to an autonomous Hamiltonian system that is linear in the action variables P, hence not amenable to Kolmogorov–Arnold–Moser (KAM) theory. The existence of invariant tori in six-dimensional phase space is investigated with the help of three-dimensional stereoscopic pictures. Both torus destruction and torus survival for finite perturbations are found.*

1. INTRODUCTION

We discuss properties of the solution of an ordinary second-order linear differential equation, Eq. (2.1), whose coefficient f is a quasi-periodic function of the independent variable t.

The investigation was motivated by a study of the "continuum" in Alfvén wave propagation[1] in nonaxisymmetric tori where the equilibrium is periodic in the poloidal and toroidal angles θ and ϕ, respectively. The wave propagates along the magnetic field, so that $\theta = \theta(t)$, $\phi = \phi(t)$, where t is a coordinate along the field, and $-\infty < t < \infty$ on irrational surfaces. To be physically acceptable, the solution must also be periodic in $\theta(t)$ and $\phi(t)$. Whether such a solution exists is unknown. Equation (2.1) represents a simplified model equation.

Independent of its origin Eq. (2.1) proves to be of more general interest. It may also be written as an autonomous Hamiltonian system, and one may ask whether it is indeed integrable, in particular for $|f| \ll 1$. The essential point is that the results of Kolmogorov, Arnol'd and Moser[2] (KAM) cannot be used here, since one of their assumptions, nonlinearity in the momenta, does not hold. In particular, as an unusual feature, both the undisturbed and the disturbed Hamiltonian are linear in the momenta. We draw attention to this class of problems, by investigating it with the "surface of section" technique[3] in six-dimensional phase space.

2. BASIC EQUATIONS AND FORMAL SOLUTION

We consider the differential equation

$$\ddot{\psi}(t) + \left[\omega_0^2 + f(t)\right]\psi = 0 \tag{2.1}$$

where

$$\begin{aligned} f(t) &\equiv f(\theta_1(t), \theta_2(t)) \\ &= \sum_{n=-\infty}^{+\infty} (-1)^n \left[F_1 \delta(t - nT_1 - c_1) + F_2 \delta(t - nT_2 - c_2)\right] \\ &= \sum_{n=-\infty}^{+\infty} C_n \delta(t - t_n) \end{aligned} \tag{2.2}$$

is a "quasi-periodic" function of t:

$$f(\theta_1, \theta_2) = f(\theta_1 + 2\pi, \theta_2) = f(\theta_1, \theta_2 + 2\pi) \tag{2.3}$$

with

$$\theta_i = \omega_i t + c_i \qquad \omega_i = \frac{\pi}{T_i} \qquad c_i = \text{const} \qquad i = 1,2 \tag{2.4}$$

Equation (2.1) is an oscillator whose frequency ω_0 is modified quasi-periodically by the δ-function terms; T_1/T_2 is assumed to be irrational. The shift $t \to t + 2T_i$ corresponds to $\theta_i \to \theta_i + 2\pi$. Equation (2.1) may be written in the form

$$\dot{\mathbf{x}}(t) = \mathsf{A}(\boldsymbol{\theta}(t)) \cdot \mathbf{x}; \qquad \mathsf{A}(\boldsymbol{\theta}) = \mathsf{A}(\boldsymbol{\theta} + 2\pi\mathbf{e}_i) \tag{2.5}$$

where in the present case

$$\mathbf{x} = \begin{pmatrix} \psi \\ \dot{\psi} \end{pmatrix}; \qquad \mathsf{A} = \begin{pmatrix} 0 & 1 \\ -\omega_0^2 - f & 0 \end{pmatrix}; \qquad \boldsymbol{\theta} = \begin{pmatrix} \theta_1 \\ \theta_2 \end{pmatrix} \tag{2.6}$$

and $\mathbf{e}_i$ are unit vectors with components 0 and 1.

Between the δ-functions Eq. (2.1) has the solution $\psi = a\cos\omega_0 t + b\sin\omega_0 t$. Integration across the δ-functions at $t = t_n$ yields the jump conditions

$$[\psi]_{t_n} = 0 \qquad \text{and} \qquad [\dot{\psi}]_{t_n} = -C_n\psi(t_n) \tag{2.7}$$

For the amplitudes a_n, b_n at $t = t_n$ one obtains the recursion relation

$$\begin{pmatrix} a_n \\ b_n \end{pmatrix} = \mathsf{T}_n \cdot \begin{pmatrix} a_{n-1} \\ b_{n-1} \end{pmatrix} = \begin{pmatrix} 1 + c_n \sin 2\omega_0 t_n & 2c_n \sin^2\omega_0 t_n \\ -2c_n \cos^2\omega_0 t_n & 1 - c_n \sin 2\omega_0 t_n \end{pmatrix} \cdot \begin{pmatrix} a_{n-1} \\ b_{n-1} \end{pmatrix} \tag{2.8}$$

where $c_n = C_n/(2\omega_0)$. This allows fast numerical calculation of the solution, in particular of $x_n = \psi\,(t = t_n)$, $y_n = \dot{\psi}\,(t = t_n)$ at the end of each jump.

3. TWO-DIMENSIONAL PLOTS

To determine the amount of stochasticity in the long-time behavior of the solution, we ran the system for $0 \leqslant n \leqslant N$, N being up to 10^6, and plotted the consecutive "states" (x_n, y_n) as points on an x–y plane. To be more precise, points belonging to full periods $2T_1$ and $2T_2$ are plotted in separate figures characterized by the indices 1 and 2, respectively (i.e., Fig. 1.1 plots points belonging to period $2T_1$). Half-period points were omitted. The origin is at the center. The values 0 and ± 1 are indicated at the margin. Each plot corresponds to a single sequence of points, usually starting at $x_0 = 1$, $y_0 = 0$. A selection of typical cases is shown in Figs. 1–8. (Stereoscopic views are given in Figs. 1.3, 2.3, etc.) The parameters T_i, F_i are normalized by setting $\omega_0 = 1$.

The amount of stochasticity varies considerably from one case to the other. Figure 1 corresponds to a high degree of stochasticity. It was obtained by a random choice of parameters. Other cases such as Fig. 5.1 show considerable correlations. Figures 6 and 7 present a set of one-dimensional curves or even discrete points. Generally speaking, it takes much more time to find strongly nonstochastic cases, by trial and error, than stochastic ones, and they may be sensitive to variations in T_i of less than 0.01%!

Figures 6.1 and 7.1, although similar in appearance, are very different. When the number N of recursions is increased, more and more segments are created in case 6.1, whereas Fig. 7 remains unchanged even for extremely large

N. Figures 3 and 4 correspond to weakly unstable cases. In case 3.2 the sequence initially converges toward the origin before its final slow exponential divergence. There also exist strongly unstable cases with just a few points on the plot. In most stable cases the region occupied reaches its final size after a few thousand iterations.

If a solution of Eq. (2.1) exists that has the same quasi-periodicity as the perturbation f, it is easily recognized on the plots. From $\psi(t) = u(\theta_1(t), \theta_2(t))$, where u is a quasi-periodic function that obeys Eq. (2.3), it follows that in the figures either θ_1 or θ_2 is constant, mod 2π, and both ψ and $\dot{\psi}$ are periodic in the complementary variable. Therefore, for $N \to \infty$, the sequence of points forms a closed curve. No such solution has been found so far. The case represented by Fig. 7, however, is a subharmonic solution with periods 2π and $\pi/4$ in θ_1 and θ_2, respectively, as closer inspection reveals. The case of Fig. 6 is difficult to interpret because of its unsteady nature with increasing N.

4. PARAMETER STUDIES AND GENERAL PROPERTIES

Equation (2.1) contains four parameters T_1, T_2, F_1, and F_2, so that an exhaustive parameter study is difficult. Some general conclusions, however, may be given.

It is useful to consider the special case of a periodic perturbation by, say, setting $F_2 = 0$. In this case the Floquet–Lyapunov theory[4] states that $\psi(t)$ is of the form

$$\psi = e^{\pm i\nu t} u(\theta_1(t)) \tag{4.1}$$

with $u(\theta_1) = (\theta_1 + 2\pi)$, ν^2 real, and ν follows from the knowledge of a fundamental solution of Eq. (2.1) after one full period, $t = 2T_1$. [From Eq. (4.1) it follows that the resulting figures (i.e., Figs. 1.1, 2.1, 4.1, 5.1, 6.1, 7.1) would be closed curves, whereas Figs. 1.2, 2.2, 3.2, and 7.2 in general would not.] With Eq. (2.8) ν may be obtained analytically. The resulting stability diagram, $\nu^2 < 0$ corresponding to instability, is shown in Fig. 9. The pattern repeats periodically in $\Delta\omega_0 T_1 = \pi$. Instability sets in at the points $(2n+1)\omega_1 = 2\omega_0$, $n = 0, \pm 1, \ldots$.

It turns out that for quasi-periodic perturbations as well the solution is (strongly) unstable if F_1 or F_2 is large enough, or if either T_1 or T_2 is inside an unstable region and F_1, F_2 are comparable in size. For $F_1/\omega_0 = F_2/\omega_0 = 0.5$ we looked for strongly unstable behavior on a coarse T_1, T_2 grid. About one-third of all cases were found to be unstable. The majority, but not all, unstable cases closely corresponded to the relation

$$m\omega_1 + n\omega_2 = r\omega_0 \tag{4.2}$$

m, n, r being integers, with absolute values $\leqslant 3$. The most conspicuous correlation is with $\omega_1 + \omega_2 = 2\omega_0$ [apart from the above-mentioned $(2n+1)\omega_i = 2\omega_0$]. The amplitudes F_i, however, also play a nontrivial role. For $\omega_0 T_1 = 0.998$, $\omega_0 T_2 = 0.7574$, for example, hence $\omega_2 - \omega_1 \approx \omega_0$, there is a sickle-shaped (weakly) unstable region around $(F_1^2 + F_2^2)^{1/2} \approx 0.6\omega_0$, with stability for both larger and smaller amplitudes (cf. cases 2 and 3).

We investigated how the distance δ_n of two neighboring starting points changes in time by calculating the Kolmogorov–Sinai (KS) entropy[5]:

$$S = \frac{1}{N} \sum_{n=1}^{N} \ln \frac{\delta_n}{\delta_{n-1}} \tag{4.3}$$

for large N. For stable cases, with all initial values tested, its absolute value converges to zero. This has to be expected since linearity of Eq. (2.1) together with stability of the solution exclude the existence of a diverging sequence of δ_n. "Local" mixing instability[5] thus does not occur.

5. HAMILTONIAN AND RELATED VIEWPOINTS

To obtain a better understanding of the properties of Eq. (2.1), more general viewpoints are helpful.

A general investigation of whether linear (and nonlinear) systems of quasi-periodic differential equations of the form of Eq. (2.5) are reducible has been made by Bogoliubov, Mitropolskii, and Samoilenko (BMS).[6] Equation (2.5) is "reducible" if an ansatz $\mathbf{x}(t) = \mathsf{U}(\boldsymbol{\theta}(t)) \cdot \mathbf{y}(t)$, $\mathsf{U}(\boldsymbol{\theta}) = \mathsf{U}(\boldsymbol{\theta} + 2\pi \mathbf{e}_i)$ with quasi-periodic matrix function U yields $\dot{\mathbf{y}} = \mathsf{K} \cdot \mathbf{y}$, where K is a constant matrix. Reducibility is thus a generalization of Floquet–Lyapunov properties to the quasi-periodic case and determines the type of solution. The BMS theory considers small quasi-periodic perturbations, $\mathsf{A} = \mathsf{A}_0 + \mathsf{A}_1(\boldsymbol{\theta})$, $\mathsf{A}_0 = \text{const}$, $\mathsf{A}_1 = 0(\varepsilon \ll 1)$, and employs the method of accelerated convergence of the perturbation series as developed by Kolmogorov, Arnol'd, and Moser (KAM).[2] Very briefly speaking, BMS show that the measure of reducible matrices A is finite and tends to 1 for $\varepsilon \to 0$, and that for any A_0 there is a slightly different reducible $\tilde{\mathsf{A}}_0$, provided the frequencies involved ($\omega_0, \omega_1, \omega_2$ in our case) are sufficiently incommensurate and the perturbation is smooth enough. In addition, BMS have proved reducibility under similar conditions, provided the eigenvalues of A_0 have a real part, which, however, does not hold in our case (ω_0 real). Consequently, BMS theory gives us no explicit results.

There is a more direct connection of Eq. (2.1) with KAM theory. By introducing the canonical variables $q_0 = \psi$, $p_0 = \dot{\psi}$, which are transformed to Q_0, P_0 by $q_0 = \sqrt{2P_0/\omega_0} \cos Q_0$, $p_0 = -\omega_0 \sqrt{2P_0/\omega_0} \sin Q_0$, and the canonical coordinates

$$Q_i = \omega_i t + c_i \qquad i = 1, 2 \tag{5.1}$$

Eq. (2.1) may be written as an autonomous Hamiltonian system with canonical variables $\mathbf{P} = (P_0, P_1, P_2)$ and $\mathbf{Q} = (Q_0, Q_1, Q_2)$ as follows:

$$H(\mathbf{Q}, \mathbf{P}) = \sum_{\alpha=0}^{2} \omega_\alpha P_\alpha + \frac{P_0}{\omega_0} \cos^2 Q_0 \cdot f(Q_1, Q_2) \tag{5.2}$$

with $f(Q_1, Q_2) = f(Q_1 + 2\pi, Q_2) = f(Q_1, Q_2 + 2\pi)$. The advantage of introducing two coordinates to make H time independent is that as a result P_α, Q_α, $\alpha = 0, 1, 2$, are conjugate action and angle variables for the undisturbed Hamiltonian $H_0 = H(f = 0)$; Q_α, mod 2π, are angles on three-dimensional "torus surfaces" $P_\alpha = \text{const}$, $\alpha = 0, 1, 2$, which are embedded in the six-dimensional phase space. KAM theory discusses the survival of invariant tori in phase space (i.e., integrability) when a small perturbation, $|f| \ll 1$ here, is applied. Unfortunately, KAM theory requires that $\det(\partial^2 H_0 / \partial P_i \, \partial P_j) \neq 0$ (or an equivalent condition). This condition, however, does not hold here, since H_0 is linear in $\mathbf{P}$. Moreover, the perturbation $\Delta H = H - H_0$ is also linear. Therefore, techniques to obtain a nonlinear $\tilde{H}_0$ to some higher order in f from the nonlinearity in ΔH[7] do not work here. Integrability of such "completely degenerate" Hamiltonians has not, to our knowledge, been investigated before. (Note that H is integrable if f is periodic in t, according to the Floquet theory; integrability is uncertain only for quasi-periodic f.)

To decide whether invariant tori exist, we use the following surface of section technique appropriate to our six-dimensional problem. Let us assume H to be completely integrable. Then, by definition, apart from $H(\mathbf{Q}, \mathbf{P}) = \text{const}$, two additional invariants of motion exist, $I_i(\mathbf{Q}, \mathbf{P}) = \text{const}$, $i = 1, 2$. They may be used in principle to express two variables (e.g., $P_2 Q_2$) in terms of the other ones. One invariant, indeed, always exists: $I_1(Q_1, Q_2) = Q_1/\omega_1 - Q_2/\omega_2$, from the definition of Eq. (5.1). In addition, we make a cut at $Q_1 = 0$, mod 2π (cf. a cut at a fixed toroidal angle of a two-dimensional torus in three-dimensional space). If invariant tori exist, the trajectories crossing the cut still form a smooth manifold on it.[3] From $H = \text{const}$ the manifold follows in the form (with Q_0, P_0 transformed into q_0, p_0)

$$h(q_0, p_0, P_1) = \text{const} \tag{5.3}$$

where $q_0 = \psi$, and $p_0 = \dot{\psi}$ and P_1 are taken at $t = n \cdot 2T_1$. Equation (5.3) represents a two-dimensional surface in q_0, p_0, P_1 space, in the limit $t \to \infty$. (A cut at $Q_2 = 0$, mod 2π is analogous.) Thus, if the points ($x_n = \psi$, $q_n = \dot{\psi}$, $z_n = P_1$) at $t = n \cdot 2T_1$ lie on a smooth surface, H is integrable, and a torus is preserved. Conversely, if the sequence of points fills a volume, H is not integrable. The method of visualizing (ψ, $\dot{\psi}$, P_1), and the results are presented in the next section.

Equation (5.3) requires a knowledge of P_1 or P_2. They follow from the canonical equations

$$\dot{P}_i = -\frac{\partial H}{\partial Q_i} = -\frac{q_0^2}{2} \frac{\partial}{\partial Q_i} f(Q_1, Q_2) \qquad i = 1, 2 \tag{5.4}$$

By integration over the δ-functions and partial integration one obtains for the jump of P_i

$$[P_i] = \int dt\, \dot{P}_i = -\frac{1}{2}\int dt\, q_0^2 \frac{\partial}{\partial \omega_i t} f(\omega_1 t, \omega_2 t)$$

$$= \frac{1}{\omega_i}\int dt\, q_0 \dot{q}_0 f = -\frac{1}{\omega_i}\int dt\, p_0 \dot{p}_0 = -\frac{1}{2\omega_i}[p_0^2] \qquad (5.5)$$

where $\dot{q}_0 = p_0, \dot{p}_0 = -(\omega_0^2 + f)q_0$ have been used and jumps at $t_n = 2nT_i$ only contribute. $[p_0^2]$ follows from Eqs. (2.7) and (2.8).

6. THREE-DIMENSIONAL PLOTS

To find out whether a numerically computed sequence of points with the Cartesian coordinates (x_n, y_n, z_n) fills a volume or lies on a smooth surface the usual methods of level lines or perspective drawing are hardly applicable because the (x_n, y_n) values are not located on a fixed grid. Therefore, we chose to represent the sequence of points direct by a stereoscopic technique. Two conjugate plots are made from the sequence, each being a projection under a slightly different viewing angle. For better inspection the whole sequence is also tilted and turned 45° counterclockwise. The double plot should be viewed not by fixing the eyes on the points but by "looking through" them and fixing at infinity. With some training a very pronounced stereoscopic impression is created. Sometimes it is helpful to put a white sheet between the eyes and the center of the double plot.

In Figs. 1.3, 2.3, and 5.3–8.3 we present a selection of cases with typical behavior. For most cases the corresponding two-dimensional figures were discussed in Section 3, Figs. 1.1, 2.1, 4.1–7.1. The ellipse is a projection of the reference unit circle. The sequence of points starts with the initial value $(1, 0, b)$, where b is an arbitrary convenient "base." The number of iterations is called N_3. To make the position of points more obvious, the z-coordinate is either slightly compressed or magnified by a factor s ranging from 0.67 to 20.

The cases represented by Figs. 8.3 and 1.3 correspond to randomly chosen parameters. The figures show a ribbonlike structure and a thin, curved disk with a hole, respectively. A cushionlike structure with holes and a fencelike structure are visible in Figs. 2.3 and 5.3. The curve segments of case 6.3 seem to lie on a "beehive." In most cases the points do not yield a surface but fill a finite volume. In these cases H is nonintegrable. The deviation from a surface, however, is relatively small in many cases, not only for nearly periodic ones. In Fig. 7.3, however, well-defined surfaces (looking like curtains) are obtained so that integrability is proved.

Similar to the horizontal dimensions, the final size of the vertical extension in stable cases is reached at about $N_3 = 5$–$10{,}000$ iterations.

Integrability should be investigated for different initial values $(\overline{\mathbf{Q}}, \overline{\mathbf{P}})$ in phase space. The effects of $\overline{P}_1$, $\overline{P}_2$ and of one of the $\overline{Q}_1$, $\overline{Q}_2$ are trivial. The absolute value of the vector $(\bar{q}_0, \bar{p}_0)$ is also irrelevant since, owing to the linearity of Eq. (2.1) and owing to Eq. (5.4), $(\alpha\psi, \alpha\dot{\psi}, \alpha^2 P_1)$ is a solution if $(\psi, \dot{\psi}, P_1)$ is. It remains to check the effects of the direction of the vector $(\bar{q}_0, \bar{p}_0)$ and of the phase difference between Q_1 and Q_2, corresponding to $c_1 - c_2$ in Eq. (2.2). In all cases investigated we found that only minor details of the solution are affected, while the type of the solution remains unchanged ($c_1 = c_2 = 0$ in Figs. 1–8).

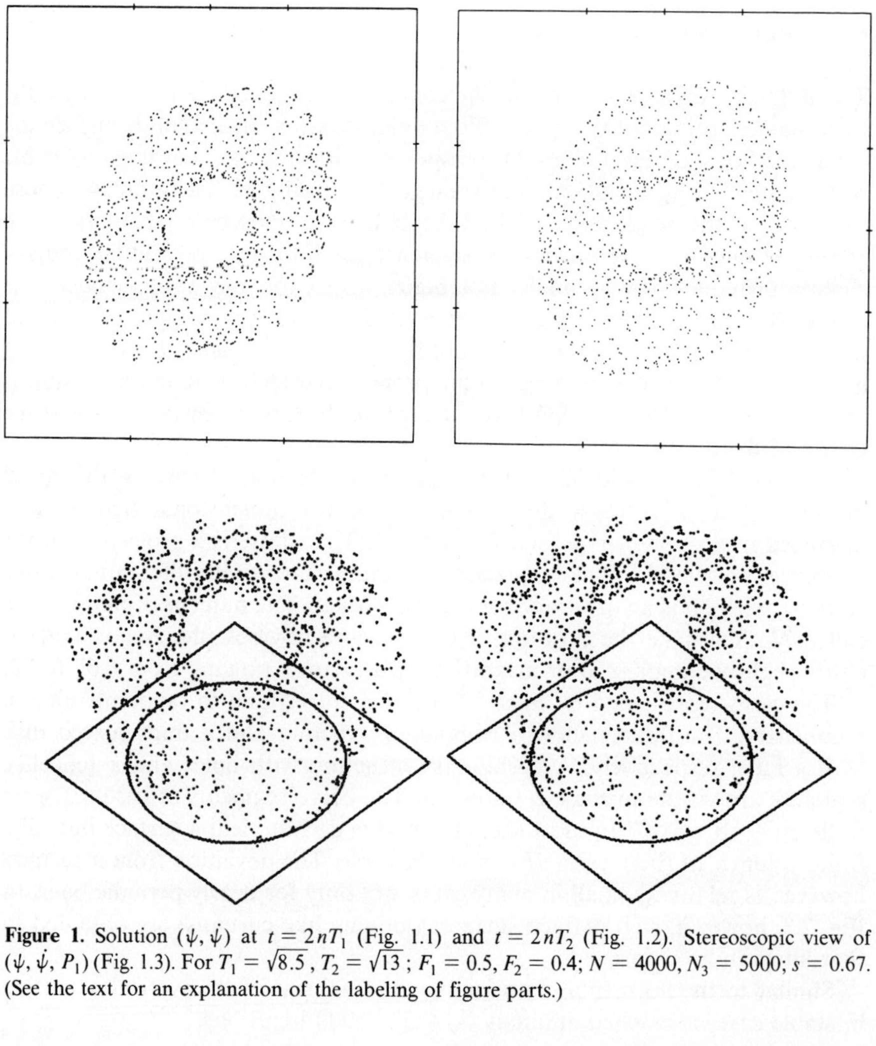

Figure 1. Solution $(\psi, \dot{\psi})$ at $t = 2nT_1$ (Fig. 1.1) and $t = 2nT_2$ (Fig. 1.2). Stereoscopic view of $(\psi, \dot{\psi}, P_1)$ (Fig. 1.3). For $T_1 = \sqrt{8.5}$, $T_2 = \sqrt{13}$; $F_1 = 0.5$, $F_2 = 0.4$; $N = 4000$, $N_3 = 5000$; $s = 0.67$. (See the text for an explanation of the labeling of figure parts.)

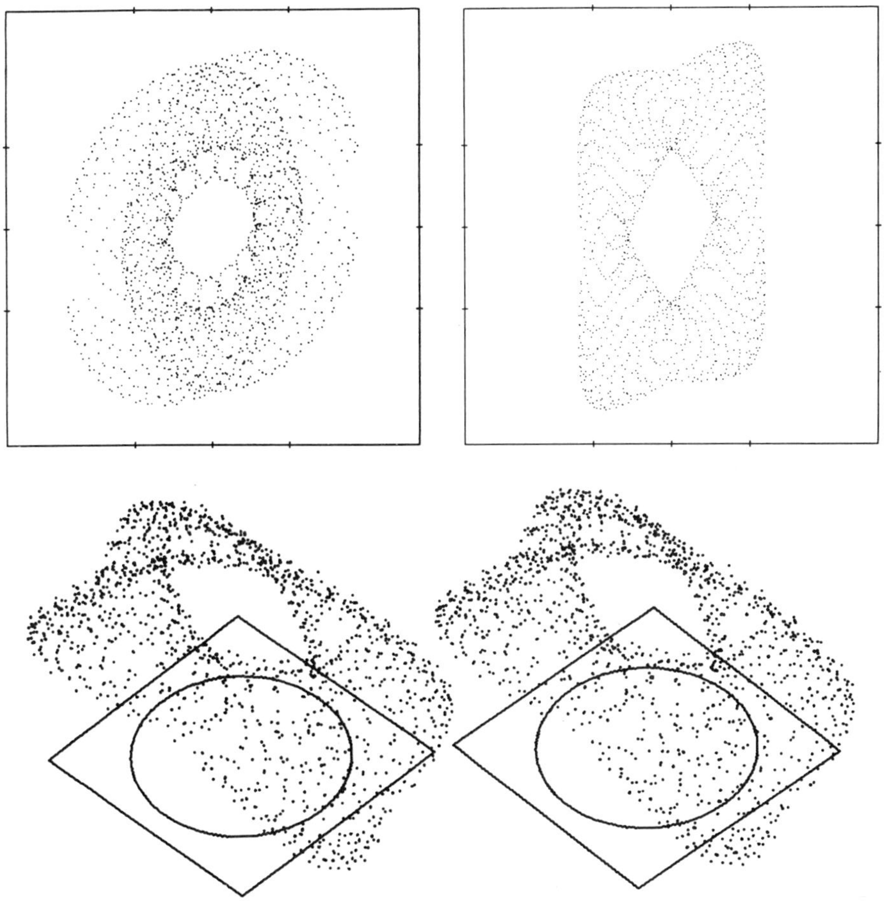

Figure 2. $T_1 = 0.998$, $T_2 = 0.7574$; $F_1 = 0.5$, $F_2 = 0.4867$; $N = 6000$, $N_3 = 6000$; $s = 0.67$. (See the text for an explanation of the labeling of figure parts.)

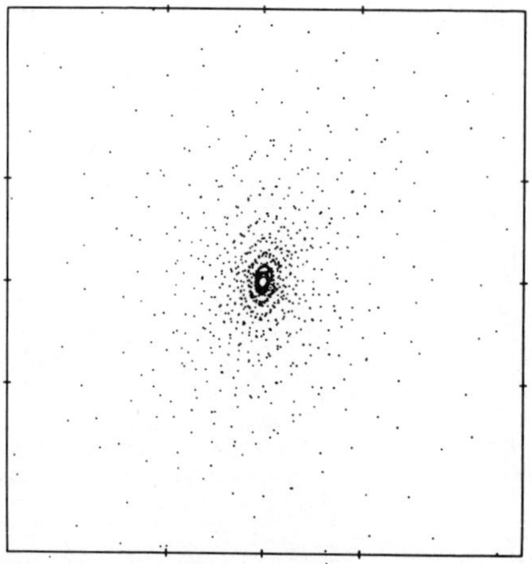

Figure 3. $T_1 = 0.998$, $T_2 = 0.7574$; $F_1 = 0.5$, $F_2 = 0.4512$; $N = 4000$. Weakly unstable. (See the text for an explanation of the labeling of figure parts.)

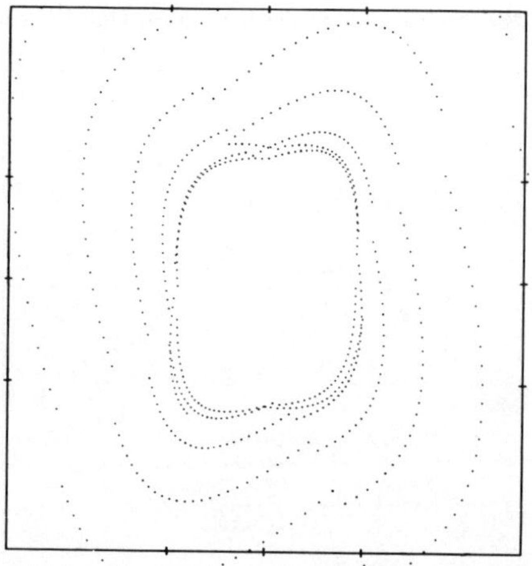

Figure 4. $T_1 = 1.027$, $T_2 = 0.77398$; $F_1 = 0.3$, $F_2 = 0.281$; $N = 6000$. Weakly unstable. (See the text for an explanation of the labeling of figure parts.)

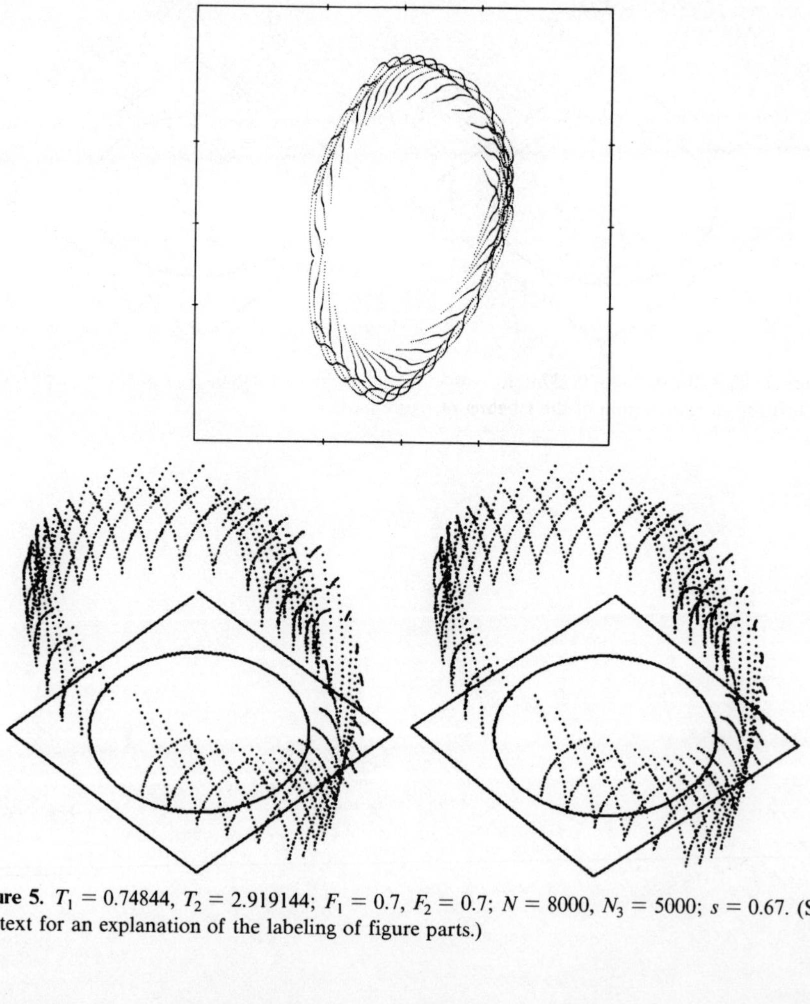

Figure 5. $T_1 = 0.74844$, $T_2 = 2.919144$; $F_1 = 0.7$, $F_2 = 0.7$; $N = 8000$, $N_3 = 5000$; $s = 0.67$. (See the text for an explanation of the labeling of figure parts.)

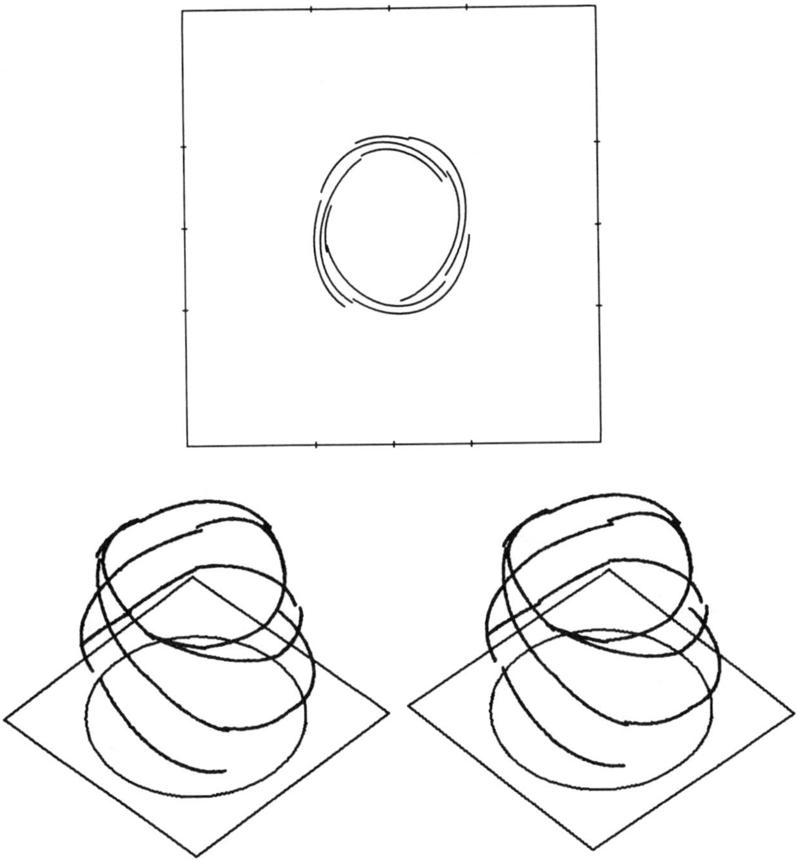

Figure 6. $T_1 = 1.045$, $T_2 = 0.7842$; $F_1 = 0.1$, $F_2 = 0.1$; $N = 8000$, $N_3 = 10\,000$; $s = 6.67$. (See the text for an explanation of the labeling of figure parts.)

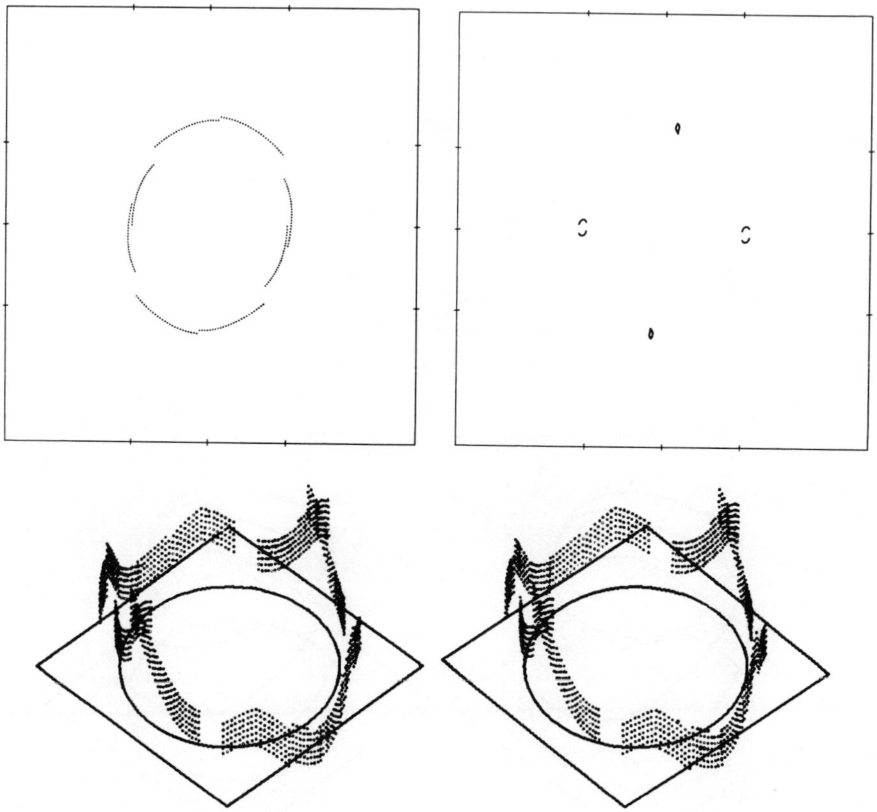

Figure 7. $T_1 = 1.027$, $T_2 = 0.773969$; $F_1 = 0.1$, $F_2 = 0.284$; $N = 8000$, $N_3 = 6000$, $s = 20$. (See the text for an explanation of the labeling of figure parts.)

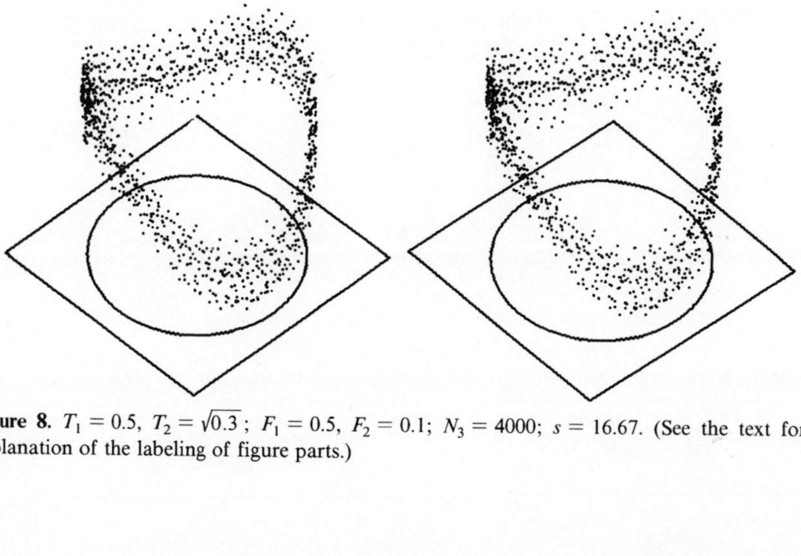

Figure 8. $T_1 = 0.5$, $T_2 = \sqrt{0.3}$; $F_1 = 0.5$, $F_2 = 0.1$; $N_3 = 4000$; $s = 16.67$. (See the text for an explanation of the labeling of figure parts.)

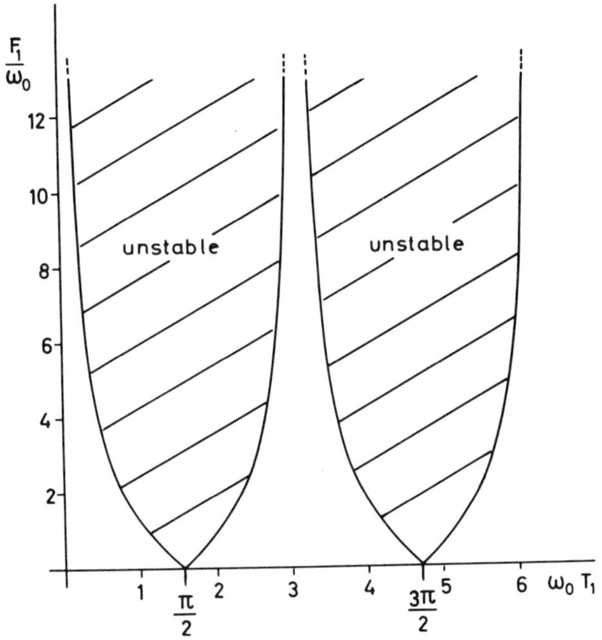

Figure 9. Stability diagram for periodic perturbation.

Thus, concerning the integrability of Hamiltonian H, Eq. (5.2), our results are as follows: depending on the parameters T_1, T_2 and F_1, F_2, H is either integrable in the whole phase space or is nonintegrable everywhere in phase space. Nonintegrability seems to occur more often.

ACKNOWLEDGMENTS

We thank H. Tasso and D. Pfirsch for helpful discussions.

This work was performed under the terms of the agreement on association between the Max-Planck-Institut für Plasmaphysik and EURATOM.

REFERENCES

1. J. A. Tataronis, J. N. Talmadge, and J. L. Shohet, "Alfvén Wave Heating in General Toroidal Geometry," University of Wisconsin Report ECE-78-10, (1978).
2. As a review, see: M. V. Berry, in *Topics in Nonlinear Dynamics*, S. Jorna, Ed., American Institute of Physics Conference Proceedings Series, Vol. 46 (AIP, New York, 1978); K. J. Whiteman, *Rep. Prog. Phys.* **40**, 1033–1069 (1977).
3. V. I. Arnol'd and A. Avez, *Ergodic Problems of Classical Mechanics* (Benjamin, New York, 1968).

4. V. A. Yakubovich and V. M. Starzhinskii, *Linear Differential Equations with Periodic Coefficients* (Wiley, New York, 1975).
5. B. V. Chirikov, *Phys. Rep.* **52**, 263–379 (1979).
6. N. N. Bogoliubov, Ju. A. Mitropolskii, and A. M. Samoilenko, *Methods of Accelerated Convergence in Nonlinear Mechanics* (Springer-Verlag, Berlin, 1976).
7. J. Ford and G. H. Lemsford, *Phys. Rev. A* **188**, 416 (1970); A. J. Lichtenberg, in *Stochastic Behavior in Classical and Quantum Hamiltonian Systems*, G. Casati and J. Ford, Eds. (Springer-Verlag, Berlin, 1979).

GRAVITATIONAL EXAMPLES OF NONDETERMINISTIC DYNAMICS

V. G. SZEBEHELY
Department of Aerospace Engineering
University of Texas at Austin, Texas

Abstract. *This chapter treats the simplest nonintegrable, nonlinear, conservative system of importance in celestial mechanics known as the restricted problem of three bodies. The differential equations of motion are presented and Birkhoff's idea of a "generating problem" is described. The existence and location of five equilibrium points are shown and their stability properties are discussed. The projection of the flow in the phase space into the configuration space is presented in the neighborhood of one of the (critically) stable Lagrangian points by studying the deformation of a line element. This is followed by establishing regions of initial conditions in the same neighborhood leading to chaotic behavior.*

1. INTRODUCTION

The classical aim of celestial mechanics is to obtain "all" solutions of the differential equations of the three- and n-body problems and to reveal the interrelations among all such solutions with various initial conditions and various masses, not only to predict the position of the celestial bodies at any instant but also to establish the universal laws that govern the motion of the system. This all-encompassing and admittedly overly ambitious aim is a generalization given by Hagihara[2] and it is partly based on Laplace's "fantasy" of deterministic dynamics. The aim is unreachable at present for several reasons.

It has been known since Poincaré's work[4] that even the simplest nontrivial dynamical system of celestial mechanics, the restricted problem of three bodies in two dimensions, is not integrable in the sense that this fourth-order system does not possess more than one uniformly valid global integral. Not unexpectedly, it can also be shown that series solutions of uniform convergence do not exist. Numerical integration efforts also fail because their randomness makes the results questionable, if not meaningless, for the long time periods of interest.

Regarding the nondeterministic nature, it is noted that the models presently used in celestial mechanics are approximations and the long-time effects of several phenomena are, consequently, not included and cannot be estimated. For the solar system accurate scientifically significant observations are available for a very short time only (as compared to its existence), consequently, there are no ways to estimate the accuracy of either the models or the initial conditions used.

Combining the foregoing properties of nonintegrability, of modeling uncertainty, of unknown initial conditions, and of random error propagations renders the dynamical systems of celestial mechanics nondeterministic in general. This chapter shows the high sensitivity to the initial conditions of the simplest, nonintegrable highly idealized model.

2. DESCRIPTION OF THE RESTRICTED PROBLEM OF THREE BODIES

Consider two point masses rotating around their center of mass according to Kepler's law of motion, that is, $n^2a^3 = GM$, where n is the mean motion, a is the semimajor axis, G is the gravitational constant, and $M = m_1 + m_2$ is the

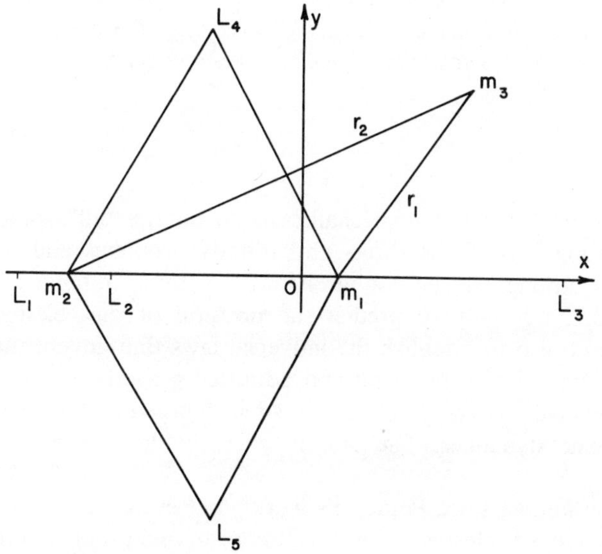

Figure 1. The restricted problem of three bodies.

total mass. Figure 1 shows the synodic or rotating coordinate system. Here m_1 and m_2, located on the x-axis of syzygies, are called the primaries, O is the fixed center of mass, r_1 and r_2 are the distances to the third body (m_3) of small mass, the motion of which is studied, and $L_1, \ldots, L_5$ are the Lagrangian or equilibrium points. Note that $m_3 \ll m_1, m_2$ so that m_1 and m_2 influence the motion of m_3 but m_3 does not influence the circular motions of m_1 and m_2.

The equations of motions are

$$\ddot{x} - 2\dot{y} = \Omega_x$$

$$\ddot{y} + 2\dot{x} = \Omega_y \tag{1}$$

where

$$\Omega = \frac{1}{2}\left[(1-\mu)r_1^2 + \mu r_2^2\right] + \frac{1-\mu}{r_1} + \frac{\mu}{r_2} \tag{2}$$

and

$$\mu = \frac{m_2}{m_1 + m_2} \qquad m_2 \leqslant m_1 \tag{3}$$

Here dots indicate derivatives with respect to time and subscripts correspond to partial derivatives of the potential function (Ω). The first term in Ω represents the centrifugal force. The second and third terms are the gravitational effects on the third body by the primaries. The Jacobian integral is given by

$$\dot{x}^2 + \dot{y}^2 = 2\Omega - C \tag{4}$$

where C is the Jacobian constant of integration. Note that in these equations the conventional nondimensional dependent and independent variables are used where the unit of distance is $\overline{m_1 m_2}$ and the dimensionless time is defined as t^*n, where t^* is the actual time in seconds if n is in radians per second.

The fourth-order system of Eq. (1) is nonintegrable and the only integral of global validity is given by Eq. (4). Therefore, chaotic motion may be expected under certain conditions. Note that the Hamiltonian of the system (corresponding essentially to the Jacobian integral) may be written as

$$H = \frac{1}{2}\left(P_1^2 + P_2^2\right) + Q_2 P_1 - Q_1 P_2 - (1-\mu)R_1^{-1} - \mu R_2^{-1} \tag{5}$$

where Q_i, P_i, and R_i are the coordinates, momenta, and time-independent distances in the synodic system, respectively.

The equilibrium points L_1, L_2, L_3 are unstable for any value of μ, while L_4 and L_5 are (critically) stable points if

$$\mu < \mu_0 = \frac{1}{2}\left[1 - \frac{(69)^{1/2}}{9}\right] = 0.038521 \tag{6}$$

For the Earth–Moon system $\mu = 0.01213$; therefore the equilateral libration points (L_4 and L_5) are stable.

Not only might the restricted problem be considered to be the simplest nonintegrable dynamical system with applications in celestial mechanics, it might be called a "generating problem,"[1] since it is a special case of the equation

$$\frac{d^2y}{dx^2} + \lambda(y)\frac{dy}{dx} = \text{grad}_y F(y) \tag{7}$$

where $y = \xi + i\eta$ is the complex dependent variable, x is a real independent variable, the functions $\lambda(y)$ and $F(y)$ are regular, and

$$\text{grad}_y F(y) = \frac{\partial F}{\partial \xi} + i\frac{\partial F}{\partial \eta} \tag{8}$$

The generalized complex differential equation above often appears when the restricted problem is regularized either for analytical or for numerical purposes.

In what follows the original formulation is used, since its behavior is of sufficient interest without introducing transformations. Nevertheless, the "generating" nature of the restricted problem is mentioned because this formulation contains no singularities. For such transformations as well as for more on the restricted problem, see, for instance, Ref. 6.

Sections 3 and 4 offer examples of deterministic and partly nondeterministic behavior of this nonintegrable dynamical system. Celestial mechanicians must be awakened from the comfortable dreams of linear and/or integrable undergraduate dynamics. Hence this chapter selects a gravitational model instead of other, often artificially created nonintegrable systems. Also emphasis is put on the dynamical aspects as opposed to the popular surface transformations, which often offer enjoyable topological exercises but are sometimes without clear dynamical implications.

3. DEFORMATION OF A LINE ELEMENT

A vertical line element is placed at L_4 with length of 0.1 (in the nondimensional system where $m_1 m_2 = 1$) with end points at $L_4(x_4 = \mu - \frac{1}{2}, y_4 = \sqrt{3}/2)$ and $P(x_4, y_4 + 0.1)$. Point L_4 is a stable and stationary equilibrium point in the synodic system, but points in its neighborhood are not stationary. Nevertheless, one expects small deformations of a line element near such a stable point. That this is not the case is demonstrated below. The initial conditions are zero velocity relative to the synodic system for all points of the line element.

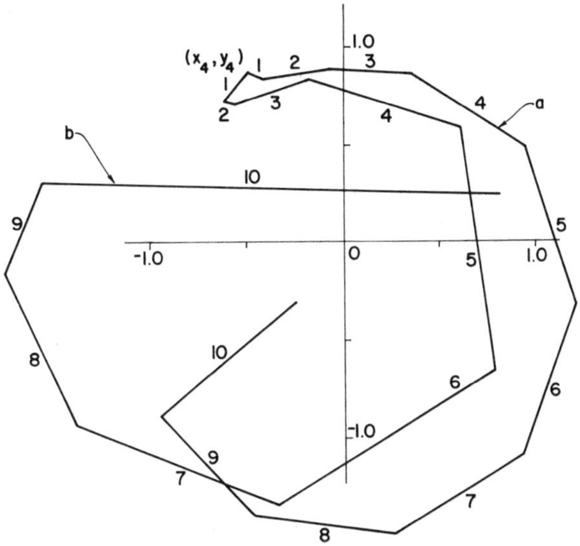

Figure 2. Deformation of the line element in 52 and 61 days.

Figure 2 shows what happens to the line element as time progresses. Note that the original length of the line element (0.1) is divided in 10 parts. Curve *a* in Fig. 2 shows what happens to the end points of every tenth of the line elements. The first part of the line element (having the original length 0.01 and denoted by 1) is between L_4 and the next point above it. The second part of the line element, which was originally located between $(y_4 + 0.01, x_4)$ and $(y_4 + 0.02, x_4)$, moved to approximately $y = 0.8$ and $x = -0.1$. The last part of the line element, located between $(y_4 + 0.09, x_4)$ and $(y_4 + 0.1, x_4)$, moved to $(x \cong -1, y \cong -1)$ and $x \cong -0.3$, $y \cong -0.3$, an increase in length from 0.01 to 1 or a factor of magnification of 100. Not only was the line element moved significantly, but it was increased in size. This occurred in 12 dimensionless time units, corresponding to approximately $12 \times 4.35 \cong 52.2$ days.

Another erratic behavior of the original small line element located near the stable point is also given in Fig. 2, showing an intersection (curve *b*). The point originally at $y_4 + 0.09$, $x_4 = \mu - \frac{1}{2}$ moved to $y \cong 0.25$ and $x \cong -1.6$, but the point originally at $y_4 + 0.1$ and $x_4 = \mu - \frac{1}{2}$ moved to $y \cong 0.2$ and $x \cong 0.8$. The original (0.01 long) vertical line element changes into a horizontal line element of length 2.4, corresponding to a stretching of 240 during a little over 2 months. For more information concerning this problem, see McKenzie and Szebehely.[3]

4. REGIONS OF REGULAR AND CHAOTIC BEHAVIOR

This section establishes the region of stability with special but well-defined conditions around the triangular libration points and shows that at the borderlines of the stability regions chaotic behavior may occur. Once again the Earth–Moon system is considered in which the points L_4 and L_5 are critically stable; that is, they have pure imaginary characteristic roots. A particle that is placed at L_4 or L_5 with zero velocity relative to the synodic system will stay there or, in case of small perturbations, will librate in a two-frequency mode. If the particle's initial relative velocity is not zero or if it is not placed at the points L_4 or L_5, once again libration will occur.

Studies of the effects of errors in the initial velocities are in progress. Here the effects of errors in the initial position are described and the sensitivity of the system to these errors is studied. For this purpose particles are placed in the neighborhood of the points L_4 and L_5 and their motion is determined by careful numerical integration. Librational motion is defined when a particle stays in the vicinity of L_4 and L_5; chaotic motion, on the other hand, is referred to when the particle leaves this region (i.e., terminates its librational motion). It might be captured and/or recaptured by the Earth and/or by the Moon, the orbit might encircle both primaries; then captured again, it might have close approaches to the primaries, or it might escape the system. Lengthy, careful, and accurate numerical studies show that once the librational amplitude is such that the orbit intersects the axis of syzygies, its motion becomes erratic.

The length of time of integration is critical in these experiments. To determine the regions of librational motion the length of time of the numerical integration should not be too short, sine the onset of stochasticity might not be reached in a short time. (Note that the long period of the linearized solution is 92 days.) The length of time of the numerical integration cannot be too long for practical reasons as well, because error accumulation can make the result meaningless by continuously changing the initial conditions. It was established after 240 nondimensional time units, corresponding to approximately 3 years, that the character of the motion did not change. Such experiments were performed for 960 nondimensional time units, corresponding to 12.4 years, without any new phenomena appearing after 3 years. As a reasonable compromise the results of this chapter show the effects for a little over 6 years, corresponding to 480 nondimensional time units or 25 long periods.

The grid size of the search for initial conditions was 5×10^{-3} nondimensional unit. The numerical integration was performed with a twelfth-order, variable mesh, multistep method. Figure 3 shows the results for the region around L_4. The coordinates (ξ_4, η_4) measure nondimensional distances from L_4 along the (x, y) coordinates. The coordinates of L_4 are $\xi_4 = \eta_4 = 0$. The outer curves determine the regions investigated. Outside these curves the motion shows irregularity. Inside these outside curves the black regions correspond to initial conditions resulting in librational motion and the white regions to irregular motions. The solid black regions around L_4 are not surprising, and sharp lines of demarcation exist between librational and irregular motions. At the extreme ends of the region considerable irregularities appear, and islands of initial conditions leading to libration are embedded in initial conditions

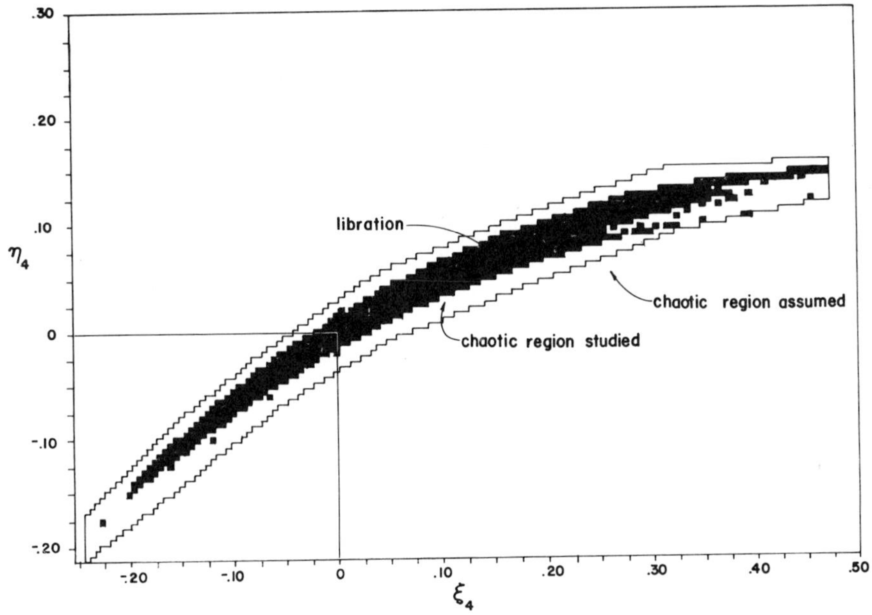

Figure 3. Regions of initial position around L_4.

associated with irregular motion. Small changes of the initial conditions in these regions result in significant changes in the character of the motion. For more information on this problem, consult McKenzie and Szebehely.[3]

5. SUMMARY

In view of the nonintegrability of the restricted problem[4] and from the point of view of Whittaker[8] concerning the frequency effects on the adelphic integrals, it is not surprising that the two examples above demonstrate discontinuous islands and irregular behavior. Prigogine's recent work[5] questions the deterministic aspects of classical mechanics. The aim of this chapter is to demonstrate such possibilities in celestial mechanics. Some of these results are most disturbing, since if celestial mechanics is not a deterministic science (and apparently it is not), is there any meaning in our attempts to predict the future of the solar system or of our galaxy?

ACKNOWLEDGMENTS

The sponsorship of the Johnson Space Center of the National Aeronautics and Space Administration, and the National Science Foundation is acknowledged.

REFERENCES

1. G. D. Birkhoff, *Rend. Circ. Mat. Palermo* **39**, 1 (1915).
2. Y. Hagihara, *Celestial Mechanics*, Vols. 1–3 (MIT Press, Cambridge, MA), Vols. 4–6 (Japan Society for the Promotion of Science, Tokyo, Japan), 1970–1976.
3. R. McKenzie and V. Szebehely, *Celestial Mech.* **23**, 223 (1981).
4. H. Poincaré, *Compt. Rend.* **123**, 1224 (1896).
5. I. Prigogine, *Celestial Mech.* **16**, 487 (1977).
6. V. Szebehely, *Theory of Orbits* (Academic Press, New York, 1967).
7. V. Szebehely and R. McKenzie, *Celestial Mech.* **23**, 131 (1981).
8. E. Whittaker, *Analytical Dynamics* (Cambridge University Press, New York, 1904).

INSTABILITIES IN PLANETARY SYSTEMS

R. O. VICENTE*
Department of Aerospace Engineering and Engineering Mechanics
University of Texas
Austin, Texas

Abstract. *One of the unsolved central problems of celestial mechanics, the stability of planetary systems, is considered, emphasizing the orbital stability of the members of our solar system. The major components (planets and satellites) are analyzed, and attention is called to the behavior of asteroids, comets, and meteors. The principal orbital elements studied are the semimajor axes, the eccentricities, and the inclinations. The physical processes considered are gravitational interactions (three-body and n-body effects), tidal effects, collisions with smaller bodies, drag effects, radiation pressure, the Poynting–Robertson effect, and electromagnetic effects.*

The tools of analysis are the investigation of the energies and angular momenta of the bodies and the method known as Hill's stability theory.[4] *In connection with the latter, a measure of stability established by Szebehely and McKenzie*[10] *is computed for members of the solar system. The final conclusions are that because of the short times of accurate observation, not enough information has been accumulated to establish the long-term stability of the members of the solar system.*

1. INTRODUCTION

The problem of stability and instability is closely linked with time intervals, and it has meaning only when the time interval we suppose to be adequate for the particular case of stability under consideration is specified.

---

*Permanent address: Department of Applied Mathematics
Faculty of Sciences
Lisbon, Portugal

We can, for instance, consider the stability of the planets corresponding to very short (10^3 years) or very long (10^9 years) time intervals. Considering the admitted age of the solar system (about 4.5×10^9 years), we can say that the orbits of the planets are stable for very short intervals. However the question of stability for very long intervals of time is much more difficult because the computational techniques employed in present-day mathematical models are not yet capable of handling stabilities or instabilities for periods of 10^9 years, and these are the time spans that are interesting. We cannot forecast much about stability for intervals of the order of 10^6 years because they correspond to short intervals in comparison with the present admitted age of the system.

The possibility of discovering instabilities in any planetary system is related to the number of bodies that describe the same type of orbit and their positions within the system. For the solar system, we can consider two categories, according to their masses and number of bodies:

Category A	Category B
Planets	Asteroids
Satellites	Comets
	Meteors

The parameters we want to analyze refer to the inclination and eccentricity of the orbit and to the motion (i.e., whether it is direct or retrograde). We are interested in a morphological study of the system, and we say that any special group of bodies exhibits regularity if a significant number of them show the same features.

The possibility of finding any instabilities in the orbits of the planets is very small because they are so few in number and, also, being the main components of the system, any unstable planets have long ago disappeared. The majority show small inclinations and eccentricities, having direct motions, which are the regular features of the planets.

The case of the natural satellites is slightly different because there are about four times more satellites than planets. They still show the same significant regularities, that is, small inclinations and eccentricities in relation to their primaries, and direct motion, with a few exceptions. One of the remarkable regularities in satellite orbits is the near-zero values of eccentricities for all close satellites.

Another distinction in the problem of studying the stability of planets and satellites refers to the time interval they take for completing a revolution, because whereas satellites have taken about 10^{12} revolutions about their primaries, the planets have gone around about 10^9 times. This fact shows again that investigations about possible instabilities of the planets are much more difficult than those involving satellites.

The mathematical models employed to explain the motions of the planetary system have been compared only with observations made in the last few centuries; the precision of astronomical observations was low even two centuries ago, and this is a very short time interval. This raises another difficulty in the study of the stability of the system.

The claim made sometimes, especially in the last century, that celestial mechanics can accurately predict the positions of the bodies of the solar system for centuries ahead, is difficult to substantiate. Our mathematical models of the dynamics of the solar system are incomplete, and the accuracy of the observations is questionable. The degree of accuracy we are interested in depends very much on the objectives of the investigations at a given epoch, that is, if we are satisfied to verify the agreement between theory and observation with an accuracy of 1″ or 0.″001. The extremely short time interval of our observations cannot support or deny the deterministic point of view of dynamics, as applied to celestial mechanics.

We are interested in finding out any possible evidence of instability in the bodies that constitute the solar system. At the same time, we notice a certain number of regularities in the planetary system that should be associated with the stability of the system.

Let us now consider some of the different physical processes that might affect the stability or instability of planetary systems:

1. Gravitational interactions.
   a. Three bodies: Sun, planet, satellite; Sun, planet, asteroid.
   b. n bodies.
2. Tides in planets and satellites.
3. Collisions with smaller bodies.
4. Gas drag—interplanetary dust, atmosphere.
5. Radiation pressure.
6. Poynting–Robertson effect.
7. Electromagnetic effects—Coulomb drag.
8. Gas cloud—collisions among the particles.

This is not an exhaustive list of all the possible physical processes that might affect the stability of planetary systems, but we can immediately see that these processes depend on two factors:

1. The mass of the bodies.
2. The time interval during which they act on the components of the system.

Some of these physical processes (e.g., the gas cloud) might have been more important at an earlier age of the planetary system than they are nowadays. Also the collisions with smaller particles have intervened more for the instability of the system during the first stages of formation of the system than at present.

We are not considering the time scale of the formation of planetary systems, but only the possible effects of these physical processes in the stability of the system.

The mass of the bodies concerned is important for the influence of some of the physical processes mentioned. We can deduce that tides are important for planets and satellites (category A), whereas physical processes 3–7 affect the stability of category B bodies.

Considering that category B bodies (asteroids, comets, and meteors) are far more numerous than those in category A, and, also, they have smaller masses, hence are more likely to be affected by a greater number of the physical processes mentioned, we should infer that the study of instabilities in planetary systems, at their point of evolution corresponding to the solar system, will be more interesting and profitable for bodies of category B.

2. ANALYTICAL CONSIDERATIONS

Let us examine the possible evolution of the orbital elements a (semimajor axis), e (eccentricity), and i (inclination) from the point of view of inferring the stability of the system from the morphological aspects of the orbits.

The total energy W contained in a satellite's orbit of semimajor axis a and mass m, moving about a fixed mass M is $W = -G(Mm/2a)$, where G is the universal gravitational constant. This result is true in the two-body problem whether m is negligible or m/M is not very small; in this case, $M/(M+m)$ of the energy is stored in the smaller and faster moving mass.

We should notice the important question of possible variations in the value of G in relation with the evolution of planetary systems. It is quite a different thing to consider time scales of 10^2 or 10^9 years. Again, our observations of the solar system cannot discriminate in favor or against any variations in the value of G. If there is a very small variation of G corresponding to a few parts in 10^{10}, we cannot detect it in short time intervals, but such a variation might be relevant for the long-time dynamics of the evolution of planetary systems.

The change in the energy of the satellite orbit, that is, the work, points out the way in which the orbit size varies

$$\Delta a = \frac{2a^2}{GMm}\Delta W$$

where ΔW is often a function of a or m.

Therefore, knowing the work done by the various forces, an integration in relation to the time will give the initial size of the orbit. But to arrive at a reliable answer ΔW must be well known, and that is one of the difficulties. The most that is frequently known is the sign of ΔW, which only gives the direction of change of a:

Work done *on* an orbit expands that orbit.
Work done *by* an orbit collapses that orbit.

It is possible to write a simple expression for the angular momentum H of a satellite orbiting a fixed planet

$$H = m\left[GMa(1 - e^2)\right]^{1/2}$$

and deduce from it the expression for the eccentricity e

$$e = \left(1 + \frac{2H^2W}{G^2M^2m^3}\right)^{1/2}$$

Differentiating this expression we obtain

$$\Delta e = \frac{e^2 - 1}{e}\left(\frac{1}{2}\frac{\Delta W}{W} + \frac{\Delta H}{H}\right)$$

and we may see that the eccentricity is affected by changes in both the orbital energy and the angular momentum. Since ΔW and ΔH can have either sign, the two terms interact to determine whether the orbit becomes more or less circularized in the evolution of the system.

The eccentricity does not change if the condition

$$\frac{\Delta W}{W} = -2\frac{\Delta H}{H}$$

is satisfied over each complete orbit. We can understand that this condition is verified for short time intervals, but it is difficult to imagine that nature will satisfy this expression for very long time intervals (e.g., 10^9 years).

A similar discussion can be considered for the inclination i and its variation. The inclination of the plane of the orbit in relation to some plane (considered to be fixed) is defined by

$$\cos i = \frac{H_n}{H}$$

where H_n is the component of angular momentum that is normal to the fixed plane. Differentiating this expression, we obtain

$$\Delta i = \frac{1}{\left[(H^2 - H_n^2)/H_n^2\right]^{1/2}}\left(\frac{\Delta H}{H} - \frac{\Delta H_n}{H_n}\right)$$

and we can see that the inclination does not change if $\Delta H/H = \Delta H_n/H_n$. Again, this is a condition that must be satisfied for any values of the time.

The problem of the tides was first discussed in a systematic way by Darwin,[1] studying the tidal forces that change the size of the orbit and affecting the other orbital elements. The important problem of tidal friction was studied by Jeffreys,[5] who applied Darwin's results to find out how the eccentricity and inclination of satellite orbits are influenced by various models of tidal friction. Jeffreys showed that in the majority of cases the eccentricity increases under the action of planetary tides. We can see that this occurs, considering the previous equation for Δe, because the energy term dominates even though the already positive angular momentum of the satellite is increased by tidal torque.

We know that all close satellites show values of the eccentricities near to zero, that is, the orbits are nearly circular, and that is one of the known regularities for this type of satellite. This observational fact seems to contradict the calculations done by Jeffreys, but the explanation of this apparent contradiction was given by Goldreich.[2]

Let us consider a satellite in an elliptic orbit whose rotation is synchronously locked to its orbital motion. It experiences an oscillating tidal strain, that is, the distortion changes with the orbital period. The structure of the satellite is heterogeneous and this oscillation dissipates energy for anelastic materials. The satellite tide corresponds to a radial tide and, because of the synchronous lock, such a tide cannot transfer angular momentum to the satellite orbit. Remembering the previous equation for Δe, we can simplify it for small values of e considering only energy loss in the satellite. We obtain

$$e\,\Delta e \approx -\frac{\Delta W}{2W}$$

and, since W is negative, the eccentricity will decrease because of the energy loss in the satellite. Small energy losses are effective considering the case of small e, but for larger e the value of $\Delta W/W$ will increase and the decay can also be appreciable.

Although we are considering very simple mathematical models, they nevertheless give us some insight about the evolution of the eccentricities of the orbit.

The tidal effects in the values of the inclination were first studied by Darwin, and his results were confirmed by Jeffreys employing an adequate tidal model. We can envisage physically the variation of the inclination with the tides, considering the previous expression for Δi, because the planetary rotation carries the tidal bulge out of the satellite orbit plane, producing a torque normal to the orbital plane. The orbital inclinations of a satellite with respect to its proper plane should change with time in the opposite sense to the change of a, and this agrees with the conservation of orbital angular momentum, neglecting the angular momentum taken from the primary.

The possible complication introduced by the precession of the primary has been shown to be negligible for the great majority of satellites, except the Moon.[2]

These simple mathematical models do not consider physical processes other than tides, and we did not even consider the tidal lag angle and the dissipation function represented by the dimensionless parameter Q. This is a very difficult subject, showing several complications for the case of the Earth–Moon system. The possibility of having reliable values for the tidal lag angle and Q for other planets and satellites is, therefore, a source of speculation at present.

Let us see whether employing an adequate definition of stability, we can justify some of the results obtained from tidal considerations and, also, explain some of the morphological aspects of the solar system.

3. HILL'S STABILITY

There are several definitions of stability in celestial mechanics, one of them originated by Hill.[4] He had the idea of applying zero-velocity surfaces to investigate the stability of the lunar system, corresponding to the restricted problem of three bodies. Hill showed that the Moon's orbit is inside a zero-velocity oval, which in turn is inside the limiting critical figure-eight curve of zero velocity. This definition has given more fruitful practical results thanks to the investigations developed by Szebehely[9] and Szebehely and McKenzie.[10].

The measure of stability S is defined by

$$S = \frac{C_{ac} - C_{cr}}{C_{cr}}$$

where C_{ac} and C_{cr} are the actual and critical values of the Jacobian constant for the orbit under consideration. The Jacobian constant is computed from the Jacobian integral of the restricted problem of three bodies.

If the satellite's Jacobian constant C_{ac} is much larger than the critical value C_{cr}, we may speak about a very stable situation that will not be influenced by the perturbing effects not included in this mathematical model. On the other hand, if $C_{ac} = C_{cr}$, the stability is questionable and external effects may introduce possible unstable behavior. Consequently, if S is a large positive number, the motion is definitely stable, but if S is zero or negative, instability may set in. We mean by stability (Hill's stability) not only that the satellite cannot leave its primary but that it cannot be captured. By instability we mean that the satellite *may* leave its primary and *could* have been captured. This corresponds to a sufficient condition.

Considering the application made by Szebehely (Fig. 1) to the motion of natural satellites, let us find a possible relationship between stability and the morphological aspects of the planetary system.

In Fig. 1 the mass parameter μ is given by $\mu = m_2/(m_1 + m_2)$, where m_1 and m_2 are the masses of the primaries P_1 and P_2, and ρ_{max} corresponds to the distance between the satellite and the primary P_2.

The results from Fig. 1 are summarized in Table 1, showing that equatorial satellites are within the stability region, except for the Moon, which is a special case. The possibility of marginal stability or noticeable instability appears only in the case of nonequatorial satellites.

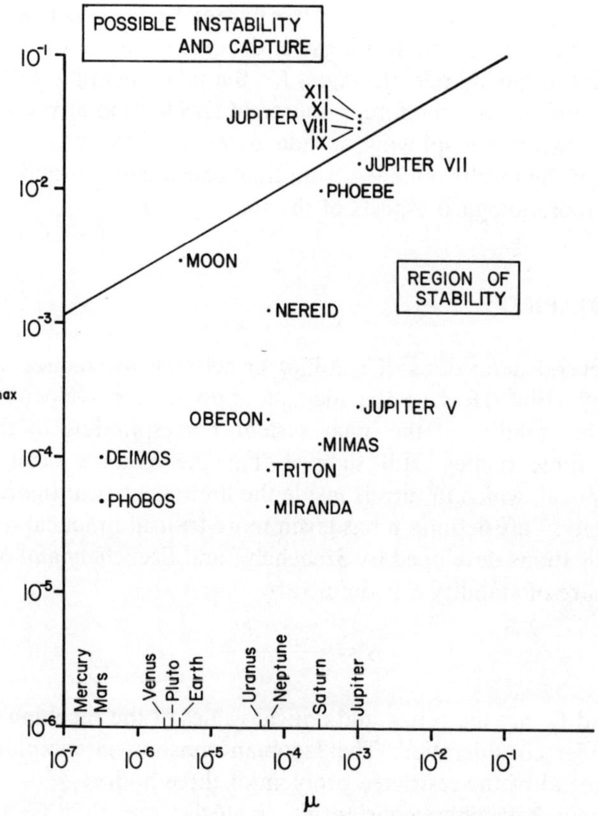

Figure 1. Plot of stability data. (Courtesy of V. G. Szebehely.)

This brief description of Hill's criterion of stability applied to the natural satellites gives us some insight into the possible evolution of the satellites. We notice some effect of the eccentricity contributing to the instability of the satellite; also, the greater inclination contributes to instability.

That the satellites are retrograde (see Table 1) does not imply, by itself, any tendency for instability. Moulton[7] showed that retrograde orbits are stable and Hénon[3] confirmed numerically these results, stating that retrograde orbits are stable for a wider range of Jacobi constants than the corresponding direct orbits.

It is difficult to generalize the results obtained (Fig. 1 and Table 1) because bodies of category A are too few and, therefore, the probability of finding some of them in a state of instability at the present time is very small, considering the time scale of the evolution of the planetary system.

The possibility of studying the stability of bodies of category B is more promising because these are far more numerous and are subject to a greater number of physical processes that might contribute to their instabilities.

Table 1. Comparison of Stability of Equatorial and Nonequatorial Satellites

Planet	Satellites Considered	Equatorial	Nonequatorial	Stability	Instability
Neptune	2		Nereid	Marginal	
			Triton (R)	Within	
Uranus	5	All		Within	
Saturn	10		Iapetus	Within	
			Phoebe (R)	Marginal	
		8		Within	
Jupiter	12	4 Galilean, JV		Within	
			3 inner group JVI, VII, X	Marginal	
			4 outer group (R) JVIII, IX, XI, XII		Within
Mars	2	2		Within	
Earth	1	Moon		Marginal	

For instance, the application of Hill's stability to the study of asteroids gives us a wider range of possibilities because statistical estimates suggest that there are about 30,000 asteroids within reach of modern instruments; unfortunately, only about 5% have their orbital elements known with some precision.

4. CONCLUSIONS

The recent application of this definition of stability to asteroids has revealed some interesting features, and the tentative conclusions obtained from the satellites about the effects of the eccentricity and inclination of the orbits have been made more certain.[11]

The measure of stability *S* for asteroids diminishes for orbits having greater eccentricity, and the same happens when orbits that have greater inclinations are considered. We have seen such an example with the outer group of Jupiter's satellites, which are within the instability region.

Comets present an interesting case because of their wide range of eccentricities and inclinations.[12] Considering these morphological aspects and following the same reasoning employed before, we should forecast that comets will present several examples of marginal stability or possible instability. This will imply, from the point of view of the evolution of comets, that because comets have a shorter evolutionary life, they will not have such long spans of evolution as do planets and satellites.

The possible shorter evolutionary life of comets has originated a number of attempts[8] to postulate a possible permanent source of material surrounding the solar system for the replenishment of comets. But this hypothesis might not be necessary, if we consider another hypothesis corresponding to the admission that comets are transient members of the solar system. They appear during a short time interval, corresponding to our present observations, because the solar system is passing through a region of our galaxy that permitted the formation of comets. Continuing our trajectory through the galaxy, they might disappear forever, or until the solar system reaches another region of the galaxy, where it would again be possible to originate another set of comets.

Some of the morphological features of the planetary system—namely, the low inclinations and small eccentricities of the majority of the orbits that impressed Laplace[6]—can be seen to correspond to a greater degree of stability in the evolution of the system. The bodies having greater inclinations and eccentricities show at least marginal instability, and, therefore, they are less numerous in the planetary system. These features correspond to instability and few bodies are, at any given time, in such a stage of their evolution.

These conclusions were obtained by applying Hill's definition of stability, but we might infer they will be supported considering less restricted cases.

Thus nearly 200 years after Laplace's remarks, it has been shown that the reason for low inclinations and small eccentricities is that orbits of this type show greater stability than any other orbits and that the less stable orbits have disappeared in the course of the evolution of the system.

REFERENCES

1. G. H. Darwin, *Phil. Trans. R. Soc. London* **171**, 713–891 (1880).
2. P. Goldreich, *Mon. Not. R. Astron. Soc.* **126**, 257–268 (1963); *Astron. J.* **70**, 5–9 (1965).
3. M. Hénon, *Astron. Astrophys.* **9**, 24–36 (1970).
4. G. W. Hill, *Am. J. Math.* **1**, 5–26, 129–147, 245–260 (1878).
5. H. Jeffreys, *Mon. Not. R. Astron. Soc.* **122**, 339–343 (1961).
6. P. S. Laplace, *Exposition du système du monde*, 2nd ed. (Paris, 1798).
7. F. R. Moulton, *Periodic Orbits* (Carnegie Institute, Washington, D.C., 1920).
8. J. H. Oort, *Bull. Astron. Inst. Neth.* **11**, 91–110 (1950).
9. V. Szebehely, *Celestial Mech.* **18**, 383–389 (1978).
10. V. Szebehely and R. McKenzie, *Astron. J.* **82**, 303–305 (1977).
11. V. Szebehely, R. O. Vicente, and J. B. Lundberg, IAU Colloquium No. 74 to be published (1982).
12. R. O. Vicente, V. Szebehely, and J. B. Lundberg, *Celestial Mech.* to be published.

Part 3

PLASMA PHYSICS

MULTIDIMENSIONAL CANONICAL/SYMPLECTIC MAPS FOR GYRORESONANCE CROSSINGS

CELSO GREBOGI AND ALLAN N. KAUFMAN
Lawrence Berkeley Laboratory and Department of Physics
University of California
Berkeley, California

Abstract. *The locally resonant interaction of a magnetically confined charged particle with a high frequency electromagnetic wave can be modeled by a mapping of the particle's phase space. This mapping represents the jump in the dynamical variables caused by each passage through the resonance zone. Two formalisms are presented for mappings that preserve the phase space structure: the Poincaré transformation, appropriate for canonical variables; and the Lie map, valid for any coordinate system on phase space. In both cases, the generating function for the map is the action integral, across the resonance, of the interaction Lagrangian.*

In the nonresonant region of phase space, the interaction is transformed to higher order by standard Lie methods. Particle motion in this (adiabatic) region can then be represented also by a mapping. Consequently, the Hamiltonian flow in phase space can be replaced by iterates of a composite map, the product of the resonance and adiabatic maps.

The key terms for the Workshop on Long-Time Prediction in Nonlinear Conservative Dynamical Systems were *conservative* and *long-time prediction*. In this chapter we interpret "conservative" to refer to the *phase space* of a Hamiltonian system (not to its energy), and we concentrate on developing a

Work supported by the Director, Office of Energy Research, Office of Fusion Energy, Applied Plasma Physics Division, of the U.S. Department of Energy under contract W-7405-ENG-48.

method, for evolution by means of iterated maps, that *preserves the structure* of that space. For "prediction," the requirement is reliability, and we maintain that any approximation method that destroys the structure of phase space cannot be reliable.

Our studies are motivated by a specific problem, of some importance in plasma physics.[1] Let there be a magnetostatic field, in which charged particles move with a complete set of three adiabatic invariants. Such fields characterize plasma confinement devices, such as tokamaks and mirror machines. One of the methods for heating the plasma is to apply an electromagnetic wave, of definite frequency ω, which resonates with particle motion in a restricted region, called the "*resonance zone*." As a particle crosses this zone, its adiabatic invariants and its energy suffer jumps, whose values depend on the relative phase of the particle motion and the wave. Outside the resonance zone, the particle oscillates nonresonantly in the field of the wave; there the invariants and Hamiltonian of the unperturbed particle motion are replaced by those of its oscillation center.[2] We call this the *adiabatic region*.

The basic idea of our study is to replace the Hamiltonian differential equations, for continuous flow in phase space, by an iterated sequence of point mappings that successively move a particle from one resonance crossing to the next. Our formalism for the mapping guarantees the preservation of phase space structure. The generating function for the resonance crossing map is obtained by physically motivated perturbation methods from the action integral representing the effect of the wave.

There are two reasons for replacing a set of differential equations by an iterated mapping. From the analytic point of view, a mapping is often more amenable to further theoretical development than its equivalent set of differential equations. This has become especially apparent in the past decade in respect to studies of intrinsic stochasticity. From the computational point of view, a mapping is far more efficient, since it represents a large finite time interval.

Previous work along these lines has usually led to maps of a two-dimensional phase plane[3] that is required to be area preserving, or, more generally, measure preserving. For the six-dimensional phase space of a particle, the corresponding requirement is that the map be *symplectic*,[4] that is, it must preserve the "*fundamental two-form*" of phase space, which reads (in canonical variables)

$$\Omega \equiv \sum_{i=1}^{3} dq_i \wedge dp_i \tag{1}$$

(The wedge $\wedge$ may be omitted by the reader unfamiliar with exterior algebra.) The use of canonical variables is sometimes inconvenient, and for a *general* phase space coordinate system, Eq. (1) becomes

$$\Omega = \sum_{i,j=1}^{6} \Omega_{ij}(z)\, dz^i \wedge dz^j \tag{2}$$

To be specific, we consider first the motion of a particle in the unperturbed field $\mathbf{B}_0(\mathbf{x})$, $\phi_0(\mathbf{x})$. It has long been customary[5] to describe this motion in terms of gyration about a guiding center (gc), which tends to follow magnetic field lines, but slowly drifts across them. The standard gc equations do not reflect the underlying Hamiltonian structure of the exact system. Only recently has Littlejohn[6] succeeded in developing a Hamiltonian theory for gc motion; here the symplectic structure is precisely preserved, even though the Hamiltonian function itself is approximate. A further advantage of Littlejohn's approach is that the dynamical variables are physical, although noncanonical. Finally, by being a Hamiltonian theory, this approach is able to utilize the powerful Lie transform methods[7] developed recently.

The gc variables of Littlejohn's formalism are $\mathbf{X}$ (gc position in physical space, in any three-dimensional coordinate system) and P (gc parallel momentum). These four are noncanonical, with the Poisson brackets:

$$[\mathbf{X},\mathbf{X}] = (B^*)^{-1}\hat{b}\times\mathbf{I}$$

$$[\mathbf{X}, P] = (B^*)^{-1}\mathbf{B}^*, \tag{3}$$

$\hat{b}(\mathbf{X})$ is the unit vector along $\mathbf{B}_0(\mathbf{X})$,

$$\mathbf{B}^*(\mathbf{X}, P) = \mathbf{B}_0(\mathbf{X}) + P\nabla \times \hat{b}(\mathbf{X})$$

$$B^*(\mathbf{X}, P) = \hat{b}(\mathbf{X}) \cdot \mathbf{B}^*(\mathbf{X}, P)$$

These brackets (which are exact, not approximations) satisfy the Jacobi condition, and thus are a legal "symplectic" structure. For applications, they must be supplemented by a Hamiltonian function $H(\mathbf{X}, P; t)$ and by a relation to the particle variables $(\mathbf{r}, \mathbf{v})$.

To illustrate the use of Eq. (3), consider the Hamiltonian $H(P) = \frac{1}{2}P^2$, representing the parallel kinetic energy alone. The equation of evolution under *any* H, for any function $g(z)$ on phase space, is $\dot{g} = [g, H]$. For this Hamiltonian, we obtain $\dot{P} = 0$ and $\dot{\mathbf{X}} = (B^*)^{-1}\mathbf{B}^*P$. Thus, $\hat{b}\cdot\dot{\mathbf{X}} = P$; the parallel component of gc velocity equals the (invariant) gc momentum, for this Hamiltonian. The perpendicular part of $\dot{\mathbf{X}}$ is the centrifugal drift.

As another example, take as the Hamiltonian $H(\mathbf{X}) = \mu B_0(\mathbf{X})$, the gyration kinetic energy alone. (We discuss μ, the magnetic moment, shortly.) Now we obtain $\dot{P} = -(B^*)^{-1}\mathbf{B}^* \cdot \nabla B_0$, the mirror force, and $\dot{\mathbf{X}} = \mu(\hat{b} \times \nabla B_0)(B^*)^{-1}$, the ∇B drift.

As the final example, we take as our Hamiltonian the electrostatic energy $\phi_0(\mathbf{X})$. (Note that we set $e = m = c = 1$ wherever convenient.) We now find $\dot{P} = -(B^*)^{-1}\mathbf{B}^* \cdot \nabla\phi_0$ and $\dot{\mathbf{X}} = (B^*)^{-1}\hat{b} \times \nabla\phi_0$, the parallel electric force and the electric drift, respectively.

In addition to the gc variables $(\mathbf{X}, P)$, Littlejohn's formalism includes the canonical pair (μ, θ), the gyromomentum (magnetic moment), and the gyrophase. These are conjugate: $[\theta, \mu] = 1$, and commute with $(\mathbf{X}, P)$. Let us examine their evolution under each of the Hamiltonians above. Since none of them depend on θ, the magnetic moment μ is invariant, as expected. However, for $H = \mu B_0(\mathbf{X})$, we obtain $\dot{\theta} = B_0(\mathbf{X})$; the gyrofrequency is given by the local magnetic field, as expected.

The total unperturbed Hamiltonian is (to lowest order) the sum of the three examples discussed above; it was instructive to examine them individually. Now we add the perturbation of an electromagnetic wave: $\delta\mathbf{E}(\mathbf{x}, t) = \tilde{\mathbf{E}}(\mathbf{x})\exp i[\psi(\mathbf{x}) - \omega t] + \text{cc}$, which we express in eikonal form. (In general, we expect a sum of such terms, but for clarity we consider only one term here.) The local wave vector is $\mathbf{k}(\mathbf{x}) = \nabla\psi(\mathbf{x})$. The perturbation Lagrangian is

$$\delta L(\mathbf{r}, \mathbf{v}; t) = \mathbf{v} \cdot \delta\mathbf{A}(\mathbf{r}, t) = (i\omega)^{-1}\mathbf{v} \cdot \tilde{\mathbf{E}}(\mathbf{r})\exp i[\psi(\mathbf{r}) - \omega t] + \text{cc}$$

At this point it is necessary to know the relation between the (noncanonical) particle variables $(\mathbf{r}, \mathbf{v})$ and the (noncanonical) gc variables $(\mathbf{X}, P, \mu, \theta)$. This is expressed by Littlejohn[6a, 6c] as a power series in the small parameter ε, representing the ratio of gyroradius to magnetic scale length. (We do not quote this relation.) Upon substituting it into δL, we find (after some algebra) $\delta L(\mathbf{X}, P, \mu, \theta, t) = \Sigma_m L_m(\mathbf{X}, P, \mu)\exp i\psi_m(\mathbf{X}, \theta, t) + \text{cc}$, where the Fourier coefficient L_m is given by

$$L_m = J_m i^m \omega^{-1}\tilde{\mathbf{E}}(\mathbf{X}) \cdot \left(P\hat{b} + \frac{mB_0}{k_\perp}\hat{k}_\perp + \frac{2iB_0}{k_\perp}\frac{\partial \ln J_m}{\partial \ln \mu}\hat{b} \times \hat{k}_\perp \right) \tag{4}$$

and the phase is

$$\psi_m(\mathbf{X}, \theta, t) = \psi(\mathbf{X}) + m\theta - \omega t \tag{5}$$

Here J_m is the Bessel function $J_m(k_\perp r_g)$ with gyroradius $r_g = [2\mu/B_0(\mathbf{X})]^{1/2}$, and $\hat{k}_\perp$ is the unit vector along $\mathbf{k}_\perp$, the projection of $\mathbf{k}(\mathbf{X})$ on the plane perpendicular to $\mathbf{B}_0(\mathbf{X})$.

Let us examine a *single* term of the perturbation Lagrangian:

$$L_m(\mathbf{X}, P, \mu)\exp[i\psi_m(\mathbf{X}, \theta, t)]$$

The crucial question is whether the phase is *rapidly varying* along an orbit or is *nearly stationary*. From Eq. (5), we have

$$\dot{\psi}_m = \mathbf{k}(\mathbf{X}) \cdot \dot{\mathbf{X}} + m\dot{\theta} - \omega$$

Using the *unperturbed* Hamiltonian for $\dot{\theta}$ and $\dot{\mathbf{X}}$, we see that the phase is stationary where

$$\omega - \mathbf{k}(\mathbf{X}) \cdot \dot{\mathbf{X}}(\mathbf{X}, P, \mu) = mB_0(\mathbf{X}) \tag{6}$$

that is, where the wave frequency in the local gc frame is an integer multiple of the local gyrofrequency. The gyroresonance condition Eq. (6) is a five-dimensional surface in the six-dimensional gc phase space $(\mathbf{X}, P, \mu, \theta)$. It is convenient to think of the unperturbed particle moving in that space and repeatedly (and rapidly) crossing the gyroresonance surface Eq. (6).

Where the phase ψ_m is rapidly varying (i.e., for the adiabatic regions of phase space, not near the resonance zone), the interaction can be transformed away.[8] This means that the gc motion is replaced by oscillation center (oc) motion: the wave field produces rapid oscillations about the oc, which moves (relatively) slowly under a new Hamiltonian, wherein the wave field appears quadratically, producing ponderomotive effects. The quadratic, or ponderomotive, term is proportional to $|L_m|^2|\dot{\psi}_m|^{-2}$; thus it can be considered to be a weak perturbation (order $|\delta E|^2$), except in the resonance zone, where $\dot{\psi}_m \to 0$. We now examine the resonance zone, where the ponderomotive term diverges. We return to the nonresonant region later.

To first order in the wave amplitude, the perturbation Hamiltonian is the negative of the perturbation Lagrangian:

$$\delta H_m(\mathbf{X}, P, \mu, \theta, t) = -L_m(\mathbf{X}, P, \mu)\exp[i\psi_m(\mathbf{X}, \theta, t)]$$

Each dynamical variable is affected by δH_m; let us examine μ, which is invariant under H_0. We have $\dot{\mu} = -\partial\delta H_m/\partial\theta = -im\,\delta H_m$; thus the invariance of μ is broken by the wave.

We wish to determine, for each dynamical variable z^i, its jump $\Delta z^i = z^{i\prime} - z^i$ to a new value $z^{i\prime}$, as a result of the resonance crossing. We work to lowest order in wave amplitude, but stress preservation of phase space structure. Denote the value *at* the crossing by $\bar{z}^i = \frac{1}{2}(z^i + z^{i\prime})$, and the time dependence relative to that value by $\delta z^i(t) = z^i(t) - \bar{z}^i$.

Returning to μ, we see that the jump is $\Delta\mu = \int dt\,\dot{\mu} = -\int dt\,\partial\delta H_m(z, t)/\partial\theta$ $= -\int dt\,\partial\delta H_m(\bar{z} + \delta z(t), t)/\partial\theta = -(\partial/\partial\bar{\theta})\int dt\,\delta H_m(z, t) = +(\partial/\partial\bar{\theta})\int \delta L\,dt$. We introduce the perturbation action integral:

$$S(\bar{z}; \bar{t}) = \int \delta L\,dt$$

and note that $\Delta\mu = \partial S/\partial\bar{\theta}$. Analogously, we calculate the resonance jump in gyrophase as $\Delta\theta = -\partial S/\partial\bar{\mu}$.

We evaluate S by the method of stationary phase, expanding the phase $\psi_m(t)$ about the stationary time $\bar{t}$, where $\dot{\psi}_m = 0$ (i.e., at the resonance surface).

With $\psi_m(t) = \psi_m(\bar{t}) + \frac{1}{2}(t-\bar{t})^2\ddot{\psi}_m(\bar{t}) + \cdots$, we obtain

$$S(\bar{z};\bar{t}) = \left(-\frac{2\pi}{\ddot{\psi}_m}\right)^{1/2} L_m(\bar{z})\exp\left[i\psi_m(\bar{\mathbf{X}},\bar{\theta},\bar{t})\right] \tag{7}$$

(Validity conditions for the approximations made are discussed below.)

Regardless of the evaluation of S, or even of its expression in terms of the perturbation Lagrangian, the formulation of the jumps $\Delta\mu$, $\Delta\theta$ in terms of the derivatives of S guarantees the preservation of area in the μ–θ plane: $d\mu' \wedge d\theta' = d\mu \wedge d\theta$. (This can be verified with a bit of straightforward exterior algebra.) We now need to extend this preservation to the full six-dimensional phase space, since each of the gc variables experiences a jump Δz^i.

For a *canonical* coordinate system (q, p), the simplest generalization is the *Poincaré transformation*.[9] For *any* function $S(\bar{q}, \bar{p})$, it is again straightforward to verify that the jump relations

$$\Delta p_i = \frac{\partial S}{\partial \bar{q}_i} \quad \text{and} \quad \Delta q_i = -\frac{\partial S}{\partial \bar{p}_i} \tag{8}$$

together with $q_i = \bar{q}_i - \frac{1}{2}\Delta q_i$, $q_i' = \bar{q}_i + \frac{1}{2}\Delta q$ (and similarly for p_i), lead to preservation of the two-form:

$$\sum dq_i' \wedge dp_i' = \sum dq_i \wedge dp_i$$

To apply these ideas in our three-degrees-of-freedom case, two alternative approaches are possible. One approach is to impose the requirement of using canonical variables, since the Poincaré method is simply invalid for our noncanonical set $(\mathbf{X}, P)$. Any canonical set will do, so one is guided by the physics of the particular problem being studied. A natural choice is the set of action-angle variables for the unperturbed gc Hamiltonian $H_0(\mathbf{X}, P)$. Another possibility (unnatural, in our opinion) is the standard set of canonical particle coordinates $(\mathbf{r}, \mathbf{p})$. One must now express S, given explicitly in Eq. (7), in terms of the canonical set chosen. This would produce such a mess that the beauty of the Littlejohn formalism would be lost. For that reason we have turned to a second approach, which works for a noncanonical coordinate system as well.

In general terms, we ask what transformation $z \to z'$ preserves the fundamental two-form. In Eq. (2), $\Omega_{ij}(z)$ is the (antisymmetric) Lagrange matrix, the reciprocal of the Poisson matrix $\sigma^{ij}(z) = [z^i, z^j]$. The Jacobi condition on Poisson brackets is equivalent to the requirement that Ω is "closed": $\partial\Omega_{ij}/\partial z^k + \partial\Omega_{jk}/\partial z^i + \partial\Omega_{ki}/\partial z^j = 0$ for all z and all (ijk). This is of course satisfied by Littlejohn's variables, for which Eq. (2) reads $\Omega = \mathbf{B}^*(\mathbf{X}, P) \cdot d\mathbf{X} \times d\mathbf{X} - \hat{b}(\mathbf{X}) \cdot d\mathbf{X}\, dP + d\mu\, d\theta$.

To answer the question above, we turn to the Lie map.[10] Again let $S(z)$ be *any* function on phase space; no particular coordinate system need yet be specified. Now define the corresponding Lie transformation operator as $R = \exp[S, \cdot]$. Let $g(z)$ be any "observable," that is, some function (on phase space) of physical interest. Then g transforms under S as $g' = Rg = g + [S, g] + \frac{1}{2}[S,[S, g]] + \cdots$. The fundamental property of R is that Poisson bracket relations are preserved; that is, if three functions g_1, g_2, and g_3, satisfy $[g_1, g_2] = g_3$ for all z, then $[g_1', g_2'] = g_3'$. Let us now apply R to each coordinate z^i. Then $\Delta z = z^{i\prime} - z^i = [S(z), z^i] +$ order S^2. Expressing z in terms of $\bar{z}$ and Δz, we obtain $\Delta z^i = [S(\bar{z}), \bar{z}^i] +$ order S^2. For *canonical* variables, this reduces to Eq. (8) to lowest order; thus we see the relation of the two methods.

The Lie method has the great advantages of being coordinate free and of allowing the use of operational methods when one proceeds to iterate. At least two disadvantages are apparent: the Lie method is asymmetric with respect to the resonance surface, at least in its present form; for numerical computation, an infinite series is called for.

Before proceeding to examine the adiabatic region, we summarize the treatment of the resonance zone. For any observable, and in particular for the dynamical variables, the jump across the resonance is given by $g' = Rg$ with $R = \exp[S, \cdot]$ and S given by Eq. (7), where the phase ψ_m is given by Eq. (5), and the amplitude L_m by Eq. (4).

In the adiabatic region we use the unperturbed Hamiltonian (in the gc variables) $H_0(\mathbf{X}, P, \mu) = \frac{1}{2}P^2 + \mu B_0(\mathbf{X}) + \phi_0(\mathbf{X})$, each term of which was discussed above. (The ponderomotive correction, quadratic in the amplitude, is neglected here but must be considered eventually.) We look for a complete set of three invariants. As discussed earlier, μ is one invariant, and H_0 (being time independent) is a second. The existence of a third invariant $I(\mathbf{X}, P, \mu)$ is guaranteed (at least locally, i.e., between resonances) by the Darboux theorem,[4,6] which also presents an algorithm for its construction. For an axisymmetric system like a tokamak, it can be identified with the angular momentum (and then is μ-independent). In a mirror machine, the bounce action or drift flux will serve as the third invariant.

In the six-dimensional phase space $(\mathbf{X}, P, \mu, \theta)$, the oscillation center orbit lies on each of the three (five-dimensional) invariant surfaces μ, $H_0(\mathbf{X}, P, \mu)$, $I(\mathbf{X}, P, \mu)$. It thus lies on the intersection $\mu \cap H_0 \cap I$, which is three dimensional.

To help us visualize the orbit, let us first project out the (μ, θ) variables, reducing the phase space to four dimensions. Then let us postulate axial symmetry (for simplicity) and project out the toroidal angle ϕ, leaving three dimensions: P and the poloidal coordinates, which we denote α, β. In this space (α, β, P), the two invariant surfaces are $H_0(\alpha, \beta, P) = E$ and $I(\alpha, \beta, P) = p_\phi$. These two-dimensional surfaces intersect in a curve $E \cap p_\phi$, along which the oc moves until it reaches the resonance surface $\dot{\psi}_m(\alpha, \beta, P) = 0$. (In practice, the latter surface usually depends also on ϕ, so one must do the analysis in a higher dimensional space, where one's geometric intuition may fail.)

For each of the three invariants of the adiabatic region, one can introduce a conjugate variable whose time derivative is constant along the orbit. One then constructs a relatively trivial symplectic map, denoted A, for the adiabatic motion. (Its explicit form should be tailored to the problem at hand.) This map then takes a point in phase space from one resonance crossing to the next. The map A depends parametrically on the three invariants of the adiabatic motion, and it advances the conjugate variables and time.

Upon crossing the next resonance, all the variables jump (both the "adiabatic" invariants and their conjugates) according to the resonance mapping R. The new values of the invariants are then inserted into A for the next adiabatic mapping.

The particle motion is thus represented as a product of alternating symplectic mappings $RARA \cdots = (RA)^n \equiv T_n$ for n resonance crossings separated by adiabatic motion. The n-fold composite T_n is itself symplectic, since symplectic maps have the group property.

With a formal expression for T_n, one may next inquire into the asymptotic behavior of its orbits, as $n \to \infty$. Since a symplectic map is the multidimensional generalization of an area preserving map, the critical questions are the same, suitably generalized: Is the motion ergodic, chaotic, mixing?[11] What are the Lyapunov characteristic exponents,[12] and how do they depend on the parameters of the problem (e.g., wave amplitude)? What are the diffusion rates for the invariants?[13] It is hoped that operational methods[10] will help provide answers to these important questions.

We now return to the validity criteria for the resonance crossing maps. In the first place, the wave amplitude must be sufficiently weak that first-order perturbation theory is adequate, since the action integral is evaluated using unperturbed orbits. Physically, this means that the crossing must be sufficiently rapid that the particle cannot get trapped by the resonance. In the second place, the eikonal conditions must allow for the evaluation of the phase integral by the stationary phase method; this requires $|\ddot{\psi}_m|^{1/2} \gg |\dddot{\psi}_m|^{1/3}$. Physically, we require sufficiently slow spatial variation of wave amplitude and wave vector on the gyroradius scale, as well as rapid passage through resonance. In the third place, the resonances should be sufficiently disjoint in phase space that the adiabatic regions are well defined. In effect, the crossing time should be short compared to the time between resonances. (Note that Chirikov's "resonance overlap" criterion[3] refers to *global* resonances, not to the local resonances discussed here. Global resonances can overlap, producing stochasticity, even though the local resonances are disjoint.)

We conclude by outlining elements of a program for future work in this area:

1. A specific model should be chosen, and the map formalism explicitly implemented.
2. The orbits for the iterated map should be compared to those for the exact differential equations of motion.[14]

3. Because the irreversible resonant diffusion is responsible for entropy production, it should be possible to extend the quasi-linear relation[16] between the diffusion tensor and the dissipative part of the dielectric susceptibility to the nonlinear regime.

4. At some stage, self-consistency for the electromagnetic perturbation should be introduced. The most promising vehicle for this is the Hamiltonian Vlasov–Maxwell theory discovered by Morrison.[17]

(The first two steps have been carried out by us in earlier work for the case of spatial variation in one dimension.[15] We found quantitative agreement for regular orbits, and qualitative agreement for chaotic orbits, as expected.)

ACKNOWLEDGMENTS

We thank F. Perkins for pointing out that our resonance crossing technique could most usefully be applied to the rf heating problem, A. Weinstein for introducing us to the Poincaré transformation, J. Cary and R. Littlejohn for educating us in the Lie transform methodology, and A. Dragt for providing his published and unpublished work on Lie maps and their applications.

REFERENCES

1. For a review of gyroresonant heating, see E. Canobbio, "Gyroresonant Particle Acceleration in a Nonuniform Magnetostatic Field," *Nucl. Fusion* **9**, 27 (1969).
2. The oscillation center concept was cast into Hamiltonian form by (a) R. L. Dewar, "Oscillation-center Quasilinear Theory," *Phys. Fluids* **13**, 2710 (1970) and further extended by (b) S. Johnston, "Oscillation Center Formalism of Classical Theory of Induced Scattering," *Phys. Fluids* **19**, 93 (1976). It is the natural entity of the standard Lie transform formalism.
3. A thorough study of area preserving maps is presented by B. Chirikov, "A Universal Instability of Many-Dimensional Oscillator Systems," *Phys. Rep.* **52**:5 (1979).
4. The standard reference for symplectic manifolds in classical physics is (a) V. Arnol'd, *Mathematical Methods of Classical Mechanics* (Springer-Verlag, Berlin, 1978). For the non-mathematician, a more accessible treatment has been prepared by (b) R. Littlejohn, "Introduction to Geometric Methods in Classical Mechanics," *Phys. Rep.* (to appear).
5. The standard reference for the traditional approach is T. G. Northrop, *The Adiabatic Motion of Charged Particles* (Interscience, New York, 1963).
6. For a full discussion, see (a) R. Littlejohn, "Hamiltonian Theory of Guiding Center Motion," Ph.D. thesis (University of California, Berkeley, 1980), available from University Microfilms International, CGM 80-29478. Parts of the thesis appear in (b) R. Littlejohn, "A Guiding Center Hamiltonian: A New Approach," *J. Math. Phys.* **20**, 2445 (1979) and (c) R. Littlejohn, "Hamiltonian Formulation of Guiding Center Motion," *Phys. Fluids* **24**, 9 (1981). The essential ideas of the Littlejohn formalism are clearly presented in (d) D. H. E. Dubin and J. A. Krommes, "Stochasticity, Superadiabaticity, and the Theory of Adiabatic Invariants and Guiding Center Motion" (following, this volume). A concise presentation is in (e) C. Grebogi, A. Kaufman, and R. Littlejohn, "Hamiltonian Theory of Ponderomotive Effects of an Electromagnetic Wave in a Nonuniform Magnetic Field," *Phys. Rev. Lett.* **43**, 1668 (1979).

7. For reviews of Lie transform methods, see (a) J. R. Cary, "Lie Transform Perturbation Theory for Hamiltonian Systems," *Phys. Rep.* **79**, 129 (1981); (b) J. R. Cary, "Lie Transforms and Their Use in Hamiltonian Perturbation Theory," U.S. Department of Energy Report ET-0074 (1978); (c) R. Littlejohn, "Pedestrian's Guide to the Lie Transform," UCB Report UCID-8091 (University of California, Berkeley, 1978).

8. For the general approach, see: (a) A. N. Kaufman, "Regular and Stochastic Particle Motion in Plasma Dynamics," in *Intrinsic Stochasticity in Plasmas*, G. Laval and D. Grésillon, Eds. (Editions de Physique, Courtaboeuf, Orsay, France, 1979). The results are presented in Ref. 6e. For details see (b) C. Grebogi and A. Kaufman, "Ponderomotive Hamiltonian in Nonuniform Magnetic Field and Local Susceptibility of Nonuniform Vlasov Magnetoplasma," Lawrence Berkeley Laboratory Report LBL-12806 (in preparation).

9. For a modern presentation see A. Weinstein, *Invent. Math.* **16**, 202 (1973).

10. The Lie map concept has been extensively developed by (a) A. J. Dragt and J. M. Finn, "Lie Series and Invariant Functions for Analytic Symplectic Maps," *J. Math. Phys.* **17**, 2215 (1976); (b) A. J. Dragt and J. M. Finn, "Normal Form for Mirror Machine Hamiltonians," *J. Math. Phys.* **20**, 2649 (1979); (c) A. Dragt, "Method of Transfer Maps for Linear and Nonlinear Beam Elements," *IEEE Trans. Nucl. Sci.* **NS-26**, 3601 (1979).

11. In addition to Ref. 3, see the survey G. M. Zaslavskii and B. V. Chirikov, "Stochastic Instability of Non-Linear Oscillations," *Sov. Phys. Usp.* **14**, 549 (1972).

12. For a clear explanation, see (a) G. Benettin, C. Froeschle, and J. P. Scheidecker, "Kolmogorov Entropy of a Dynamical System with an Increasing Number of Degrees of Freedom," *Phys. Rev. A* **19**, 2454 (1979); (b) G. Benettin, L. Galgani, A. Giorgilli, and J. P. Strelcyn, "Tous les nombres caractéristiques de Lyapunov sont effectivement calculables," *C.R. Acad. Sci. Paris* **286A**, 431 (1978); (c) G. Benettin, L. Galgani, A. Giorgilli, and J. P. Strelcyn, "Lyapunov Characteristic Exponents for Smooth Dynamical Systems and for Hamiltonian Systems; A Method for Computing All of Them," *Meccanica*, **15**, 9, 21 (1980).

13. R. H. Cohen and G. Rowlands, "Calculation of Resonant Transport Coefficients from Mappings," *Phys. Fluids* **24**, 2295 (1981).

14. This comparison should take account of the important studies of G. Benettin, M. Casartelli, L. Galgani, A. Giorgilli, and J. P. Strelcyn, "On the Reliability of Numerical Studies of Stochasticity," *Nuovo Cimento* **44B**, 183 (1978); **50B**, 211 (1979).

15. C. Grebogi and A. Kaufman, "Multiple Resonance Crossing in the Eikonal Model of Plasma Turbulence," Lawrence Berkeley Laboratory Report LBL-12805, in preparation.

16. This relation is explored in: A. N. Kaufman, "Quasilinear Diffusion of an Axisymmetric Toroidal Plasma," *Phys. Fluids* **15**, 1063 (1972).

17. (a) P. J. Morrison, "Maxwell–Vlasov Equation as a Continuous Hamiltonian System," *Phys. Lett.* **80A**, 383 (1980). The mathematical basis is presented by: (b) J. Marsden and A. Weinstein, "Hamiltonian Structure of the Maxwell–Vlasov Equations," *Physica* **4D**, 394 (1982).

STOCHASTICITY, SUPERADIABATICITY, AND THE THEORY OF ADIABATIC INVARIANTS AND GUIDING CENTER MOTION

DANIEL H. E. DUBIN
JOHN A. KROMMES
Plasma Physics Laboratory
Princeton University
Princeton, New Jersey

Abstract. *The theory of adiabatic invariants is discussed within the modern framework of symplectic Hamiltonian dynamics. The distinctions between exact, adiabatic, and superadiabatic invariants are clarified. The intimate connection between adiabatic (as opposed to exact) invariance and resonant interactions between motions on disparate time scales is elucidated. For the important case of charged particle motion in a strong magnetic field, resonances between gyration, bounce motion, and an external sinusoidal perturbation are described explicitly by introducing a time-dependent symplectic formulation of the guiding center motion. Destruction of invariance is discussed for quite general situations of physical interest, including the case of a trapped particle in a tokamak.*

1. INTRODUCTION

The roots of the concept of adiabatic invariance extend far back into the history of physics. We do not attempt to provide a complete historical introduction, which would be worthy of a treatise in itself. However, the idea was familiar to, and used by, the founding fathers of statistical and quantum mechanics. In the 1911 Solvay Conference, Einstein considered a simple pendulum and argued that "If one alters the length of the pendulum infinitely slowly, the energy of oscillation will remain always $h\nu$ if it was initially $h\nu$; the energy of oscillation changes only through ν. The same holds for undamped electrical oscillations and for free radiation."[1] According to Born,[2] Ehrenfest considered in 1914 general classical mechanical systems in which certain dynamical variables took on only integral values, and argued that if such systems were disturbed sufficiently slowly, the initial integral value would remain unchanged—"adiabatically invariant." This principle led to the method of quantizing classical systems in the old quantum mechanics.[2] It persists in modern quantum mechanics in the form of the so-called adiabatic theorem, which states[2-4] that under certain technical restrictions, the eigenstates of a sufficiently slowly perturbed quantum mechanical system will be those that evolve continuously from the unperturbed system.

Though applications of adiabatic invariance could be traced through the development of many modern physics specialties, we concentrate here on the application to the motion of charged particles and, thus, to plasma physics. Both practical and theoretical reasons foster an intense interest in the dynamics of charged particles trapped in the terrestrial magnetic field; extensive reviews have been given.[5] In the laboratory, the quest for controlled fusion has provided important examples of the utility of adiabatic invariants. The theory of one of the major confinement geometries, the magnetic mirror,[6] depends fundamentally on the adiabatic invariance of the magnetic moment of gyration[7] $\bar{\mu}$, when the gyroradius of the particle is much smaller than the macroscopic scale of variation of the magnetic field seen by the particle in the course of its motion, to reflect particles from the open ends of the container. Laboratory experiments[8] have verified adiabatic confinement of trapped particles for more than 10^{10} gyroperiods. Toroidal geometry, as in the case of the tokamak,[9] eliminates the need for such reflection; however, it also leads to an important component of trapped particles. The conventional single-particle picture of confinement in a tokamak[10] relies critically on the theory of the magnetic and electric drifts[11] that exist in the limit of small gyroradius, the same limit in which $\bar{\mu}$ is (adiabatically) conserved. Furthermore, free energy arguments[12] suggest that the dominant instabilities of confined tokamak plasmas will be low frequency, $\omega \ll \Omega_i \equiv eB/m_i c$, again the limit in which $\bar{\mu}$ should be a good invariant. This idea is used extensively in the analyses of microfluctuations in tokamaks. Space precludes a discussion of the important role of the second, or longitudinal invariant[13,14] $\bar{J}$, the adiabatic conservation of which was apparently first suggested by Rosenbluth on the basis of quantum mechanical analogies.

In the early historical discussions of adiabatic invariance, the concept of "sufficiently slowly" remained quantitatively undefined. Study of special models[11,15] led to the conclusion that the exact relative change of an adiabatic invariant after a slow perturbation was often of the form $\exp(-\alpha/\varepsilon)$, where α

is a constant and ε is the small ratio of slow perturbation frequency to fast unperturbed oscillation frequency. Because the asymptotic expansion in powers of ε of $\exp(-\alpha/\varepsilon)$ vanishes, it is often said, and sometimes taken as a definition, that adiabatic invariants are *constant to all orders*[5,16] in ε. In fundamental work, Kruskal[17] gave an elegant description of the construction of the asymptotic series that represent the adiabatic invariants. Chirikov[18,19] emphasized that the basic concept was not $\varepsilon \ll 1$, but rather the presence or absence of *resonances*[20] of the perturbation with the periodic motion of the unperturbed system. If resonances existed in a particular region of phase space, the invariant was destroyed *in that region*; otherwise, it was preserved.[21] The inequality $\varepsilon \ll 1$ does not preclude resonances, since nonlinearity can create high harmonics from a slow fundamental. We shall return to this point.

We now understand that in the region where the invariant is destroyed, stochastic motion[22] ensues. Loosely, in that region trajectories depend exponentially sensitively on initial conditions. The theory of stochastic regions has been pursued with various degrees of sophistication.[23-25] In work of considerable intuitive and practical importance, Chirikov[25] has developed the theory of resonances to the point where one now understands how to estimate the dimensions of the stochastic regions for a wide variety of conservative systems, including certain models of particle motion in a magnetic mirror.

As is clear from the work of Chirikov and many others, the appropriate framework in which to discuss questions of invariance and stochasticity is Hamiltonian dynamics. However, in the application to adiabatic invariance the usual Hamiltonian formulation meets with certain difficulties, and historically the theory of adiabatic invariants was generally developed from some version of the method of averaging[17,26] applied directly to the equations of motion. Some of these approaches are reviewed in the monograph by Northrop.[27]

To illustrate the difficulty with the Hamiltonian procedures, let us consider the theory of motion of a charged particle.[28] The Hamiltonian thereof is well known[29]:

$$H(\mathbf{p}, \mathbf{q}; t) = \left|\mathbf{p} - \frac{e\mathbf{A}(\mathbf{q}, t)}{c}\right|^2 \Big/ 2m + e\phi(\mathbf{q}, t) \tag{1}$$

Here $\mathbf{A}$ and ϕ are the vector and scalar potentials, $\mathbf{q}$ is the spatial coordinate $\mathbf{x}$, and $\mathbf{p}$ is the canonical momentum conjugate to $\mathbf{q}$. The problem is concerned with writing Eq. (1) in the standard form, appropriate for perturbation theory,

$$H = H_0 + \varepsilon H_1 + \cdots \tag{2}$$

where ε is a small parameter and H_0 is soluble, and becomes apparent when Eq. (1) is written in dimensionless variables. Normalize velocities to a typical velocity v_0, introduce a typical magnetic field B_0 and a macroscopic scale length L, and define a gyroradius ρ_0 by $\rho_0 \equiv v_0/\Omega_0$, where $\Omega_0 \equiv eB_0/mc$. If we then introduce a dimensionless momentum $\hat{\mathbf{p}} \equiv \mathbf{p}/mv_0$, a dimensionless vector potential $\hat{\mathbf{A}} \equiv \mathbf{A}/LB_0$, and a dimensionless scalar potential $\hat{\Phi} \equiv e\phi/(\frac{1}{2}mv_0^2)$, we find

$$\frac{H}{\frac{1}{2}mv_0^2} = \tfrac{1}{2}|\hat{\mathbf{p}} - \varepsilon^{-1}\hat{\mathbf{A}}|^2 + \hat{\Phi} \tag{3}$$

where $\varepsilon \equiv \rho_0/L$ is assumed to be small. Then, if the dimensionless velocity $\hat{\mathbf{v}} \equiv \hat{\mathbf{p}} - \varepsilon^{-1}\hat{\mathbf{A}}$ is to be of order unity, we see that $\hat{\mathbf{p}}$ must be of order ε^{-1}. It is not excluded—and, in fact, must be true—that $\hat{\mathbf{p}}$ also contains terms of higher order in ε, but the forms of these terms are by no means apparent. Thus, although it is sometimes justifiable to set $\hat{\Phi} = O(\varepsilon)$, it is not obvious how to extract the $O(\varepsilon)$ part related to the magnetic field, and we have not succeeded in writing the Hamiltonian in the form of Eq. (2).

In an attempt to deal with this difficulty, Gardner[30] introduced an ingenious canonical transformation and indicated how one could apply canonical perturbation theory to his transformed Hamiltonian. Stern[31] studied the same method in much greater detail. However, the variables of Gardner and Stern were neither completely intuitive nor systematic, and interpretation of the resulting perturbation theory was somewhat opaque. Dragt and Finn[32] studied Hamiltonians of a special form appropriate to mirror machines with the aid of Lie transforms. Mynick[33] studied more general Hamiltonians, employing a novel combination of mixed-variable generating functions and Lie transforms to give a systematical procedure for doing perturbation theory to all orders. All these authors worked exclusively with canonical variables.

Recently Littlejohn[34] reconsidered the possibility of finding an efficient Hamiltonian theory of guiding center motion and, thereby, of the magnetic moment. By introducing relatively unfamiliar yet strikingly simple techniques—in particular, the use of noncanonical phase space coordinates—he succeeded in providing a beautiful, mathematically elegant, and physically clear Hamiltonian formulation of charged particle motion that in ease of use and interpretation far superseded all previous work and made accessible to the study of adiabatic invariance a number of important techniques and theorems from the general theory of stochasticity. Part of this chapter is devoted to a review, brief of necessity, of Littlejohn's procedure, on which our work depends heavily. We emphasize that Littlejohn deserves the fullest credit for his insight and fundamental contributions and calculations. Our modest contributions are primarily to discuss certain physics points that Littlejohn did not completely interpret, and to touch on questions of resonances and stochasticity, which Littlejohn did not address (though he emphasized that his method affords a powerful tool in this direction). Whereas Littlejohn stressed the guiding center equations of motion, we are concerned more with the adiabatic invariants and the conditions under which they are destroyed.

The picture that emerges of charged particle motion in a strong magnetic field is essentially Chirikov's[18, 19] coupled to the modern symplectic techniques that Littlejohn introduced. Consider, for definiteness, the case of a single magnetically trapped particle in time-independent magnetic and electric fields. The phase space of the particle can then be described in terms of a pair of field line coordinates $\{\alpha, \beta\}$ and two action-angle pairs $\{\mu, \theta\}$ and $\{J, \psi\}$. The latter pairs (which are not unique) describe, respectively, gyration about and bouncing along a field line. These variables are not canonical, but are rather semicanonical in a technical sense to be described. In general, there are resonances between low harmonics of the gyration and higher harmonics of the bounce motion. In regions of phase space where the resonance condition is not satisfied, μ and J are connected, by a transformation that differs from the identity by an amount of $O(\varepsilon)$, to quantities $\bar{\mu}$ and $\bar{J}$ that label *invariant tori*[22] —Kolmogorov–Arnol'd–Moser (KAM) surfaces—embedded in the phase space. Thus, away from resonance $\bar{\mu}$ and $\bar{J}$ are *exactly* (not merely to all orders) conserved. Near resonances, the KAM tori do not exist—colloquially, they "are destroyed." They are replaced by stochastic regions in which μ and J diffuse. For analytic Hamiltonians and small ε it can be shown that the amplitudes of the resonant harmonics, and thus the width of the stochastic regions and the rates of change of μ and J, are exponentially small [of the form $\exp(-\alpha/\varepsilon)$], in agreement with the historical picture based on special cases and with Kruskal's asymptotic theory. The quantities $\bar{\mu}$ and $\bar{J}$ are called adiabatic invariants, the adjective "adiabatic" reminding us that the barred quantities are not defined in the stochastic region.

More precisely, the resonant and nonresonant regions are not isolated. Rather, there are stochastic regions near any point of the phase space, densely intermixed with the KAM tori. However, for small ε, the fraction of the space that is stochastic is very small, and the intuition presented in the preceding paragraph is useful for many purposes. As another caveat, when a third periodicity is present Arnol'd diffusion[25] can occur. Lack of space precludes us from discussing this interesting possibility.

If one now adds a sinusoidal perturbation—for example, $\phi \sim \exp(-i\omega t)$—one of two things can happen depending on the point in phase space: the KAM tori can shift and distort relative to fixed phase space axes, but remain topologically intact, or the invariants may be destroyed because of new resonances between the perturbation and the unperturbed motion. In the first case, the invariants change their functional form [e.g., $\bar{\mu} \to \bar{\mu}(\omega)$]. The ω-dependent functions have been called *superadiabatic invariants*.[35] However, it is too restrictive to reserve the term "superadiabatic" for invariants topologically preserved under temporal perturbation, since we will show that time dependence can be placed on a footing equal to that of the other phase space coordinates. To us, the term "superadiabatic invariant" is synonymous with a KAM torus—that is, with an exact invariant in a certain part of phase space.

Common usage contains some confusing imprecision. The invariants we find will emerge as power series in the small parameter ε; away from resonance, the series converge. Although sometimes the lowest order term (e.g., $\mu_0 \equiv v_\perp^2/2\Omega$) is loosely called the adiabatic invariant, to us the adiabatic invariant is the full series, whose value *does not change* in the nonresonant region where it is defined, even though μ_0 does change. We call the gyration-related quantity $\bar{\mu}$ the "first adiabatic invariant" or the "magnetic moment"; we call the bounce-related quantity $\bar{J}$ the "second adiabatic invariant," the "longitudinal invariant," or the "bounce action." The quantity μ_0 would be called the "lowest order magnetic moment."

Along with the adiabatic invariants, the various magnetic and electric drifts emerge naturally from the Hamiltonian formalism. The magnetic drifts are relatively straightforward and have been adequately discussed by Littlejohn.[34,36] However, the way in which the $\mathbf{E} \times \mathbf{B}$ and polarization drifts enter the theory is somewhat subtle, particularly when $\mathbf{E}_\perp$ is ordered small. Careful, though straightforward, applications of Lie transformations are needed to describe these effects. We describe this part of the theory in some detail, since it demonstrates a beautiful link between fundamental, intuitive physics and the somewhat formal mathematical apparatus of Lie transformations.

The chapter is organized as follows. Section 2 briefly reviews relevant notions of symplectic Hamiltonian dynamics and of Lie transforms, giving many references to the literature. Section 3 formulates and, to some extent, discusses a time-dependent symplectic Hamiltonian theory of guiding center motion and the magnetic moment for a particular ordering of the electric and magnetic fields; possible interactions with the longitudinal motion arc ignored. Section 4 treats gyration and bouncing on equal footing, so that the resonant interaction between them emerges naturally; we discuss with the aid of an example how one estimates the extent of the nonadiabatic regions. Section 5 contains a brief discussion. Intermediate algebraic results are collected in several appendices.

2. SYMPLECTIC HAMILTONIAN DYNAMICS AND TRANSFORMATIONS

We review here the modern, coordinate-free representation of Hamiltonian dynamics and sketch Littlejohn's procedure for applying this formalism to Eq. (1). The key ingredients of his method are the preparatory use of noncanonical variables to place the Hamiltonian into the form of Eq. (2), the construction of semicanonical variables to isolate the (nearly) periodic motion, then the use of near-identity symplectic (Lie) transformations to remove angle dependence order by order.

It is, perhaps, not widely enough appreciated that Hamiltonian dynamics does not require the use of canonical coordinates, but can be couched in a coordinate-free representation. Once understood, such a formalism is both more transparent and often more useful than the original representation in terms of p's and q's, just as the concept of a vector $\mathbf{v}$ as an abstract entity transcends in importance and utility the idea of the collection of numbers $\{v^i\}$, the components of the vector with respect to a particular basis. Now for the purpose of this review, though not for most of the rest of the chapter, we must be prepared to deal with simple ideas about p-forms, exterior calculus, and the like. These concepts, perhaps unfamiliar, are beautifully introduced at a level appropriate to our needs in Ref. 37, a pedagogical masterpiece. A more compressed and rigorous treatment can be found in Ref. 38. Also summarized in Ex. 4.11 of Ref. 37 is the germ of the application to Hamiltonian dynamics. Namely, define[37] the action integral

$$I \equiv \int_A^B [p^i \underline{d}q^i - H(p^i, q^i; t)\, \underline{d}t] \equiv \int_A^B \underline{\Omega}$$

(We denote p-forms, functions of p arguments, by an underbar and vectors by boldface type; we use the summation convention.) The principle of least action[29,37–39] states that the variation of I between two adjacent paths in the $(2n+1)$-dimensional space $\{p^i, q^i, t\}$ vanishes for the physical path $\mathbf{P}$. An application of Stokes's theorem[37,38] then leads to

$$\underline{d\Omega}\left(\ldots, \frac{d\mathbf{P}}{dt}\right) = 0 \tag{4}$$

where $\underline{d}$ signifies exterior differentiation and the dots signify[37] that the first argument of $\underline{\Omega}$ remain unspecified. Thus, by definition, $\underline{\Omega}$ is a closed differential two-form.[37,38] The component of Eq. (4) in the phase space is

$$\underline{\omega}\left(\ldots, \frac{d\mathbf{P}}{dt}\right) = \underline{d}H \tag{5}$$

where

$$\underline{\omega} \equiv \underline{d}p^i \wedge \underline{d}q^i = \underline{d}(p^i \underline{d}q^i) \tag{6}$$

is called the fundamental or distinguished (differential) two-form, or the Lagrange tensor.[34] It is also closed ($\underline{d\omega} = 0$) and nondegenerate.[38] This important result means, by definition,[38] that the phase space together with the fundamental two-form is a *symplectic manifold* M^{2n}.

Equation (5) is one form of the desired abstract expression of Hamiltonian dynamics. It establishes an isomorphism[38] $\vec{\sigma}$ between the phase space velocity (tangent vector) $d\mathbf{P}/dt$ and the differential one-form $\underline{d}H$:

$$\frac{d\mathbf{P}}{dt} = \vec{\sigma}(\underline{d}H) \tag{7}$$

which is another useful abstract expression of the Hamiltonian equations of motion.

Loosely, the isomorphism $\vec{\sigma}$ can be thought of as the inverse of the fundamental two-form $\underline{\omega}$. More precisely, expand $\underline{\omega}$ (covariantly) with respect to an arbitrary basis $\underline{d}z^i$: $\underline{\omega} = \frac{1}{2}\omega_{ij}\, \underline{d}z^i \wedge \underline{d}z^j$. Similarly, expand (contravariantly) $d\mathbf{P}/dt = \dot{z}^i\, d\mathbf{z}_i$ and note that one has $\underline{d}H = (\partial H/\partial z^i)\, \underline{d}z^i$. The covariant components of Eq. (5) are then readily shown to be $\omega_{ij}\dot{z}^j = \partial H/\partial z^i$. Defining the contravariant tensor σ^{ij} by $\omega_{ij}\sigma^{jk} = \delta_i^k$, one also has

$$\frac{dz^i}{dt} = \sigma^{ij}\frac{\partial H}{\partial z^j} \tag{8a}$$

or

$$\frac{d\mathbf{z}}{dt} = \overleftrightarrow{\sigma}\cdot\frac{\partial H}{\partial \mathbf{z}} \tag{8b}$$

the desired covariant expression of Hamiltonian dynamics; $\overleftrightarrow{\sigma}$ is called the Poisson tensor.

Following Arnol'd, define the Poisson bracket $[F, H]$ of two phase functions F and H as the derivative of F in the direction of the flow U^t: $M^{2n} \to M^{2n}$ generated by H:

$$[F, H] = \frac{d}{dt}F(U^t\mathbf{z})|_{t=0} \tag{9}$$

A corollary[38] of this definition is that

$$[F, H] = \underline{\omega}[\overleftrightarrow{\sigma}(\underline{d}H), \overleftrightarrow{\sigma}(\underline{d}F)] \tag{10}$$

from which follow the important results that

$$[F, H] = \frac{\partial F}{\partial z^i}\sigma^{ij}\frac{\partial H}{\partial z^j} \tag{11}$$

or

$$\sigma^{ij} = [z^i, z^j] \equiv [\mathbf{z}, \mathbf{z}]^{ij} \tag{12}$$

With Eq. (12), Eq. (8) becomes

$$\frac{dz^i}{dt} = [z^i, z^j]\frac{\partial H}{\partial z^j} \tag{13}$$

which is the form we actually use.

There is a shorter, if less intuitive, route to Eq. (13). It can be readily verified that the components of $\underline{\omega}$ with respect to an arbitrary basis $\underline{d}z^i$ are Lagrange brackets[29]:

$$\omega_{ij} = \{z^i, z^j\} \equiv \{\mathbf{z}, \mathbf{z}\}_{ij} = -\sum_l\left(\frac{\partial q^l}{\partial z^i}\frac{\partial p^l}{\partial z^j} - \frac{\partial q^l}{\partial z^j}\frac{\partial p^l}{\partial z^i}\right) \tag{14}$$

Since it is well known[29] that the Lagrange and Poisson brackets are inverses,

$$\{\mathbf{z},\mathbf{z}\}_{ij}[\mathbf{z},\mathbf{z}]^{jk} = \delta_i^k \tag{15}$$

Eq. (13) follows immediately from Eq. (5). Indeed, it is interesting to note that Goldstein remarks that "[Eq. (15)] is invariant under *all* transformations of the coordinates, *without restriction even to canonical coordinates*" (last emphasis added). Unfortunately, Goldstein did not pursue the implications of this observation.

It is assumed that the Hamiltonian describes rapid, nearly periodic motion in an angle θ. Then, with the appropriate choice of noncanonical coordinates, it is possible to cast the Hamiltonian into the standard form Eq. (2), and various averaging techniques are available for effecting the time-honored procedure of removing angle dependence order by order. However, unlike the very special case of canonical coordinates, *such averaging does not*, in general, *remove angle dependence from the equations of motion* for any of the variables, because in noncanonical coordinates the Poisson tensor may depend on θ. To isolate the rapid motion, an intermediate set of ("semicanonical") variables must be constructed for which the fast angle and its conjugate action are decoupled from—have zero Poisson brackets with—the remaining variables. That such a transformation can be found is a consequence of the Darboux theorem,[38] which states that given a closed, nondegenerate, differential two-form $\underline{\omega}$ in a certain neighborhood in R^{2n}, a local coordinate system $\{p^i, q^i\}$ can be chosen such that $\underline{\omega}$ has the standard form Eq. (6). Littlejohn[34] has discussed the Darboux theorem in detail. The proof of the theorem is constructive, so that explicit equations for the desired transformations are available.

The Hamiltonian written in Darboux variables can now be subjected to a near-identity symplectic transformation[38] that removes angle dependence to any desired order. The functional form of the Poisson tensor is preserved under such a transformation.[34] The most efficient technique for this purpose is Lie transformation, the theory and application of which has been thoroughly discussed.[34,40–42] A Lie transformation T to new variables $\bar{\mathbf{z}}$ and to a new Hamiltonian K is effected by $\bar{\mathbf{z}} = T\mathbf{z}$ and $K = T^{-1}H$; if $L_n \equiv \{w_n, \ldots\} = \sum_{l=0}^{\infty} \varepsilon^l L_n^{(l)}$, then[34,36]

$$T = 1 - \varepsilon L_1^{(0)} + \tfrac{1}{2}\varepsilon^2\left[-L_2^{(0)} + \left(L_1^{(0)}\right)^2 - 2L_1^{(1)}\right] + O(\varepsilon^3) \tag{16}$$

The expression $K = T^{-1}H$ is expanded order by order, and the generating functions w_n are chosen to make K_n independent of angle.

The function $\bar{\mu} = T\mu$, the Lie transform of the Darboux action variable μ conjugate to the angle θ, is the adiabatic invariant associated with the nearly periodic motion described by θ. If the Lie transformation, constructed as a power series in ε, converges, $\bar{\mu}$ exists and is an exact invariant. The Lie series will not converge everywhere if the system supports one or more additional periodicities that can resonate with the first. This effect can be isolated by first introducing Darboux action-angle variables for all periodicities, and by designing T to remove dependence on all angles simultaneously. Potential loss of convergence will be signified by the appearance of resonant denominators in the formal expressions for the w_n's, and can be analyzed with the aid of the KAM and related theorems. See Section 4.

3. A NONRELATIVISTIC, TIME-DEPENDENT, GUIDING CENTER HAMILTONIAN

Littlejohn's original work[34] on the guiding center problem ignored time dependence and electric fields. Here we indicate a possible generalization that includes these effects.[43] The results of the formalism will be needed in the next section where we study resonances. We use the well-known technique[29] of extending the phase space by introducing the energy and the reversed time as a new canonical momentum–coordinate pair. Thus, defining $s \equiv -t$, we introduce a variable w such that $[s, w] = 1$, while $[s, s] = [w, w] = [s, z^i] = [w, z^i] = 0$. Then, we define an extended Hamiltonian Λ by

$$\Lambda(\mathbf{q}, \mathbf{p}, s, w; \tau) \equiv H(\mathbf{q}, \mathbf{p}; t) - w \tag{17}$$

where τ now plays the role of the time and does not appear on the right-hand side of Eq. (17). We have $\partial\Lambda/\partial w = -1 = ds/d\tau$, so $\tau = -s = t$ (choosing the origin of time to vanish). Then

$$\frac{\partial \Lambda}{\partial \mathbf{p}} = \frac{\partial H}{\partial \mathbf{p}} = \frac{d\mathbf{q}}{dt} = \frac{d\mathbf{q}}{d\tau}$$

and similarly $\partial\Lambda/\partial\mathbf{q} = -d\mathbf{p}/d\tau$; thus, the original canonical dynamics are preserved in the extended phase space. Finally, from

$$\frac{dw}{d\tau} = -\frac{\partial \Lambda}{\partial s} = \frac{\partial H}{\partial t} = \frac{dH}{dt} = \frac{dH}{d\tau}$$

we conclude that we can choose $w = H$, consistent with $\partial\Lambda/\partial\tau = d\Lambda/d\tau = 0$. The fact that the value of Λ is identically zero with this choice does not, of course, affect the canonical apparatus.

We must now decide how to order the electric field. It is conventional to order $E_\parallel = O(\varepsilon)$; we will make the same choice for $\mathbf{E}_\perp$. In the (drift-kinetic) ordering we shall enforce, where gradients and time derivatives are taken to be $O(\varepsilon)$, this implies that the potentials $\mathbf{A}$ and ϕ that generate the electric field are $O(1)$ [while the potential that generates the equilibrium magnetic field is $O(\varepsilon^{-1})$]. In fact, it is often true that $\phi = O(\varepsilon)$. For example, in the theory of drift-type fluctuations in tokamaks, various arguments[44] predict saturation at fluctuation levels $e\langle\delta\phi^2\rangle^{1/2}/T_e \lesssim (k_\perp L_n)^{-1}$, where $k_\perp$ is a typical wave number of the fluctuations and L_n is the (macroscopic) density scale length. We take $(k_\perp L_n)^{-1} = O(\varepsilon)$, thus precluding discussion of the very longest wavelengths. In the laboratory, measured fluctuation levels are on the order of several percent. Littlejohn[43] has recently treated the Magnetohydrodynamic (MHD) ordering $\mathbf{E}_\perp = O(1)$, $E_\parallel = O(\varepsilon)$.

Following Littlejohn[34,36] closely, we now effect a succession of variable changes designed to bring the Hamiltonian into the standard form, Eq. (2). We first pass to the set $\{z^i\} = \{\mathbf{x}, \mathbf{v}, s, w\}$, where $\mathbf{v} \equiv \mathbf{p} - e\mathbf{A}(\mathbf{x}, s)/c$, and find

$$\Lambda(\mathbf{x}, \mathbf{v}, s, w; \tau) = \tfrac{1}{2}mv^2 + e\phi(\mathbf{x}, s) - w \tag{18}$$

In these variables, the Poisson bracket of two arbitrary functions F and G, which defines the Poisson tensor $\vec{\sigma}$, is given in Appendix A. Next one introduces a spatial, orthonormal basis $\{\hat{b}, \hat{\tau}_1, \hat{\tau}_2\}$, where $\hat{b}(\mathbf{x}, s) \equiv \mathbf{B}/B$ and $\hat{\tau}_1$ and $\hat{\tau}_2$ are space–time fields, arbitrary except for the constraint of orthonormality. One then introduces the coordinates $\{x, v_\parallel, \theta, v_\perp, s, w\}$, where $v_\parallel \equiv \hat{b}\cdot\mathbf{v}$, $v_\perp \equiv |\hat{b}\times(\mathbf{v}\times\hat{b})|$, and $\theta \equiv \tan^{-1}(\mathbf{v}\cdot\hat{\tau}_1/\mathbf{v}\cdot\hat{\tau}_2)$. The Hamiltonian becomes

$$\Lambda(\mathbf{x}, v_\parallel, \theta, v_\perp, s, w; \tau) = \tfrac{1}{2}m\left(v_\parallel^2 + v_\perp^2\right) + e\phi(\mathbf{x}, s) - w \tag{19}$$

The nonvanishing Poisson brackets in these coordinates are also recorded in Appendix A.

To isolate the fast θ-dependence, we design a Darboux transformation $\{\mathbf{x}, v_\parallel, \theta, v_\perp, s, w\} \to \{\mathbf{X}, U, \theta, \mu, T, W\}$ by requiring

$$[\theta, \mu] = (m\varepsilon)^{-1} \tag{20a}$$

$$[X, \theta] = [U, \theta] = [W, \theta] = [T, \theta] = 0 \tag{20b}$$

$$[X, \mu] = [U, \mu] = [W, \mu] = [T, \mu] = 0 \tag{20c}$$

That is, we determine the variable μ semicanonically conjugate to θ. The solution of these equations subject to the conditions $\mu = 0$, $\mathbf{X} = \mathbf{x}$, $U = v_\parallel$, $W = w$, and $T = s$ on the $v_\perp = 0$ surface is recorded in Appendix B. The Hamiltonian that results can be written in the form

$$\begin{aligned}\Lambda(\mathbf{X}, U, \theta, \mu, T, W) &= m\left(\tfrac{1}{2}U^2 + \mu\Omega\right) + e\phi_0 - W \\ &\quad + \varepsilon\left\{e\phi_1 + m\rho\left[\tfrac{1}{3}(F_3 + \hat{a}\cdot\nabla\ln\Omega)v_\perp^2\right.\right. \\ &\quad \left.\left. + \tfrac{1}{2}\left(\tfrac{1}{2}Z_0 + Z_2 + S_0\right)v_\perp U + \left(U^2F_0 - e\mathbf{E}\cdot\frac{\hat{a}}{m}\right)\right]\right\} \\ &\quad + O(\varepsilon^2)\end{aligned} \tag{21}$$

where we have redefined $v_\perp$ as $v_\perp \equiv (2\Omega\mu)^{1/2}$; also, $\rho \equiv v_\perp/\Omega$, $\mathbf{E}(\mathbf{X}, T) \equiv -\partial\phi/\partial\mathbf{X} + c^{-1}\partial\mathbf{A}/\partial T$ is the electric field, and the remaining symbols are defined in Appendix A. Equation (21) is used in Section 4, where we introduce bounce motion explicitly.

The next step is to (attempt to) construct a Lie transformation to a set of barred variables such that $\bar{\theta}$ dependence is removed from the Hamiltonian to any desired order. The barred variables can be reasonably called guiding center variables; $\bar{\mu}$ will be the magnetic moment associated with the guiding center. In general, carrying the Lie transformation through second order is complicated, though straightforward, and is not of particular interest here, since Littlejohn has discussed the magnetic effects in detail.[34,36] It is, however, instructive to consider the special case where the magnetic field is independent of space and time, while the electric field is left arbitrary. Here, the relations (B1)–(B4) simplify considerably:

$$\mathbf{X} = \mathbf{x} - \varepsilon\rho\hat{a} \tag{22a}$$

$$U = v_\parallel \tag{22b}$$

$$\mu = \frac{v_\perp^2}{2\Omega} \tag{22c}$$

$$W = w - \varepsilon\left(\frac{e}{c}\right)\rho\hat{a}\cdot\frac{\partial\mathbf{A}}{\partial s} + \frac{1}{2}\varepsilon^2\left(\frac{e}{c}\right)\rho^2\hat{a}\cdot\nabla\left(\hat{a}\cdot\frac{\partial\mathbf{A}}{\partial s}\right) \tag{22d}$$

The Lie operations described in Section 2 are also straightforward; one finds, for the low-β approximation where $\mathbf{E}_\perp = -\nabla_\perp\phi \to i\mathbf{k}_\perp\phi$,

$$\Lambda(\overline{\mathbf{X}}, \overline{U}, \theta, \bar{\mu}, T, \overline{W}) = m\left(\tfrac{1}{2}\overline{U}^2 + \overline{\mu\Omega}\right) + \left[1 - \tfrac{1}{4}(k_\perp\bar{\rho})^2\right]e\bar{\phi}$$
$$- \left(\overline{W} + \tfrac{1}{2}m\bar{u}_E^2\right) + O(\varepsilon^3) \tag{23}$$

where $\mathbf{u}_E \equiv c\mathbf{E}\times\hat{b}/B$ is the familiar electric drift, and

$$\overline{\mathbf{X}} = \mathbf{X} + O(\varepsilon^3) \tag{24a}$$

$$\bar{\theta} = \theta - \varepsilon(\Omega v_\perp)^{-1}e\hat{c}\cdot\frac{\mathbf{E}}{m} + O(\varepsilon^2) \tag{24b}$$

$$\overline{U} = U + \varepsilon^2\left(\frac{c}{B}\right)\rho\hat{b}\cdot\nabla(\hat{c}\cdot\mathbf{E}) + O(\varepsilon^3) \tag{24c}$$

$$\bar{\mu} = \mu - \varepsilon\rho\frac{c}{B}\hat{a}\cdot\mathbf{E} + \frac{1}{2}\varepsilon^2\left\{\frac{c}{B}\left(\frac{\rho}{\Omega}\right)\left[-\frac{1}{2}v_\perp(\hat{a}\hat{a} - \hat{c}\hat{c}):\nabla\mathbf{E} + 2\frac{d}{dT}(\hat{c}\cdot\mathbf{E})\right]\right.$$
$$\left. + \frac{u_E^2}{\Omega}\right\} \tag{24d}$$

where $d/dT \equiv \partial/\partial T - U\hat{b} \cdot \nabla$. The factor $1 - \frac{1}{4}(k_\perp \rho)^2$ in Eq. (23) is the long wavelength limit of the well-known factor $J_0(k_\perp \rho)$, which describes the effective field felt by the guiding center. (A gyrokinetic ordering, in which terms of all orders in $k_\perp \rho$ are retained, is straightforward, but is not discussed here.)

The equations of motion that follow from Eq. (23) are

$$\frac{d\bar{\mathbf{X}}}{d\tau} = \bar{U}\hat{b} + \varepsilon\bar{\mathbf{u}}_E + O(\varepsilon^2) \tag{25a}$$

$$\frac{d\bar{U}}{d\tau} = \frac{e\bar{E}_\parallel}{m} + O(\varepsilon) \tag{25b}$$

$$\frac{d\bar{\theta}}{d\tau} = \bar{\Omega} + O(\varepsilon^2) \tag{25c}$$

$$\frac{d\bar{\mu}}{d\tau} = 0 \qquad \text{(to all orders)} \tag{25d}$$

The physical interpretations are obvious. However, it is important to note that Eq. (25a) contains *no hint of the polarization drift*. There is no paradox; the polarization drift is an effect seen in laboratory, not guiding center, coordinates. Indeed, using $d\hat{a}/d\tau = \hat{c}\, d\theta/d\tau$ and $\mathbf{x} = \bar{\mathbf{X}} + \varepsilon\rho\hat{a}$, we have

$$\frac{d\mathbf{x}_\perp}{d\tau} = \frac{d\bar{\mathbf{X}}_\perp}{d\tau} + \varepsilon\left(\frac{1}{v_\perp}\frac{d\mu}{d\tau}\hat{a} + \rho\frac{d\theta}{d\tau}\hat{c}\right)$$

Using Eqs. (24) and (25), we then find

$$\frac{d\mathbf{x}_\perp}{d\tau} = \mathbf{v}_\perp + \varepsilon\mathbf{u}_E + \frac{\varepsilon^2}{\Omega}\frac{d}{d\tau}\left(\frac{c}{B}\mathbf{E}_\perp\right) + O(\varepsilon^3) \tag{26}$$

the last term of which is the well-known polarization drift.[11]

Further insight into the role of the barred, guiding center variables and of the Lie transformation is gained by considering the form of the magnetic moment μ. For constant $\mathbf{E}$, this can be written correct to $O(\varepsilon^2)$ as

$$\bar{\mu} = \frac{|\mathbf{v}_\perp - \mathbf{u}_E|^2}{2\Omega} \tag{27}$$

the well-known result[27] for the adiabatic invariant in the presence of electric fields. Thus, among other things the Lie transformation effects a Galilean transformation to the drift frame. If this interpretation is to be consistent, the new variable $\bar{\theta}$ should describe gyration in the drift frame. That is, it should be true that $\bar{v}_\perp \sin\bar{\theta} = (\mathbf{v}_\perp - \mathbf{u}_E) \cdot \hat{\tau}_1$. Writing $\bar{\theta} = \theta + \delta\theta$ and noting that $\bar{v}_\perp \simeq v_\perp - \hat{c} \cdot \mathbf{u}_E$, one finds that $\delta\theta = \hat{a} \cdot \mathbf{u}_E/v_\perp = -(c/B)\hat{c} \cdot \mathbf{E}/v_\perp$, in agreement with the formal result, Eq. (24).

4. THE LONGITUDINAL INVARIANT, RESONANCES, AND DESTRUCTION OF INVARIANCE

The procedure of Section 3 produces a magnetic moment $\bar{\mu}$ that is formally constant to all orders in ε. However, because the choice of coordinates took no explicit account of possible near-periodicities other than the gyration, it is not guaranteed that the Lie series converges. We study this issue with the aid of a specific example, and consider the effects on the invariance of μ of the longitudinal periodicity (bouncing) of a particle trapped in a magnetic well, as well as the effects of temporal resonance.

In unpublished work of considerable importance, Littlejohn[45] has discussed the longitudinal invariant J (for the time-independent case). However, he transformed the *gyration-averaged* Hamiltonian into bounce coordinates, and thus was unable to examine the resonant interaction between bouncing and gyration. As he indicated, the appropriate technique is to introduce bounce coordinates *before* Lie transformation is applied, and this is what we do. We begin with the Hamiltonian Eq. (21), and assume that the magnetic structure is such that particles with sufficiently small $v_\parallel$ are trapped in a magnetic well. We assume that the bounce frequency is $O(\varepsilon)$ relative to the gyrofrequency. To lowest order, particles bounce along a field line. The natural spatial coordinates are therefore the field line coordinates $\{\mathbf{y}, l\}$, where

$$l(\mathbf{X}, T) = \int^{\mathbf{X}} d\mathbf{X}' \cdot \hat{b}(\mathbf{X}', T)$$

is the distance along a field line (computed with time frozen), $y^1 \equiv \beta(\mathbf{X}, T)$, $y^2 \equiv e\alpha(\mathbf{X}, T)/mc$, where α and β are the Euler potentials[46] for the field: $\mathbf{B}(\mathbf{X}, T) = \nabla\alpha \times \nabla\beta$. One can choose a gauge such that $\mathbf{B} = \nabla \times \mathbf{A}$ with

$$\mathbf{A} = \frac{1}{2}(\alpha\nabla\beta - \beta\nabla\alpha) \qquad \text{or} \qquad A_i = \frac{1}{2}\left(\frac{mc}{e}\right)\gamma_{ab}\, y^b \frac{\partial y^a}{\partial z^i}$$

where γ is a two-dimensional antisymmetric matrix with $\gamma_{12} = 1$. As an intermediate variable, it is convenient to introduce the zeroth-order particle energy E_0:

$$E_0 \equiv \frac{1}{2}U^2 + \mu\Omega + \frac{e\phi_0}{m} \tag{28}$$

The zeroth-order bounce angle of the particle is then

$$\psi = \omega_b \int_{l_0}^{l} \frac{dl'}{U(\mathbf{y}, l', E_0, \mu, T)} \tag{29}$$

where

$$\frac{2\pi}{\omega_b(\mathbf{y}, E_0, \mu, T)} \equiv \int_b \frac{dl'}{U(\mathbf{y}, l', E_0, \mu, T)} \tag{30}$$

is the period of the nonlinear bounce motion ($\int_b$ denotes the line integral over a complete period), and l_0 is a turning point ($U = 0$) of the zeroth-order motion. For later use, it is convenient to introduce the quantity (the zeroth-order longitudinal action)

$$I(\mathbf{y}, E_0, \mu, T) \equiv \frac{1}{2\pi} \int_b dl' U$$

in terms of which Eq. (30) can be written

$$\omega_b(\mathbf{y}, E_0, \mu, T) = \left(\frac{\partial I}{\partial E_0} \right)^{-1} \equiv \Omega_b(\mathbf{y}, I, \mu, T) \tag{31}$$

Rather than determine the Poisson brackets of the new variables directly, it is convenient to note[45] that since ω is closed, it can be written as the exterior derivative of a distinguished one-form ρ: $\underline{\omega} = d\underline{\rho}$. The one-form ρ, being isomorphic to a vector $\boldsymbol{\rho}$, can be computed simply, $\underline{\omega}$ can be easily determined by computing the exterior derivative, then $\overleftrightarrow{\sigma}$ found by inverting $\underline{\omega}$. If one defines $\xi^1 \equiv T$, $\xi^2 \equiv W$, it can be shown that

$$\rho_i = \frac{1}{2} \varepsilon^{-1} \gamma_{ab} y^b \frac{\partial y^a}{\partial z^i} + \frac{1}{2} \gamma_{ab} \xi^b \frac{\partial \xi^a}{\partial z^i} + U\hat{b} \cdot \frac{\partial \mathbf{X}}{\partial z^i} \tag{32}$$

Next, a Darboux transformation taking $\{\mathbf{y}, l, E_0, T, W\}$ to $\{\mathbf{Y}, \psi, J, T, \Upsilon\}$ is designed by requiring

$$[\psi, J] = 1 \tag{33a}$$

$$[\mathbf{Y}, \psi] = [T, \psi] = [\Upsilon, \psi] = 0 \tag{33b}$$

$$[\mathbf{Y}, J] = [T, J] = [\Upsilon, J] = 0 \tag{33c}$$

and $\Upsilon = W$, $\mathbf{Y} = \mathbf{y}$, and $J = 0$ on the initial value surface $U = 0$, $E_0 = \kappa_0$, where κ_0 is defined in Appendix C. The details of solution of Eqs. (33) are not entirely trivial and cannot be recorded here. One finds that the Hamiltonian takes the form

$$\begin{aligned}
\Lambda(\mathbf{Y}, \psi, J, \theta, \mu, T, \Upsilon) = {} & \kappa(\mathbf{Y}, J, \mu, T) - \Upsilon \\
& + m\varepsilon \Big\{ D(\mathbf{Y}, \psi, T, \kappa) - \omega_b(\mathbf{Y}, J, \mu, T) \\
& \times \Big[\gamma^{ab} F_a(\mathbf{Y}, \psi, T, \kappa) \frac{\partial}{\partial Y^b} I(\mathbf{Y}, \mu, T, \kappa) \\
& + G(\mathbf{Y}, \psi, T, \kappa) \Big] \\
& + m^{-1} \delta\Lambda(\mathbf{Y}, \psi, J, \theta, \mu, T) \Big\} + O(\varepsilon^2)
\end{aligned} \tag{34}$$

Here κ is the lowest order energy E_0 expressed in terms of the action I,

$$\kappa(\mathbf{Y}, J, \mu, T) \equiv \int_0^J dJ' \, \Omega_b(\mathbf{Y}, J', \mu, T) \tag{35}$$

and $\delta\Lambda$ is the $O(\varepsilon)$ part of Eq. (21) expressed in the longitudinal Darboux coordinates; to lowest order, $\mathbf{Y} = \mathbf{y}$ and $J = I$. The θ-independent functions D, F_a, and G and the remaining Poisson brackets of the Darboux variables are recorded in Appendix C.

We are now prepared to remove angle dependence by Lie transformation. Several ordering schemes are possible. We first assume that all angle derivatives maintain their nominal ordering. As we will see, such ordering misses the resonant interaction between multiple periodicities. It also necessitates a modification of Eq. (16). Because the fundamental Poisson bracket $[\theta, \mu]$ is of order ε^{-1}, the Lie operator $L_1 \equiv [w_1, \ldots]$ contains a term $\varepsilon^{-1} L_1^\theta$, where

$$L_n^\theta \equiv \frac{\partial w_n}{\partial \theta} \frac{\partial}{\partial \mu} - \frac{\partial w_n}{\partial \mu} \frac{\partial}{\partial \theta}$$

functions L_n^ψ and L_n^T are defined similarly. Unless the term in L_1^θ is removed, the transformation will not be near identity. Littlejohn[34] removed it by setting $w_1 \equiv 0$. Here, however, a more general procedure is necessary, and it is sufficient to choose

$$L_1^\theta \Lambda_0 = 0 \qquad \text{or} \qquad \frac{\partial w_1}{\partial \theta} = 0 \tag{36}$$

The first two equations defining the Lie transformation are then

$$\Gamma_0 = \Lambda_0 \tag{37}$$

$$\Gamma_1 = \Lambda_1 + L_2^\theta \Lambda_0 + L_1^\theta \Lambda_1 + \left(L_1^\psi + L_1^T\right)\Lambda_0 \tag{38}$$

Explicitly, Eq. (38) is

$$\Gamma_1 = \Lambda_1 + \bar{\Omega}_c \frac{\partial w_2}{\partial \theta} - \frac{\partial w_1}{\partial \mu} \frac{\partial \Lambda_1}{\partial \theta} + \Omega_b \frac{\partial w_1}{\partial \psi} - \frac{\partial w_1}{\partial T} - \frac{\partial w_1}{\partial \Upsilon} \frac{\partial \Lambda_0}{\partial T} \tag{39}$$

Here

$$\bar{\Omega}_c \equiv \left.\frac{\partial \kappa}{\partial \mu}\right|_{\mathbf{Y}, J, T} = \frac{\int dl' \, \Omega(l') / [2(\kappa - \mu\Omega(l') - e\phi_0]^{1/2}}{\int dl' / [2(\kappa - \mu\Omega(l') - e\phi_0]^{1/2}} \tag{40}$$

is the bounce-averaged gyrofrequency. Equation (39) is unusual in that the second-order generating function appears in the first-order transformation equation. Nevertheless, Eq. (39) can be solved for w_1. It is consistent to take $\partial w_1/\partial \Upsilon = 0$. We choose Γ_1 to be independent of θ, ψ, and τ. Averaging Eq. (39) successively over θ, ψ, and T gives

$$\Gamma_1 = \langle \Lambda_1 \rangle_\theta + \Omega_b \frac{\partial w_1}{\partial \psi} - \frac{\partial w_1}{\partial T} \tag{41a}$$

$$\Gamma_1 = \langle \Lambda_1 \rangle_{\theta, \psi} - \frac{\partial}{\partial T} \langle w_1 \rangle_\psi \tag{41b}$$

$$\Gamma_1 = \langle \Lambda_1 \rangle_{\theta, \psi, T} \tag{41c}$$

Choosing Fourier decomposition of the form $\exp[i(l\theta + m\psi + n\omega T)]$, we see that Eq. (41b) defines the $(0, 0, n \neq 0)$ component of w_1, and that Eq. (41a) (formally) defines the $(0, m \neq 0, n \neq 0)$ component of w_1. Because of Eq. (36), w_1 is completely determined if we demand $\langle w_1 \rangle_{\theta, \psi, T} = 0$. Combining Eq. (41a) with Eq. (39) gives

$$\bar{\Omega}_c \frac{\partial w_2}{\partial \theta} = -\left(1 - \frac{\partial w_1}{\partial \mu} \frac{\partial}{\partial \theta}\right)(\Lambda_1 - \langle \Lambda_1 \rangle_\theta)$$

We see that w_2 is responsible for removing θ-dependence from Γ_1, just as in Littlejohn's theory.

The function w_1 removes ψ-dependence. The formal solution of Eq. (41a) is

$$w_1 = -\sum_{\substack{m \neq 0 \\ n \neq 0}} \frac{(\Lambda_1)_{0, m, n}}{i(m\Omega_b - n\omega)} \exp[i(m\psi + n\omega T)] \tag{42}$$

where $(\Lambda_1)_{l, m, n}$ is the Fourier amplitude of Λ_1. Away from resonance, w_1 is well defined; the Lie transformation thus constructed generates the first correction to the longitudinal invariant.[45,47] Resonance overlap in some part of phase space signifies development of stochastic regions and destruction of μ and J in that region.

An important example of destruction of J has been given by Smith[48] in his study of the trapped ion mode. In our notation, Smith computed, in a simple tokamak geometry, the amplitudes $(\Lambda_1)_{0, m, \pm 1}$ for the Hamiltonian Eq. (21), taking $\phi_0 = 0$ and retaining only the ϕ_1 part of Λ_1. He then used the Chirikov criterion for resonance overlap[23,25] to estimate the extent of the stochastic region that exists around the separatrix between trapped and passing particles. He found, for mode frequencies of the order of the bounce frequency, that destruction of the adiabatic invariant occurred in quite a large region around the separatrix. This effect would greatly affect the dispersion relation for trapped ion modes and is a subject of continuing investigation.

By taking $\partial/\partial\psi$ to be $O(1)$, we cannot correctly describe the effects of high harmonics of the bounce frequency; we thus miss the resonant interaction between gyration and bouncing. To describe this, we may take L_1^ψ and L_1^T to be $O(\varepsilon^{-1})$. One must then choose $w_1 = 0$, and more conventional Lie equations emerge:

$$\Gamma_0 = \Lambda_0 \tag{43a}$$

$$\Gamma_1 = \Lambda_1 + \left(L_2^\theta + L_2^\psi + L_2^T\right)\Lambda_0 \tag{43b}$$

The solution

$$w_2 = -\sum_{(l,m,n)\neq\mathbf{0}} \frac{(\Lambda_1)_{l,m,n}\exp[i(l\theta + m\psi + n\omega T)]}{i\left(l\bar{\Omega}_c + m\Omega_b - n\omega\right)} \tag{44}$$

displays the characteristic resonant denominators.

We may illustrate the use of Eq. (44) by estimating the degree of destruction of μ and J for a trapped particle in an axisymmetric tokamak. We consider the "theorist's equilibrium"[49] of static, nested flux surfaces concentric about a toroidal magnetic axis of radius R_0. In terms of the radial, poloidal, and toroidal unit vectors $\{\hat{r}, \hat{\Theta}, \hat{\phi}\}$, we take $\mathbf{B} = [(r/Rq)\hat{\Theta} + (R_0/R)\hat{\phi}]B_0$, where $R \equiv R_0(1 + \delta\cos\Theta)$, $\delta \equiv r/R_0 < 1$, and $q(r) = O(1)$; we ignore electric fields. Defining $\Omega_0 \equiv eB_0/mc$, we find an approximate Hamiltonian from the time-independent version of Eq. (21):

$$H = m\mu\Omega_0(1 - \delta\cos\Theta) + \frac{1}{2}mU^2$$

$$-m\varepsilon\left[\frac{(2\mu\Omega_0)^{3/2}}{3\Omega_0 R_0}\right]\left(\frac{3}{2}\cos\theta\cos\Theta + \sin\theta\sin\Theta\right) + O(\varepsilon^2) \tag{45}$$

In arriving at the form Eq. (45), a canonical transformation was used to remove an $O(\varepsilon)$ term (related to the first correction to the magnetic moment) independent of Θ. The zeroth-order Hamiltonian describes a standard pendulum, characterized by the trapping parameter $M \equiv [E_0 - \mu\Omega_0(1 - \delta)]/(2\mu\Omega_0\delta)$, E_0 being the constant value of the energy. Trapped particles satisfy $0 \leqslant M \leqslant 1$. Equation (45) can be written straightforwardly in terms of the lowest order angle-action variables $\{\psi, J\}$, which are Jacobi elliptic functions. Fourier analysis of the perturbation reveals that a dense set of resonances "pile up" and overlap near the separatrix $M = 1$. Although a resonance overlap criterion can be used to estimate the width of the stochastic region, it is more rigorous to follow Chirikov[25] and look directly at the evolution of the action near the separatrix. Since the bounce frequency and gyrofrequency are well separated, it can be shown that the change in J occurs principally near $\psi = 0$ because of kicks rapid compared to the bounce time, and that the change in μ is small compared to the change in J. The differential equations for J over each half-bounce period can then be approximated by a standard mapping,[25]

$$I_{n+1} = I_n + K \sin \theta_n \tag{46a}$$

$$\theta_{n+1} = \theta_n + I_{n+1} \tag{46b}$$

where K is a known function. Using the well-known result[25,50] that global stochasticity ensues for Eqs. (46) when $K \gtrsim 1$ leads to the result[51] that stochasticity exists in the region

$$|1 - M| \lesssim \left(\frac{\rho}{r}\right)\left(\frac{\overline{\Omega}_c}{\omega_0}\right)^3 \exp\left(-\frac{\pi \overline{\Omega}_c}{2\omega_0}\right) \tag{47}$$

where $\overline{\Omega}_c \equiv \Omega_0(1 + \delta)$ and $\omega_0 \equiv (\mu \Omega_0 \delta)^{1/2}/qR_0$. To within logarithmic corrections, this result has the familiar form $\exp(-\alpha/\varepsilon)$. This is, in fact, a quite general result for analytic Hamiltonians, by virtue of an uncertainty principle of Fourier analysis.

5. DISCUSSION

Our principal contributions have been to introduce a time-dependent Hamiltonian formulation of guiding center motion in the drift-kinetic ordering and to show from the Hamiltonian point of view how resonant interactions between multiple periodicities lead to adiabatic, as opposed to exact, invariants. As an example, we computed the extent of the stochastic region for a trapped particle in an axisymmetric tokamak. In itself, this calculation is academic, since the stochastic region is extremely small for typical device parameters. However, it serves as a very useful prototype for more relevant (and much more difficult) calculations of particle stochasticity in the magnetic fields of realistic devices.

Others have analyzed nonadiabatic effects by studying directly the equations of evolution for μ_0 or μ (see, e.g., Ref. 52). The well-known feature of the Hamiltonian method is that it collects all the physics into one scalar function, an advantage not substantially vitiated by the noncanonical formulation. We believe that the techniques and point of view discussed here will have wide applications to further studies of regular and stochastic particle motion.

ACKNOWLEDGMENTS

We are very grateful for informative and stimulating discussions with J. Cary, R. Cohen, R. Dewar, H. Grad, S. Johnston, C. Karney, A. Kaufman, R. Kulsrud, R. Littlejohn, P. Morrison, H. Mynick, C. Oberman, G. Smith, and B. Taylor. We are indebted to Robert Littlejohn for kindly providing us with Refs. 43 and 45. One of us (JAK) especially thanks Allan Kaufman for the hospitality he has shown on several visits to Berkeley, during which time some of this work was performed.

This work was supported in part by the U.S. Air Force Office of Scientific Research Contract No. PR # 80-00656 and in part by the U.S. Department of Energy Contract No. DE-AC02-76-CH03073.

APPENDIX A: POISSON BRACKETS

In the coordinates $\{\mathbf{x}, \mathbf{v}, s, w\}$,

$$[F, G] = \frac{1}{m}\left(\frac{\partial F}{\partial \mathbf{x}} \cdot \frac{\partial G}{\partial \mathbf{v}} - \frac{\partial F}{\partial \mathbf{v}} \cdot \frac{\partial G}{\partial \mathbf{x}} + \mathbf{\Omega} \cdot \frac{\partial F}{\partial \mathbf{v}} \times \frac{\partial G}{\partial \mathbf{v}}\right)$$
$$- \frac{e}{mc}\frac{\partial \mathbf{A}}{\partial s} \cdot \left(\frac{\partial F}{\partial \mathbf{v}}\frac{\partial G}{\partial w} - \frac{\partial F}{\partial w}\frac{\partial F}{\partial \mathbf{v}}\right) + \frac{\partial F}{\partial s}\frac{\partial G}{\partial w} - \frac{\partial F}{\partial w}\frac{\partial G}{\partial s} \quad \text{(A1)}$$

In the coordinates $\{\mathbf{x}, v_{\parallel}, \theta, v_{\perp}, s, w\}$, all Poisson brackets except the following vanish:

$$[\mathbf{x}, v_{\parallel}] = \frac{\hat{b}}{m} \quad \text{(A2a)}$$

$$[\mathbf{x}, \theta] = -\frac{\hat{a}}{mv_{\perp}} \quad \text{(A2b)}$$

$$[\mathbf{x}, v_{\perp}] = \frac{\hat{c}}{m} \quad \text{(A2c)}$$

$$[\theta, v_{\perp}] = m^{-1}\left[\frac{\Omega}{v_{\perp}} + F_3 + (v_{\parallel}/v_{\perp})Z_0\right] \quad \text{(A2d)}$$

$$[v_{\parallel}, v_{\perp}] = m^{-1}\left[v_{\perp}\left(\frac{1}{2}Z_1 - S_1\right) + v_{\parallel}F_1\right] \quad \text{(A2e)}$$

$$[v_{\perp}, w] = -\hat{c} \cdot \left(\frac{e}{mc}\frac{\partial \mathbf{A}}{\partial s} + v_{\parallel}\frac{\partial \hat{b}}{\partial s}\right) \quad \text{(A2f)}$$

$$[v_{\parallel}, \theta] = m^{-1}\left[\frac{1}{2}Z_0 - S_0 - Z_2 - \left(\frac{v_{\parallel}}{v_{\perp}}\right)F_0\right] \quad \text{(A2g)}$$

$$[\theta, w] = \hat{a} \cdot \left\{\left(\frac{e}{mcv_{\perp}}\right)\frac{\partial \mathbf{A}}{\partial s} + \left[\left(\frac{v_{\parallel}}{v_{\perp}}\right)\frac{\partial \hat{b}}{\partial s} + \frac{\partial \hat{c}}{\partial s}\right]\right\}, \quad \text{(A2h)}$$

$$[v_{\parallel}, w] = -\left(\frac{e}{mc}\right)\hat{b} \cdot \frac{\partial \mathbf{A}}{\partial s} + v_{\perp}\hat{c} \cdot \frac{\partial \hat{b}}{\partial s} \quad \text{(A2i)}$$

$$[s, w] = 1 \quad \text{(A2j)}$$

where

$$\hat{a} \equiv \cos\Theta\hat{\tau}_1 - \sin\Theta\hat{\tau}_2 \tag{A3a}$$

$$\hat{c} \equiv -\sin\Theta\hat{\tau}_1 - \cos\Theta\hat{\tau}_2 \tag{A3b}$$

$$Z_0 \equiv \hat{b}\cdot\nabla\times\hat{b} \tag{A4a}$$

$$Z_1 \equiv \nabla\cdot\hat{b} \tag{A4b}$$

$$Z_2 \equiv \hat{b}\cdot\mathbf{R} \tag{A4c}$$

$$\mathbf{R} \equiv (\nabla\hat{c})\cdot\hat{a} = (\nabla\hat{\tau}_1)\cdot\hat{\tau}_2 \tag{A5}$$

$$F_0 \equiv \hat{b}\cdot(\nabla b)\cdot\hat{a} \tag{A6a}$$

$$F_1 \equiv \hat{b}\cdot(\nabla\hat{b})\cdot c \tag{A6b}$$

$$F_2 \equiv \hat{a}\cdot\mathbf{R} \tag{A6c}$$

$$F_3 \equiv \hat{c}\cdot\mathbf{R} \tag{A6d}$$

$$S_0 \equiv \frac{1}{2}[\hat{a}\cdot(\nabla\hat{b})\cdot\hat{c} + \hat{c}\cdot(\nabla\hat{b})\cdot\hat{a}] \tag{A7a}$$

$$S_1 \equiv \frac{1}{2}[\hat{a}\cdot(\nabla\hat{b})\cdot\hat{a} - \hat{c}\cdot(\nabla\hat{b})\cdot\hat{c}] \tag{A7b}$$

APPENDIX B: DARBOUX TRANSFORM FOR GUIDING CENTER PROBLEM

The solutions of Eqs. (20) are

$$\mathbf{X} = \mathbf{x} - \varepsilon\rho\hat{a} - \varepsilon^2\rho^2\hat{a}\left\{\frac{1}{2}\hat{a}\cdot\nabla\ln\Omega - \left[\frac{1}{2}F_3 + \left[\frac{v_\parallel}{v_\perp}Z_0\right]\right]\right\} + O(\varepsilon^3) \tag{B1}$$

$$U = v_\parallel - \varepsilon\rho\left[\frac{1}{2}v_\perp\left(-\frac{1}{2}Z_0 + S_0 + Z_2\right) + v_\parallel F_0\right] + O(\varepsilon^2) \tag{B2}$$

$$\begin{aligned} W = w &- \varepsilon\frac{e}{c}\rho\hat{a}\cdot\frac{\partial\mathbf{A}}{\partial s} \\ &+ \varepsilon^2\rho^2\left\{m\Omega\frac{v_\parallel}{v_\perp}\hat{a}\cdot\frac{\partial\hat{b}}{\partial s} + \frac{1}{2}m\Omega\hat{a}\cdot\frac{\partial\hat{c}}{\partial s} + \frac{e}{c}\hat{a}\right. \\ &\quad\cdot\frac{\partial\mathbf{A}}{\partial s}\left[\frac{1}{2}F_3 + \left(\frac{v_\parallel}{v_\perp}\right)Z_0 - \frac{1}{2}\hat{a}\cdot\nabla\ln\Omega\right] \\ &\quad\left. + \frac{1}{2}\frac{e}{c}\hat{a}\cdot\nabla\left(\hat{a}\cdot\frac{\partial\mathbf{A}}{\partial s}\right)\right\} + O(\varepsilon^3) \end{aligned} \tag{B3}$$

$$\mu = \frac{v_\perp^2}{2\Omega} + \varepsilon\rho^2\left[\frac{1}{3}v_\perp\left(F_3 + \frac{1}{2}\hat{a}\cdot\nabla\ln\Omega\right) - \frac{1}{2}v_\parallel Z_0\right] + O(\varepsilon^2) \tag{B4}$$

$$T = s \tag{B5}$$

where

$$\rho \equiv \frac{v_\perp}{\Omega(\mathbf{x}, s)} \tag{B6}$$

The Poisson brackets of the Darboux variables not specified by Eqs. (20) are *exactly*

$$[\mathbf{X}, \mathbf{X}] = \varepsilon\hat{b} \times \frac{\overleftrightarrow{1}}{m\Omega^*} \tag{B7a}$$

$$[\mathbf{X}, U] = m\left\{\hat{b} + \frac{\varepsilon U[\hat{b}\times(\hat{b}\cdot\nabla)\hat{b}]}{\Omega^*}\right\} \tag{B7b}$$

$$[X, W] = \frac{\varepsilon b \times \left(\frac{e}{mc}\frac{\partial\mathbf{A}}{\partial T} + \varepsilon U\frac{\partial\hat{b}}{\partial T}\right)}{\Omega^*} \tag{B7c}$$

$$[U, W] = -\frac{e}{mc}\hat{b}\cdot\frac{\partial\mathbf{A}}{\partial T} - \frac{\varepsilon U[\hat{b}\times(\hat{b}\cdot\nabla)\hat{b}]\cdot\left(\frac{e}{mc}\frac{\partial\mathbf{A}}{\partial T} + \varepsilon U\frac{\partial\hat{b}}{\partial T}\right)}{\Omega^*}$$

$$[T, W] = 1 \qquad \text{(B7d) and (B7e)}$$

where

$$\Omega^* \equiv \Omega + \varepsilon U\hat{b}\cdot(\nabla\times\hat{b}) \tag{B8}$$

APPENDIX C: QUANTITIES RELATED TO THE LONGITUDINAL INVARIANT

The following quantities appear in Eq. (34):

$$D(\mathbf{Y}, \psi, T, \kappa) \equiv \int_{\kappa_0}^{\kappa} dE_0\left[\frac{\partial}{\partial E_0}\left(U\hat{b}\cdot\frac{\partial\mathbf{X}}{\partial T}\right) - \frac{\partial}{\partial T}\left(U\frac{\partial s}{\partial E_0}\right)\right]$$

$$F_a(\mathbf{Y}, \psi, T, \kappa) \equiv \int_{\kappa_0}^{\kappa} dE_0\left[\frac{\partial}{\partial Y_a}\left(U\frac{\partial s}{\partial E_0}\right) - \frac{\partial}{\partial E_0}\left(U\hat{b}\cdot\frac{\partial\mathbf{X}}{\partial Y_a}\right)\right]$$

$$G(\mathbf{Y}, \psi, T, \kappa) \equiv \gamma_{ab}\int_{\kappa_0}^{\kappa} dE_0\frac{\partial F_a}{\partial E_0}\frac{\partial F_b}{\partial\psi}$$

where $\kappa_0 \equiv \mu\Omega(\mathbf{Y}, l_{\min}) + e\phi_0(\mathbf{Y}, l_{\min})$ and $l_{\min}$ labels the position of minimum field strength along the line labeled by **Y**.

The Poisson brackets unspecified by Eqs. (33) are

$$[Y_a, Y_b] = \varepsilon\gamma_{ab} \qquad [Y_a, \Upsilon] = 0 \qquad [Y_a, T] = 0 \qquad [T, \Upsilon] = 1$$

REFERENCES

1. Verhandlungen des Conseil Solvay 1911: *Die Theorie der Strahlung und der Quanten*, A. Eucken, Ed. (Knapp, Halle an Saale, Germany, 1914), p. 364. Quotation translated by the authors.
2. M. Born, *Atomic Physics*, 6th ed. (Hafner, New York, 1957), Section V.2.
3. M. Born and V. Fock, *Z. Phys.* **51**, 165 (1928); A. Messiah, *Quantum Mechanics*, Vol. II, translated by J. Potter (North-Holland, Amsterdam, 1966), Chap. XVII.
4. A. Lenard, *Ann. Phys. (N.Y.)* **6**, 261 (1959).
5. *Earth's Particles and Fields*, B. M. McCormac, Ed. (Reinhold, New York, 1968), and references therein.
6. G. I. Budker, in *Plasma Physics and the Problems of Controlled Thermonuclear Reactions*, Vol. 3 (Pergamon Press, New York, 1959), p. 1; D. E. Baldwin, *Rev. Mod. Phys.* **49**, 317 (1977).
7. H. Alfvén, *Cosmical Electrodynamics* (Clarendon Press, Oxford, England, 1950).
8. G. Gibson, W. C. Jordan, and E. J. Lauer, *Phys. Rev. Lett.* **5**, 141 (1960); S. N. Rodionov, *Plasma Phys.* **1**, 247 (1960).
9. L. I. Artsimovich, *Nuclear Fusion* **12**, 215 (1972); H. P. Furth, *Nuclear Fusion* **15**, 487 (1975).
10. The ideas can be traced back to the earliest days of stellarator research: L. Spitzer, Jr., U.S. Atomic Energy Commission Report No. NYO-993 (1951).
11. S. Chandrasekhar, *Plasma Physics*, notes compiled by S. K. Trehan (University of Chicago Press, Chicago, 1960), Chap. III.
12. M. N. Rosenbluth, in *Advances in Plasma Physics*, vol. 5, A. Simon and W. B. Thompson, Eds. (Wiley-Interscience, New York, 1974), p. 75.
13. G. F. Chew, M. L. Goldberger, and F. E. Low, in *Series of Lectures on Physics of Ionized Gases*, J. L. Tuck, Ed., Los Alamos Scientific Laboratory Report No. LA-2055 (1956), p. T-759.
14. M. N. Rosenbluth, U.S. Atomic Energy Commission Report No. LA-2030 (1956).
15. F. Hertweck and A. Schlüter, *Z. Naturforsch.* **12A**, 844 (1957).
16. The first proof of invariance to all orders (for a special case) was apparently given by R. M. Kulsrud, *Phys. Rev.* **106**, 205 (1957).
17. M. Kruskal, *J. Math. Phys.* **3**, 806 (1962).
18. B. V. Chirikov, *Plasma Phys.* **1**, 253 (1960).
19. B. V. Chirikov, *Fiz. Plazmy* **4**, 521 (1978); Engl. transl. in *Sov. J. Plasma Phys.* **4**, 289 (1978).
20. L. I. Mandel'shtam, A. A. Andronov, and M. A. Leontovich, *J. Phys. Chem. (Moscow)* **60**, 413 (1928).
21. V. I. Arnol'd, *Usp. Mat. Nauk* **18**, 91 (1963).
22. M. Berry, in *Topics in Nonlinear Dynamics*, S. Jorna, Ed., American Institute of Physics Conference Proceedings Series, Vol. 46 (AIP, New York, 1978), p. 16.
23. G. M. Zaslavskii and B. V. Chirikov, *Usp. Fiz. Nauk* **105**, 3 (1971); Engl. transl. in *Sov. Phys. Usp.* **14**, 549 (1972).
24. V. I. Arnol'd and A. Avez, *Ergodic Problems of Classical Mechanics* (Benjamin, New York, 1968).

25. B. V. Chirikov, *Phys. Rep.* **52**, 265 (1979).
26. H. Poincaré, *Les Méthodes Nouvelles de la Méchanique Céleste* (Gauthier-Villars, Paris, 1892).
27. T. G. Northrop, *The Adiabatic Motion of Charged Particles* (Interscience, New York, 1963).
28. M. Kruskal, in *Plasma Physics* (International Atomic Energy Agency, Vienna, 1965), p. 91.
29. H. Goldstein, *Classical Mechanics* (Addison-Wesley, Reading, MA, 1950).
30. C. S. Gardner, *Phys. Rev.* **115**, 791 (1959).
31. D. P. Stern, Goddard Space Flight Center Report No. X-641-71-56 (1977).
32. A. J. Dragt and J. M. Finn, *J. Math. Phys.* **20**, 2649 (1979).
33. H. E. Mynick, Ph.D. Thesis, University of California at Berkeley (1979), unpublished; *Phys. Fluids* **23**, 1888 (1980).
34. R. G. Littlejohn, *J. Math. Phys.* **20**, 2445 (1979).
35. M. N. Rosenbluth, *Phys. Rev. Lett.* **29**, 408 (1972).
36. R. G. Littlejohn, Lawrence Berkeley Laboratory Report No. LBL-9141 (1979).
37. C. W. Misner, K. S. Thorne, and J. A. Wheeler, *Gravitation* (Freeman, San Francisco, 1973).
38. V. I. Arnol'd, *Mathematical Methods of Classical Mechanics*, transl. K. Vogtmann and A. Weinstein (Springer-Verlag, New York, 1978).
39. C. Lanczos, *The Variational Principles of Classical Mechanics*, 4th ed. (University of Toronto Press, Toronto, 1970).
40. A. Deprit, *Celestial Mech.* **1**, 12 (1969); R. L. Dewar, *J. Phys. A* **9**, 2043 (1976).
41. J. R. Cary, U.S. Department of Energy Report No. DOE/ET-0074 (1979); R. G. Littlejohn, University of California Report No. UCID-8091 (1978).
42. A. N. Kaufman, in *Intrinsic Stochasticity in Plasmas*, G. Laval and D. Grésillon, Eds. (Editions de Physique Courtaboeuf, Orsay, France, 1979), p. 131.
43. After our work was completed, we received from Littlejohn a preprint [R. G. Littlejohn, La Jolla Institute Report No. TN-81-113 (1981)] in which an alternative generalization is effected. The primary difference is that Littlejohn takes the perpendicular electric field to be of order unity and we take it to be $O(\varepsilon)$. Also, Littlejohn transforms immediately to gauge-independent variables, whereas we do not. Both procedures have merit. Otherwise, the formalisms are quite similar; it is instructive to compare both the similarities and the differences.
44. W. M. Tang, *Nuclear Fusion* **18**, 1089 (1978).
45. R. G. Littlejohn, Ph.D. thesis, University of California at Berkeley (1980), unpublished.
46. D. P. Stern, *Am. J. Phys.* **38**, 494 (1970).
47. T. G. Northrop, C. S. Liu, and M. D. Kruskal, *Phys. Fluids* **9**, 1503 (1966).
48. G. R. Smith, *Phys. Rev. Lett.* **38**, 970 (1977).
49. G. Knorr, *Phys. Fluids* **8**, 1334 (1965).
50. J. M. Greene, *J. Math. Phys.* **20**, 1183 (1979).
51. The details of this calculation will be presented elsewhere: D. H. E. Dubin and J. A. Krommes (1981), in preparation.
52. R. J. Hastie, G. D. Hobbs, and J. B. Taylor, in *Plasma Physics and Controlled Nuclear Fusion Research*, Vol. I (International Atomic Energy Agency, Vienna, 1969), p. 389; R. H. Cohen, G. Rowlands, and J. H. Foote, *Phys. Fluids* **21**, 627 (1978).

RAY ERGODICITY AND ITS CONSEQUENCES FOR PLASMA HEATING, STABILITY, AND EMISSION

EDWARD OTT
Laboratory for Plasma and Fusion Energy Studies
University of Maryland
College Park, Maryland

Abstract. *The theory of stochasticity in Hamiltonian systems is applied to the propagation of plasma waves. Illustrative applications include the accessibility of waves to the plasma interior, and the application of the ergodic hypothesis to determine plasma stability, thermal radiation properties and wall absorption. Finite wavelength diffraction effects are estimated by developing the properties of the quantum Chirikov Taylor map.*

1. INTRODUCTION

In many situations of interest in plasma physics one is concerned with the propagation of waves in inhomogeneous plasmas that are possibly surrounded by reflecting boundaries. When the plasma inhomogeneity scale length and the typical radius of curvature of the reflecting boundaries are large compared to the typical wavelength of the wave in question, the ray approximation may be useful. We consider a wave packet with total energy E and a frequency ω that satisfies a local dispersion relation $\omega = \omega_r(\mathbf{k}, \mathbf{r}) + i\gamma(\mathbf{k}, \mathbf{r})$, where the real part of the wave frequency greatly exceeds the imaginary part, $\omega_r \gg |\gamma|$ ($\gamma > 0$ for

Work supported by the U.S. Department of Energy.

unstable waves and $\gamma < 0$ for damped waves). In this case the trajectory of the wave packet is given by the ray equations

$$\frac{d\mathbf{r}(t)}{dt} = \frac{\partial \omega_r}{\partial \mathbf{k}} \tag{1.1a}$$

$$\frac{d\mathbf{k}(t)}{dt} = -\frac{\partial \omega_r}{\partial \mathbf{r}} \tag{1.1b}$$

and the variation of the wave packet energy along the ray can be obtained from

$$\frac{dE(t)}{dt} = 2\gamma E \tag{1.2}$$

It is evident from Eqs. (1.1) that the ray equations are a Hamiltonian system for which $\omega_r(\mathbf{k}, \mathbf{r})$ plays the role of the Hamiltonian, and $(\mathbf{k}, \mathbf{r})$ are analogous to the momenta and coordinate variables (usually denoted $(\mathbf{p}, \mathbf{q})$). Since Eqs. (1.1) constitute a Hamiltonian system, they are subject to the same phenomena of ergodicity onset, ergodic motions, and so on, of other Hamiltonian systems. We discuss several applications in plasma physics for which this point of view may be useful.

2. ACCESSIBILITY OF LOWER HYBRID WAVES IN TOROIDAL PLASMAS

One of the central problems in creating a controlled thermonuclear reactor lies in raising the temperature of the confined plasma sufficiently to permit fusion reactions to take place. One way of doing this is by launching from outside the plasma waves that then propagate to the plasma interior, where they dissipate their energy to heat. Clearly, conditions must be such that the wave is able to reach the plasma interior. In this case, in the terminology of the field, the wave is said to be "accessible." Of all the various types of wave that can be used for plasma heating, the lower hybrid wave is perhaps the most attractive from a technological point of view. The accessibility problem for this wave was originally considered by Stix[1] for the case in which there are two symmetry directions. For example, in a straight cylinder k_z and $m = k_\theta r$ are constants of the ray equations due to translation symmetry along the axis of the cylinder (the z-axis) and to rotation symmetry around the cylinder (in θ). The accessibility situation for this case is illustrated in Fig. 1 for a cylinder with an applied magnetic field $\mathbf{B} = B_0\mathbf{z}_0 + B_\theta\boldsymbol{\theta}_0$, $B_0^2 \gg B_\theta^2$, and a wave launched from vacuum with $m = 0$. Let $n_\parallel = k_z c/\omega_r$ (where c is the speed of light) and $n_\perp = k_r c/\omega$. Figure 1 shows plots of $n_\perp^2$ (obtained from the dispersion relation) as a function of plasma density, N. For $n_\parallel < n_a$, $n_\parallel = n_a$, and $n_\parallel > n_a$, Figs. 1a, 1b, and 1c apply, respectively, where n_a is a certain critical value.[1] Between $N = 0$ and $N = N_s$ there is a narrow cutoff region through which a slow wave (i.e., lower hybrid wave), launched from the vacuum region, typically has little

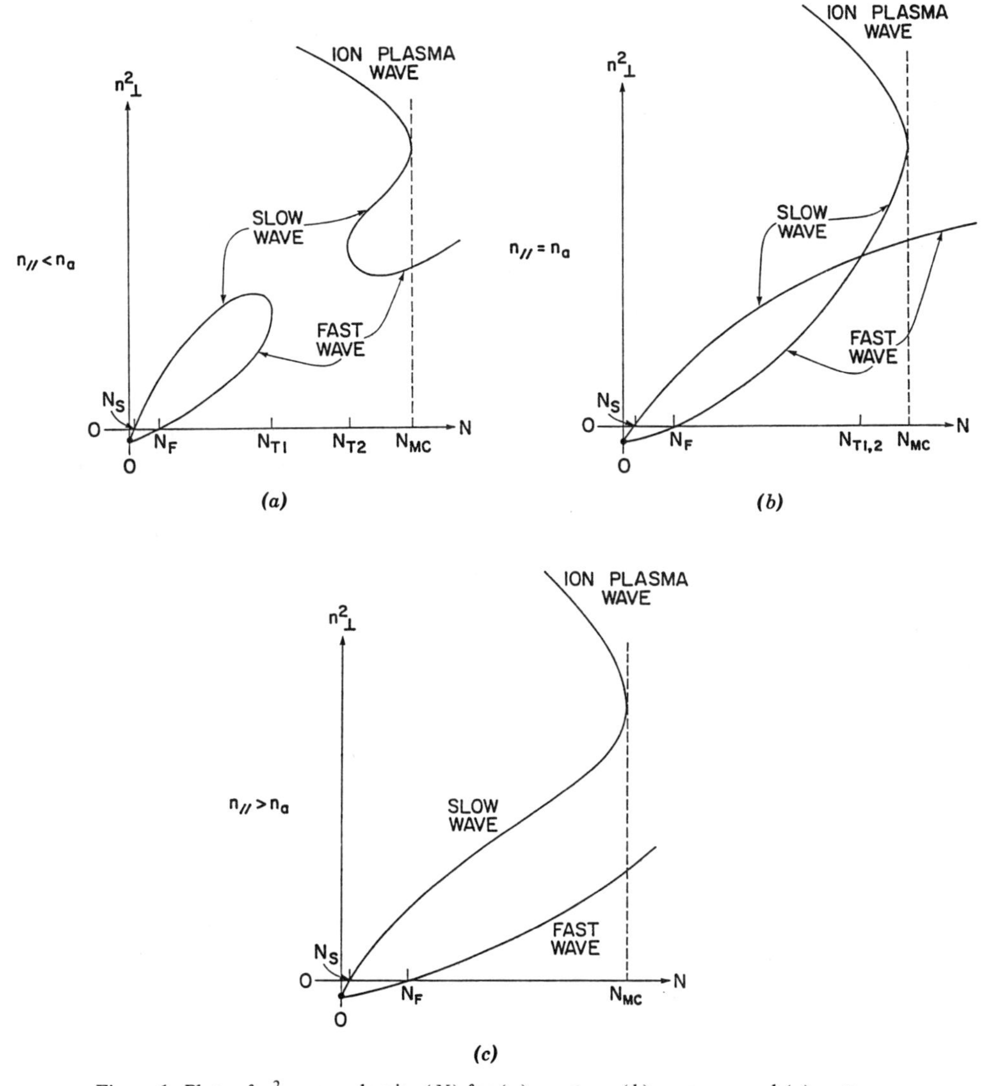

Figure 1. Plots of $n_\perp^2$ versus density (N) for (a) $n_\parallel < n_a$, (b) $n_\parallel = n_a$, and (c) $n_\parallel > n_a$.

trouble in tunneling. Figure 1a shows that for $n_\parallel < n_a$ an additional, effectively much wider, cutoff region between $N = N_{T1}$ and $N = N_{T2}$ exists. This cutoff region presents a barrier for propagation to the plasma center and prevents accessibility. Figure 1c shows that for $n_\parallel > n_a$ this barrier is absent, and the lower hybrid (slow) waves becomes accessible.

Now we consider heating a circular cross section toroidally symmetric plasma (a tokamak). We use toroidal coordinates wherein r, θ are circular polar coordinates centered in the circular cross section of the tokamak plasma such that the distance of a point from the major axis of the torus is

$R = R_0 + r\cos\theta$, ϕ is the toroidal angle, and R_0 is the distance from the major axis of the torus to the center of the plasma cross section. Let $\varepsilon = a/R_0$, where $r = a$ denotes the plasma boundary. As $\varepsilon \to 0$ with a fixed, the straight cylinder limit is approached. However, for finite ε the plasma equilibrium depends on θ. Thus it is no longer expected that $m = rk_\theta$ is a constant of the motion, although the toroidal symmetry still guarantees that a constant of the motion analogous to k_z in the cylinder still exists; namely, $n = Rk_\phi$ is a constant. The questions that now arise are what happens to the constant m, and how is the accessibility condition for lower hybrid waves effected? As is known from previous work on Hamiltonian systems, for finite ε there may still be some other constant, $\tilde{m} = \tilde{m}(r, \theta, k_r, m)$, which takes the place of m. For small ε, regions where $\tilde{m}$ exists occupy most of the phase space; but there can also exist small regions in which the trajectory wanders ergodically. As ε increases the regions occupied by ergodic trajectories increase, until almost all regions where $\tilde{m}$ exists are gone. A ray in the region with no $\tilde{m}$ may eventually approach the plasma interior and be absorbed even if $n_\parallel$ at launch does not satisfy the straight cylinder accessibility condition. Thus we need to know at what value of ε most of the rays have no constant $\tilde{m}$.

Figures 2–5 show numerical results testing for the existence of $\tilde{m}$ by the surface of section method with $\theta = 0 \pmod{2\pi}$ as the surface of section (cf. also Refs. 2 and 3). Further details of these calculations will appear in Ref. 2. Figure 2 shows that for $\varepsilon = 0.10$, $\tilde{m}$ exists, and initially inaccessible rays (i.e., $n_\parallel < n_a$ at launch) do not reach the plasma interior. Figures 3 and 4 show a case for $\varepsilon = 0.15$, illustrating the coexistence of ergodic and integrable orbits including (Fig. 4) higher order island structures. For $\varepsilon = 0.25$, $\tilde{m}$ is apparently

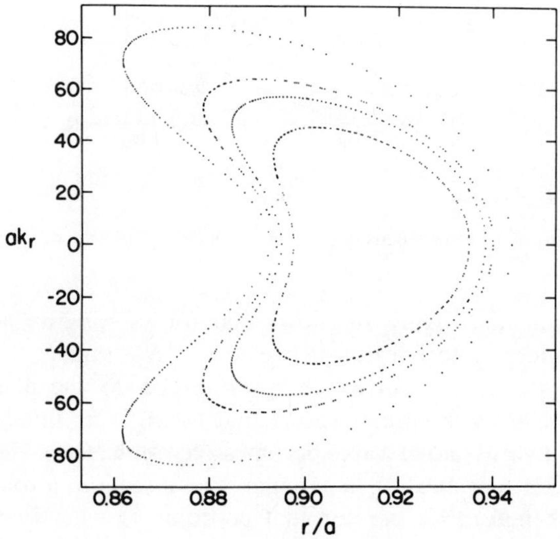

Figure 2. Surface of section ($\theta = 0$) plots for four different initial conditions for a torus with inverse aspect ratio $a/R_0 = 0.10$: $n_a = 2.0$, $1.3 \leqslant n_\parallel \leqslant 1.4$.

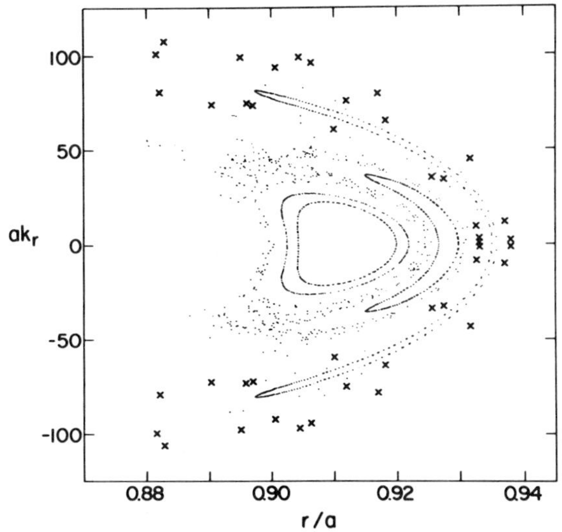

Figure 3. Surface of section plots for several different initial conditions for a torus with inverse aspect ratio $a/R_0 = 0.15$. Integrable and nonintegrable (ergodic) regions of phase space are evident: $n_a = 2.0$; $1.25 \leqslant n_{\|} \leqslant 1.4$.

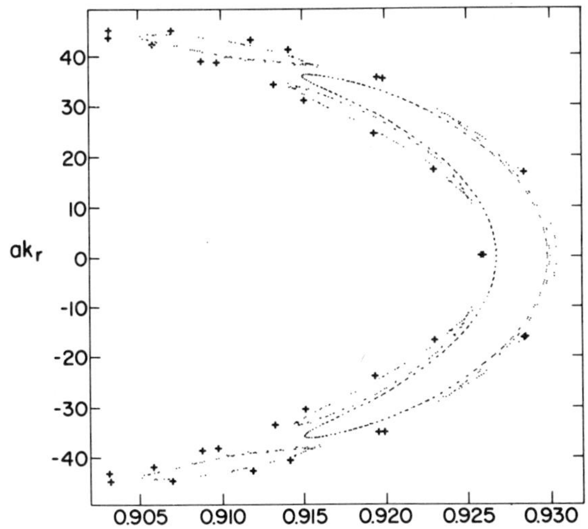

Figure 4. Same conditions as in Fig. 3 but for an initial condition yielding a fifth-order island.

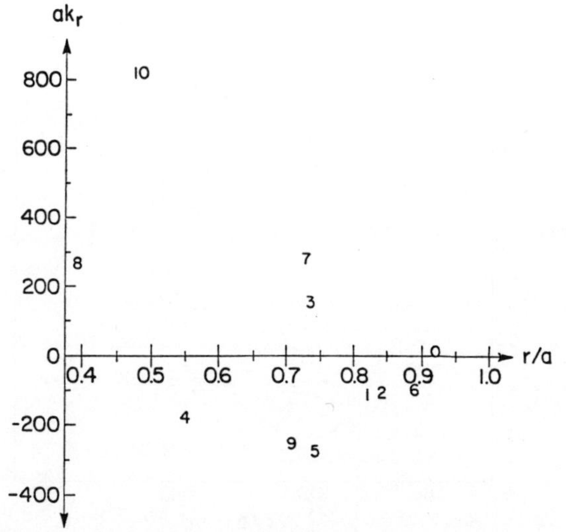

Figure 5. Surface of section for a single orbit with $n_{\parallel} < n_a$ in a torus of inverse aspect ratio $a/R_0 = 0.25$; numbers denote sequence in which the surface of section was pierced. After about 10 piercings of the surface of section, the wave reaches the interior of the plasma, where it is strongly absorbed by ion Landau damp.

completely destroyed, and even waves launched with $n_{\parallel}$ substantially below n_a are absorbed in the plasma interior after a few piercings of the surface of section (Fig. 5).

3. APPLICATION OF THE ERGODIC HYPOTHESIS

As shown in Section 2 and in Refs. 2 and 3, ergodic wandering of plasma wave rays can occur in toroidal plasmas of circular cross section (tokamaks). This results because the toroidal geometry breaks the (poloidal) circular symmetry, thus leading to ergodic wandering in the poloidal plane. Even stronger poloidal plane asymmetry can occur in other toroidally symmetric devices such as multipoles and doublets. Furthermore, other plasma devices (e.g., bumpy tori and stellarators) are highly asymmetric and possess no symmetry direction. It is natural to expect waves to wander ergodically in such situations. Sections 4–6 examine the consequences of ray ergodicity for some problems in plasma physics. The time average of some functions $g(\mathbf{k}, \mathbf{r})$ over the trajectory of a wave packet,

$$\bar{g} \equiv \lim_{t \to \infty} t^{-1} \int_0^t g[\mathbf{k}(t), \mathbf{r}(t)]\, dt,$$

shall prove to be a quantity of great interest for subsequent considerations. By the ergodic hypothesis

$$\bar{g} = \langle g \rangle \tag{3.1}$$

where $\langle g \rangle$ is a suitable phase space average. For example, consider a case with toroidal symmetry (with toroidal angle ϕ). Then $n = Rk_\phi$ is a constant of the motion for the ray equations (where R denotes the major toroidal radius coordinate). In this case, assuming ergodic motion in the poloidal plane, we have, in analogy with the usual microcanonical distribution,

$$\langle g \rangle = \frac{\int\int g(\mathbf{k},\mathbf{r})\delta(k_\phi R - n)\delta[\Omega - \omega_r(\mathbf{k},\mathbf{r})]\, d\mathbf{k}\, d\mathbf{r}}{\int\int \delta(k_\phi R - n)\delta[\Omega - \omega_r(\mathbf{k},\mathbf{r})]\, d\mathbf{k}\, d\mathbf{r}} \tag{3.2}$$

where the real part of the wave frequency Ω plays the role analogous to the system energy in the microcanonical distribution. In devices with no symmetry direction, the appropriate phase space average in an ergodic situation is

$$\langle g \rangle = \frac{\int\int g(\mathbf{k},\mathbf{r})\delta[\Omega - \omega_r(\mathbf{k},\mathbf{r})]\, d\mathbf{k}\, d\mathbf{r}}{\int\int \delta[\Omega - \omega_r(\mathbf{k},\mathbf{r})]\, d\mathbf{k}\, d\mathbf{r}} \tag{3.3}$$

One might argue that a wave packet is subject to diffractive spreading and that it thus cannot be regarded as well localized for all time. In this case one might question the relevance of the limit $t \to \infty$ in the definition of $\bar{g}$. An alternate point of view, which does not suffer from this weakness, is to consider a nonlocalized solution, namely, an eigenmode. In the case of an eigenmode with sufficiently short wavelength, we introduce the ordering

$$L \gg \delta \gg \lambda \tag{3.4}$$

where L is the scale size of the eigenmode (e.g., in a cavity with reflecting walls, L would be a typical dimension of the cavity), λ is a typical wavelength, and δ is some intermediate scale. With the ordering of Eq. (3.4) one can look at the wave number power spectrum within a cell of size δ centered about some point $\mathbf{r}$, which we denote $W(\mathbf{k},\mathbf{r})$. Thus δ represents a coarse graining scale length. If one is dealing with a quantum mechanical wave equation, then $W(\mathbf{k},\mathbf{r})$ can be identified with a coarse-grained Wigner function, and the rays are just the classical particle orbit trajectories. Voros[4] and Berry[5] have argued on the basis of the correspondence principle that in the case of ergodic rays, $W(\mathbf{k},\mathbf{r})$ should be given by (in our notation)

$$W(\mathbf{k},\mathbf{r}) = \left[\int\!\!\int W(\mathbf{k},\mathbf{r})\,d\mathbf{k}\,d\mathbf{r}\right]\frac{\delta[\Omega-\omega_r(\mathbf{k},\mathbf{r})]}{\int\!\!\int \delta[\Omega-\omega_r(\mathbf{k},\mathbf{r})]\,d\mathbf{k}\,d\mathbf{r}} \tag{3.5}$$

for a case where ω_r is the only constant of the motion [as is the case for Eq. (3.3)]. If there are additional constants of the motion due to symmetry but the motion is otherwise ergodic, Eq. (3.5) will be modified. For example, for a case with toroidal symmetry $\delta[\Omega-\omega_r(\mathbf{k},\mathbf{r})]$ in Eq. (3.5) is replaced by $\delta[\Omega-\omega_r(\mathbf{k},\mathbf{r})]\delta[Rk_\phi - n]$ [cf. Eq. (3.2)]. Although the arguments of Voros and Berry are in the context of the quantum mechanical wave equation, it seems clear that, if the result [Eq. (3.5)] applies in this case, it should apply equally well for other systems of equations with an associated group velocity and system of ray equations [Eqs. (1.1)]. One expects that Eq. (3.5) should be true in the limit of $(\lambda/L)\to 0$; however, it is not yet known how good Eq. (3.5) is for finite (λ/L). In this connection, recent numerical experiments may be of interest[6–9] (typically $(\lambda/L)\sim 10^{-1}$ in these experiments). For the purposes of obtaining the results of Sections 4–6, one may employ *either* the wave packet point of view, Eq. (3.1), or the eigenfunction point of view, Eq. (3.5). In writing Eq. (3.5) we have in mind a cavity filled with a magnetized plasma or anisotropic dielectric. In this case the two polarizations of plane electromagnetic waves propagating in a given direction, with given frequency are, in general, nondegenerate. That is, they have different values of $|\mathbf{k}|$. Thus the dispersion relation $\Omega = \omega_r(\mathbf{k},\mathbf{r})$ will possess these two solutions. Hence, in this case Eq. (3.5) automatically gives the distribution of energy in the two polarizations. In the case of an unmagnetized plasma, or vacuum, the two independent polarizations are degenerate, that is, they have the same value of $|\mathbf{k}|$. In this case we supplement Eq. (3.5) with the information that both polarizations are equally likely.

In the next three sections we discuss some specific applications of Eqs. (3.1)–(3.5):

1. Plasma stability[10] (Section 4 and Appendix A).
2. Determination of wall absorption by conducting boundaries[11] (Section 5).
3. Thermal emission from plasmas[11] (Section 6).

Parenthetically we note that a result analogous to Eq. (3.5) can also be obtained for many situations independent of the existence of a group velocity. This conclusion follows from noting that the equations for the characteristics of the eikonal equation are Hamiltonian (ray equations).

We shall be interested in applying Eqs. (3.1) and (3.5) to cases of the wave experiencing damping and/or growth within the plasma [given by $\gamma(\mathbf{k},\mathbf{r})$] and possibly absorption at the boundaries upon reflection. Under these circumstances we shall require for the validity of our applications the following two conditions:

Condition i. The ray trajectory of the wave is confined to a finite region of **k**–**r** phase space.

Condition ii. The growth and/or damping is small enough that a ray can wander over a representative region of the $\omega_r(\mathbf{k},\mathbf{r}) = $ constant surface before the energy content of a wave packet is appreciably modified.

Condition i is necessary for the integrals in Eqs. (3.2) and (3.3) to exist. This condition would be violated, for example, for waves that experience a resonance (i.e., $k \to \infty$ at some point in space). [Also note that the ergodic waves in Section 2 do not satisfy Eq. (3.5), since they experience strong absorption once they reach the plasma interior (i.e. Condition ii is violated).]

4. PLASMA STABILITY CRITERION FOR WAVES WITH ERGODIC RAY TRAJECTORIES

Consider a case such that the ray approximation applies, and waves are locally unstable with local growth rate $\gamma(\mathbf{k},\mathbf{r}) > 0$ in some regions of **k**–**r** phase space and locally damped, $\gamma(\mathbf{k},\mathbf{r}) < 0$, in other regions. Under these circumstances a reasonable stability condition is that the exponentiation of a wave packet

$$I \equiv \int_0^T \gamma[\mathbf{k}(t),\mathbf{r}(t)]\, dt \tag{4.1}$$

can become large (> 10, say) over some time interval 0 to T with some initial condition $[\mathbf{k}(o),\mathbf{r}(o)]$. We now consider the possibility of evaluating Eq. (4.1) by making use of the ergodic theorem, Eq. (3.1).

Making use of Conditions i and ii of Section 3, it makes sense to express the number of exponentiations in Eq. (4.1) as

$$I \cong \langle \gamma(\mathbf{k},\mathbf{r}) \rangle T \tag{4.2}$$

Since $\gamma(\mathbf{k},\mathbf{r})$ can be either positive or negative in different regions of **k**–**r** phase space, $\langle \gamma \rangle$ can be either positive or negative. From Eq. (4.2) the instability condition is

$$\langle \gamma \rangle > 0 \tag{4.3}$$

Equation (4.3) offers a definite prescription for evaluating stability in the case of ergodic rays, without having to calculate the integral $\int \gamma[\mathbf{k}(t),\mathbf{r}(t)]\, dt$. An example, in which Eq. (4.3) can be used for a particular plasma mode in tokamak geometry, is discussed in Appendix A.

We now offer some additional discussion concerning the validity of this approach. The total exponentiation of the wave packet is $\exp(I)$. For this approach to be reasonable, the observation of a growing or decaying wave packet should be independent of initial conditions. We therefore introduce the quantity Δ defined by

$$\Delta = \frac{\left\langle \left(e^{I} - \langle e^{I} \rangle_0\right)^2 \right\rangle_0}{\left(\langle e^{I} \rangle_0\right)^2} \tag{4.4}$$

where $\langle \cdots \rangle_0$ denotes the phase space average [e.g., Eqs. (3.2) or (3.3)] over initial conditions, and we require $\Delta \ll 1$. Defining $\delta I = I - \langle I \rangle_0$ and assuming that δI may be treated as an approximately Gaussian random variable, we have $\Delta \cong \exp[\langle (\delta I)^2 \rangle_0] - 1$. Thus the condition $\Delta \ll 1$ yields $\Delta \cong \langle (\delta I)^2 \rangle_0 \ll 1$ or

$$\Delta \cong \left\langle \int_0^T \int_0^T \delta\gamma(t')\, \delta\gamma(t'')\, dt'\, dt'' \right\rangle_0 \ll 1 \tag{4.5}$$

where $\delta\gamma(t) = \gamma[\mathbf{k}(t), \mathbf{r}(t)] - \langle \gamma \rangle$. Equation (4.5) gives

$$\Delta \cong \int_0^T \int_0^T \left\langle (\delta\gamma)^2 \right\rangle C(t' - t'')\, dt'\, dt'' \ll 1 \tag{4.6}$$

where $C(t' - t'') \equiv \langle \delta\gamma(t')\, \delta\gamma(t'') \rangle_0 / \langle (\delta\gamma)^2 \rangle$. Defining a correlation time $\tau_c = \int_0^\infty C(s)\, ds$ and assuming $T \gg \tau_c$, we have

$$\Delta \cong 2\left\langle (\delta\gamma)^2 \right\rangle \tau_c T \tag{4.7}$$

Equation (4.7) shows that for large enough T the criterion $\Delta \ll 1$ will be violated. However, if we assume that $I \sim 10$ puts the system in a strongly nonlinear regime, we are not interested in times T greater than $\langle \gamma \rangle T \sim 10$. Thus the validity conditions become

$$\frac{\left\langle (\delta\gamma)^2 \right\rangle}{\langle \gamma \rangle} \tau_c \ll 5 \tag{4.8a}$$

$$\langle \gamma \rangle \tau_c \ll 10 \tag{4.8b}$$

where the latter condition is necessary to obtain Eq. (4.7) (i.e., we required $T \gg \tau_c$). Having established the validity conditions Eqs. (4.8), which apply to "typical" rays, there still remains a problem. Namely, there may be exceptional rays that have trajectories that happen to place them in the more unstable regions of phase space much more often than is typical. For such rays the exponentiation I may become large over some time interval even though $\langle\gamma\rangle < 0$. Presumably, under conditions in Eqs. (4.8), the number of such rays will be very small. In addition, diffraction spreads a wave packet over a finite region of phase space so that the existence of such exceptional rays may not be relevant. The problem discussed in Section 7 is meant to shed some light on this question.

5. ABSORPTION BY CONDUCTING WALLS

We consider a high order mode in a cavity containing a plasma and assume that the rays are ergodic with ω_r the only constant of the motion. From energy conservation Γ, the mode damping rate per unit time due to the conducting walls, is given by

$$2\Gamma W_t = \eta_s \langle |\mathbf{J}_s|^2 \rangle_s A \tag{5.1}$$

where W_t is the total mode energy,

$$W_t = \int\int W(\mathbf{k}, \mathbf{r})\, d\mathbf{k}\, d\mathbf{r}$$

η_s is the surface skin resistivity [$\eta_s = (\mu_0 \omega_r / 2\sigma)^{1/2}$, where σ is the wall conductivity], A is the total area of the cavity wall, $\mathbf{J}_s$ is the (root-mean-square) surface current density vector, and $\langle \cdots \rangle_s$ denotes an average over the boundary surface, $\langle |\mathbf{J}_s|^2 \rangle_s \equiv A^{-1} \int |\mathbf{J}_s|^2\, dA$. The cavity Q for an eigenmode is $Q \equiv \omega_r/(2\Gamma)$. Here we wish to calculate $\langle |\mathbf{J}_s|^2 \rangle_s$ (hence Γ).

To utilize Eq. (3.5), the wall absorption must be a small perturbation on the cavity mode structure. This is true if Condition ii of Section 3 is satisfied. For simplicity we assume that near the cavity walls the plasma density is zero. In this region $\omega_r(\mathbf{k}, \mathbf{r}) = |\mathbf{k}|c$, where c is the speed of light, and from Eq. (3.5) $W(\mathbf{k}, \mathbf{r}) \sim \delta[\Omega - |\mathbf{k}|c]$. Thus, near the walls, $W(\mathbf{k}, \mathbf{r})$ is independent of $\mathbf{r}$ and the direction of $\mathbf{k}$. To treat this case we recall that $\lambda \ll L$ [cf. Eq. (3.4)] and consider a plane wave in vacuum that is incident on a plane conductor. Let θ be the angle of incidence, and ϕ the angle that the magnetic field vector $\mathbf{H}$ makes with the plane of incidence. Then, from Maxwell's equations and the condition that $\mathbf{E}$ tangential vanish (approximately true for large σ), the current $\mathbf{J}_s$ created by this plane wave is

$$\mathbf{J}_s = 2H \cos\phi\, \boldsymbol{\alpha} + 2H \sin\phi \cos\theta\, \boldsymbol{\beta}$$

where $\boldsymbol{\alpha}$ and $\boldsymbol{\beta}$ are unit vectors in the surface of the conductor, respectively, perpendicular to and in the plane of incidence. Averaging over ϕ and the solid angle of the incident waves we have

$$\langle|\mathbf{J}_s|^2\rangle_s = \int_0^{2\pi} \frac{d\phi}{2\pi} \int \frac{d\Omega}{4\pi} \left[4H^2(\cos^2\phi + \sin^2\phi\cos^2\phi)\right] \tag{5.2}$$

where $d\Omega = 2\pi \sin\theta\, d\theta$, the integral over $d\Omega$ is from $\theta = 0$ to $\theta = \pi/2$ (i.e., only over incident waves), and H^2 may be related to the average energy density of the eigenmode in the vacuum near the wall [from Eq. (3.5)]

$$\mu_0 H^2 = \int\int W(\mathbf{k},\mathbf{r})\, d\mathbf{k}\, d\mathbf{r} \frac{\int \delta[\Omega - |\mathbf{k}|c]\, d\mathbf{k}}{\int\int \delta[\Omega - \omega_r(\mathbf{k},\mathbf{r})]\, d\mathbf{k}\, d\mathbf{r}} \tag{5.3}$$

From Eq. (5.2),

$$\langle|\mathbf{J}_s|^2\rangle_s = \left(\tfrac{4}{3}\right)H^2 \tag{5.4}$$

From Eqs. (5.1)–(5.4)

$$\Gamma = \frac{8\pi\eta_s\omega_r^2 A}{3\mu_0 c^3} \left\{\int\int \delta[\Omega - \omega_r(\mathbf{k},\mathbf{r})]\, d\mathbf{k}\, d\mathbf{r}\right\}^{-1} \tag{5.5}$$

Equation (5.5) is the main result of this section and is used in the following discussion of thermal emission from plasmas (Section 6). [In the case of a vacuum-filled cavity of volume V, Eq. (5.5) becomes $\Gamma = (2\eta_s A)(3\mu_0 V)^{-1}$.]

Similar arguments to those above can be applied to Helmholtz's equation in two dimensions $\{\partial^2/\partial x^2 + \partial^2/\partial y^2 + \omega^2/c^2\}\xi = 0$, with $\xi = 0$ on the boundary, with the result that Eq. (3.5) yields ($k = \omega/c$)

$$\left\langle \left(\frac{\partial \xi}{\partial n}\right)^2 \right\rangle_s = \frac{k^2 \int \xi^2\, dx\, dy}{\int dx\, dy} \tag{5.6}$$

This expression has been checked in numerical experiments by McDonald and Manheimer[12] using a racetrack-shaped boundary.[13] Reasonable agreement was obtained, thus giving an indication of the correctness of our approach. To obtain Eq. (5.6) consider a plane wave of amplitude $\tilde{\xi}$ incident on a planar boundary where $\xi = 0$. This leads to a normal derivative of ξ at the boundary of amplitude $2\tilde{\xi}k\cos\theta$, where θ is the angle of incidence. Averaging $(2\tilde{\xi}k\cos\theta)^2$ over θ for incident waves ($0 \leqslant \theta \leqslant \pi/2$) then yields Eq. (5.6).

6. THERMAL EMISSION FROM PLASMAS

As in Section 5, we consider a plasma bounded by conducting walls and assume an ergodic eigenmode such that Eq. (3.5) applies. We assume that the wall temperature can be neglected compared to the plasma temperature. Let $S(\mathbf{k}, \mathbf{r})$ be the local energy emission rate. Since the local dispersion relation is satisfied we may write $S(\mathbf{k}, \mathbf{r})$ in the form

$$S(\mathbf{k},\mathbf{r}) = s(\mathbf{k},\mathbf{r})\delta[\Omega - \omega_r(\mathbf{k},\mathbf{r})] \tag{6.1}$$

where Ω is the resonant frequency for the particular eigenmode in question. From conservation of energy, we have for perfectly conducting walls (i.e., $\Gamma = 0$)

$$\int S(\mathbf{k},\mathbf{r})\, d\mathbf{k}\, d\mathbf{r} = \int 2\gamma(\mathbf{k},\mathbf{r}) W(\mathbf{k},\mathbf{r})\, d\mathbf{k}\, d\mathbf{r} \tag{6.2}$$

where $W(\mathbf{k}, \mathbf{r})$ is given by Eq. (3.5). Again, we emphasize that (as in Section 5) application of Eq. (3.5) requires that a typical ray be able to wander over the $\Omega = \omega_r(\mathbf{k}, \mathbf{r})$ surface before it experiences significant damping (i.e., an optically thin plasma is required). If the plasma temperature $\bar{T}$ inside the cavity is uniform, the total mode energy is

$$W_t = \frac{\kappa \bar{T}}{2} \tag{6.3}$$

From Eqs. (3.5), (6.2), and (6.3)

$$\int\!\!\int S(\mathbf{k},\mathbf{r})\; dk\; dr = \int\!\!\int \gamma(\mathbf{k},\mathbf{r})\kappa\bar{T} \frac{\delta[\Omega - \omega_r(\mathbf{k},\mathbf{r})]}{\int\!\!\int \delta[\Omega - \omega_r(\mathbf{k},\mathbf{r})]\, d\mathbf{k}\, d\mathbf{r}}\, d\mathbf{k}\, d\mathbf{r}$$

which together with Eq. (6.1) implies

$$s(\mathbf{k},\mathbf{r}) = \frac{\kappa\bar{T}\gamma(\mathbf{k},\mathbf{r})}{\int\!\!\int \delta[\Omega - \omega_r(\mathbf{k},\mathbf{r})]\, d\mathbf{k}\, d\mathbf{r}}$$

Since this result should hold locally and for electrons and ions separately, we have the following generalization for nonuniform electron and ion temperatures, $T_e(\mathbf{r})$ and $T_i(\mathbf{r})$:

$$s(\mathbf{k},\ \mathbf{r}) = \frac{\kappa[T_e(\mathbf{r})\gamma_e(\mathbf{k},\mathbf{r}) + T_i(\mathbf{r})\gamma_i(\mathbf{k},\mathbf{r})]}{\int\!\!\int \delta[\Omega - \omega_r(\mathbf{k},\mathbf{r})]\, d\mathbf{k}\, d\mathbf{r}} \tag{6.4}$$

where γ_e and γ_i are the contributions to the damping rate from energy absorption by the electrons and ions, respectively, $\gamma = \gamma_e + \gamma_i$. [One can also use Eq. (6.7) and the test particle emission result for a homogeneous medium to calculate $S(\mathbf{k}, \mathbf{r})$. This method would also allow one to deal with non-Maxwellian particle distributions.] To take into account finite wall resistivity we add the term $2\Gamma W_t$ to the right-hand side of Eq. (6.2). Equations (6.1)–(6.4) yield the following result for the mode energy:

$$W_t = \frac{1}{2}\,\frac{\int\int[\gamma_e(\mathbf{k},\mathbf{r})\kappa T_e(\mathbf{r}) + \gamma_i(\mathbf{k},\mathbf{r})\kappa T_i(\mathbf{r})]\,\delta[\Omega - \omega_r(\mathbf{k},\mathbf{r})]\,d\mathbf{k}\,d\mathbf{r}}{\int\int\gamma(\mathbf{k},\mathbf{r})\delta[\Omega - \omega_r(\mathbf{k},\mathbf{r})]\,d\mathbf{k}\,d\mathbf{r} + \Gamma\int\int\delta[\Omega - \omega_r(\mathbf{k},\mathbf{r})]\,d\mathbf{k}\,d\mathbf{r}} \tag{6.5}$$

which generalizes the result $W_t = \kappa\bar{T}/2$ to include unequal and nonuniform species temperatures and absorption by the walls.

A quantity of some interest in the study of fusion reactors is the power absorbed by the walls. Let $P(\omega)\,\Delta\omega$ be the total power absorbed by the walls due to modes with resonant frequencies between ω and $\omega + \Delta\omega$. Then

$$P(\omega) = 2\Gamma W_t N(\omega) \tag{6.6}$$

where $N(\omega)$ is the number of modes with resonant frequencies between ω and $\omega + \Delta\omega$, which is given by (cf. Appendix B)

$$N(\Omega) = \pi^{-3}\int\int\delta[\Omega - \omega_r(\mathbf{k},\mathbf{r})]\,d\mathbf{k}\,d\mathbf{r} \tag{6.7}$$

As an example, wall absorption of thermal cyclotron radiation is currently thought to be a major plasma energy loss channel for nuclear fusion reactors using advanced fuels. For the part of the frequency spectrum for which the plasma is not optically thin, the wall absorption may be calculated fairly economically by numerically following rays. However, for the optically thin portion of the spectrum, such calculations may be difficult because an emitted wave would have to be followed for a long time before its energy is absorbed. Fortunately, it is just in this regime where the ergodic results of this section and Section 5 can be applied.

Finally even in devices, such as tokamaks, where the plasma has a high degree of axisymmetry, the conducting walls surrounding the plasma commonly have corrugations and are partly composed of baffles and limiters. Thus the assumption that the rays are ergodic with ω_r the only constant of the motion is probably well justified for waves such as those considered in this section and in Section 5, which experience reflections from the walls.

7. QUANTUM CHIRIKOV–TAYLOR MAP[18,25]

In this section we consider a particular model problem that includes finite wavelength. The motivation is to compare ray approximation results in cases with ergodic rays to corresponding results with finite wavelength. In particular, we use this model to investigate our contention at the end of Section 4 that the effect of exceptional rays can be mitigated by a small amount of diffraction.

Our discussion is in the context of the quantum mechanical Schrödinger equation. However, we emphasize that the same mathematical problem arises in classical physics under various guises. For example, the propagation of electron plasma waves in a plasma of density $N_0 + \delta N(x, t)$ with $N_0 \gg \delta N$ is governed by an equation with the same form as the Schrödinger equation,

$$2i\omega_p \frac{\partial \hat{E}}{\partial t} = -3v_e^2 \frac{\partial^2 \hat{E}}{\partial x^2} + \omega_p^2 \frac{\delta N(x, t)}{N_0} \hat{E}$$

where ω_p is the plasma frequency in the density N_0, v_e is the electron thermal speed, the wave electric field is $E(x, t) = \exp(-i\omega_p t)\hat{E}(x, t)$, and it is assumed that $\omega_p \gg |\hat{E}^{-1} \partial \hat{E}/\partial t|$. Other contexts in which the problem to be treated in this section [embodied in Eq. (7.7)] arises are waves bouncing between corrugated walls and wave propagation through a system of periodically spaced lenses in the paraxial approximation.

References 7 and 25 discuss quantization of two-dimensional maps. If one considers Newton's equations in one dimension, $dp/dt = -\partial \bar{V}(q, t)/\partial q$ and $dq/dt = p$ with $\bar{V}(q, t) = V(q)\Sigma_n \delta(t - n)$, integration leads to a family of two-dimensional area preserving maps,

$$q_{n+1} = q_n + p_n \tag{7.1}$$

$$p_{n+1} = p_n - V'(q_{n+1}) \tag{7.2}$$

where the subscript n denotes the time $t = n + 0^+$. Following Refs. 7 and 25, we insert this potential in Schrödinger's equation to obtain

$$i\hbar \frac{\partial \psi}{\partial t} = -\frac{\hbar^2}{2} \frac{\partial^2 \psi}{\partial q^2} + \bar{V}(q, t)\psi \tag{7.3}$$

which, upon integration, yields a discrete time advancement equation for ψ,

$$\psi_{n+1}(q) = (2\pi\hbar)^{-1/2} \exp\left[\frac{-i\pi}{4} - \frac{iV(q)}{\hbar}\right] \int_{-\infty}^{+\infty} dq' \exp i\left[\frac{(q - q')^2}{2\hbar}\right] \psi_n(q') \tag{7.4}$$

For a state $\psi(q)$ the Wigner function is defined by

$$W(q, p) \equiv (\pi\hbar)^{-1} \int_{-\infty}^{+\infty} dx\, e^{2ipx/\hbar} \psi(q + x)\psi^*(q - x) \tag{7.5}$$

The Wigner function has the important feature that its integral along q yields the probability density in p, and vice versa. Insertion of Eq. (7.4) in Eq. (7.5) yields a time advancement equation for W,

$$W_{n+1}(q, p) = \int_{-\infty}^{+\infty} dq' \int_{-\infty}^{+\infty} dp'\, K(q, p; q', p') W_n(q', p') \tag{7.6}$$

where

$$K(q, p; q', p') = \frac{2\pi}{\hbar} \delta(q' - q + p') \int_{-\infty}^{+\infty} dx$$

$$\times \exp \frac{i}{\hbar} \{2x(p' - p) + V(q - x) - V(q + x)\}$$

Here we consider Eq. (7.6) for the case of the Chirikov–Taylor map.[18,25] The resulting problem will turn out to be very similar to that in the classical case ($\hbar \to 0$) for which successful analytical techniques have already been found.[19–24] By using these techniques it will be possible to obtain the diffusion coefficient in p and to compare it with the classical limit. In this way we hope to shed some light on how good the ray approximation is for finite $\hbar$.

To obtain the Chirikov–Taylor map from Eqs. (7.1) and (7.2), let $V(q) = \varepsilon \cos q$. Inserting this potential into Eq. (7.6) we have

$$W_{(n+1)}(q, p) = \sum_l J_l(-2\varepsilon\hbar^{-1} \sin q) W_n\left(q - p - \frac{l\hbar}{2}, p + \frac{l\hbar}{2}\right) \tag{7.7}$$

where J_l denotes the Bessel function of order l. Expressing $W_n(q, p)$ as a Fourier transform in p and a Fourier series in q,

$$W_j(q, p) = \sum_m \int \frac{dk}{2\pi} P_m(k, j) \exp i[mq + kp]$$

we obtain a time advancement equation for $P_m(k, j)$,

$$P_m(k, j + 1) - \sum_l J_l[c\kappa(k)] P_{m'}(k', j) \tag{7.8}$$

where $m' = m + l$, $k' = m + l + k$ and $\kappa(k) = [\sin(\hbar k/2)](2/\hbar)$. Note that for $\hbar \to 0$, $\kappa = k$, and the classical limit of the advancement equation for the Fourier-transformed distribution function for the Chirikov–Taylor map is recovered. Following the techniques developed for calculating the diffusion coefficient resulting from classical maps like the Chirikov–Taylor map (in particular, Refs. 19–23), one can obtain a formal series expansion for large ε. The first few terms of this expansion for the diffusion coefficient are

$$D = \frac{\varepsilon^2}{4}\Big\{1 - 2J_2[\varepsilon\kappa(1)] + 2J_3^2[\varepsilon\kappa(1)] - 2J_1^2[\varepsilon\kappa(1)]$$

$$+2J_2^2[\varepsilon\kappa(1)] + \sum_{p\neq 0} J_{p-2}[\varepsilon\kappa(1)]\,J_{p+2}[-\varepsilon\kappa(1)]\,J_{2p}[\varepsilon\kappa(p)]$$

$$-\sum_p J_{p+1}^2[\varepsilon\kappa(1)]\,J_{2p}[\varepsilon\kappa(p-1)]\Big\} + (\text{higher order terms}) \quad (7.9)$$

It has been pointed out[21] that in the $\hbar \to 0$ case, accelerator modes of the Chirikov–Taylor map prevent the convergence of the series above. However, it was also noted that the addition of a small amount of external noise can lead to convergence. For large ε the leading term in D is $\varepsilon^2/4$, which is precisely the classical result even though $\hbar \neq 0$. In the next two orders (i.e., $O(\varepsilon^{3/2})$ and $O(\varepsilon)$), which result from the second and the third through fifth terms in Eq. (7.9), there is $\hbar$ dependence through $\kappa(1) \equiv (2/\hbar)\sin(\hbar/2)$. This difference between the classical and quantum problems is clearly small for $\hbar$ small. Each term in the two summations in Eq. (7.9) contributes to D to order $\varepsilon^{1/2}$. However, there is the possibility that when considering the summation the result will be larger than order $\varepsilon^{1/2}$. As shown in Ref. 24 this is indeed the case. In fact, for $\hbar \to 0$ and $\varepsilon = 2\pi(N + \frac{1}{2})$ (where N is an integer) there are narrow resonances of width in ε of the order of ε^{-1}. Inside these resonances, terms in the summation add coherently to produce a contribution to D of order ε, that is, of the same order as the third, fourth, and fifth terms in Eq. (7.9). Furthermore, these resonances may be associated with the appearance of stable fixed points. Proceeding with an analysis similar to that in Ref. 24 but including a small finite $\hbar$, it can be shown that the coherent resonant addition of terms in the summation that occurs for $\hbar = 0$, is spoiled, if $\hbar$ becomes large enough; namely, if $\hbar \gtrsim O(\varepsilon^{-2})$. Thus a small $\hbar$ can remove the effect of a small region of nontypical orbits, in much the same way that the inclusion of noise does in the classical ($\hbar = 0$) problem (cf. Ref. 21). This lends support to the idea that diffraction (i.e., finite $\hbar$) can mitigate the effects of exceptional orbits (cf. Section 4). Further details and analytical and numerical results on this problem will be reported elsewhere.[26]

8. CONCLUSIONS

Several possible applications of stochasticity in Hamiltonian systems have been outlined for the ray equations applied to plasma waves. In Section 2 it was shown that lower hybrid waves used for heating tokamaks can become accessible to the plasma interior even when the straight cylinder theory of accessibility says that this cannot happen; and this phenomenon was shown to result from the onset of stochastic ray wandering when ε, the inverse aspect ratio of the torus, became large enough. Sections 3–6 considered the possibility of applying the ergodic hypothesis to problems in plasma stability, the determination of wall energy absorption, and the determination of thermal emission from plasmas. Many problems for future research remain. Some of these are: estimation of the size of the error in the approximation of Eq. (3.5) for finite (λ/L), the role of exceptional rays in determining stability in an ergodic case (Section 4), and the amount of damping or growth that can be tolerated before Eq. (3.5) becomes modified.

ACKNOWLEDGMENTS

This work was supported by the U.S. Department of Energy. The work reported here has benefited from useful discussions with T. M. Antonsen, P. T. Bonoli, Y. C. Lee, and W. M. Manheimer.

APPENDIX A: A POSSIBLE APPLICATION OF EQ. (4.3)

We consider an electrostatic magnetized electron plasma wave in a toroidally symmetric tokamak plasma. Such waves can be driven unstable by parametric effects during wave heating of plasmas[14] or by a population of runaway electrons.[15,16] The scope for consideration of the latter possibility was recently expanded when it was found that runaways can be greatly enhanced during lower hybrid heating.[17] The dispersion relation is $(\omega_{pi}^2 \ll \omega^2 \ll \Omega_e^2)$

$$1 + \frac{\omega_{pe}^2}{\Omega_e^2}\frac{k_\perp^2}{k^2} - \frac{\omega_p^2}{\omega^2}\frac{k_\parallel^2}{k^2} = 0$$

where $\omega_{pe}(\omega_{pi})$ is the electron (ion) plasma frequency, Ω_e is the electron cyclotron frequency, $k_\parallel = \mathbf{k}\cdot\mathbf{B}|\mathbf{B}|^{-1}$, $k_\perp^2 = k^2 - k_\parallel^2$, $\mathbf{B}$ is the equilibrium magnetic field, $\mathbf{B} = \mathbf{B}_p + B_\phi \boldsymbol{\phi}_0$, $\boldsymbol{\phi}_0$ is a unit vector in the toroidal direction, and wave damping mechanisms have been neglected. From toroidal symmetry $k_\phi = n/R$, where R is the major toroidal radius coordinate and n the toroidal mode number, is a constant of the motion for the ray equations. We now verify that these waves satisfy Condition i of Section 3, namely, that the ray trajectory is confined to a finite region of $\mathbf{k}$–$\mathbf{r}$ phase space [i.e., $\int\int \delta(\Omega - \omega_r(\mathbf{k},\mathbf{r}))\delta(Rk_\phi - n)\, d\mathbf{k}\, d\mathbf{r} < \infty$]. Writing the dispersion relation as

$$\omega^2 = \omega_{pe}^2 \frac{k_\parallel^2}{k^2 + k_\perp^2\left(\omega_{pe}^2/\Omega_e^2\right)}$$

we see that ω^2 is always less than ω_{pe}^2. Thus the waves are confined to the high density region of the plasma, $\omega^2 < \omega_{pe}^2(\mathbf{r})$. Now, introducing k_χ and k_ψ, the components of the poloidal wave number, respectively, parallel and perpendicular to $\mathbf{B}_p$, the dispersion relation can be written in the form

$$k_\chi^2\left[1+\frac{\omega_{pe}^2}{\Omega_e^2}-\frac{\omega_{pe}^2}{\omega^2}\frac{B_p^2}{B^2}\right]+k_\psi^2\left[1+\frac{\omega_{pe}^2}{\Omega_e^2}\right]\cong k_\chi\left(\frac{n}{R}\right)\left(\frac{B_pB_\phi}{B^2}\right)\left(\frac{\omega_{pe}^2}{\omega^2}\right)$$

$$+\left(\frac{n}{R}\right)^2\left[\left(\omega_{pe}^2/\omega^2\right)-1\right]$$

where we have utilized the approximations $B_\phi^2/B^2 \cong 1$ and $\omega^2 \ll \Omega_e^2$. From the form above it is seen that k_χ versus k_ψ is an ellipse (confined in k) rather than a hyperbola (unconfined), provided

$$\omega^2 > \frac{\omega_{pe}^2}{1+\left(\omega_{pe}/\Omega_e\right)^2}\frac{B_p^2}{B^2}$$

Since $B_p^2 \ll B^2$ for tokamaks, this condition can be easily satisfied. Thus the wave can be confined both in $\mathbf{r}$ and $\mathbf{k}$, hence satisfies Condition i of Section 3.

APPENDIX B: DENSITY OF MODES IN ω

For a homogeneous cubical cavity of side L, the $\mathbf{k}$-space volume of a mode is $(\Delta k)^3 = (\pi/L)^3 = \pi^3/V$, where V is the cavity volume. For a homogeneous medium, $\omega_r = \omega_r(\mathbf{k}) = \omega_r(k_x, k_y, k_z)$. Thus

$$dk_x\,dk_y\,dk_z = dk_x\,dk_y\sum_j\left[\frac{\partial\omega_r}{\partial k_z}\right]_{k_z=k_{zj}}^{-1}d\omega$$

where k_{zj} denote solutions of the dispersion relation for k_z as a function of ω_r, k_x, and k_y. Let $N(\omega_r)\,\Delta\omega$ be the number of modes with resonant frequencies between ω_r and $\omega_r + \Delta\omega$. Then

$$N(\omega_r) = \sum_j\int\frac{dk_x\,dk_y}{[\partial\omega_r/\partial k_z]_{k_{zj}}}\left(\frac{\pi^3}{V}\right)^{-1}$$

hence

$$N(\Omega) = \left(\frac{V}{\pi^3}\right)\int d\mathbf{k}\,\delta[\Omega - \omega_r(\mathbf{k})]$$

In the case of an inhomogeneous plasma this generalizes to Eq. (6.7).

REFERENCES

1. T. H. Stix, *Phys. Rev. Lett.* **15**, 878 (1965).
2. P. T. Bonoli and E. Ott, *Phys. Fluids* (to be published).
3. J.-M. Wersinger, J. M. Finn, and E. Ott, *Phys. Fluids* **21**, 2263 (1978).
4. A. Voros, in *Stochastic Behavior in Classical and Quantum Systems*, Lecture Notes in Physics Vol. 93, Volta Memorial Conference, 1977, G. Casati and J. Ford, Eds. (Springer-Verlag, Berlin, 1979).
5. M. V. Berry, *J. Phys. A* **10**, 2083 (1977).
6. J. S. Hutchinson and R. E. Wyatt, *Chem. Phys. Lett.* **72**, 378, (1980).
7. M. V. Berry, N. L. Balazs, M. Tabor, and A. Voros, *Ann. Phys.* **122**, 26 (1979).
8. D. W. Noid, M. L. Koszykowski, and R. A. Marcus, *J. Chem. Phys.* **71**, 2864 (1979).
9. S. W. McDonald, A. N. Kaufman, and M. V. Berry, *Bull. Am. Phys. Soc.* **24**:8, 942 (1979).
10. E. Ott, *Phys. Fluids* **22**, 2246 (1979).
11. E. Ott and W. M. Manheimer, *Phys. Rev.* **A25**, 1808 (1982).
12. S. W. McDonald and W. M. Manheimer, *Bull. Am. Phys. Soc.* **25**:8, 988 (1980); and to be published.
13. S. W. McDonald and A. N. Kaufman, *Phys. Rev. Lett.* **42**, 1189 (1979).
14. E. Ott, B. H. Hui, and K. R. Chu, *Phys. Fluids* **23**, 1031 (1980).
15. C. S. Liu and Y. Mok, *Phys. Rev. Lett.* **38**, 169 (1977).
16. K. Molvig, M. S. Tekula, and A. Bers, *Phys. Rev. Lett.* **38**, 1404 (1977).
17. C. S. Liu, L. Muschietti, K. Appert, J. Vaclavik, and D. A. Boyd, "Enhanced Runaway Production Rate by Waves in Plasmas," preprint (October 1980).
18. In collaboration with T. M. Antonsen.
19. A. B. Rechester and R. B. White, *Phys. Rev. Lett.* **14**, 1586 (1980).
20. A. B. Rechester, M. N. Rosenbluth, and R. B. White, *Phys. Rev.* **A23**, 2664 (1981).
21. C. F. F. Karney, A. B. Rechester, and R. B. White, *Physica D*, to be published.
22. J. R. Cary, J. D. Meiss, and A. Bhattacharjee, *Phys. Rev.* **A23**, 2744 (1981).
23. H. D. I. Abarbanel and J. D. Crawford, Report LBL-11889 (Lawrence Berkeley Laboratory, Berkeley, CA, 1980).
24. T. M. Antonsen and E. Ott, *Phys. Fluids* **24**, 1635 (1981).
25. The quantum Chirikov–Taylor map has been previously considered by G. Casati, B. V. Chirikov, F. M. Izraelev, and J. Ford, in *Stochastic Behavior in Classical and Quantum Systems* Lecture Notes in Physics Vol. 93, Volta Memorial Conference, 1977, G. Casati and J. Ford, Eds. (Springer-Verlag, Berlin, 1979). One of the interesting results of this paper is that changes from diffusive behavior (i.e., increase of $\langle p^2 \rangle$ linearly with t) can occur for large t as a result of finite h.
26. E. Ott, J. D. Hanson, and T. M. Antonsen, *Bull. Am. Phys. Soc.* **26**, (September 1981); and to be published.

RENORMALIZED PLASMA TURBULENCE THEORY

C. W. HORTON, JR.
Institute for Fusion Studies
University of Texas
Austin, Texas

Abstract. *Recent developments in the renormalized turbulence theory of collisionless plasmas described by the Vlasov–Poisson system of equations are presented. The need for renormalization in plasma turbulence is discussed in terms of the historical problem of anomalous resistivity due to ion–acoustic turbulence.*

1. THE VLASOV–POISSON SYSTEM

The Vlasov–Poisson description of a collisionless plasma is a continuous Hamiltonian system with a volume preserving flow in the noncanonical variables of the particle distribution function and the electric field. The fields $f(x, v, t)$ and $E(x, t) = -\partial_x \phi(x, t)$ are defined over the six-dimensional phase space of the single-particle trajectories. The nonlinear evolution of the fields is given by the bilinear coupling $E\, \partial_v f$, which may be expressed as a Poisson bracket operator with respect to the Hamiltonian functional

$$\mathcal{H} = \int \frac{1}{2} m v^2 f\, d^d x\, d^d v + (8\pi)^{-1} \int E^2\, d^d x$$

as shown by Morrison.[1] The basic problem of nonlinear plasma theory is to obtain a reduced dynamical description of this conservative system suitable for predicting the observables in fully developed plasma turbulence.

In many plasma problems it is clear from the outset that the self-consistent electric field is such as to give rise to the universal instability of particle orbits.[2] Even in one-dimensional models where $\phi = \phi(x, t)$, we know that the generic form of plasma oscillations $\phi(x, t) = \Sigma_k \phi_k \exp(ikx - i\omega_k t)$ in the later stages of evolution in laser–pellet interactions, particle beam–plasma interactions, and turbulent heating, is a spectrum of waves ϕ_k, ω_k satisfying the Chirikov criterion for overlapping nonlinear resonances, namely,

$$K_j(k, v) = \frac{\left|\dfrac{e_j}{m_j}\phi_k\right|^{1/2}}{\Delta k\left|\dfrac{d\omega}{dk} - v\right|} > 1 \tag{1}$$

where Δk is the spacing between the wave amplitudes ϕ_k. The particle motion is described by the $1\frac{1}{2}D$ Hamiltonian $H = \frac{1}{2}m_j v^2 + e_j\phi(x, t)$, where the initial phase of the time-dependent wave potential adds the third degree of freedom to the phase space that destroys integrability. The onset of large-scale stochasticity through the destruction of the Kolmogorov–Arnol'd–Moser (KAM) tori in the case of two plasma waves is perhaps best calculated by the renormalization method given by Escande and Doveil.[3]

Increasing the dimensionality of the wave potential from $\phi(x, t)$ to $\phi(x, y, t)$ and $\phi(x, y, z, t)$ rapidly increases the stochasticity of the particle motion at a fixed ratio of the root-mean-square wave amplitude to the particle kinetic energy, $\epsilon = e_j\langle\phi^2\rangle^{1/2}/m_j v^2 \ll 1$. For a Boltzmann-like distribution of particle energies with a mean kinetic energy $(d/2)T_j$, we estimate the rate of decrease of the fraction of quasi-trapped particles by a given wave amplitude with increasing dimensionality $d = 1, 2, 3$ as $\epsilon^{d/2}$. For ion–acoustic waves and drift waves where the energy density W in the fluctuation $W \simeq e^2 n_e\langle\phi^2\rangle/T_e$ is a small fraction of the electron thermal energy density $n_e T_e$, the fraction of phase space in which a trapping oscillation may occur decreases with increasing dimensionality as $(W/n_e T_e)^{d/4}$, where $W/n_e T_e$ is the basic expansion parameter of plasma turbulence theory. The rapid decrease of the quasi-trapped particle fraction with increasing d suggests that the phase space correlations observed at $d = 1$ in particle simulations are less important, and perhaps insignificant, at higher dimensionality.

In regions of phase space where the neighboring orbits diverge exponentially because of the overlapping of resonances from $K_j(k, v) > 1$ in Eq. (1), every small region of orbits undergoes strong mixing. After several e-foldings in the separation, the local orbit x, v, t is extremely sensitive to its initial conditions. Neglecting the exponentially small correlations with the initial data required to maintain the "in principle deterministic systems" as advocated by Chirikov (Ref. 2, p. 326), leads to a reduced statistical description of the dynamics based on renormalization kinetic equations. Furthermore, Misra and Prigogine[4] argue that in the presence of a suitably strong form of trajectory instability, the very concept of phase space trajectories becomes an unphysical idealization. They demonstrate that in this situation the dynamical evolution is "equivalent," in a well-defined sense, to the stochastic evolution of Markov processes.

In the stochastic regime the renormalized kinetic equations describe an increase of entropy of the system and wave–particle dynamics described by semigroups. The renormalized kinetic equations are derived by truncating the chain of phase space correlations, and the principal differences in the theories reflect the different arguments invoked to truncate the chain of correlations.

2. QUASI-LINEAR DIFFUSION IN THE STOCHASTIC DOMAIN AND PHASE SPACE CORRELATIONS

The renormalized background velocity distribution function $f_j(v, t)$ must satisfy a diffusionlike equation as is easily shown by iteration of the Chirikov map or by direct integration of the orbits for the $1\frac{1}{2}D$ Hamiltonian. Rechester and White[5] investigate convergence of the velocity distribution to the quasi-linear diffusion law as a function of wave parameters concluding that for $K_j(k, v) \gg 1$ the distribution $f_j(\mathbf{v}, t)$ rapidly converges to quasi-linear diffusion. In the general system, the quasi-linear diffusion tensor is

$$\mathbf{D}_j(\mathbf{v}) = \left(\frac{e_j}{m_j}\right)^2 \int d^dk\, d\omega\, I(\mathbf{k}, \omega)\mathbf{kk}\pi\delta(\omega - \mathbf{k}\cdot\mathbf{v}) \tag{2}$$

where $I(\mathbf{k}, \omega)$ is the spectral distribution of the electric potential given by $\langle\phi^2(\mathbf{x}, t)\rangle = \int I(\mathbf{k}, \omega)\, d^dk\, d\omega$. By giving up the exponentially small correlations required by the concept of single-particle trajectories in the mixing system, the dynamics becomes irreversible. A simple statistical calculation by Rechester et al.[6] shows that in the stochastic region of x, v the average rate of divergence of the orbits is given by $(k^2D)^{1/3}$, where $D(v)$ is the quasi-linear diffusion coefficient in Eq. (2).

Renormalized turbulence theories predict the fluctuating part of the particle distribution $\delta f_j(\mathbf{k}, \mathbf{v}, \omega)$ induced by the collective field $\phi_{\mathbf{k},\omega}$ as a function of the spectral distribution $|\phi_k|^2$. The distribution δf_j evolves according to the renormalized particle propagator $g^j_{\mathbf{k},\omega}(\mathbf{v})$. This propagator controls the transfer of energy and momentum between the fluctuations and the particles. Associated with the resonant transfer proportional to $\mathrm{Im}\, g_k$ of energy and momentum is a change in the entropy of the system. In the extreme case of a wide spectrum of low amplitude fluctuations, the propagator describes evolution according to a Markovian semigroup. For general spectral distributions $I(\mathbf{k}, \omega)$ we show that the energy and momentum conservation laws are contained in the renormalized turbulence theory. As the level of turbulence increases, new regions of $\mathbf{k}, \mathbf{v}$ are opened up to the emission and absorption of wave energy. Frequency sum rules given in Sec. 4 constrain the total strength of the transfer of energy and momentum. These properties of the renormalized turbulence theory are important in the problem of ion–acoustic turbulence and other problems of plasma turbulence.

In statistical theories of plasma turbulence, it must be kept in mind that the collective modes from the long-range Coulomb forces

$$\phi(\mathbf{x}, t) = \sum_j e_j \int\int K_c^d(|\mathbf{x} - \mathbf{x}'|)\delta f_j(\mathbf{x}', \mathbf{v}', t)\, d^d x'\, d^d v'$$

create correlations in the turbulent state. The assumption regarding the strength of these correlations is one of the principal differences between the renormalized turbulence theories. In the simpler theories, such as that of Choi and Horton,[7] the fields are assumed to be sufficiently close to Gaussian statistics to drop cumulants of fourth and higher order. In contrast, DuBois and Espedal[8] have developed the direct interaction approximation (DIA), as originally suggested by Orszag and Kraichnan,[9] to provide a systematic higher order expansion in the derivations away from Gaussian statistics. The plasma DIA is considerably more complicated than the conventional renormalized turbulence theory of Horton and Choi[7] and Misguich and Balescu,[10] because polarization effects associated with a so-called incoherent contribution to $\delta f_k(\mathbf{v})$ are included. The plasma DIA is based on a truncation of the equation for the two-point correlation function $\langle \delta f(1)\, \delta f(2) \rangle$ as given by DuBois and Espedal[8,11] and Krommes and Kleva.[12] A test particle approximation within the DIA recovers the conventional renormalized turbulence theory as shown by Rose.[13]

The creation of phase space correlations has been considered by Dupree[14] and Kadomtsev and Pogutse.[15] Recently, considerable effort has been directed at the development of two phase space–point correlation function $\langle \delta f(1)\, \delta f(2) \rangle$ theories by DuBois[8,11] and Boutros-Gali and Dupree.[16] The general importance of these correlations remains unsettled, although some special $d = 1$ regimes near the onset of turbulent trapping $(k^2 D)^{1/3} > \gamma_k$ have been shown theoretically to contain strong correlations associated with higher order irreducible correlations. In the classical problem of the gentle bump instability where the quasi-linear conditions are well satisfied, recent work by Pesme and Laval[17] solves the mode coupling equation for the two-point equal time correlation function

$$\begin{aligned}\left[-i(\omega - kv_-) - 2D(v_+)\partial_{v_-}^2\right]&\langle \delta f\, \delta f' \rangle_k \\ &= D(v_+)\left[\delta(k - k_+) + \delta(k + k_+)\right]\left(\partial_{v_+} f\right)^2 \\ &\quad - 2D(v_+)\partial_{v_-}^2\left[\langle \delta f\, \delta f' \rangle_{k-k_+} + \langle \delta f\, \delta f' \rangle_{k+k_+}\right] \qquad (3)\end{aligned}$$

where $v_+ = \frac{1}{2}(v + v')$, $v_- = v - v'$ and k_+ is the average resonant wave number defined by $\omega(k_+) = k_+ v_+$. Solutions of Eq. (3) are obtained for close orbits where $|k_+ v_-| < (k_+^2 D)^{1/3}$ and $|x_-| < v_+ (k_+^2 D)^{-1/3}$. Iterating on the mode coupling terms and summing the divergent contributions led Pesme and

Laval to conclude that the irreducible parts of the correlation functions are comparable to the reducible contributions when $(k^2 D)^{1/3} > \gamma_k$. Mathematically, the enhanced correlations arise from large contributions in the mode coupling terms due to resonant particles. In this $d = 1$ example, the onset of the irreducible correlations may signal the ability of the Coulomb system to self-organize into strongly correlated structures in the nonlinear regime.

Historically, the need to renormalize plasma turbulence theory was discovered principally by the failure of weak turbulence theory to describe a number of simple features of ion–acoustic turbulence. Weak turbulence theory based on a third-order expansion of $\delta f_k(\mathbf{v})$ in powers of $g_k^\circ E_k \partial_v$ where $g_k^\circ(\mathbf{v}) = (\omega - \mathbf{k} \cdot \mathbf{v} + io^+)^{-1}$ fails to correctly describe the fluctuating electron charge density when the electric field amplitudes exceed an almost trivially low level, as first found by Sizonenko and Stepanov[18] in applying weak turbulence theory to ion–acoustic turbulence. The breakdown of the expansion can be understood from estimating the amplitude of an ion–acoustic wave $\omega_k = kc_s/(1 + k^2\lambda_{De}^2)^{1/2}$ with $c_s = (T_e/m_i)^{1/2}$ required to accelerate a resonant electron $\omega = \mathbf{k} \cdot \mathbf{v}$ out of the phase velocity resonance during one wave period. From $\Delta v = (e/m_e)\, g_k^\circ \mathbf{k}\phi_k$ with $g_k^\circ \sim 1/\omega_k$, we find that $\mathbf{k} \cdot \Delta\mathbf{v} > \omega_k$ when $|e\phi_k/T_e| > (\omega/kv_e)^2 \sim m_e/m_i$. Evidently, for wave amplitudes above this low critical level, resonant contributions from all orders in $g_k^\circ E_k \partial_v$ must be collected to correctly describe the resonant electron–wave interaction. In terms of the critical turbulence level[18–22] for a $d = 2$ or 3 spectrum, the breakdown level is $W/nT_e \sim m_e/m_i$, where observations from particle simulations show no appreciable change in the rate of exponentiation of $W(t)$ in sharp contrast to the prediction of weak turbulence theory. The weak turbulence theory calculation for the critical turbulence level W/nT_e requires an evaluation of the pole pinching contributions[19] to the nonlinear mode coupling matrix elements. Other prominent failures of weak turbulence theory are discussed in the review article on ion–acoustic turbulence by Horton and Choi.[22]

3. RENORMALIZED TURBULENCE THEORY

Now we summarize some of the principal results of renormalized turbulence theory following the infinite-order perturbation expansion of Choi and Horton.[7, 22] The fluctuating part of the distribution function $\delta f_k(v)$ is expanded in powers of $(e/m) g_k^\circ E_k \partial_v$ to obtain, in the notation of Choi and Horton,

$$\delta f_j(v) = \frac{k\!\downarrow}{k} f(v) + \frac{k_1|k - k_1\!\downarrow}{k \quad k - k_1} f(v) + \cdots$$

$$+ \frac{k_1|k_2| \cdots k - k_1 - \cdots - k_{n-1}\!\downarrow}{k \qquad\qquad k - k_1 - \cdots} f(v) \tag{4}$$

where $g_k^\circ(\mathbf{v}) = -$, and $(e/m)\phi_k \mathbf{k}\cdot\partial_v = k_1\downarrow$ and $\Sigma_{k_1}(e/m)\phi_{k_1}\mathbf{k}_1\cdot\partial_v = k_1|$ and in Eq. (4), all free $k_n = \mathbf{k}_n$, ω_n are summed over without restriction. Asking for the highest order pole in the resonance $\omega = \mathbf{k}\cdot\mathbf{v}$ at order $(g_k E_k)^{2n+1}$ in the expansion leads to the simply renormalized subseries

$$\delta f_k(\mathbf{v}) = \frac{\quad}{k}\left(1 + \frac{k_1\,\square\,-k_1}{k-k_1 \quad k} + \frac{k_1\,\square\,-k_1\;\;k_1\,\square\,-k_1}{k \qquad\qquad k} + \cdots\right)^{k}\Big\downarrow f(\mathbf{v})$$

$$= \frac{\quad}{k}\left(1 - \frac{k_1\,\square\,-k_1}{k-k_1 \quad k}\right)^{-1}{}^{k}\Big\downarrow f(\mathbf{v}) \tag{5}$$

where the repeated operator

$$\frac{k_1\,\square\,-k_1}{k-k_1 \quad k} = \sum_{k_1}\left(\frac{e}{m}\right)\phi_{k_1}\mathbf{k}_1\cdot\frac{\partial}{\partial\mathbf{v}}g^\circ_{k-k_1}\left(\frac{e}{m}\right)\phi_{-k_1}\left(-\mathbf{k}\cdot\frac{\partial}{\partial\mathbf{v}}\right)g_k^\circ(\mathbf{v}) \tag{6}$$

arises from terms containing poles of order $(\omega - \mathbf{k}\cdot\mathbf{v})^{-n-1}$. In the simply renormalized theory that sums these contributions to all orders, the wave–particle propagator $\tilde{g}_k^j(\mathbf{v})$ is

$$\left[\tilde{g}_k^j(\mathbf{v})\right]^{-1} = \omega - \mathbf{k}\cdot\mathbf{v} + \frac{\partial}{\partial\mathbf{v}}\left(\frac{e}{m}\right)^2\cdot\int\frac{d\omega_1\,d\mathbf{k}_1\,\mathbf{k}_1\mathbf{k}_1 I(\mathbf{k}_1\omega_1)}{\omega-\omega_1-(\mathbf{k}-\mathbf{k}_1)\cdot\mathbf{v}+io^+}\cdot\frac{\partial}{\partial\mathbf{v}} \tag{7}$$

and

$$\delta f_k^j(\mathbf{v}) = -\frac{e_j}{m_j}\phi_k\tilde{g}_k^j(\mathbf{v})\mathbf{k}_1\cdot\frac{\partial f^j}{\partial\mathbf{v}} \tag{8}$$

where the solution Eq. (8) requires inverting the second-order differential operator Eq. (7) in velocity space. The simply renormalized propagator $\tilde{g}_k(\mathbf{v})$ is a nonlocal operator $\tilde{g}_k F(\mathbf{v}) = \int\tilde{g}_k(\mathbf{v},\mathbf{v}')F(\mathbf{v}')\,d\mathbf{v}'$ with range of order $\Delta v \sim (D/k)^{1/3}$, where $D(\mathbf{v})$ is the quasi-linear diffusion coefficient in Eq. (2). The magnitude of $\tilde{g}_k(\mathbf{v},\mathbf{v}')$ is $|\omega - \mathbf{k}\mathbf{v}_+ + i(k^2D)^{1/3}|^{-1}$.

The simply renormalized propagator $\tilde{g}_k(\mathbf{v},\mathbf{v}')$ can be calculated for rather general fluctuation spectra $I(\mathbf{k},\omega)$ and consequently plays a central role in practical applications of the theory. For example, a broad spectrum of ion–acoustic waves produces quasi-elastic scattering of the electrons described by the Lorentz gas form of the propagator $[\tilde{g}_k^e(\mathbf{v})]^{-1} = \omega - kv\mu - i\nu_{\text{eff}}\partial_\mu(1 -$

$\mu^2)\partial_\mu$, where $\mu = \hat{\mathbf{k}} \cdot \hat{\mathbf{v}}$ and $\nu_{\text{eff}} = (\pi/4)\int dk_1 |\mathbf{k}_1| I(\mathbf{k}_1, \omega_1)$. On the other hand, for the low velocity ions, another explicit form of the propagator $\tilde{g}_k(\mathbf{v})$ can be worked out, which is useful for practical calculations.

The simply renormalized theory, although valuable for practical calculations, has the formal difficulty of still containing the singular propagator $g^\circ_{k-k_1}(\mathbf{v})$. A higher order selection of the "next most secular" terms in the subseries for $\delta f_k(v)$ can be summed to yield a "doubly renormalized" propagator $\tilde{\tilde{g}}_k(\mathbf{v})$ given by Eq. (7) with $\tilde{g}_k \to \tilde{\tilde{g}}_k$ on the left-hand side and $g^\circ_{k-k_1} \to \tilde{g}_{k-k_1}$ on the right-hand side of the equation. The series for $\tilde{\tilde{g}}_k$ contains an infinite-order set of new terms over the terms in $\tilde{g}_k$ with the first new term appearing at fourth order

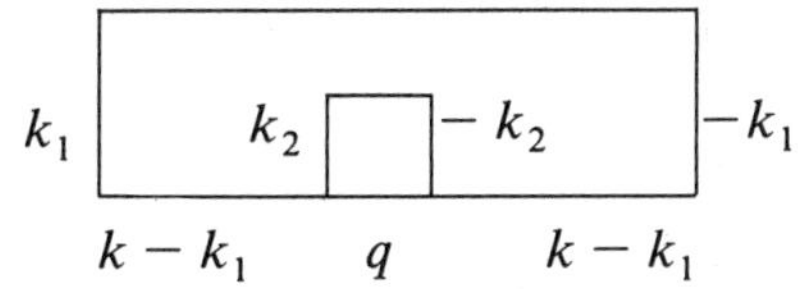

being added to

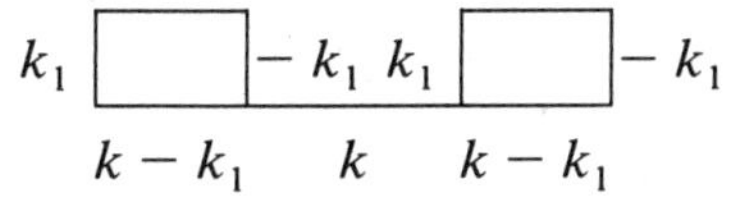

contained in $\tilde{g}_k(\mathbf{v})$. Continuing this selection of higher order subseries to eliminate the dependence on the singular propagator $g^\circ_k(\mathbf{v})$ generates a continued fraction representation of the propagator. At finite order in the sequence we have a Padé approximant for the propagator expressed as a ratio of finite-order polynomials in the operator $g_k E_k \partial_v$. Assuming that the sequence of iterations converges to $g_k(\mathbf{v})$, we obtain the nonlinear integral equation

$$\left(\omega - \mathbf{k}\cdot\mathbf{v} + \frac{\partial}{\partial \mathbf{v}} \cdot \left(\frac{e}{m}\right)^2 \int dk_1\, \mathbf{k}_1 \mathbf{k}_1 I(k_1) g_{k-k_1}(\mathbf{v}) \cdot \frac{\partial}{\partial \mathbf{v}}\right) g_k(\mathbf{v}) = 1 \qquad (9)$$

for the fully renormalized propagator. The fluctuating distribution is $\delta f_k = -(e/m)\phi_{\mathbf{k}} g^j_k \mathbf{k} \cdot \partial_v f^j$, and the turbulent response function $\varepsilon_{\mathbf{k}}(\omega, I)$ is

$$\varepsilon_{\mathbf{k}}(\omega) = 1 + \sum_j \varepsilon^j_{\mathbf{k}}(\omega)$$

$$= 1 + \sum_j \frac{4\pi e_j^2}{k^2 m_j} \int d\mathbf{v} \int d\mathbf{v}'\, g_k(\mathbf{v}, \mathbf{v}')\mathbf{k} \cdot \frac{\partial f^j}{\partial \mathbf{v}'} \qquad (10)$$

The quasi-linear equation is

$$\frac{\partial f^j}{\partial t} = \frac{\partial}{\partial \mathbf{v}} \cdot \int \mathsf{D}_j(\mathbf{v}, \mathbf{v}') \cdot \frac{\partial f^j}{\partial \mathbf{v}'}(\mathbf{v}', t)\, d\mathbf{v}' \qquad (11)$$

with the generalization of Eq. (1) given by

$$\mathsf{D}_j(\mathbf{v},\mathbf{v}') = -\left(\frac{e_j}{m_j}\right)^2 \int d\mathbf{k}\, d\omega\, \mathbf{k}\mathbf{k} I(\mathbf{k},\omega) \operatorname{Im} g_k^j(\mathbf{v},\mathbf{v}') \tag{12}$$

In the limit $W/nT \to 0$ the propagator's range $\Delta v = (D/k)^{1/3} \to 0$ and $-\operatorname{Im} g_k(\mathbf{v},\mathbf{v}') \to \delta(\mathbf{v}-\mathbf{v}')\delta(\omega - \mathbf{k}\cdot\mathbf{v})$, so that weak turbulence theory is recovered.

The theory of Padé approximants[23] suggests that the sequence of renormalized propagators may well represent the exact propagator outside the limited circle of convergences of the original power series expansion in $(e/m)g_k^\circ E_k\,\partial_v$. At a given order in the sequence of Padé approximants a new mode coupling perturbation series for $\delta f_k(\mathbf{v})$ is generated by expanding in the renormalized propagator. The expansion parameter

$$\epsilon = \left|\left(\frac{e}{m}\right) g_k E_k\,\partial_v\right| = \left|\frac{e}{m}\right| |g_k|^2 k^2 |\phi_k|$$

estimated with max $g_k = [|\omega - \mathbf{k}\cdot\mathbf{v}| + i(k^2D)^{1/3}]^{-1}$ and $D = (e/m)^2 k^2 \phi_k^2 g_k$ is found to be bounded by

$$\epsilon \leqslant \frac{kD^{1/2}}{\left[\gamma + (k^2D)^{1/3}\right]^{3/2}} < 1$$

for all $|\phi_k|^2$. Thus, the renormalized perturbation expansion is more convergent than the original expansion, although little is known mathematically about the convergence properties of the renormalized series. The second-order renormalized perturbation expansion, for example, gives

$$\delta f_k^{(2)}(\mathbf{v}) = \left(\frac{e}{m}\right)^2 g_k \int dk_1\, \phi_{k_1}\mathbf{k}_1 \cdot \frac{\partial}{\partial \mathbf{v}} g_{k-k_1}\phi_{k-k_1}(\mathbf{k}-\mathbf{k}_1)\cdot\frac{\partial f}{\partial \mathbf{v}}$$

which is used to calculate

$$\rho_k^{(2)} = \left(\frac{k^2}{4\pi}\right) \sum_{k_1+k_2=k} \varepsilon^{(2)}(k_1,k_2)\phi_{k_1}\phi_{k_2} = \sum_j e_j \int f_k^{(2)}(\mathbf{v})\, d\mathbf{v}$$

Computing the second- and third-order renormalized charge densities and using the Poisson equation yields

$$\varepsilon_k\phi_k + \sum_{k_1+k_2=k} \varepsilon^{(2)}_{k_1,k_2}\phi_{k_1}\phi_{k_2} + \sum_{k_1+k_2+k_3=k} \varepsilon^{(3)}_{k_1,k_2,k_3}\phi_{k_1}\phi_{k_2}\phi_{k_3} + \cdots = 0 \tag{13}$$

where the definitions of $\varepsilon^{(n)}(k_1 \cdots k_n)$ are given by the velocity integral of the $\phi_{k_1}\phi_{k_2} \cdots \phi_{k_n}$ coefficient of $\delta f_k^{(n)}(\mathbf{v})$. In Eq. (13) we must recognize that $\varepsilon_k^{(n)} = \varepsilon_k^{(n)}[g_{k_n}(I)]$ eliminates the apparent divergence problem of series (13) as $|\phi_k| \to \infty$.

The mode coupling Eq. (13) in principle determines the spectral distribution $I(\mathbf{k}, \omega)$ required to close the theory. In practice, however, an assumption on the statistics of ϕ_k must be introduced to obtain a reduced equation for $I(k)$. The standard procedure is to assume that the fields are not too far from Gaussian statistics, to calculate the three-field correlation function $\langle \phi_{-k_2}\phi_{-k_1}\phi_k \rangle$ perturbatively, and to close the hierarchy by dropping the irreducible part of the four-field correlation function. The resulting theory for the two-field correlation function $\langle \phi_k \phi_{k'} \rangle = \delta(k + k')I(k)$ has a certain obvious lack of symmetry, described below after Eq. (15), which can be overcome by summing higher order terms in the nonlinear response function $\varepsilon^{nl}(k)$. An alternative procedure is to introduce the direct interaction approximation of fluid turbulence theory to truncate the chain of correlations. The reduced mode coupling equations determining ε^{nl} and $I(k)$ are given by

$$\varepsilon^{nl}(k) = \varepsilon(k) - \int \left[\frac{4\varepsilon^{(2)}(k_1, k - k_1)\varepsilon^{(2)}(-k_1, k)}{\varepsilon^{nl}(k - k_1)} - 2\varepsilon^{(3)}(k_1, k, -k_1) \right] I(k_1)\, dk_1 \tag{14}$$

and

$$\varepsilon^{nl}(k)I(k) = \frac{2}{[\varepsilon^{nl}(k)]^*} \times \int\int |\varepsilon^{(2)}(k_1, k_2)|^2 \delta(k - k_1 - k_2) I(k_1) I(k_2)\, dk_1\, dk_2 \tag{15}$$

In the simple perturbative derivation, the lowest order linear response function $\varepsilon^{nl}(k) \to \varepsilon(k)$ appears on the right-hand side of Eqs. (14) and (15). From Eq. (15) we see that one feature of the higher order contributions in the DIA is to guarantee the positive-definite values of $I(\mathbf{k}, \omega)$, which as noted by DuBois and Espedal[8] is not assured in the earlier renormalized turbulence equations of Choi and Horton.[7] Tsytovich[24] has also given an alternative derivation of the higher terms contained on the right-hand side of Eqs. (14) and (15).

Assuming the existence of a turbulent steady state, we may approximate the correlation function at each $\mathbf{k}$ by frequency profile peaked at the mode $\omega \simeq \omega_{\mathbf{k}}$ determined from the roots of the nonlinear response function $\varepsilon_{\mathbf{k}}^{nl}(\omega) = 0$. Provided the turbulence is not too strong, we separate $\varepsilon_{\mathbf{k}}^{nl}(\omega)$ into its real $\varepsilon'_{\mathbf{k}}(\omega)$ and imaginary $\varepsilon''_{\mathbf{k}}(\omega)$ parts for real ω. Expanding about $\varepsilon'_{\mathbf{k}}(\omega_{\mathbf{k}}) = 0$ the lifetime of the nonlinear mode is governed by $\omega = \omega_{\mathbf{k}} - i\nu_{\mathbf{k}}$ where

$$\nu_{\mathbf{k}} = \frac{\operatorname{Im} \varepsilon_{\mathbf{k}}^{nl}(\omega_{\mathbf{k}})}{[\partial \varepsilon_{\mathbf{k}}^{nl}(\omega_{\mathbf{k}})]/\partial \omega_{\mathbf{k}}} \tag{16}$$

and from Eq. (15) the line width is approximately Lorentzian

$$I(\mathbf{k}, \omega) = \frac{I(\mathbf{k})}{\pi} \frac{\nu_{\mathbf{k}}}{(\omega - \omega_{\mathbf{k}})^2 + \nu_{\mathbf{k}}^2} \tag{17}$$

with the frequency-integrated spectrum $I(\mathbf{k}) = \int I(\mathbf{k}, \omega)\, d\omega$ satisfying the balance equation

$$\nu_{\mathbf{k}} I(\mathbf{k}) = \frac{2\pi}{|\partial \varepsilon_{\mathbf{k}}^{nl}/\partial \omega_{\mathbf{k}}|^2} \int |\varepsilon_{\mathbf{k}_1, \mathbf{k}-\mathbf{k}_1}^{(2)}(\omega_{\mathbf{k}_1}, \omega_{\mathbf{k}-\mathbf{k}_1})|^2 \delta(\omega_{\mathbf{k}} - \omega_{\mathbf{k}_1} - \omega_{\mathbf{k}-\mathbf{k}_1}) \times I(\mathbf{k}_1) I(\mathbf{k} - \mathbf{k}_1)\, d\mathbf{k}_1 \tag{18}$$

Equation (18) describes a balance of the decay of fluctuations at $\mathbf{k}$ with their production from $\mathbf{k}_1$ and $\mathbf{k} - \mathbf{k}_1$.

A major simplification occurs when the dispersion relation $\omega = \omega(\mathbf{k})$ is of the nondecay type where $\Delta\omega = \omega_{\mathbf{k}} - \omega_{\mathbf{k}_1} - \omega_{\mathbf{k}-\mathbf{k}_1} \neq 0$, as is the case for ion–acoustic waves, for example. The balance equations Eqs. (14) and (15) are then satisfied to the lowest order by $\varepsilon^{nl}(k)I(k) = 0$, which requires that nontrivial $I(k)$ satisfy the marginal stability equation $\varepsilon^{nl}(k, I) = 0$, giving the integral equation

$$\int \left[\frac{4\varepsilon^{(2)}(k_1, k - k_1)\varepsilon^{(2)}(-k_1, k)}{\varepsilon(k - k_1)} - 2\varepsilon^{(3)}(k_1, k, -k_1) \right] I(k_1)\, dk_1 = \varepsilon(k) \tag{19}$$

The large matrix equation Eq. (19) is most practical to solve dynamically by restoring the two time scale variations $T = \frac{1}{2}(t_1 + t_2)$ for the slow nonlinear evolution, and $\tau = t_2 - t_1$ for the fluctuation time scale. Integrating the imaginary part of Eq. (19) over the frequency ω associated with τ, the dynamical equation for the slowly evolving wave number spectrum $I(\mathbf{k}, T)$ is derived in the form $dI/dT = (2\gamma_{\mathbf{k}}^{ql} + 2\gamma_{\mathbf{k}}^{nl})I(\mathbf{k}, T)$. These dynamical equations have been studied extensively for the problem of ion–acoustic turbulence by Horton et al.[25]

4. PROPERTIES OF THE RENORMALIZED PROPAGATOR AND DIELECTRIC

The renormalized propagator $g_k(\mathbf{v})$ has a number of useful properties that apply at each level of renormalization, that is, to the simply renormalized, doubly renormalized,..., and fully renormalized series. The properties reflect conservation laws and causality properties contained in the original Vlasov–Poisson system and are preserved by the partial summations.

1. The propagator $g_k(\mathbf{v})$ is analytic in the upper-half ω-plane. This property arises from the causality of the system and ensures that unstable oscillations in $\delta f_{\mathbf{k}}(\mathbf{v}, t)$ arise only from the nonequilibrium features contained in the collective modes $\varepsilon_{\mathbf{k}}(\omega) = 0$. Using Cauchy integrals to relate the real and imaginary parts of $g_{\mathbf{k}\omega}(\mathbf{v})$, it is straightforward to derive a number of frequency sum rules, two of which are

$$\int_{-\infty}^{+\infty} \operatorname{Im} g_{\mathbf{k}\omega}(\mathbf{v})\, d\omega = -\pi \qquad \text{for} \quad \text{all } \mathbf{k}, \mathbf{v}, I_k \tag{20}$$

and

$$\int_{-\infty}^{+\infty} \omega \varepsilon_{\mathbf{k}}^{j}(\omega)\, d\omega = -\pi\omega_{pj}^2 \qquad \text{for} \quad \text{all } \mathbf{k} \text{ and } I_k \tag{21}$$

The first sum rule Eq. (20) shows that as the turbulence opens new regions of emission and absorption in $\operatorname{Im} g_{\mathbf{k}\omega}(\mathbf{v}, I)$, the net amount of absorption over the entire frequency range is invariant. This is the basis for the often-made approximation that the broadening of the response function is unimportant under integrals with smoothly varying distributions.

2. The renormalized propagator conserves particle number density and thus charge density ρ_k through the divergence of the particle current $\mathbf{j}_k$. This property follows because at each level of renormalization the turbulent collision operator is proportional to $\partial_{\mathbf{v}} \cdot \mathbf{J}(\mathbf{v})$ where $\mathbf{J}(\mathbf{v})$ is the flux of particles due to the electric field. Taking the velocity moment of $(g_k^j)^{-1}\delta f_{\mathbf{k}}$ and assuming that the $\lim v \to \infty$ of $v^2 J(v) = 0$, we obtain

$$\omega \rho_{\mathbf{k}\omega} = \mathbf{k} \cdot \mathbf{j}_{\mathbf{k}\omega}$$

for each species j and all W/nT. This property is used in the demonstration of energy conservation, as noted by Pelletier and Pomot,[26] in transforming the work done on the particles $\langle \mathbf{j} \cdot \mathbf{E} \rangle$ into the imaginary part of $\varepsilon_{\mathbf{k}}(\omega, I)$.

3. The renormalized propagator replaces the singular unitary evolution of $g_k^{\circ}(\mathbf{v})$ by an evolution described by semigroups. This property represents the diffusion of orbits about the unperturbed trajectory. Mathematically, we can show that for a broad frequency spectrum where $D_{\mathbf{k}\omega}(\mathbf{v}, \mathbf{v}') \to D_{\mathbf{k}}(\mathbf{v}_+)$ independent of ω and $\mathbf{v}_- = \mathbf{v} - \mathbf{v}'$, the inverse of

$$g_{\mathbf{k}}(\tau, \mathbf{v}, \mathbf{v}') = \int_{-\infty}^{+\infty} g_{\mathbf{k}\omega}(\mathbf{v}, \mathbf{v}') \exp(-i\omega\tau)\, d\omega \tag{22}$$

can be calculated and is

$$\tilde{g}_{\mathbf{k}}(\tau, \mathbf{v}, \mathbf{v}') = -i(4\pi D_k \tau)^{-3/2} \times \exp\left[-\frac{(\mathbf{v} - \mathbf{v}')^2}{4D_k\tau} - \frac{k^2 D_k}{12}\tau^3 - \frac{i}{2}\mathbf{k} \cdot (\mathbf{v} + \mathbf{v}')\tau \right] \tag{23}$$

for $\tau \geqslant 0$.

The operator $g_k(\tau, \mathbf{v}, \mathbf{v}')$ can be shown to form a semigroup, using explicit calculations in the case of Eq. (23), to show that

$$\int d\mathbf{v}'\, g_k(\tau_1, \mathbf{v}, \mathbf{v}') g_k(\tau_2, \mathbf{v}', \mathbf{v}'') = g_k(\tau_1 + \tau_2, \mathbf{v}, \mathbf{v}'')$$

$$g_k(\tau \to 0^+, \mathbf{v}, \mathbf{v}') = \delta(\mathbf{v} - \mathbf{v}') \tag{24}$$

$$g_k(\tau < 0, \mathbf{v}, \mathbf{v}') \equiv 0$$

In the case of a localized spectrum of width $\Delta\omega$ the propagator is both non-Markovian and nonunitary, describing a complicated dynamics.

4. The propagator describes an irreversible dynamic. Once the exponential separation of neighboring particle trajectories sets in and the physical description gives up the exponentially small correlations present after many *e*-foldings, the evolution becomes irreversible. In complicated multispecies systems, energy and momentum can be transferred from particles to fluctuations and back to particles, making an effective Carnot cycle in the system. The requirement that there be no net gain of energy and momentum in the cycle then must be contained in the Im g_k that controls this transfer. We find that the semidefinite form of the interaction operator

$$-\text{Im}\, g_{\mathbf{k}\omega}(\mathbf{v}, \mathbf{v}') \geqslant 0 \tag{25}$$

guarantees a positive definite entropy production functional for the turbulent system.

5. ION–ACOUSTIC TURBULENCE AND THE NEED FOR RENORMALIZATION

The problem of ion–acoustic turbulence has been investigated in detail over the years with both weak turbulence theory[20] and renormalized turbulence theory.[22] As comparisons with experiments and particle simulations have been developed, it has been found that a number of simple observations, now understood in terms of renormalized turbulence theory, could not be explained on the basis of weak turbulence theory. The basic difficulty was that at the values of W/nT of interest, the nonlinear decorrelation of the wave–particle resonances is faster than the quasi-linear decorrelation given by the width of the spectrum. Rather than attempt to review the predictions of renormalized turbulence theory and the comparison with computer and laboratory experiments (see, e.g., Refs. 20, 22, and 27), let us simply define the basic quantities given by theory and measured in the experiments. The theory and experiments are for $d = 2$ and $d = 3$ and are characterized by the electron drift velocity u/c_s, the electron–ion temperature ratio T_e/T_i, and the mass ratio m_i/m_e.

Using θ as the polar angle measured from $\mathbf{u}$, the basic theoretical distributions are

$$\frac{e^2 I(\mathbf{k}, t)}{T_e^2} = \frac{A}{k^d} \mathfrak{f}(k\lambda_D, \theta, t\omega_{pi})$$

$$f_e(v, t) = \frac{B n_e}{v_e^d} \exp\left[-\left(\frac{v}{v_e}\right)^s\right]\left[1 + \frac{\mathbf{u} \cdot \mathbf{v}}{v_e^2} G\left(\frac{v}{v_e}\right)\right]$$

$$f_i(v, t) = \frac{C n_e}{v_i^d} \exp\left(\frac{-v^2}{v_i^2}\right) + \frac{D n_t}{v_t^d} f\left(\frac{v}{v_t}, \theta\right) \tag{26}$$

where the spectrum is peaked in the direction of the current carried by the electron distribution as are also a small fractional density $n_t/n_e \sim (m_e/m_i)^{1/4}$ for high energy ions $T_t = m_i v_t^2 \sim 5T_e$. The quasi-linear evolution of $f_e(v, t)$ from an initial Maxwellian distribution with $s = 2$ leads to final heated distribution $T_e(t) \gg T_e(0)$ with a much flatter lower energy distribution and a sharp dropoff of high energy electrons described by $s \leqslant 5$. These features of the time-asymptotic electron and ion distribution functions are predicted by the self-similar theory of Vekshtein and Sagdeev[28] and Vedenov and Ryutov.[29] The dynamics of the spectrum $I(\mathbf{k}, t)$ is complicated and has been the subject of many investigations (see Refs. 20, 22, and 25). Generally, the theory predicts a transformation of energy from the region of maximum growth rate $k\lambda_D \lesssim 1$ to long wavelengths $k\lambda_D \ll 1$ where the spectrum $I(k)$ reaches a maximum intensity. An underlying basis of the results is the conservation of energy and momentum as it passes from the drifting electrons to the waves and ultimately to the ions. The rate of transfer of energy and momentum is controlled by Im $g_{\mathbf{k}}^e(\mathbf{v})$ and Im $g_{\mathbf{k}}^i(\mathbf{v})$, respectively.

6. THE CONSERVATION LAWS

The macroscopic conservation laws for the mean kinetic energy and the momentum densities are one of the most important aspects of the theory in regard to describing a complicated plasma experiment. The renormalized turbulence theory yields a detailed balance between each species, denoted by j, and the energy–momentum densities in the fluctuating fields. The conservation laws can be established for the simply renormalized and fully renormalized theories. We consider the renormalized quasi-linear theory and then briefly describe the situation in the renormalized mode coupling equations. The energy density in the fluctuating fields is identified by separating the response function $\varepsilon_{\mathbf{k}}(\omega)$ into its real $\varepsilon'_{\mathbf{k}}(\omega)$ and imaginary $\varepsilon''_{\mathbf{k}}(\omega)$ parts for real ω. The energy density in the fluctuations when $I(\mathbf{k}, \omega)$ is peaked around $\omega \simeq \omega_{\mathbf{k}}$ is given by $W(\mathbf{k})$, where

$$W(\mathbf{k}) = \frac{k^2 I(\mathbf{k}, T)}{8\pi} \omega_{\mathbf{k}} \frac{\partial \varepsilon'_{\mathbf{k}}(\omega_{\mathbf{k}})}{\partial \omega_{\mathbf{k}}} \tag{27}$$

and the local rate of transfer between species j and the fluctuation $I(\mathbf{k})$ is controlled by $\gamma_{\mathbf{k}}^j$ defined by

$$\frac{\partial \varepsilon'_k}{\partial \omega_{\mathbf{k}}} \gamma_{\mathbf{k}}^j[g(I)] = -\varepsilon''_{\mathbf{k}}(\omega_{\mathbf{k}})$$

$$= \frac{4\pi e_j^2}{k^2 m_j} \int d^d v \int d^d v' \operatorname{Im} g_k^j(\mathbf{v}, \mathbf{v}') \left(\mathbf{k} \cdot \frac{\partial f^j}{\partial \mathbf{v}'} \right) \tag{28}$$

The background distribution $f^j(\mathbf{v}, t)$ evolves due to the particle transport in velocity space according to

$$\frac{\partial f^j}{\partial t} = \frac{\partial}{\partial \mathbf{v}} \cdot \int d^d v' \, \mathsf{D}_j(\mathbf{v}, \mathbf{v}') \cdot \frac{\partial f^j}{\partial \mathbf{v}'} \tag{29}$$

with

$$\mathsf{D}_j(\mathbf{v}, \mathbf{v}') = -\left(\frac{e_j}{m_j} \right)^2 \int d^d k \, \mathbf{k}\mathbf{k} I(\mathbf{k}, t) \operatorname{Im} g_{\mathbf{k}\omega_{\mathbf{k}}}^j(\mathbf{v}, \mathbf{v}') \tag{30}$$

describing the scattering with finite $\Delta\mathbf{v}$ in velocity space due to the finite amplitude electric fields.

With the identification of the renormalized wave energy and momentum density and the use of properties 2 and 3 for the renormalized propagator, we can show that

$$\frac{d}{dt} \int m_j \mathbf{v} f^j \, d^d v = -\int \frac{\mathbf{k}}{\omega_{\mathbf{k}}} \gamma_{\mathbf{k}}^j W(\mathbf{k}, t) \, d^d k \tag{31}$$

$$\frac{d}{dt} \int \tfrac{1}{2} m_j v^2 f^j \, d^d v = -\int \gamma_{\mathbf{k}}^j W(\mathbf{k}, t) \, d^d k \tag{32}$$

for the flow of momentum and energy between each species and the waves. The total energy and momentum conservation laws follow from summing Eqs. (31) and (32) over all species and using that $\gamma_{\mathbf{k}} = \Sigma_j \gamma_{\mathbf{k}}^j$ controls the evolution of $W(\mathbf{k}, t)$ in the absence of mode coupling. The conservation laws are then

$$\frac{d}{dt} \left[\sum_j \int m_j \mathbf{v} f^j \, d^d v + \int \frac{\mathbf{k}}{\omega_{\mathbf{k}}} W(\mathbf{k}, t) \, d^d k \right] = 0 \tag{33}$$

$$\frac{d}{dt} \left[\sum_j \int \tfrac{1}{2} m_j v^2 f^j \, d^d v + \int W(\mathbf{k}, t) \, d^d k \right] = 0 \tag{34}$$

with $W(\mathbf{k}, t)$ being the frequency integrated distribution of fluctuation energy density. In the ion–acoustic problem and in particle beam-driven plasma turbulence it is necessary to divide the kinetic energy into streaming energy $\frac{1}{2}mu^2$ and thermal energy density $\frac{3}{2}T$ with $\int \frac{1}{2}mv^2 f\,dv = \frac{1}{2}nmu^2 + \frac{3}{2}nT$. Using Eq. (31) to compute the change in streaming energy yields the conservation law for turbulent heating

$$\frac{d}{dt}\left(\frac{3}{2}n_j T_j\right) = \int (\mathbf{k}\cdot\mathbf{u}_j - \omega_{\mathbf{k}})\gamma^j_{\mathbf{k}} \frac{\partial \varepsilon'_{\mathbf{k}}}{\partial \omega_{\mathbf{k}}} \frac{k^2 I(\mathbf{k})}{8\pi} d^d k \tag{35}$$

For example, in ion–acoustic turbulence with $j = e$ and $\gamma^e_{\mathbf{k}} \cong (\pi/8)^{1/2}\omega_{\mathbf{k}}(\mathbf{k}\cdot\mathbf{u} - \omega_{\mathbf{k}})/|\mathbf{k}|v_e$, Eq. (35) becomes the law of anomalous ohmic heating, implying a rapid increase in the electron temperature proportional to $\nu_* u^2$ for $u \gg c_s$, where ν_* is given by an appropriate integral over the fluctuation spectrum. In addition, the momentum balance law Eq. (31) for the electrons implies a well-known formula for the anomalous turbulent collision frequency.

In addition to the renormalized quasi-linear processes described by $\varepsilon_{\mathbf{k}}[\omega, g(I)]$ there are changes in energy and momentum associated with the mode coupling processes given by Eqs. (14) and (15). The most typical process is that in which the induced scattering of the wave is the dominant transfer mechanism. The induced scattering occurs in $\mathbf{k}, \mathbf{v}$ regions where $\omega_{\mathbf{k}} - \omega_{\mathbf{k}_1} \cong (\mathbf{k} - \mathbf{k}_1)\cdot\mathbf{v}$ at a rate controlled by the mode coupling matrix elements and the $\operatorname{Im} g_{k-k_1}(\mathbf{v})$. It can be shown from the renormalized kinetic equation that the evolution for this process reduces to the form

$$k^2 \frac{\partial \varepsilon'_{\mathbf{k}}}{\partial \omega_{\mathbf{k}}} \frac{d^{nl} I(\mathbf{k}, t)}{dt} = I(\mathbf{k}, t) \sum_j \omega^2_{pj} \int I(\mathbf{k}_1, t) |M_j(\mathbf{k}, \mathbf{k}_1, \mathbf{v}, \mathbf{v}')|^2$$

$$\times \operatorname{Im} g^j_{\mathbf{k}-\mathbf{k}_1}(\mathbf{v}, \mathbf{v}')(\mathbf{k} - \mathbf{k}_1)\cdot \frac{\partial f^j}{\partial \mathbf{v}'} d^d k_1\, d^d v\, d^d v' \tag{36}$$

with the associated change in $f^j(\mathbf{v}, t)$ given by

$$\frac{\partial f^j(\mathbf{v}, t)}{\partial t} = -\frac{\partial}{\partial \mathbf{v}} \cdot \sum_j \left(\frac{e_j}{m_j}\right)^2 \int |M_j(\mathbf{k}, \mathbf{k}_1, \mathbf{v}, \mathbf{v}')|^2 I(\mathbf{k}, t) I(\mathbf{k}_1, t)$$

$$\times (\mathbf{k} - \mathbf{k}_1) \operatorname{Im} g_{\mathbf{k}-\mathbf{k}_1}(\mathbf{v}, \mathbf{v}')(\mathbf{k} - \mathbf{k}_1)\cdot \frac{\partial f^j(\mathbf{v}')}{\partial \mathbf{v}'} d^d k_1\, d^d k\, d^d v' \tag{37}$$

where $|M_j|^2$ is related to $\varepsilon(k - k_1)$, $\varepsilon^{(2)}(-k_1, k)$, and $\varepsilon^{(3)}(k_1, -k_1, k)$ in a complicated manner[20,22] that describes the scattering of waves from shielded test particles. The evolution in Eqs. (36) and (37) describes the scattering of $\omega_{\mathbf{k}}$, $I(\mathbf{k})$ into $\omega_{\mathbf{k}_1}$, $I(\mathbf{k}_1)$, with the particle picking up the difference in the wave energy and momentum. As a consequence, it can be shown that

$$\left(\frac{d}{dt}\right)^{nl}\left[\int m_j \mathbf{v} f^j \, d^d v + \int \mathbf{k} \frac{\partial \varepsilon'_{\mathbf{k}}}{\partial \omega_{\mathbf{k}}} \frac{k^2 I(\mathbf{k}, t)}{8\pi} d^d k\right] = 0 \tag{38}$$

$$\left(\frac{d}{dt}\right)^{nl}\left[\int \frac{1}{2} m_j v^2 f^j \, d^d v + \int \omega_{\mathbf{k}} \frac{\partial \varepsilon'_{\mathbf{k}}}{\partial \omega_{\mathbf{k}}} \frac{k^2 I(\mathbf{k}, t)}{8\pi} d^d k\right] = 0 \tag{39}$$

and that

$$\left(\frac{d}{dt}\right)^{nl} \int \frac{\partial \varepsilon'_{\mathbf{k}}}{\partial \omega_{\mathbf{k}}} \frac{k^2 I(\mathbf{k}, t)}{8\pi} d^d k = 0 \tag{40}$$

which states that the action or number density of the plasma waves $N(\mathbf{k}, t) = W(\mathbf{k}, t)/\omega_{\mathbf{k}}$ is not changed in the scattering process. Typically, the waves are scattered from the regions $\gamma_{\mathbf{k}_0}[g(I)] > 0$ to regions of absorption $\gamma_{\mathbf{k}}[g(I)] < 0$ where the transfer $\gamma_{\mathbf{k}}^{nl}$ increases with W/nT until the lifetime of the waves balances the rate of input $2\gamma_{\mathbf{k}_0} W(\mathbf{k}_0)$ of wave energy density.

Finally, we consider the entropy production function for the evolution of our reduced system of equations. Since the Boltzmann collision operator plays no role in this theory, it is natural to replace the entropy function from $-f(\mathbf{x}, \mathbf{v}, t) \ln f(\mathbf{x}, \mathbf{v}, t)$ to $-f^2(\mathbf{x}, \mathbf{v}, t)$. Thus, we consider the change in the entropy functional

$$\sigma(f_j) = -h(f_j) = -\tfrac{1}{2} \int f_j^2(\mathbf{x}, \mathbf{v}, t) \, d^d x \, d^d v \tag{41}$$

in the renormalized turbulence equations. In the original Vlasov–Poisson system we have $d\sigma_j/dt = 0$. As noted earlier, however, the exponential filamentation of the phase space volumes introduces arbitrarily small scale oscillations in the phase space, which are dropped in the renormalized theory. Thus, we compute for this theory that

$$\frac{d\sigma_j}{dt} = -\left(\frac{e_j}{m_j}\right)^2 \int\int\int I(k) \mathbf{k} \cdot \frac{\partial f^j}{\partial \mathbf{v}} \operatorname{Im} g_k^j(\mathbf{v}, \mathbf{v}') \mathbf{k} \cdot \frac{\partial f^j}{\partial \mathbf{v}'} dk \, d^d v \, d^d v' \tag{42}$$

From the negative definite property of the propagator (item 4 in Section 4), it follows that

$$\frac{d\sigma_j}{dt} \geqslant 0 \tag{43}$$

throughout the renormalized kinetic theory.

The entropy production principle Eq. (43) provides the basis for the existence of a thermodynamic transport theory in terms of $n_j(\mathbf{R}, T)$, $\mathbf{u}_j(\mathbf{R}, T)$

and $T_j(\mathbf{R}, T)$ where **R** and T are the macroscopic space–time variables. In the case of turbulent heating, it follows that consistent transport coefficients such as the anomalous collision frequency $\nu_{\text{eff}}(T_e/T_i, u/c_s, m_e/m_i)$ and anomalous thermal conductivity κ_{eff}, which are the properties measured in the laboratory, can be derived from the background kinetic equations. Finally, we note that from the renormalized mode coupling equation we can also expect to establish that $\partial_t \int \langle \delta f^2 \rangle_{\mathbf{k}\omega} \, d^d k \, d\omega \, dv \leqslant 0$. Special cases of this result can be given as discussed by DuBois[8, 11] and Boutros-Ghali and Dupree.[16]

REFERENCES

1. P. J. Morrison, *Phys. Lett. A* **80a**, 383 (1980).
2. B. V. Chirikov, *Phys. Rep.* **52**, 263 (1979).
3. D. F. Escande and F. Doveil, *J. Stat. Phys.* **26**, 257 (1981); *Phys. Lett. A* **83**, 307 (1981); *Phys. Lett. A* **84**, 399 (1981). See also chapter by D. F. Escande in Part 2 pp. 149–177 of this volume.
4. B. Misra, I. Prigogine, and M. Courbage, *Physica A* **98**, 1 (1979); B. Misra and I. Prigogine. See also Ch. 2, pp. 21–43 of this volume.
5. A. B. Rechester and R. B. White, *Phys. Rev. Lett.* **44**, 1586 (1980).
6. A. B. Rechester, M. N. Rosenbluth, and R. B. White, *Phys. Rev. Lett.* **42**, 1247 (1979) pp. 471–483.
7. D. I. Choi and W. Horton, *Phys. Fluids* **17**, 2048 (1974).
8. D. F. DuBois and M. Espedal, *Plasma Physics* **20**, 1209 (1978).
9. S. A. Orszag and R. H. Kraichnan, *Phys. Fluids* **10**, 1720 (1967).
10. J. H. Misguich and R. Balescu, *J. Plasma Phys.* **13**, 385 (1975).
11. D. F. DuBois, *Phys. Rev. A* **23**, 865 (1981).
12. J. A. Krommes and R. G. Kleva, *Phys. Fluids* **22**, 2168 (1979).
13. H. A. Rose, private communication (1981); *J. Stat. Phys.* **20**, 441 (1978).
14. T. H. Dupree, *Phys. Fluids* **15**, 334 (1972); *Phys. Rev. Lett.* **25**, 789 (1970).
15. B. B. Kadomtsev and O. P. Pogutse, *Phys. Rev. Lett.* **25**, 1115, (1970).
16. T. Boutros-Ghali and T. H. Dupree, *Phys. Fluids* **24**, 1839 (1981).
17. D. Pesme and G. Laval, in Proceedings of CNLS Workshop, Los Alamos Scientific Laboratory, May (1981) and *Phys. Rev. Lett.* **43**, 1671 (1979).
18. V. L. Sizonenko and K. N. Stepanov, *Zh. Eksp. Theor. Fiz. Pis. Red.* **9**, 282 (1969) [*J. Exp. Theor. Phys. Lett.* **9**, 165 (1969)]; *Nuclear Fusion* **19**, 155 (1970).
19. D. I. Choi and W. Horton, *Phys. Fluids* **20**, 628 (1977).
20. A. A. Galeev and R. Z. Sagdeev, in *Reviews of Plasma Physics*, Vol. 7, M. A. Leontovich, Ed. (Consultants Bureau, New York, 1979), pp. 141–180.
21. L. I. Rudakov and V. N. Tsytovich, *Plasma Phys.* **13**, 213 (1971).
22. W. Horton and D. I. Choi, *Phys. Rep.* **49**, 273–410 (1979).
23. G. Baker, *The Essentials of Padé Approximants* (Academic Press, New York 1975).
24. V. N. Tsytovich, *Theory of Turbulent Plasmas*, (Consultants Bureau, New York, 1977) pp. 84–110.
25. W. Horton and D. Brock, *Phys. Fluids* **24**, 509 (1981); R. A. Koch and W. Horton, *Phys. Fluids* **18**, 861 (1975); see also Ref. 22.

26. G. Pelletier and C. Pomot, *J. Plasma Phys.* **14**, 153 (1975); G. Pelletier, *J. Plasma Phys.* **18**, 49 (1977).

27. K. Papadopoulos, *Rev. Geophys. Space Phys.* **15**, 125 (1976).

28. G. E. Vekshtein and R. Z. Sagdeev, *Zh. Eksp. Teor. Fiz. Pis. Red.* **11**, 297 (1970) [*J. Exp. Theor. Phys. Lett.* **9**, 165 (1970)].

29. A. A. Vedenov and D. D. Ryutov, in *Reviews of Plasma Physics*, Vol. 6, M. A. Leontovich, Ed. (Consultants Bureau, New York, 1972), pp. 40–55.

TURBULENT PLASMA RESPONSE IN A STOCHASTIC ORBIT REGIME

K. MOLVIG, J. P. FREIDBERG, AND R. POTOK
Nuclear Engineering Department and Plasma Fusion Center,
Massachusetts Institute of Technology,
Cambridge, Massachusetts

S. P. HIRSHMAN AND J. C. WHITSON
Oak Ridge National Laboratory,
Oak Ridge, Tennessee

T. TAJIMA
Institute for Fusion Studies,
University of Texas,
Austin, Texas

Abstract. *The theory for the nonlinear, turbulent response in a system with intrinsic stochasticity and long-lived fluctuations is considered. It is argued that perturbative Eulerian theories, such as the direct interaction approximation (DIA) are inherently unsuited to describe such a system. The exponentiation and chaotic properties that characterize stochasticity appear in the Lagrangian picture and cannot even be defined in the Eulerian (one-point) representation. An approximation for stochastic systems—the normal stochastic approximation (NSA)—is developed; it states that the perturbed orbit functions (Lagrangian fluctuations) behave as normally distributed random variables. This is independent of the Eulerian statistics and, in fact, these Eulerian fluctuations are treated as fixed. A simple model problem (appropriate for the electron response in the drift wave) is subjected to a series of computer experiments. To within numerical noise, the results are in agreement with the NSA. The predictions of the DIA for this model show substantial qualitative and quantitative differences.*

1. INTRODUCTION

The phenomenon of intrinsic stochasticity has received considerable attention in recent years.[1] To a large extent this has been motivated by its presumed relevance to anomalous transport in tokamaks.[2] Indeed, stochasticity for the electrons is difficult to avoid, since at the fluctuation levels needed to account for the observed anomalies, the island overlap or Chirikov condition is strongly satisfied. Thus the tokamak transport problem suggests a new branch of turbulence theory—namely, that of incorporating orbit stochasticity into the nonlinear wave dynamics in a self-consistent way. The stochasticity studies referenced above[1] do not address this problem but concentrate on specific orbit properties, exponentiation, Kolmogorov entropy, phase space maps, and related phenomena, which have exhibited an incredible variety of structure and pathological behavior. In fact, the occurrence of so much structure in the orbits alone has suggested a kind of futility in attempting a theory for the self-consistent wave dynamics. To the contrary, our basic result is that stochasticity, in spite of its myriad complex and bewildering properties, actually *simplifies* the part of the theory that we need to know to calculate the nonlinear electron response and the drift wave dynamics. This claim has been made previously,[3] in an abbreviated form. Here we detail the theory and validate the assertion with a series of numerical experiments.

Although the problem we address has quite general implications for turbulence theories of stochastic systems, we use a specific system appropriate for the drift wave response. This is chosen both for its practical applications and for its theoretical convenience. The relevant phase space is everywhere stochastic. It is not complicated by division into integrable and stochastic regions, which would require separate treatment. The lifetime of the long-scale (Eulerian) fluctuations associated with the waves, τ_{AC}, is assumed to be long compared to the Kolmogorov time τ_K, characterizing the rate of the randomization in the particle orbits.* From a formal integration of the kinetic equation we motivate an approximation based on the properties of the stochastic orbits. In essence, for the purpose of computing the fluctuations associated with the waves, the orbit perturbation can be treated as a normally distributed random variable with variance $\langle \delta x^2 \rangle = 2Dt$. Hence, we use the term "normal stochastic approximation" (NSA) to describe this method. We then have the single parameter D (only weakly dependent on the wave spectrum) as the manifestation of the nonlinear response. The result is simply a broadened resonance response. This differs from previous resonance broadening theories[4] in several important respects. Resonance broadening here is tied inextricably with stochasticity. It is not a separate effect added to the usual panoply of nonlinear processes,

*This problem actually has three distinct characteristic times, which are often termed *correlation* times. The τ_{AC} used here would be the half-width, in τ, of the correlation function $\langle \tilde{H}(J, \theta, t + \tau)\tilde{H}(J, \theta, t)\rangle$. The third time that is sometimes referred to as a correlation time is the half-width, in τ, of the function $\langle \tilde{H}(J, \theta + J\tau, t + \tau)\tilde{H}(J, \theta, t)\rangle$.

wave–wave coupling, nonlinear Landau damping, and so on. Rather, for fluctuations on the scale of the waves, the entire hierarchy of nonlinear processes dissolves into the single complicated phenomenon of stochasticity. The parameter D is then the embodiment of all these processes, which have combined to produce radial diffusion. In short, for the wave response, there is resonance broadening, nothing else. This argument is based on averaging over the microscale structure resulting from stochasticity, or effectively coarse graining. This leads to a closure in which the wave fluctuation statistics are arbitrary and are determined by a higher level of dynamics.

Our objective is to make, for a specific system, a deductive turbulence theory in which the physical basis for the approximations is clear. Our viewpoint is that a valid closure is based on a underlying chaotic process arising from the nonlinear dynamics. Therefore, to be plausible, a turbulence theory must have three features:

1. *Identification of the Underlying Source of Chaos.* Or irreversible relaxation process. In the present example, this is intrinsic orbital stochasticity.
2. *Formulation of a Statistical Ansatz or Closure.* To replace the nonlinear dynamical random process and to make an explicit connection with the destruction of high order correlations to justify the closure.
3. *Testing the Closure Scheme Numerically.* One is generally interested in a complex nonlinear dynamics that cannot be calculated analytically, but hopefully has tractable average properties. The theory developed herein, albeit limited and incomplete, follows this philosophy for a system of fusion interest.

In contrast the Eulerian schemes (which we loosely lump under the category "DIA") view closure as essentially a mathematical problem. The approach is systematic and, at least in the case of the DIA, is complete and has many desirable features.[6,7] However, it leaves open the question of what underlying physical process, if any, is responsible for the irreversible behavior generally found in such a closure, and it casts the validity criterion for the theory in purely mathematical terms, such as "smallness of the skewness parameter." The applicability of such a theory for any given physical system is then rather obscure.

In the present problem, to identify the underlying random process, we use a mixed representation in which the Eulerian fluctuations associated with the waves are expressed as an integral over the orbital characteristics, or Lagrangian fluctuations. With intrinsic stochasticity, these Lagrangian orbit perturbations develop much finer scales than the wavelengths of the Eulerian fluctuations. This scale disparity is exploited by applying what amounts physically to a coarse graining operation. A closure is affected in which the Eulerian wave dynamics remain deterministic. Thus, as in the drift wave,[3] one can analyze the stability of these fluctuations. The validity of this closure is confirmed by numerical experiments.

2. DISCUSSION OF THE MODEL AND STATISTICAL PROCEDURES

Consider the simple model system with Hamiltonian $H(J, \theta, t) = H_0(J) + \tilde{H}(J, \theta, t)$, and perturbing Hamiltonian in the form

$$\begin{aligned}\tilde{H}(J, \theta, t) &= \sum_{m,n} H_{mn}(J)\exp[im\theta - i\omega_{mn}t] \\ &\equiv \sum_{m,n} |H_{mn}|\exp[im\theta - i\omega_{mn}t + i\phi_{mn}(t)]\end{aligned} \tag{1}$$

We assume that H_{00} is not present, so that $\int_0^{2\pi} d\theta\, \tilde{H} = 0$. For completeness and clarification later, we denote the phase of the complex Fourier coefficient, the *Eulerian phase*, by ϕ_{mn}. In general this depends on time, with a characteristic scale τ_{AC}, since in a turbulent system the waves have a finite lifetime. However, in the limit we consider, $\tau_{AC} \gg \tau_K$, this time dependence is negligible for the period during which the τ integral in Eq. (14) is significant. The equilibrium frequencies for the particle motion are given by $\omega_0(J) = \partial H_0/\partial J$. Shear, or dependence of the frequency on J, occurs in general and is necessary to produce the phase space islands associated with individual resonances. For small perturbations δJ about some equilibrium values J_0, the frequency can be linearized, $\omega_0(J_0 + \delta J) = \omega_0(J_0) + (\partial\omega_0/\partial J_0)\,\delta J$. The measure of this approximation is the ratio of the correlation length J_c (in units of action) to the global scale $J_s \equiv |d\ln\omega_0/dJ|^{-1}$, which is generally small in practice. For $J_c/J_s \ll 1$, we can, without loss of generality take $H_0 = \frac{1}{2}J^2$ in our model, since ω_0 can always be absorbed into ω_{mn}.

The distribution function $f(J, \theta, t)$ is described by the equation

$$\frac{\partial f}{\partial t} + \frac{\partial H}{\partial J}\frac{\partial f}{\partial \theta} - \frac{\partial H}{\partial \theta}\frac{\partial f}{\partial J} = 0 \tag{2}$$

which, according to the interpretation of f is either the Vlasov or the Klimontovich equation. Finally, we have a condition (e.g., Poisson's equation) relating f and $\tilde{H}$, $\tilde{H}(J, \theta, t) = \int dJ'\, d\theta'\, K(J, J'; \theta, \theta')f(J', \theta', t)$ which will make Eq. (2) nonlinear if $\tilde{H}$ is eliminated in favor of f. This gives a closed system evolving f self-consistently.

In spite of its simplicity there are many practical problems that can be reduced to essentially this model: electron drift wave turbulence (**E × B**), stochastic magnetic field fluctuations, ion cyclotron resonance, velocity space Langmuir turbulence, and so on. The $\tilde{H}$ we use below is intended to represent the electron drift wave response.

The approach is to integrate Eq. (2) generally to obtain f as a nonlinear functional of $\tilde{H}$. Self-consistency can then be enforced as a final step in determining the nonlinear dynamics. In this chapter we consider only the first step of solving for f, the turbulence response. However, since the results are not dependent on the details of $\tilde{H}$ (in a stochastic regime), the requirement of self-consistency can be invoked later without difficulty.

Equation (2) can be cast in an integral form that is a convenient starting point for the nonlinear theory. We define an angular average as $\langle F \rangle = (1/2\pi)\int_0^{2\pi} d\theta\, F$, and decompose f into an average part $f_0(J) = \langle f \rangle$, and a fluctuating part $\tilde{f}(J, \theta, t) = f - f_0$. The angular average of Eq. (2) is

$$\frac{\partial f_0}{\partial t} = \frac{\partial}{\partial J}\left\langle \frac{\partial \tilde{H}}{\partial \theta}\tilde{f}\right\rangle \tag{3}$$

and the fluctuating part is

$$\left(\frac{\partial}{\partial t} + \frac{\partial H}{\partial J}\frac{\partial}{\partial \theta} - \frac{\partial \tilde{H}}{\partial \theta}\frac{\partial}{\partial J}\right)\tilde{f} = \frac{\partial \tilde{H}}{\partial \theta}\frac{\partial \bar{f}}{\partial J} - \frac{\partial}{\partial J}\left\langle \frac{\partial \tilde{H}}{\partial \theta}\tilde{f}\right\rangle \tag{4}$$

Equation (4) can be formally solved for $\tilde{f}$ by the method of characteristics, yielding

$$\tilde{f}(J, \theta, t) = \int_{-\infty}^{t} dt' \left(\frac{\partial \tilde{H}}{\partial \theta}\frac{\partial f_0}{\partial J} - \frac{\partial}{\partial J}\left\langle \frac{\partial \tilde{H}}{\partial \theta}\tilde{f}\right\rangle\right)\Bigg|_{\substack{J = J + \delta J(J, \theta, t, t') \\ \theta = \theta + J(t - t') + \delta\theta(J, \theta, t, t')}} \tag{5}$$

where the perturbed orbits or characteristics are given by

$$\frac{d\delta J}{dt'} = -\frac{\partial \tilde{H}}{\partial \delta\theta}$$

$$\frac{d\delta\theta}{dt'} = \delta J + \frac{\partial \tilde{H}}{\partial \delta J} \tag{6}$$

with boundary conditions $\delta\theta = \delta J = 0$ at $t' = t$. Equations (3), (5), and (6) [which are equivalent to the Eulerian form Eq. (2)] together with the self-consistency condition, form a closed system. The right-hand side of Eq. (3) is a turbulent collision operator, which is zero—determining f_0—in the steady state. Generally we consider f_0 to be slowly varying, consistent with $J_c/J_s \ll 1$, and ignore Eq. (3). To the same approximation the term $\partial/\partial J\langle(\partial\tilde{H}/\partial\theta)\tilde{f}\rangle$ in Eq. (5) may be neglected. Our considerations are then reduced to Eqs. (5) and (6).

Equation (5) expresses the Eulerian perturbation $\tilde{f}$ in terms of the perturbed Hamiltonian $\tilde{H}$, also Eulerian, and the characteristics δJ, $\delta\theta$. The characteristics are essentially a Lagrangian representation of the perturbations $\tilde{H}$, determined by integration of Eq. (6). Thus Eqs. (5) and (6) form a mixed Eulerian–Lagrangian representation.

In certain applications it is convenient to write Eq. (5) in terms of a propagator $\tilde{G}$,

$$\tilde{f}(J, \theta, t) = \int_{-\infty}^{t} dt' \int dJ' \, d\theta' \, \tilde{G}(J, \theta, t; J', \theta', t') \left(\frac{\partial \tilde{H}}{\partial \theta'} \frac{\partial f_0}{\partial J'} - \frac{\partial}{\partial J'} \left\langle \frac{\partial \tilde{H}}{\partial \theta} \tilde{f} \right\rangle \right) \tag{7}$$

The propagator can be expressed directly in terms of the Lagrangian characteristics, $\tilde{G}(J, \theta, t; J', \theta', t') = \delta(J' - J - \delta J(J, \theta, t, t'))\, \delta(\theta' - \theta - \dot{J}(t - t') - \delta\theta(J, \theta, t, t'))$, which implies the integral form,

$$\tilde{G}(J, \theta, t; J', \theta', t') = \frac{1}{(2\pi)^2} \int dk \, dm \exp\left[ik(J' - J) - ik \, \delta J + im(\theta - \theta') - imJ(t - t') - im \, \delta\theta \right] \tag{8}$$

Alternately, $\tilde{G}$ can be determined from the differential equation

$$\left(\frac{\partial}{\partial t} + \frac{\partial H}{\partial J} \frac{\partial}{\partial \theta} - \frac{\partial \tilde{H}}{\partial \theta} \frac{\partial}{\partial J} \right) \tilde{G}(J, \theta, t; J', \theta', t') = 0 \tag{9}$$

with boundary condition $\tilde{G}(J, \theta, t; J', \theta', t) = \delta(J - J')\, \delta(\theta - \theta')$.

We define all statistical averages in terms of the Liouville density and, accordingly, regard f as the Klimontovich distribution. This is completely general. It encompasses the more restricted statistical averages and can be utilized directly in the two-point, two-time theory for the calculation of fluctuations. All functions now depend parametrically on the particle variables J_i, θ_i, at time t in addition to the usual independent variables J, θ. For performing the statistical averages the notation can be abbreviated by writing $f = f(\mathbf{X}, t; \mathbf{X}_1, \mathbf{X}_2, \ldots, \mathbf{X}_N)$, where $\mathbf{X} = (J, \theta)$ is the phase space variable and $\mathbf{X}_i = (J_i, \theta_i)$ is the parametric variable corresponding to the particle phase space position at time t. The Liouville density Γ depends on the particle variables, $\Gamma = \Gamma(\mathbf{X}_1, \mathbf{X}_2, \ldots, \mathbf{X}_N; t)$, and gives the joint probability that, at time t, particle 1 is at $\mathbf{X}_1$, particle 2 at $\mathbf{X}_2$, and so on. The statistical average of any function F is given by

$$\{F\} = \bar{F}(\mathbf{X}, t) = \int d\mathbf{X}_1 \, d\mathbf{X}_2 \cdots d\mathbf{X}_N F(\mathbf{X}, t; \mathbf{X}_1, \mathbf{X}_2, \ldots, \mathbf{X}_N) \times \Gamma(\mathbf{X}_1, \mathbf{X}_2, \ldots, \mathbf{X}_N; t) \tag{10}$$

This is the complete formal statistical average, and does not restrict $\bar{F}$ in any way. In particular, $\bar{F}$ may be an arbitrarily complex function of J, θ. Formulated this way there is no distinction, statistically, between the various scales of F.

When the physics on the various scales differ (or in some cases when the physics is independent of scale[5]), it is desirable to have a statistical procedure that recognizes the different scales. To this end we decompose the perturbation $\tilde{H}$ as follows

$$\tilde{H}(\mathbf{X}) = \sum_{\mathbf{k}=\mathbf{k}_1}^{\mathbf{k}_m} H_{\mathbf{k}} e^{i\mathbf{k}\cdot\mathbf{X}} + \tilde{H}_0,$$

where $\tilde{H}_0$ is the fine scale (microscale) fluctuation, corresponding to $|\mathbf{k}| > |\mathbf{k}_m|$. The identity

$$1 = \int dH_{\mathbf{k}_1}\, dH_{\mathbf{k}_2} \cdots dH_{\mathbf{k}_m}$$

$$\times \prod_{\mathbf{k}=\mathbf{k}_1}^{\mathbf{k}_m} \delta\left(H_{\mathbf{k}} - \int d\mathbf{X}\, e^{-i\mathbf{k}\cdot\mathbf{X}} \int d\mathbf{X}'\, K(\mathbf{X},\mathbf{X}') \times f(\mathbf{X}', t; \mathbf{X}_1, \mathbf{X}_2, \ldots, \mathbf{X}_N)\right)$$

is then inserted in the integrand of Eq. (10) and, upon reversing the integrations, the average becomes

$$\{F\} = \int dH_{\mathbf{k}_1}\, dH_{\mathbf{k}_2} \cdots dH_{\mathbf{k}_m} \int d\mathbf{X}_1\, d\mathbf{X}_2 \cdots d\mathbf{X}_N\, F(\mathbf{X}, t; \mathbf{X}_1 \cdots \mathbf{X}_N)$$

$$\times \Gamma(\mathbf{X}_1, \ldots, \mathbf{X}_N, t)$$

$$\times \prod_{\mathbf{k}=\mathbf{k}_1}^{\mathbf{k}_m} \delta\left(H_{\mathbf{k}} - \int d\mathbf{X}\, d\mathbf{X}'\, e^{-i\mathbf{k}\cdot\mathbf{X}} K(\mathbf{X},\mathbf{X}') f(\mathbf{X}', t; \mathbf{X}_1, \ldots, \mathbf{X}_N)\right) \quad (11)$$

The integration over the particle coordinates in Eq. (11) is now restricted by the constraint that the fluctuations from $\mathbf{k}_1$ to $\mathbf{k}_m$ are all fixed. This is the microscale average. Denoting this by $[\ ]_0$, Eq. (11) can be written

$$\{F\} = \int dH_{\mathbf{k}_1}\, dH_{\mathbf{k}_2} \cdots dH_{\mathbf{k}_m} [F]_0(\mathbf{X}, t;\, H_{\mathbf{k}_1} \cdots H_{\mathbf{k}_m}) \Gamma'(H_{\mathbf{k}_1}, \ldots, H_{\mathbf{k}_m}; t) \quad (12)$$

where the remainder of the ensemble is parametrized by the Fourier coefficients, and Γ' is their associated probability density. Now, we can repeat the procedure and average over some subset of the $H_{\mathbf{k}}$'s, fixing the remainder, and so on. This will define a hierarchy of scales and associated averages, so that $\{F\}$ can be written

$$\{F\} = [\cdots [[F]_0]_1 \cdots]_n \quad (13)$$

The complete average is now performed as a sequence over ever-increasing scales, where the long scale fluctuations are fixed while the average over the fine scale is performed. In this chapter we distinguish only two scales, so that $\{F\} = [[F_0]]_1$.

In many cases, this sequential averaging procedure is of very little advantage. For example, suppose there is only one relevant scale and the fine scale average $[\ ]_0$ has a negligible effect. Then $[F]_0$ will depend sensitively on the perturbations $H_{\mathbf{k}}$ and the coarse scale average, $[\ ]_1$, will be just as difficult as the full average $\{\ \}$.

The utility of the decomposition, Eq. (13), is evident in problems with disparate scales. Then the fine scale average $[\ ]_0$, can eliminate significant information and simplify the long-scale dynamics. Thus, in the stochasticity problem of present concern, the average propagator $[\tilde{G}]_0$ depends on the $H_{\mathbf{k}}$ only through D, Eq. (22), which is essentially constant over the realizations of interest, and the averages $[\ ]_0$ and $[\ ]_1$ can be done independently. This simplifies enormously the calculation of the long-scale fluctuations.

We consider only the first step, the average $[\ ]_0$, for a stochastic regime; thus the sequential averaging procedure is not required. The discussion shows the connection to the larger picture and demonstrates that since the long-scale fluctuations are rigorously fixed, this step in the calculation can be performed before requiring self-consistency on the long scale.

3. NORMAL STOCHASTIC APPROXIMATION: HEURISTIC THEORY

In a typical plasma turbulence problem, the source of the turbulence is instability, localized in $\mathbf{k}$, ω-space (m, n in our model). There may be many modes or m and n values present, but the fractional spectral width $\Delta m/m$, $\Delta n/n$ is generally small, so that there is a single scale characterizing the Eulerian perturbations $\tilde{H}$. Eulerian theories like the DIA[6,7] are designed for a single scale and tacitly assume that the Lagrangian perturbations have the same scale. However, the defining characteristic of stochasticity is the exponentiation of neighboring trajectories. In this case, the perturbed orbit function $\delta\theta$ is both *secular* with respect to $t - t'$ and *unstable* with regard to the initial conditions θ, J. Not only do the deviations from the unperturbed orbits grow in time, but initially neighboring particles suffer radically different large perturbations. In fact, after a time τ_K, the Kolmogorov time, a very small element, $m\,\Delta\theta \ll 2\pi$, gets mapped to cover the full range 2π. This has been called the *mixing* property by Zaslavskii and Chirikov.[8] Its implications for the Lagrangian orbit function $\delta\theta$ are shown in Fig. 1. A plot like Fig. 1 for larger τ would show larger excursions in the amplitude of $\delta\theta$ *and* more rapid fluctuations as a function of θ. In short, it is a consequence of stochasticity that the Lagrangian perturbations develop much finer spatial scales than their Eulerian counterparts. We exploit this disparity of scales with the microscale average to develop the NSA.

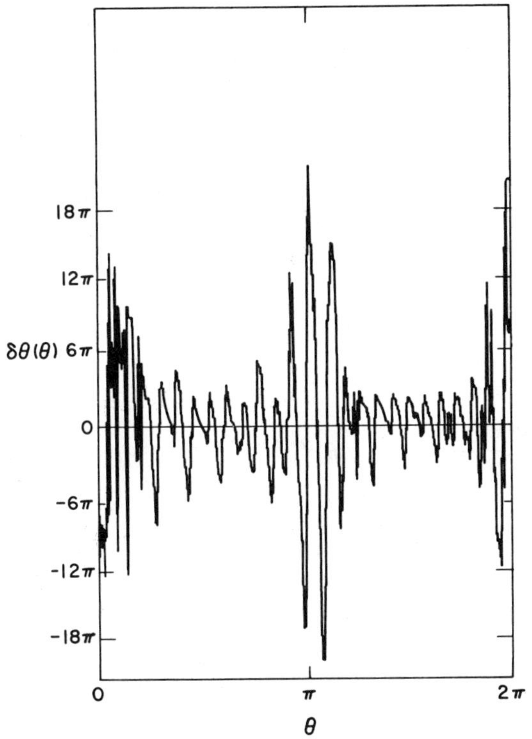

Figure 1. Characteristic orbit functions in stochastic case: for $t - t' > \tau_c$, where τ_c is the correlation time and approximation Kolmogorov entropy. For large $t - t'$, the oscillations have larger amplitude and finer scale.

To be more explicit, we use the form of Eq. (1) for the perturbation and compute the mth Fourier component of Eq. (5). There results

$$f_{m'}(J, t) = \sum_{m, n} e^{-i\omega_{mn}t} \int_0^{\infty} d\tau\, e^{i(\omega_{mn} - mJ)\tau} \times \langle e^{i(m-m')\theta + im\,\delta\theta(J,\theta;\tau)} im|H_{mn}(J + \delta J)|\rangle e^{i\varphi_{mn}(t+\tau)} \tag{14}$$

We focus on values of m and m' typical of the normal modes. That is, we assume that the spectrum of fluctuations is dominated by the normal modes and compute here only these Fourier components. Of course, the stochastic mapping produces much higher values of m, presumably incoherent fluctuations, but these belong to the microscale and require a separate treatment. We also assume that all phase space (J, θ) is stochastic, without isolated integrable curves. In this case $\delta\theta$ is secular with respect to time, and the exponent cannot be expanded as a perturbation series. Rather, we evaluate Eq. (14) asymptotically using the properties of $\delta\theta$ implied by stochasticity. Furthermore, for simplicity, we treat the case where H_{mn} is independent of J (the results are readily generalized) so that we must evaluate

$$A_{mm'} = \langle \exp(i(m - m')\theta + im\,\delta\theta(J, \theta, \tau)) \rangle \tag{15}$$

For small τ, $\delta\theta \ll 1$, and $A_{mm'} \simeq \delta_{m,m'}$. For large τ/τ_K, the θ-dependence of $\delta\theta$ is the dominant variation in the integrand, and $A_{mm'}$ can be computed as a series of integrals over small intervals $2\pi/M$. Thus letting $\theta_j = 2\pi j/M$,

$$\begin{aligned} A_{mm'} &= \sum_{j=1}^{M} \frac{1}{2\pi} \int_0^{2\pi/M} d\theta \exp\left[i(m - m')(\theta_j + \theta) + im\delta\theta(\theta_j + \theta, \tau)\right] \\ &= \sum_{j=1}^{M} \frac{e^{i(m-m')\theta_j}}{2\pi} \int_0^{2\pi/M} d\theta\, e^{im\,\delta\theta(\theta_j+\theta,\tau)}[1 + i(m - m')\theta + \cdots] \end{aligned} \tag{16}$$

The essential feature of stochasticity is that an M can be chosen such that $\delta\theta$ undergoes many oscillations with respect to θ in period $2\pi/M$, but $(m - m')/M$ is small. Thus, the integral above will approach an average, independent of θ_j, and the sum gives a Kronecker delta, yielding,

$$A_{mm'} \simeq \delta_{mm'} \langle e^{im\,\delta\theta} \rangle \tag{17}$$

with an error of order $((m - m')/M)\langle e^{im\,\delta\theta}\rangle$. For stochastic orbits, M increases with τ, such that $M \to \infty$, so that Eq. (17) is asymptotically exact in the limit of large τ/τ_K. The disparity between the wavelength and the spatial scale of the Lagrangian orbit functions is the key element here. By restricting m values, however, a separate microscale average was not required.

The basic assertion of the NSA can now be stated. The microscale average $A_{mm'} \to [A_{mm'}]_0$ is equivalent to averaging with $\delta\theta$ as a normally distributed random variable. With δJ a diffusion or Wiener process $[\delta J^2]_0 = 2D\tau$, $\delta\theta$ is the integrated Wiener process or $[\delta\theta^2]_0 = \frac{2}{3}D\tau^3$, and therefore

$$[A_{mm'}]_0 = \delta_{m,m'}[\langle e^{im\,\delta\theta}\rangle]_0 = \delta_{m,m'} e^{-(1/2)m^2[\delta\theta^2]_0} = \delta_{m,m'} e^{-(1/3)m^2 D\tau^3} \tag{18}$$

More formally, the relation between the moments $[\delta\theta^n]_0$ of $\delta\theta$ is normal,

$$[\delta\theta^{2p}]_0 = \frac{(2p)!}{p!2p}[\delta\theta^2]_0^p$$

Note that since ϕ_{mn} and $|H_{mn}|$ belong to the long-scale statistical ensemble, the microscale average of Eq. (14) acts only on $A_{mm'}$. We now show how the stochastic properties of the orbits lead to this result. The perturbed orbit function is expanded in a series with τ- and J-dependent coefficients

$$\delta\theta(J,\theta;\tau)=\sum_{l}\theta(J,\tau)e^{il\theta} \tag{19}$$

Since $\delta\theta$ has zero mean, the $l=0$ term is not present. Using Eq. (19) yields

$$\langle e^{im\,\delta\theta}\rangle=\sum_{n=0}^{\infty}\frac{(im)^n}{n!}\frac{1}{2\pi}\int\left\langle d\theta\left(\sum_{l_1}\theta_{l_1}e^{il_1\theta}\right)\cdots\left(\sum_{l_n}\theta_{l_n}e^{il_n\theta}\right)\right\rangle \tag{20}$$

The θ integral yields a selection rule for the l sums, $l_1+l_2+\cdots+l_n=0$. The cumulant expansion results by assuming a hierarchy where the dominant contribution arises from combining the l_n in pairs, and so on. Regardless of whether a hierarchy exists, Eq. (20) can always be expressed in terms of this rearrangement. Thus the pairwise combinations, requiring $n=2p$, can be made $(2p)!/p!2^p$ ways, and this leads to

$$\begin{aligned}\langle e^{im\,\delta\theta}\rangle_{(2)}&=\sum_{p=0}^{\infty}\frac{(-m^2)^p}{(2p)!}\frac{(2p)!}{p!2^p}\left(\sum_{l_1}|\theta_{l_1}|^2\right)\cdots\left(\sum_{l_p}|\theta_{l_p}|^2\right)\\&=\exp(-\tfrac{1}{2}m^2\langle\delta\theta^2\rangle)\end{aligned} \tag{21}$$

where $\langle\delta\theta^2\rangle=\Sigma_l|\theta_l|^2$. The pairwise combinations are equivalent to a normal distribution.

The next order may be computed in a similar manner in terms of the triple moment

$$\langle\delta\theta^3\rangle\equiv\sum_{l_1,l_2}\theta_{l_1}\theta_{l_2}\theta_{-l_1-l_2}$$

Here, however, as in all the higher cumulants, there is dependence on the phases of the Fourier coefficients, which implies sensitivity to other variables, like J and τ. The type of behavior indicated in Fig. 1 also occurs with respect to J in a stochastic situation. This implies rapid oscillatory behavior of the triplet (and higher cumulants) with respect to J in a stochastic situation. Thus integration over J tends to annihilate the higher order cumulants, leaving only the pairs. It is this phase independence of the pair combinations that makes the normal distribution dominant. The essential point about stochasticity is that the scale of oscillations in J decreases rapidly with τ, so that the *microscale* interval goes to zero for $\tau\gg\tau_K$.

The foregoing is basically a duplication of the central limit theorem, in a physical context. There is, in effect, a random phase principle acting when stochasticity is present. We emphasize that it is the phases of the Fourier coefficients of the perturbed orbit functions or the Lagrangian representation, *not the Eulerian wave phases*, that are relevant here.

It remains to compute the diffusion coefficient from the orbit equations Eqs. (6), by using the microscale average,

$$D = \frac{[\delta J^2]_0}{2t}$$

$$= \frac{1}{2t}\int_0^t dt_1 \int_0^t dt_2 \left[\tilde{H}(\theta + Jt_1 + \delta\theta(t_1), t_1) \cdot \tilde{H}(\theta + Jt_2 + \delta\theta(t_2), t_2)\right]_0$$

Substituting for $\tilde{H}$, letting $t_2 = t_1 - \tau$, and, to order $\tau_K/t \ll 1$, extending the τ-integral to infinity, this becomes

$$D = \frac{1}{t}\int_0^t dt_1 \int_0^\infty d\tau \sum_{\substack{m,n \\ m',n'}} H_{mn} H_{m'n'}(-1) mm' \exp(i(m+m')\theta$$

$$+ i((m+m')J - (\omega_{mn} + \omega_{m'n'})t_1 + i(\omega_{m'n'} - mJ)\tau))$$

$$\times \left[\exp(im\,\delta\theta(t_1) + im'\,\delta\theta(t_1 - \tau))\right]_0$$

To order τ_K/t, we can assume $t_1 \gg \tau_K$, then, provided τ_K^{-1} exceeds the frequency separation of modes—which is essentially the Chirikov overlap condition—the t_1 oscillations reduce the double sum to the diagonal $m' = -m$, $n' = -n$, leaving

$$D = \frac{1}{t}\int_0^t dt_1 \int_0^\infty d\tau \sum_{m,n} m^2 |H_{mn}|^2 e^{i(mJ - \omega_{mn})\tau} \left[\exp(im\,\delta\theta'(\tau))\right]_0$$

where $\delta\theta'(\tau) = \delta\theta(t_1) - \delta\theta(t_1 - \tau)$. This is the usual characteristic, but with fields evaluated at position $J = J + \delta J(t_1)$, $\theta = \theta + Jt_1 + \delta\theta(t_1)$ for $\tau = 0$. The exponential factor is then the same form as above, and we get

$$D = \sum_{m,n} m^2 |H_{mn}|^2 \int_0^\infty d\tau \exp\left[i(mJ - \omega_{mn})\tau - \tfrac{1}{3}m^2 D\tau^3\right] \qquad (22)$$

or the usual Fokker–Planck form with a broadened resonance instead of $\pi\delta(mJ - \omega_{mn})$.

To summarize, we have formulated the theory from the mixed Eulerian–Lagrangian representation, Eqs. (5) aned (6), together with a microscale average. In a stochastic regime, the Lagrangian orbit functions δJ, $\delta\theta$ tend to become normally distributed under the microscale average, with variance, $[\delta J^2]_0 = 2D\tau$, $[\delta J\,\delta\theta]_0 = D\tau^2$, $[\delta\theta^2]_0 = \frac{2}{3}D\tau^3$. The NSA (for this model) can then be succinctly expressed as the rule

$$[f_{m'}]_0 = \sum_{m,n} e^{-i\omega_{mn}t} im|H_{mn}| e^{i\phi_{mn}} \int_0^\infty d\tau \exp\left[i(\omega_{mn} - mJ)\tau - \tfrac{1}{3}mn^2 D\tau^3\right]$$

or, more formally,

$$\left[\tilde{H}(\theta + Jt + \delta\theta(t))\right]_0 = \int d\delta\theta \frac{\exp\left(-\delta\theta^2/\frac{4}{3}Dt^3\right)}{\sqrt{4\pi Dt^3/3}} \tilde{H}(\theta + Jt + \delta\theta) \quad (23)$$

This can be readily generalized to include δJ variations. Using the rule of Eq. (23), the Eulerian responses in Eq. (5) can be immediately averaged to obtain Eq. (18). Notice that the time dependence of $\phi_{mn}(t)$ is irrelevant, since the integrand in Eq. (14) decays in a time, $\tau_K \sim (\frac{1}{3}m^2 D)^{-1/3}$, much less than the time scale τ_{AC} on which $\phi_{mn}(t)$ varies. This precludes the possibility that correlations (in the long-scale ensemble) between ϕ_{mn} and $\delta\theta$ could modify the result.

The NSA is the leading order in an expansion for the distribution of Lagrangian orbit perturbations. It does not depend on the smallness of the Eulerian field amplitudes H_{mn}, which must be *above* some threshold to give stochasticity. The small parameter is kurtosis, of the distribution, measuring nearness to normal. This contrasts the DIA, which maximizes normality of the Eulerian field fluctuations $\tilde{H}$, and as shown below, cannot properly account for the secularities, scale disparity, and distribution of the Lagrangian fluctuations that characterize stochasticity.

4. FORMAL THEORY: COMPARISON WITH DIRECT INTERACTION APPROXIMATION

We now use a common framework within which both the DIA and the NSA can be simply expressed and reduced to a few explicit assumptions. The object of present concern is the propagator $\tilde{G}$, and the different forms proposed for the ensemble average of it. Of course the averaging procedures differ markedly between the two theories. In the DIA the complete statistical average is performed at the outset and one attempts to truncate the chain of statistically averaged equations. In the NSA the statistical averages are performed sequentially, to utilize the disparity of scales, and become an essential part of the argument justifying the approximation. For comparing the different results for the propagator in the present section, however, the statistical distinctions are secondary and are not emphasized. We consider the DIA first.

The starting point is Eq. (9) for the propagator. For $\tilde{H}$ independent of J, and defining $\tilde{E}(\theta, t) = -\partial\tilde{H}/\partial\theta$, this becomes

$$\left(\frac{\partial}{\partial t} + J\frac{\partial}{\partial\theta} + \tilde{E}(\theta, t)\frac{\partial}{\partial J}\right)\tilde{G}(J, \theta, t; J', \theta', t') = 0 \quad (24)$$

with $\tilde{G}(J, \theta, t; J', \theta', t) = \delta(J - J')\,\delta(\theta - \theta')$. The procedure of the DIA, according to Orszag and Kraichnan,[6] is as follows. The propagator is developed as a powers series in the Eulerian fluctuation amplitude $\tilde{E}$. This is done by using the unperturbed propagator G_0, obeying $[\partial/\partial t + J(\partial/\partial\theta)]G_0 = 0$, to construct the integral equation

$$\tilde{G}(J,\theta,t;\,J',\theta',t') = \int_{t'}^{t} dt'' \int dJ''\, d\theta''\, G_0(J,\theta,t;\,J'',\theta'',t'') \times \tilde{E}(\theta'',t'')\frac{\partial}{\partial J''}\tilde{G}(J'',\theta'',t'';\,J',\theta',t') \quad (25)$$

which is then iterated in the obvious way. The average is then performed, retaining only pair correlations* and the reduced series resumed to give $\{\tilde{G}\} = G^{\mathrm{DIA}}$. This can be done simply with diagrammatic methods and leads to an integral equation for the DIA propagator

$$\left(\frac{\partial}{\partial t} + J\frac{\partial}{\partial\theta}\right)G^{\mathrm{DIA}}(J,\theta,t;\,J',\theta',t') - \frac{\partial}{\partial J}\int_{t'}^{t} dt''\, dJ''\, d\theta''\, G^{\mathrm{DIA}}(J,\theta,t;\,J'',\theta'',t'')\{\tilde{E}(\theta,t)\hat{E}(\theta'',t'')\} \times \frac{\partial}{\partial J''}G^{\mathrm{DIA}}(J'',\theta'',t'';\,J',\theta',t') = 0 \quad (26)$$

Thus, the DIA makes two assumptions. First, the perturbation expansion of $\tilde{G}$ in powers of the Eulerian amplitude $\tilde{E}$ is assumed to converge. Second, a statistical assumption, is that pair correlations of the $\tilde{E}$ dominate, or in other words, that the Eulerian fluctuations are normally distributed. Although the second assumption is plausible, the convergence assumption is crucial and highly dubious in a stochastic regime. Since the propagator is just a formal expression of the orbit, this is equivalent to assuming convergence of the orbit expansion in powers of the Eulerian fields. But the resonant behavior underlying stochasticity is *not* analytic, even for an isolated resonance. For sufficiently small amplitude, well below the stochasticity threshold, the phase space volume occupied by the resonances is small, and the perturbation expansion converges in the remaining, dominant, volume of phase space. Here the DIA might be appropriate. For this reason, even though certain select terms are summed to all orders, the DIA is inherently a perturbative theory. Furthermore, the physical effect represented by the nonlinear (broadening) term in Eq. (26) corresponds to the third-order, two-wave, one-particle, perturbative process of Compton scattering. One might also argue, above the stochasticity threshold, that although the series for the exact propagator $\tilde{G}$ does not converge, the statistically averaged series does and gives the correct $\{\tilde{G}\}$. In a sense the statistical average salvages the divergent series. This becomes plausible when the Eulerian fluctuations relax to their normal distribution before the orbits

*Actually certain *crossing diagrams* are also neglected in the resummation.

become stochastic, or in other words, in the limit $\tau_{AC} < \tau_K$. Below the stochasticity threshold, τ_K goes to infinity, so that this condition is trivially satisfied.

Equation (26) can be reduced to a diffusion form, commonly used in applications[9] where the operand of G^{DIA} depends only on θ. In this case, one needs only $G^{\mathrm{DIA}}(\theta, t; \theta', t'; J) = \int dJ' \, G^{\mathrm{DIA}}(J, \theta, t; J', \theta', t')$. Noting that G^{DIA} depends rapidly on the difference variable $J - J'$, but weakly on the sum $\frac{1}{2}(J + J')$, integration over J' in Eq. (26) gives

$$\begin{aligned}\left(\frac{\partial}{\partial t} + J\frac{\partial}{\partial \theta}\right) G^{\mathrm{DIA}}(J) &= \frac{\partial}{\partial J}\int_{t'}^{t} dt'' \, d\theta'' \, dJ' \, dJ'' G^{\mathrm{DIA}}(J, J'')\{\tilde{E}\tilde{E}\} \\ &\quad \times \frac{\partial}{\partial J''} G^{\mathrm{DIA}}(J'', J') \\ &= \frac{\partial}{\partial J}\int_{t'}^{t} dt'' \, d\theta'' \, d\delta J'' G^{\mathrm{DIA}}(J, J + \delta J'')\{\hat{E}\hat{E}\} \\ &\quad \times \frac{\partial}{\partial J} G^{\mathrm{DIA}}(J + \delta J'') \\ &\cong \frac{\partial}{\partial J}\int_{t'}^{t} dt'' \, d\theta'' \, G^{\mathrm{DIA}}(\theta, t; \theta'', t''; J) \\ &\quad \times \{\tilde{E}(\theta, t)\tilde{E}(\theta'', t'')\} \frac{\partial}{\partial J} G^{\mathrm{DIA}}(\theta'', t''; \theta', t'; J)\end{aligned}$$

This can be transformed in t and θ, assuming a quasi-stationary fluctuation spectrum and causal propagators to finally give

$$\begin{aligned}&-i(\omega - mJ)G^{\mathrm{DIA}}_{m,\omega}(J) \\ &\qquad - \frac{\partial}{\partial J}\sum_{m',\omega'} G^{\mathrm{DIA}}_{m-m',\omega-\omega'}(J)\{|E|^2\}_{m',\omega'}\frac{\partial}{\partial J}G^{\mathrm{DIA}}_{m,\omega}(J) = 1 \qquad (27)\end{aligned}$$

Equation (27) is in the form of a diffusion equation for $G^{\mathrm{DIA}}_{m,\omega}$, where the diffusion coefficient

$$D_{m,\omega} = \sum_{m',\omega'} G^{\mathrm{DIA}}_{m-m',\omega-\omega'}\{|E|^2\}_{m',\omega'} \qquad (28)$$

depends on the frequency and wave number. Apart from the broadening of the resonance function G, Eq. (28) has the form of diffusion resulting from Compton scattering in the standard weak turbulence theory. It corresponds to diffusion arising from the beat wave resonance, $\omega - \omega' = (m - m')J$. The resonance width is not of practical importance for the evaluation of $D_{m,\omega}$. The resonance function is invariably narrow (in m', ω') compared to the spectrum

$\{|E|^2\}_{m',\omega'}$, so that one may take $G^{\text{DIA}}_{m-m',\omega-\omega'}(J) \simeq \pi\delta[\omega - \omega' - (m - m')J]$ in Eq. (28), and pass from the sum over m' to an integral. Treatment of the sum as an integral is legitimized by the resonance width of G^{DIA}.

The DIA departs from the standard weak turbulence theory by including $D_{m,\omega}$ in the equation for the propagator, and accordingly deleting some of the standard terms appearing in the equations for the one-particle distribution function and the spectrum. This is the familiar *renormalization*, basically a rearrangement of the perturbation series. In a strict sense it requires that the series converge. If the series does not converge, the procedure may still be valid in an asymptotic sense, but becomes largely an article of faith. Some aspects of the renormalization are then more reasonable than others. For example, the elimination of certain terms from the standard perturbation theory remains plausible because it does not require infinite summation. This part of the procedure was the main concern in the renormalization of quantum field theory, where certain divergent diagrams had to be eliminated. In the quantum theory the deleted diagrams never appear explicitly, but, when summed to all orders, serve to redefine the mass and charge. One can never observe, separately, the renormalized terms. By contrast in the DIA, these terms—requiring infinite summation—give an explicit form for the broadening of the propagator. This form is the issue of this chapter and is observable in our numerical experiments.

The NSA begins with the integral form for the propagator, Eq. (8). This is averaged over the microscale directly assuming normality of the Lagrangian fluctuations. We use the result that $[\exp(ik\delta J + im\delta\theta)]_0 = \exp(-\frac{1}{2}k^2[\delta J^2]_0 - km[\delta J\delta\theta]_0 - \frac{1}{2}m^2[\delta\theta^2]_0)$ for normal δJ and $\delta\theta$. This is shown as in Section 3, by writing the series, averaging, and resumming. Note, however, that no convergence assumption is required, since the exponential function is analytic. This then gives, from Eq. (8),

$$G^{\text{NSA}}(J,\theta,t;J',\theta',t') = \frac{1}{(2\pi)^2}\int dk\,dm \exp\Big(ik(J-J') + im(\theta-\theta') + imJ(t-t') - \tfrac{1}{2}k^2[\delta J^2]_0 - km[\delta J\delta\theta]_0 - \tfrac{1}{2}m^2[\delta\theta^2]_0\Big) \tag{29}$$

The variance $[\delta J^2]_0$, $[\delta\theta^2]_0$, and $[\delta J\,\delta\theta]_0$ can be computed under the same approximation, as indicated above, and are independent of J and θ. Thus, integration over J' gives

$$G^{\text{NSA}}(\theta,t;\theta',t';J) = \frac{1}{2\pi}\int dm \exp\Big(im(\theta-\theta') - imJ(t-t') - \tfrac{1}{2}m^2[\delta\theta^2]_0\Big)$$

Finally, this can be transformed, and using the expression for $[\delta\theta^2]_0$ from Eqs. (18) and (22), is equivalent to

$$-i(\omega - mJ)G_{m,\omega}^{\text{NSA}}(J) - \frac{\partial}{\partial J}$$

$$\sum_{m',\omega'} G_{m',\omega'}^{\text{NSA}}(J)\big[|E|_{m',\omega'}^2\big]_0 \frac{\partial}{\partial J} G_{m,\omega}^{\text{NSA}}(J) = 1 \tag{30}$$

This is to be compared with Eq. (27) for the DIA. It is a diffusion equation with $D = \Sigma_{m',\omega'} G_{m',\omega'}^{\text{NSA}}(J)[|E|_{m',\omega'}^2]_0$ independent of frequency and wave number, corresponding to diffusion arising from the primary wave–particle resonance $\omega' = m'J$. Thus the two approximations differ in physical content on the processes causing diffusion in the propagator.

To summarize, the DIA assumes normality of the Eulerian field fluctuations and convergence of the series expansion for the propagator in powers of the Eulerian amplitudes. The latter assumption is untenable in a stochastic regime. The NSA assumes normality of the Lagrangian fluctuations and does not require a convergence assumption. In practical terms the two theories predict different forms for the broadening of the propagator. The regimes of expected validity are also distinct. Thus the DIA is basically an amplitude expansion, presumably valid for $\tau_{AC} < \tau_K$. The NSA is essentially a multiple-scale expansion valid for amplitudes above the stochasticity threshold, with $\tau_K < \tau_{AC}$.

5. NUMERICAL EXPERIMENT

From a spectrum of known fields, chosen to be in a stochastic regime, one can numerically compute a family of orbits to generate the Lagrangian fluctuations $\delta\theta$, δJ, and evaluate the expression $A_{mm'}$ of Eq. (15). This, together with the theoretical formulation of Sections 2 and 3, provides the propagator for the limit considered, $\tau_K \ll \tau_{AC}$. The validity of the NSA propagator, Eq. (30), is then confirmed. Comparison with the DIA predictions is somewhat confusing, since the propagator is developed from an Eulerian expansion, and the functions $A_{mm'}$ and $\langle\exp(im\,\delta\theta)\rangle$ never appear in the theory. Thus we must directly compare the propagator deduced from the numerical experiment with that predicted by the DIA.

The propagators from the two theories can be written in the same characteristic forms as shown in Eqs. (27) and (30) or, equivalently in the integral representation

$$G_{mn}(J) = \int_0^\infty d\tau\, e^{i(\omega_{mn} - mJ)\tau} I_{mn}(\tau) \tag{31}$$

The decay function $I_{mn}(\tau)$ takes the same form in both theories; namely,

$$I_{mn} = \exp\left(-\tfrac{1}{3}m^2 D_{mn}\tau^3\right) \tag{32}$$

The theories differ in the forms claimed for D_{mn}. The NSA has $D_{mn} = D$, or simply the diffusion coefficient, independent of m and n, whereas the DIA has D_{mn} given by Eq. (20) with $\omega = nq$. Also, in the NSA, I_{mn} is related to the Lagrangian orbital characteristics by $I_{mn} = [\langle \exp(im\,\delta\theta)\rangle]_0$, where it should be recalled that $\langle\ \rangle$ denotes a θ-integral average, and $[\ ]_0$ is the microscale statistical average. In the DIA, the prediction for I_{mn} is not so clearly connected to the orbital characteristics and as we find below, is *inconsistent* with $[\langle \exp(im\,\delta\theta)\rangle]_0$.

Parameters are chosen to replicate the electron response for the drift wave. The m's correspond to poloidal mode numbers, typically in the several hundred range. The actual mode frequency for the drift wave is small, and $\omega_{mn} \equiv nq$. Resonances then occur at the action locations, $J = nq/m$. The m-spectrum ranges from $M - \Delta m/2$ to $M + \Delta m/2$, whereas the n-spectrum is sufficiently large that boundaries are not encountered in the simulation. To estimate the resonance spacing we take q-order unity, consider the region near $J = 1$, and take the n values from $M - \Delta m/2$ to $M + \Delta m/2$ that fill this region. There are then Δm^2 resonances from $J = 1 - \Delta m/M$ to $J = 1 + \Delta m/M$ for an *average* resonance spacing of $\Delta J_R \simeq 1/M\Delta m$. The island width of the individual resonances is $\Delta J_1 \simeq \sqrt{H_{mn}}$. A crude estimate of the island overlap criterion is then

$$\left(\frac{\Delta J_I}{\Delta J_R}\right)^4 \simeq |H_{mn}|^2 m^4 \Delta m^4 > 1 \tag{33}$$

The diffusion coefficient can be estimated (see below) as $D \simeq |H_{mn}|^2 M^2 \Delta m$, so the Kolmogorov time is expressed as

$$\tau_K^{-3} \simeq |H_{mn}|^2 M^4 \Delta m \tag{34}$$

The value M has been chosen from practical considerations. Parameters H_{mn} and Δm are determined from Eqs. (33) and (34) by requiring strong inequality in Eq. (33) and $\tau_K > 1$ in Eq. (34) so that the decay in time of $\langle \exp(im\,\delta\theta)\rangle$ is observable on the unit time scale of our model. These two conditions require $\Delta m > 1$, so that spectral width is important.

These considerations can become invalid near low order resonances (m/n approximating a low order rational or integer) where the resonances are not evenly spaced but cluster together and leave gaps. For example, near $J = 1$ and $q = 1$, the resonances cluster in groups of Δm modes, the groups being spaced by $1/m$ in J but within the cluster resonance spacing is $1/m^2$. Then, if $\Delta m/m \ll 1$, there will obviously be large gaps. To avoid this difficulty, we have used an irrational J (clearly q can be scaled out of the problem, so we take $q = 1$, without loss of generality). In addition the resonances were tabulated numerically and compared to the island widths to be sure that no such gaps occurred.

Although the integration of Eqs. (6) is straightforward, it can be time-consuming numerically because of the accuracy required to resolve the fine scale structure. Many of the relevant orbit properties can be studied using a simplified model where Eqs. (6) reduce to a simple mapping. The model is obtained by making the further simplification of an infinite n spectrum with constant amplitude and phase. Specifically, we take

$$\hat{H}(J,\theta,t) = \sum_{m=M_1}^{M_2} \sum_{n=-\infty}^{+\infty} H_m \cos[(mJ - n)t + m\theta + \varphi_m]$$

The identity $\Sigma_{n=-\omega}^{+\infty} e^{inqt} = 2\pi\Sigma_{l=-\infty}^{+\infty}\delta(t - 2\pi l)$ is used to express the force as a series of impulses; Eqs. (6) are then integrated to give the mapping

$$\Delta J_{l+1} = \delta J_l + 2\pi \sum_m mH_m \sin[2\pi l m J + m(\theta + \delta\theta_l) + \varphi_m] \tag{35}$$

$$\delta\theta_{l+1} = \delta\theta_l + 2\pi\delta J_l$$

where the lth step occurs at time $2\pi l$. The orbits were computed using both the mapping, Eq. (35), and differential form of Eqs. (6). Although the results are virtually indistinguishable and in good agreement with the NSA, the mapping is somewhat degenerate in that Eqs. (22) and (28) become identical. The differential form Eqs. (6) is necessary to resolve the differences with the DIA discussed in the preceding section.

With the mapping, orbits for 3000 particles distributed uniformly in θ at $J = 1 + 12/\pi$ are computed. For these experiments, the m spectrum extended from 250 to 500, with the amplitudes constant at $H_m = 10^{-8}$ and phases φ_m chosen at random, but, of course, fixed through the integration.

With the differential form, orbits were computed for 8400 particles distributed uniformly in θ at $J = 1.4$. (The same results were obtained with 25,600 particles). The m spectrum ranged from 90 to 111, while the n spectrum ran from 132 to 153 with amplitudes $H_{mn} = 3 \times 10^{-5}$ and phases φ_{mn} also random. Particles remained within the resonance region for several Kolmogorov times. The inevitable computer truncation errors play a natural coarse graining role as the orbits become excessively fine scaled.

The results are displayed in Figs. 2–6. For comparison purposes the data are plotted in normalized form relative to the diffusion coefficient D, evaluated from Eq. (22), for each model. The broadened resonance allows the n sum to be done as an integral, since it is wide on the scale of the resonance spacing but narrow compared to the spectral width in n. There is an additional factor of $\frac{1}{2}$ owing to the cosine form used in the model. Thus

$$D = \frac{\pi}{2} \sum_{m=M_1}^{M_2} m^2 H^2_{m,mJ/q} \simeq \frac{\pi}{6} H^2_{m,mJ/q}(M_2^3 - M_1^3) \tag{36}$$

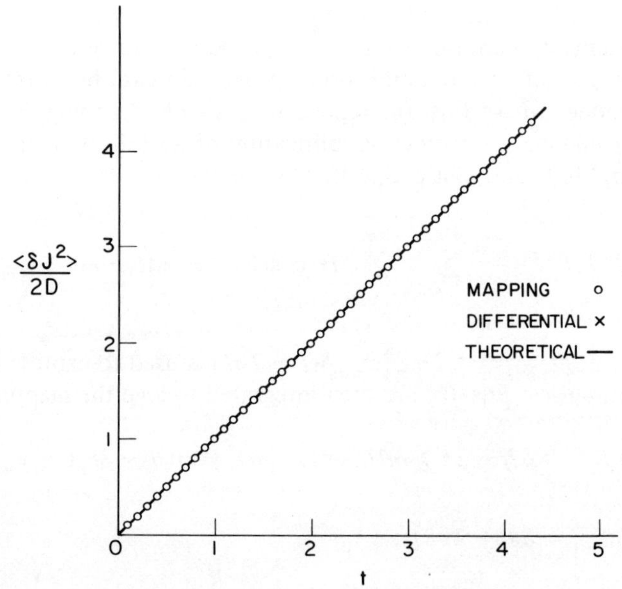

Figure 2. Observed variance $\langle \delta J^2 \rangle / 2D$ for the mapping and differential models compared to the theoretical value.

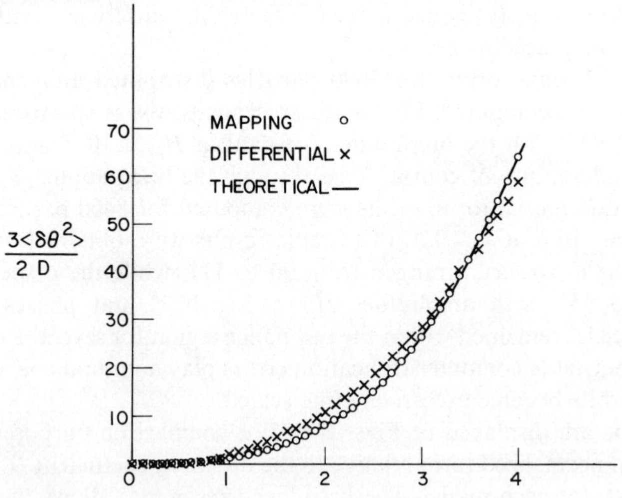

Figure 3. Observed variance $3\langle \delta\theta \rangle / 2D$ for the mapping and differential models compared to the theoretical value.

The observed variances, $\langle \delta J^2 \rangle / 2D$ and $3\langle \delta\theta^2 \rangle / 2D$, plotted in Figs. 2 and 3 show the expected t and t^3 time dependencies, respectively, for both the mapping and differential models. Actually, there is a slight discrepancy in the short-time behavior $t < \tau_K$ of $\langle \delta\theta^2 \rangle$ for the differential model. It shows up more clearly in a plot of $(\langle \delta\theta^2 \rangle 3/2D)^{1/2}$ where the initial slope is somewhat larger than one. This may be due to the relatively small number of modes in the differential model, since the short-time behavior is dependent more on the random initial phases than the randomness in the Lagrangian phases when stochasticity develops for $t > \tau_K$.

A typical decay function, $I_{mn} = \langle \exp(im\, \delta\theta) \rangle$, where n is the toroidal mode multiplied, is shown in Fig. 4. These exhibit the same behavior for both models and all m. They decay down to noise with roughly a t^3 exponent, while the off-diagonal terms $A_{mm'}$, for $m \neq m'$ (not shown), start at zero and build up to noise. The noise is produced by the finite number N of particles, and no smoothing has been done to remove it. By varying N we have verified that the level scales like $N^{-1/2}$ as consistent with white noise.

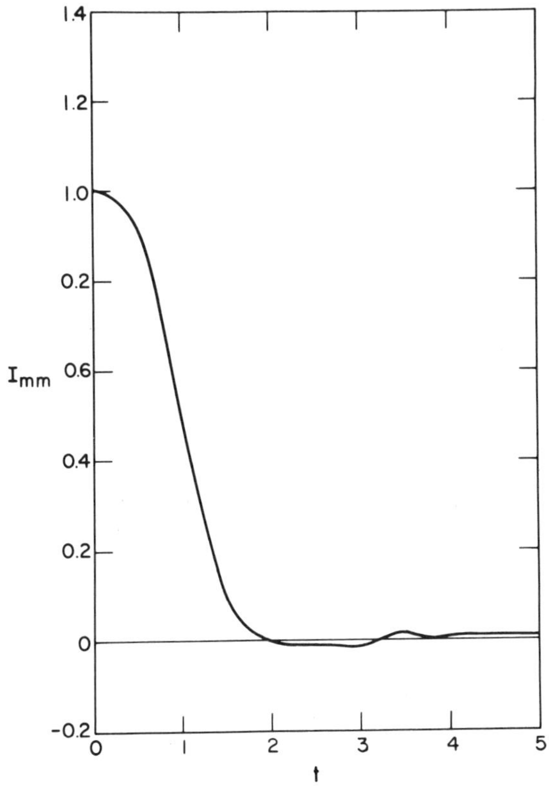

Figure 4. Typical decay function I_{mn} versus time showing characteristic decay to noise.

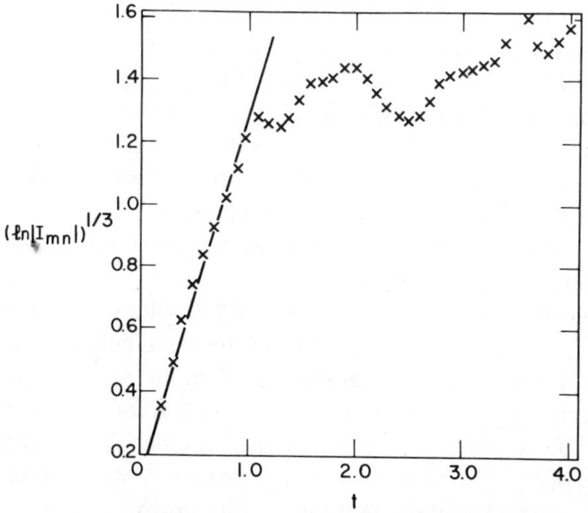

Figure 5. Plot of $(\ln|I_{mn}|)^{1/3}$ to display characteristic time dependence of correlation function. The linear fit to the initial decay, from which D_{mn} is determined, is also shown. The form of Eq. (35) is a reasonable description of the decay until the noise level is reached.

Comparison of the decay function with the I_{mn} of the propagators is indicated in Fig. 5, (for the differential model). The predictions are compared by determining the observed D_{mn}, fitting the form of Eq. (32) to the numerical data using least squares to compute the slope as indicated in Fig. 5. The results are shown in Fig. 6, where D_{mn}/D is plotted as a function of m, using the parameters of the differential equation model.

The observations show clearly that D_{mn} is independent of m, consistent with the NSA. The quantitative agreement is good, although there appears to be a slight systematic error. We interpret this as an error in the variance $\langle \delta\theta^2 \rangle$ as noted earlier, not as a deviation from normal in the distribution of δJ. This interpretation was confirmed by directly measuring the kurtosis, $\langle\langle \delta J^4 \rangle\rangle \equiv (\langle \delta J^4 \rangle - 3\langle \delta J^2 \rangle^2)/3\langle \delta J^2 \rangle^2$, which was found to be small, of order 0.1, independent of time. This measurement can be improved by reducing the noise level (increasing the number of particles) so that the short-time behavior contributes less to the value of the fitted slope. In the present experiments the noise level was reached after a few correlation times so that discrepancies for $\tau < \tau_K$ (which are expected) effect the determination of D_{mn}.

In contrast, the DIA predictions have some characteristic features that are quite difficult to reconcile with these observations. First, there is the dependence of the decay function on n and the related resonance function dependence on the frequency difference $\omega_{mn} - \omega_{m'n'}$. These are natural features of the Eulerian representation (arising from the beat wave resonance) but difficult to understand in the Lagrangian picture, where a secular dependence of $\delta\theta$ on t, in Eq. (14), would be implied. This would be a rather obvious feature in the numerics and nothing of the kind was observed. Second, even for fixed n, there is a strong m dependence of D_{mn} in the DIA, and this is not observed either.

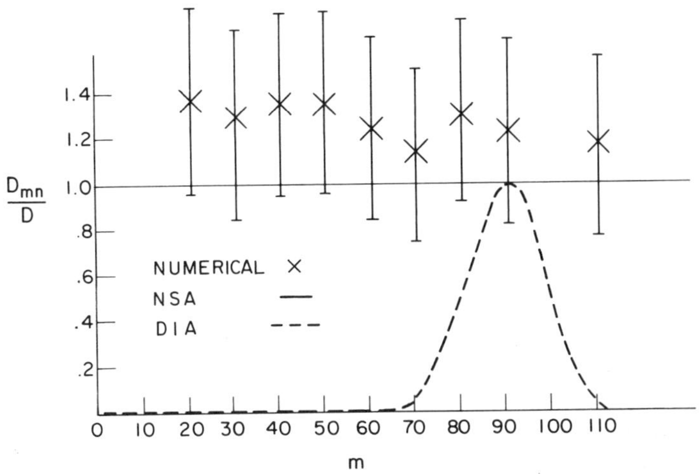

Figure 6. Comparison of the decay parameter, D_{mn} in the propagation function. The predictions of the DIA are plotted for fixed $n = 130$.

The magnitude of this discrepancy is illustrated in Fig. 6, where D_{mn}/D is plotted versus m for $n = 130$. It should be emphasized that D_{mn} is observed in an exponent. In fact, the DIA would predict no observable decay of I_{mn} during the course of the experiment for most values of m. Note that the DIA and NSA predictions in Fig. 6 agree at a point. This is where m is such that $\omega = nq - mJ$ and corresponds to the particle (J-value) being in resonance with the wave (m, n-value) whose response is being computed. In fact, it is evident from Eqs. (28) and following that $D_{m,mJ} = D$. The discrepancies arise when $\omega \neq mJ$, corresponding to the particle not being resonant with the fluctuation m, n. However, the particle *is* resonant with some other wave (m', n') in the spectrum and *is* diffusing with diffusion coefficient D, such that $\langle \delta\theta^2 \rangle = \frac{2}{3} D\tau^3$. Thus, in the context of the Lagrangian formulation, the DIA is predicting that $\langle \exp(im\,\delta\theta) \rangle$ is not equal to $\exp(-\frac{1}{2}m^2\langle \delta\theta^2 \rangle)$ even though the particle in question is diffusing, a clearly incongruous result. The only way this could occur is by allowing more rapid time dependence in the Eulerian phase φ_{mn} of Eq. (14), so that the decay of the time integral is not simply determined by the orbit characteristic $\delta\theta$. Such a possibility is precluded in the limit $\tau_K \ll \tau_{AC}$, considered here, and again suggests that the DIA closure is more appropriate in the limit $\tau_{AC} < \tau_K$, where the irreversibility arises more from the waves than from the particle orbits.

To summarize, the diffusion coefficient determined from the observed variances $\langle \delta\theta^2 \rangle, \langle \delta J^2 \rangle$ and the decay functions I_{mn} was the same in all cases with the value, Eq. (36), as predicted by the NSA. The discrepancies with the DIA predictions or the propagator are evident, amounting to orders of magnitude in a measured exponent. These observations strongly support the view that the physical process responsible for the resonance broadening in a stochastic regime ($\tau_{AC} \gg \tau_K$) is diffusion, not Compton scattering.

6. CONCLUSION

The main points of this chapter follow from an examination of the implications of intrinsic stochasticity for collisionless turbulence theory. Most of the consequences stem from stochasticity's most striking property, namely, the pathological complex, fine-scale, phase space structures that can result from relatively simple, long-scale, perturbations. The fine-scale structures result from integrating the long-scale perturbations along a particle orbit and thus suggest, at least in part, a Lagrangian representation. The fine-scale structures cannot be obtained by a perturbative treatment of the orbit (implicit in any perturbative Eulerian theory). This casts doubt on the applicability of theories like the DIA, for a stochastic system. The basic features that distinguish a stochastic system are thus the inherent nonanalyticity of the orbits and the inevitable multiplicity of scales.

The theory described exploits these features by using a Lagrangian representation and a statistical procedure of averaging sequentially over increasing scales. The dynamics of the long-scale fluctuations (associated with the waves or instabilities) can be simplified by averaging over the fine scales to give, in effect, a diffusing orbit response. We have termed this limit the "normal stochastic approximation." It is essentially a multiple-scale theory, appropriate for long-lived waves and amplitudes *above* the stochasticity threshold. Below the stochasticity threshold (or when $\tau_{AC} < \tau_K$), one might expect Eulerian amplitude expansion theories, like the DIA, to be valid. Not surprisingly, the two theories give different predictions with different underlying physical causes.

A qualifying comment is in order here. There is actually a third regime at still higher amplitude when the individual islands overlap the entire resonant region. One then gets a slowly modulated single island, and no evident stochasticity.[8] Trapping and diffusion phenomena appear together, but one is not clearly dominant. This regime was observed numerically for a different model but has not yet been studied extensively.

The situation is somewhat reminiscent of that in fluid turbulence, where the DIA has had its only established success. At moderate Reynolds number, where the relevant phenomenon are limited to a single scale, the Eulerian DIA works very well. However, as the Reynolds number gets larger, the scales multiply and discrepancies appear. The DIA does not give the Kolmogorov, $k^{-5/3}$ spectrum for inertial range turbulence.[10] The difficulty is that the Eulerian DIA treats all scales identically. Any interacting triplet of fluctuations irreversibly creates or destroys correlations. The theory does not allow for the small scales to be convected without being destroyed by the large scales as required for the Kolmogorov spectrum.

Attempts to remedy these deficiencies have utilized a Lagrangian representation,[11] and physical arguments to treat the large scales differently.[12] Fluid turbulence is, of course, quite different from the kinetic theory turbulence problem addressed in this chapter. The comparison here is intended simply to underscore the limits of the DIA for dealing with problems involving disparate scales. We are echoing Kraichnan's 18-year old refrain[11] and making a plea for Lagrangian theories.

ACKNOWLEDGMENTS

Fruitful conversations with D. F. DuBois, J. A. Krommes, D. Pesme, D. J. Sigmar, and P. Similion are gratefully acknowledged.

This work partially supported by U.S. Department of Defense contracts DE-ACO2-78ET-51013 (Massachusetts Institute of Technology) and W7405 ENG.26 (Oak Ridge National Laboratories).

REFERENCES

1. G. Laval and D. Grésillon, Eds. *International Workshop on Intrinsic Stochasticity in Plasmas* (Institut d'Etudes Scientifiques de Cargese, Corsica, France, 1979).
2. T. H. Stix, *Phys. Rev. Lett.* **30**, 833 (1973); J. D. Callen, *Phys. Rev. Lett.* **39**, 1540 (1977); A. B. Rechester and M. N. Rosenbluth, *Phys. Rev. Lett.* **40**, 38 (1978); K. Molvig, J. E. Rice, and M. S. Tekula, *Phys. Rev. Lett.* **41**, 1240 (1978).
3. K. Molvig, paper presented at the International Conference on Plasma Physics, Nagoya, Japan (1980); S. P. Hirshman and K. Molvig, *Phys. Rev. Lett.* **42**, 648 (1979); K. Molvig, S. P. Hirshman, and J. C. Whitson, *Phys. Rev. Lett.* **43**, 582 (1979).
4. T. H. Dupree, *Phys. Fluids* **9**, 1773 (1966), G. Benford and J. J. Thomson, *Phys. Fluids* **15**, 1496, J. J. Birmingham and M. Bornatici, *Phys. Fluids* **14**, 2234 (1971); **15**, 1778 (1972).
5. K. G. Wilson, *Rev. Mod. Phys.* **47**, 773 (1975).
6. S. A. Orszag and R. H. Kraichnan, *Phys. Fluids* **10**, 1720 (1967).
7. D. F. DuBois and M. Espedal, *Plasma Phys.* **20**, 1209 (1978); J. A. Krommes, in *Theoretical and Computational Plasma Physics* (International Atomic Energy Agency, Vienna, 1978), p. 405.
8. G. M. Zaslavskii and B. V. Chirikov, *Sov. Phys. Usp.* **14**, 549 (1972) [*Usp. Fiz. Nauk* **105**, 3 (1971)].
9. T. H. Dupree and D. J. Tetrault, *Phys. Fluids* **21**, 425 (1978); P. H. Diamond and M. N. Rosenbluth, *Phys. Fluids* **24**, 1641 (1981).
10. R. H. Kraichnan, *J. Math. Phys.* **2**, 124 (1961).
11. R. H. Kraichnan, *Phys. Fluids* **7**, 1723 (1964); **8**, 525 (1965); **8**, 1385 (1965).
12. T. Nakano, *Ann. Phys.* **73**, 326 (1972); T. Tajima, S. Ichimaru, and T. Nakano, *J. Plasma Phys.* **12**, 381 (1974).

NEW INTEGRABLE NONLINEAR EVOLUTION EQUATIONS LEADING TO EXOTIC SOLITONS

YOSHI H. ICHIKAWA
Institute of Plasma Physics
Nagoya University
Nagoya, Japan

KIMIAKI KONNO
Department of Physics
College of Science and Technology
Nihon University
Tokyo, Japan

MIKI WADATI
Institute of Physics
College of General Education
University of Tokyo
Tokyo, Japan

Abstract. *A generalization of the Ablowitz–Kaup–Newell–Segur inverse scattering transformation has confirmed the soliton property for spiky modulations of the traveling wave solution of the derivative nonlinear Schrödinger equation, which describes the circular polarized Alfvén wave. The generalized inverse scattering transformation leads to two types of new integrable nonlinear evolution equation, which bear exotic soliton solutions such as the cusp soliton and the loop soliton.*

1. INTRODUCTION

During the past two decades, great advances have been achieved in the exploration of the chaotic behavior that underlies the dynamics of most nonlinear Hamiltonian systems on long time scales. Self-generation of chaotic behavior is now fully acknowledged in nonlinear mechanics.[1] On the other hand, Fermi, Pasta and Ulam[2] failed to observe the commonly expected chaos in a one-dimensional nonlinear lattice. Their observation of recurrence phenomena led to the discovery of soliton solutions of the Korteweg–de Vries equation,[3] the Toda lattice,[4] and a number of other integrable systems.[5] The Toda lattice has played a key role in the history of studies of nonlinear dynamical systems; it serves as a canonical model in the search for chaos, as well as for integrals of motion.

While carrying out a detailed analysis of the Fermi–Pasta–Ulam recurrence phenomena, Zabusky and Kruskal[3] discovered the novel stability of solitary wave solutions of the Korteweg–de Vries equation and coined the term "soliton" to designate this remarkable stability and individuality. Before the discovery of the soliton, Gardner and Morikawa[6] rediscovered the Korteweg–de Vries equation, which was studied by fluid dynamists a century ago in connection with investigations of the nonlinear propagation of magnetohydrodynamic waves in plasma. The essence of the Gardner–Morikawa approach lies in linking a stationary state of the exact nonlinear system with the large time-asymptotic evolution of the approximately linear system. Generalizing this approach, Taniuti and collaborators[7] formulated the reductive perturbation method and developed a systematic analysis of nonlinear wave propagation in plasmas, lattices, and fluids.

A plasma in the presence of external magnetic fields is a unique medium sustaining many kinds of oscillations, all of which easily attain such large amplitudes that nonlinear effects are as important as dispersive effects. Based on the reductive perturbation theory, one can derive the Korteweg–de Vries equation for the ion–acoustic wave, the cubic nonlinear Schrödinger equation for the electron Langmuir wave, and the modified Korteweg–de Vries equation for the Alfvén wave. Yet, these do not exhaust the possible varieties of nonlinear evolution equations describing nonlinear wave propagation in plasmas. For example, the derivative nonlinear Schrödinger equation describes the propagation of an Alfvén wave parallel to the magnetic field.[8,9]

It was the great success of Ablowitz, Kaup, Newell, and Segur[10] (hereafter referred to as AKNS) to have found a unified scheme called the inverse scattering transformation for solving the first three of the above-named nonlinear evolution equations. The derivative nonlinear Schrödinger equation is not solved by the AKNS scheme; nevertheless, we found its traveling wave solutions.[11,12] Kaup and Newell[13] (hereafter referred to as KN) then discovered a new inverse scattering transform for the new type of nonlinear evolution equation.

According to our analysis,[12] however, our solutions are stationary solutions of a generalized nonlinear Schrödinger equation whose nonlinear term is the sum of the cubic nonlinear term and the derivative nonlinear term. Through a further generalization of the inverse scattering transformation, we[14] showed explicitly that the generalized nonlinear Schrödinger equation is solved by a linear superposition of the AKNS scheme and the KN[13] scheme. Furthermore, within the scheme of our generalization of the inverse scattering transformation, we discovered two new integrable nonlinear evolution equations that have saturating nonlinear terms.[15]

This chapter summarizes recent developments of our studies and discusses novel features of the soliton solutions of these new integrable nonlinear evolution equations.

2. SPIKY SOLITARY ALFVÉN WAVES

The circularly polarized Alfvén wave propagating along the magnetic field obeys the derivative nonlinear Schrödinger equation[8,9]

$$i\frac{\partial}{\partial t}q \pm \mu\frac{\partial^2}{\partial \xi^2}q + i\frac{1}{4}\frac{\partial}{\partial \xi}\{|q|^2 q\} = 0 \tag{1}$$

where q is the complex magnetic field

$$q = B_y \mp iB_z \tag{2}$$

and ξ is the moving coordinate with the Alfvén velocity in the x-direction.

We seek stationary solutions to Eq. (1) describing the nonlinear self-modulation of a large amplitude plane wave.[12] Letting q be

$$q(\xi, t) = \sqrt{8}\,\psi(\xi, t)\exp[i\chi(\xi, \tau)] \tag{3a}$$

$$\chi(\xi, t) = (k\xi - \Omega t) + \theta(y) \tag{3b}$$

$$\psi(\xi, t) = \psi(y) \tag{3c}$$

with

$$y \equiv \xi - \lambda t \tag{3d}$$

we obtain

$$\psi^2(y) \equiv \Phi(y) = \Phi_0 + 8\kappa\gamma^2\beta^{-1}\{\kappa m + \cosh[2\gamma|\mu|^{-1/2}(y - y_0)]\}^{-1} \tag{4a}$$

$$\theta(y) = \theta(y_0) + 3\kappa\frac{|\mu|^{1/2}}{\mu}\arctan\left\{\left(\frac{1-\kappa m}{1+\kappa m}\right)^{1/2}\tanh[\gamma|\mu|^{-1/2}(y - y_0)]\right\}$$
$$+\kappa\delta\frac{|\mu|^{1/2}}{\mu}\arctan\left\{\left(\frac{1-\kappa l}{1+\kappa l}\right)^{1/2}\tanh[\gamma|\mu|^{-1/2}(y - y_0)]\right\} \tag{4b}$$

where

$$\kappa = \pm 1 \tag{5a}$$

$$\delta = \text{sign of } (3\Phi_0 - \lambda - 2\mu k) \tag{5b}$$

$$\alpha = 2\{2\Phi_0 - \lambda - 2(1+\mu)k\} \tag{5c}$$

$$\beta = 4[(\Phi_0 + k)[\lambda + (1+2\mu)k - 2\Phi_0]]^{1/2} \tag{5d}$$

$$\gamma^2 = \tfrac{1}{4}(\lambda - \lambda_1)(\lambda_2 - \lambda) \tag{5e}$$

$$l = \frac{\alpha}{\beta} + \frac{8\gamma^2}{(\beta\Phi_0)} \tag{5f}$$

$$m = \frac{\alpha}{\beta}$$

The allowed range of propagation velocity λ

$$\lambda_1 < \lambda < \lambda_2 \tag{6}$$

where

$$\lambda_1 \equiv 2(2\Phi_0 - \mu k) - 2[\Phi_0(\Phi_0 + k)]^{1/2} \tag{7a}$$

$$\lambda_2 \equiv 2(2\Phi_0 - \mu k) + 2[\Phi_0(\Phi_0 + k)]^{1/2} \tag{7b}$$

Considering the left-polarized Alfvén wave, Fig. 1 illustrates the characteristic feature of the solitary wave solutions given by Eqs. (4a) and (4b). The parameters are chosen arbitrarily as $\kappa = +1$, $\sqrt{\Phi} = 0.5/\sqrt{8}$, $k = 0.01$, and $\mu = 0.5$, $\lambda = 2(2\Phi_0 - k)$. This represents the bright hyperbolic solitary wave, resulting from the self-modulation of a large amplitude plane wave, due to a strong coupling of the amplitude and the phase. When the propagation velocity λ takes the limiting velocities λ_1 or λ_2, Eqs. (7), the hyperbolic solitary wave solution reduces to the algebraic form

$$\psi^2(y) \equiv \Phi(y) = \Phi_0 + \frac{4\rho}{4 + \rho^2|\mu|^{-1}(y - y_0)^2} \tag{8a}$$

$$\theta(y) = \theta(y_0) + \frac{|\mu|^{1/2}}{\mu}\varepsilon \arctan\left(\frac{1}{2}\rho\nu|\mu|^{-1/2}(y - y_0)\right)$$

$$+ \frac{|\mu|^{1/2}}{\mu} 3 \arctan\left(\frac{1}{2}\rho|\mu|^{-1/2}(y - y_0)\right) \tag{8b}$$

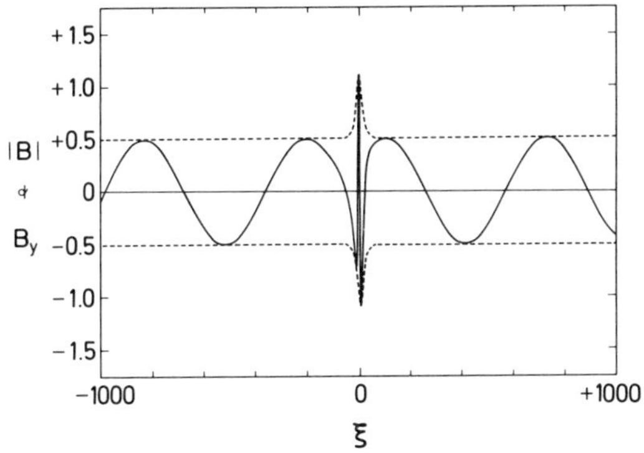

Figure 1. The bright hyperbolic solitary wave of the left-polarized Alfvén wave for a value of $\lambda = 2(2\Phi_0 - \mu k)$ with an arbitrarily chosen phase constant $\theta(y_0) = \pi/6$: solid curve, magnetic field component B_y; dashed line, $|B|$.

where

$$\rho = 4(\Phi_0 + k) + \varepsilon 4\left[\Phi_0(\Phi_0 + k)\right]^{1/2} \tag{9a}$$

$$\nu = \frac{\sqrt{\Phi_0}}{|2\sqrt{\Phi_0 + k} + \varepsilon\sqrt{\Phi_0}|} \tag{9b}$$

$$\varepsilon = +1 \qquad \text{for} \quad \lambda = \lambda_2 \tag{9c}$$

$$\varepsilon = -1 \qquad \text{for} \quad \lambda = \lambda_1 \tag{9c$'$}$$

Figure 2 illustrates the fast algebraic solitary wave moving with the velocity λ_2 for the same values of ψ_0, k, and μ as in Fig. 1.

To show physical differences between the derivative nonlinear Schrödinger equation for the Alfvén wave and the cubic nonlinear Schrödinger equation, we substitute

$$q(\xi, t) = Q(\eta, t)\exp\left[i(k\eta - \mu k^2 t)\right] \tag{10a}$$

with

$$\eta = \xi - 2\mu k t \tag{10b}$$

into Eq. (1), obtaining an equation for the complex amplitude $Q(\eta, t)$:

$$i\frac{\partial}{\partial t}Q + \mu\frac{\partial^2}{\partial \eta^2}Q - \frac{k}{4}|Q|^2 Q + \frac{i}{4}\frac{\partial}{\partial \eta}\{|Q|^2 Q\} = 0 \tag{11}$$

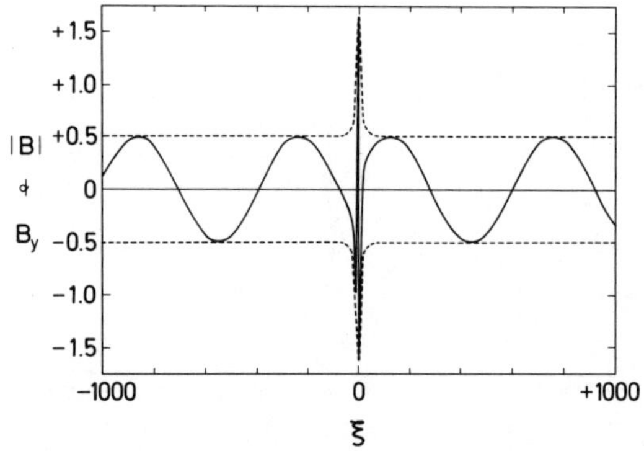

Figure 2. The fast algebraic solitary wave of the left-polarized Alfvén wave with a propagation velocity λ_2; B_y and $|B|$ as in Fig. 1.

Here, the nonlinearity contains both the usual cubic nonlinear term and the derivative nonlinear term. The spiky solitary waves presented as Eqs. (4a), (4b), and (8a), (8b) are the stationary state solution of Eq. (11). At this stage of analysis, however, we cannot call them spiky solitons because Eq. (11) has not yet been shown to be a completely integrable nonlinear evolution equation. We turn to this question in the next section.

3. A GENERALIZATION OF THE INVERSE SCATTERING METHOD

In spite of the success of AKNS[10] in presenting a unified scheme of the inverse scattering transformation for a certain class of nonlinear evolution equations, their scheme does not appear to be general enough to cover large varieties of the integrable nonlinear evolution equation. Indeed, Kaup and Newell[13] have presented another scheme for the derivative nonlinear Schrödinger equation and the massive Thirring equation. Inspired by the solitary wave solutions of the generalized nonlinear Schrödinger equation, Eq. (11), we seek a generalization of the inverse scattering transformation to cover a wider class of nonlinear evolution equations.[14]

We consider the eigenvalue problem

$$\frac{\partial}{\partial x}v_1 + F(\lambda)v_1 = G(\lambda)q(x,t)v_2 \tag{12a}$$

$$\frac{\partial}{\partial x}v_2 - F(\lambda)v_2 = G(\lambda)r(x,t)v_1 \tag{12b}$$

where $F(\lambda)$ and $G(\lambda)$ are functions of the eigenvalue λ. The AKNS scheme is a special case of our generalization, $F(\lambda) = i\lambda$ and $G(\lambda) = 1$. The time dependence of the eigenfunctions is chosen to be

$$\frac{\partial}{\partial t}v_1 = A(\lambda, q, r)v_1 + B(\lambda, q, r)v_2 \tag{13a}$$

$$\frac{\partial}{\partial t}v_2 = C(\lambda, q, r)v_1 - A(\lambda, q, r)v_2 \tag{13b}$$

Noting that $(v_{ix})_t = (v_{it})_x$, $i = 1, 2$, and assuming that the eigenvalues λ are time invariant, we readily find that $A(\lambda, q, r)$, $B(\lambda, q, r)$, and $C(\lambda, q, r)$ satisfy

$$A_x + G(rB - qC) = 0 \tag{14a}$$

$$Gq_t - B_x - 2FB - 2GqA = 0 \tag{14b}$$

$$Gr_t - C_x + 2FC + 2GrA = 0 \tag{14c}$$

The proper choices of A, B, C, F, and G yield various integrable nonlinear evolution equations.

Referring to the AKNS scheme and the KN scheme, we choose

$$F(\lambda) = i\alpha\lambda^2 - \sqrt{2\beta}\lambda \tag{15a}$$

$$G(\lambda) = \alpha\lambda + i\sqrt{\beta/2} \tag{15b}$$

Namely, we consider the superposition of these two schemes of the inverse scattering transformation. Here, α and β are positive constants. Then, with the choice of

$$\begin{aligned} A(\lambda, q, r) = {} & -2i\alpha^2\lambda^4 + 4\alpha\sqrt{2\beta}\,\lambda^3 + (4i\beta - i\alpha^2 rq)\lambda^2 \\ & + \sqrt{2\beta}\,\alpha rq\lambda + i\left(\frac{\beta}{2}\right)rq \end{aligned} \tag{16a}$$

$$\begin{aligned} B(\lambda, q, r) = {} & 2\alpha^2 q\lambda^3 + 3i\sqrt{2\beta}\,\alpha q\lambda^2 + \left(-2\beta q + i\alpha q_x + \alpha^2 rq^2\right)\lambda \\ & + \left(-\sqrt{\beta/2}\,q_x + i\alpha\sqrt{\beta/2}\,rq^2\right) \end{aligned} \tag{16b}$$

$$\begin{aligned} C(\lambda, q, r) = {} & 2\alpha^2 r\lambda^3 + 3i\sqrt{2\beta}\,\alpha r\lambda^2 + \left(-2\beta r - i\alpha r_x + \alpha^2 r^2 q\right)\lambda \\ & + \left(\sqrt{\beta/2}\,r_x + i\alpha\sqrt{\beta/2}\,r^2 q\right) \end{aligned} \tag{16c}$$

we obtain from Eqs. (14b) and (14c) the set of nonlinear evolution equations

$$iq_t + q_{xx} - i\alpha(rq^2)_x + \beta rq^2 = 0 \tag{17a}$$

$$ir_t - r_{xx} - i\alpha(r^2q)_x - \beta r^2q = 0 \tag{17b}$$

If we take $r = \pm q^*$, the set of equations Eqs. (17) is reduced to the generalized nonlinear Schrödinger equation

$$iq_t + q_{xx} \mp i\alpha(|q|^2q)_x \pm \beta|q|^2q = 0 \tag{18}$$

Having established the inverse scattering transformation for Eq. (18), we can show explicitly that Eq. (18) poses a set of conservation laws of infinite number. Thus, we conclude that the spiky solitary waves illustrated in the preceding section are indeed solitons. The analysis above is the first illustration of the solution of the nonlinear evolution equation with the superposition of nonlinear terms by linear superposition of two different schemes of the inverse scattering transformation.

4. NEW INTEGRABLE NONLINEAR EVOLUTION EQUATIONS

Encouraged by the success of our generalization of the inverse scattering transformation, we can explore new types of integrable nonlinear evolution equations.[15]

We assign for $F(\lambda)$ and $G(\lambda)$ of Eqs. (12) the following expressions:

$$F(\lambda) = i\lambda \tag{19a}$$

$$G(\lambda) = \lambda \tag{19b}$$

Then, choosing

$$A = -\frac{2i}{(1 - rq)^{1/2}}\lambda^2 \tag{20a}$$

$$B = 2\frac{q}{(1 - rq)^{1/2}}\lambda^2 + i\left[\frac{q}{(1 - rq)^{1/2}}\right]_x \lambda \tag{20b}$$

$$C = 2\frac{r}{(1 - rq)^{1/2}}\lambda^2 - i\left[\frac{r}{(1 - rq)^{1/2}}\right]_x \lambda \tag{20c}$$

we obtain from Eqs. (14b) and (14c)

$$q_t - i\left[\frac{q}{(1-rq)^{1/2}}\right]_{xx} = 0 \tag{21a}$$

$$r_t + i\left[\frac{r}{(1-rq)^{1/2}}\right]_{xx} = 0 \tag{21b}$$

If we take $r = \pm q^*$, Eqs. (21) are reduced to

$$iq_t + \frac{\partial^2}{\partial x^2}\left\{\frac{q}{(1 \mp |q|^2)^{1/2}}\right\} = 0 \tag{22}$$

Another choice of A, B, and C

$$A = -\frac{4i}{(1-rq)^{1/2}}\lambda^3 + \frac{rq_x - qr_x}{(1-rq)^{3/2}}\lambda^2 \tag{23a}$$

$$B = \frac{4q}{(1-rq)^{1/2}}\lambda^3 + \frac{2iq_x}{(1-rq)^{3/2}}\lambda^2 - \left(\frac{q_x}{(1-rq)^{3/2}}\right)_x \lambda \tag{23b}$$

$$C = \frac{4r}{(1-rq)^{1/2}}\lambda^3 - \frac{2ir_x}{(1-rq)^{3/2}}\lambda^2 - \left(\frac{r_x}{(1-rq)^{3/2}}\right)_x \lambda \tag{23c}$$

leads to the following set of equations:

$$q_t + \frac{\partial^2}{\partial x^2}\left(\frac{1}{(1-rq)^{3/2}}\frac{\partial q}{\partial x}\right) = 0 \tag{24a}$$

$$r_t + \frac{\partial^2}{\partial x^2}\left(\frac{1}{(1-rq)^{3/2}}\frac{\partial r}{\partial x}\right) = 0 \tag{24b}$$

If we take $r = -q$, the set of Eqs. (24) is reduced to

$$\frac{\partial}{\partial t}q + \frac{\partial^2}{\partial x^2}\left(\frac{1}{(1+q^2)^{2/3}}\frac{\partial r}{\partial x}\right) = 0. \tag{25}$$

We notice also, for $r = -1$ and $q = u - 1$, Eqs. (24a) and (24b) are reduced to

$$\frac{\partial}{\partial t}u = 2\frac{\partial^3}{\partial x^3}u^{-1/2} \tag{26}$$

which is the equation discovered by Harry Dym (see Ref. 16).

5. CUSP SOLITONS

As an explicit illustration of the soliton solution of these new integrable nonlinear evolution equations, we discuss a special case of Eqs. (24) with the choice of $r = -1$.[17] The eigenvalue problem of Eqs. (12) reduces to

$$\psi_{xx} + \lambda^2(1+q)\psi = 0 \tag{27}$$

and the temporal evolution of the eigenfunctions becomes

$$\psi_t = 2\lambda^2\left\{\frac{2}{(1+q)^{1/2}}\frac{\partial}{\partial x} - \left[\frac{1}{(1+q)^{1/2}}\right]_x\right\}\psi \tag{28}$$

By assuming $\partial\lambda/\partial t = 0$, the compatibility condition of Eqs. (27) and (28) gives rise to the nonlinear evolution equation

$$\frac{\partial}{\partial t}q = 2\frac{\partial^3}{\partial x^3}(1+q)^{-1/2} \tag{29}$$

which is equivalent to Eq. (26).

Introducing the Jost functions by

$$\phi(\lambda, x) \to e^{-i\lambda x} \quad \text{as} \quad x \to -\infty \tag{30a}$$

$$\psi(\lambda, x) \to e^{i\lambda x} \quad \text{as} \quad x \to +\infty \tag{30b}$$

and the scattering coefficients by

$$\phi(\lambda, x) = a(\lambda)\psi(-\lambda, x) + b(\lambda)\psi(\lambda, x) \tag{31}$$

we examine the analytic properties of $a(\lambda)$ and the Jost functions for large $|\lambda|$. Substituting the expression

$$\phi(\lambda, x) = \exp\left\{-i\lambda x + \int_{-\infty}^{x}\sigma(\lambda, x')\,dx'\right\} \tag{32}$$

into Eq. (27), we get

$$\sigma_x + \sigma^2 - 2i\lambda\sigma + \lambda^2 q = 0 \tag{33}$$

Expanding σ in a power series of λ, we obtain the first two conserved densities as

$$\sigma_{-1} = 1 - (1+q)^{1/2} \tag{34a}$$

$$\sigma_0 = -\frac{1}{4}\frac{\partial}{\partial x}\log(1+q) \tag{34b}$$

Then, we have

$$\log a = \int_{-\infty}^{+\infty} \sigma \, dx = i\lambda\varepsilon + O(\lambda^{-1}) \tag{35}$$

where

$$\varepsilon = \int_{-\infty}^{+\infty} \sigma_{-1} \, dx \tag{36}$$

As $|x| \to \infty$, we have

$$\phi e^{i\lambda(x-\varepsilon_-)} = (1+q)^{-1/4} + O(\lambda^{-1}) \tag{37a}$$

$$\psi e^{-i\lambda(x+\varepsilon_+)} = (1+q)^{-1/4} + O(\lambda^{-1}) \tag{37b}$$

where

$$\varepsilon_+(x) = \int_x^{\infty} \sigma_{-1} \, dx \tag{38a}$$

$$\varepsilon_-(x) = \int_{-\infty}^{x} \sigma_{-1} \, dx \tag{38b}$$

The Gelfand–Levitan equation for the eigenvalue problem Eq. (27) is constructed as

$$K(x, y) - F(x+y) - \int_x^{\infty} K(x, s) F'(s+y) \, ds = 0 \tag{39}$$

for $x \leqslant y$, where $F(z)$ and $F'(z)$ are defined by

$$F(z) = \frac{1}{2\pi} \int_c \frac{b(\lambda)}{a(\lambda)} \frac{1}{i\lambda} \exp(i\lambda(z + 2\varepsilon_+(x)) \, d\lambda \tag{40a}$$

$$F'(z) = \frac{1}{2\pi} \int_c \frac{b(\lambda)}{a(\lambda)} \exp[i\lambda(z + 2\varepsilon_+(x)] \, d\lambda \tag{40b}$$

The potential $q(x)$ is given by a solution $K(x, y)$ as

$$1 + q = [1 - K(x, x)]^{-4} \tag{41}$$

Now, the time dependence of the scattering coefficients are determined from Eq. (28) as

$$a(\lambda, t) = a(\lambda, 0) \tag{42a}$$

$$b(\lambda, t) = b(\lambda, 0) \exp(8i\lambda^3 t) \tag{42b}$$

The zeros of $a(\lambda)$ in the upper-half λ-plane are the bound state eigenvalues. To construct a one-soliton solution, we assume that $a(\lambda)$ has only one simple zero in the upper-half λ-plane and $\rho(\lambda) = a(\lambda)/b(\lambda) = 0$ for real λ. Then, Eqs. (40a) and (40b) become

$$F(z) = \frac{1}{i\lambda_1} C_1(t)\exp\left[i\lambda_1(z + 2\varepsilon_+(x)\right] \tag{43a}$$

$$F'(z) = C_1(t)\exp\left[i\lambda_1(z + 2\varepsilon_+(x)\right] \tag{43b}$$

With the substitution of Eqs. (43) into Eq. (39), we get a solution $K(x, y)$, and finally can write down the potential $q(x, t)$ as (setting $i\lambda_1 = \kappa < 0$)

$$q(x, t) = \tanh^{-4}\left[\kappa\left(x - x_0 - 4\kappa^2 t + \varepsilon_+(x)\right)\right] - 1 \tag{44}$$

Differentiating Eq. (38a) and using Eq. (44), we have

$$\frac{\partial}{\partial x}\left[x - x_0 - 4\kappa^2 t + \varepsilon_+(x)\right] = \tanh^{-2}\left[\kappa\left(x - x_0 - 4\kappa^2 t + \varepsilon_+(x)\right)\right] \tag{45}$$

Comparing Eqs. (44), (45), and (38a), we find that $\varepsilon_+(x)$ should satisfy

$$\kappa\varepsilon_+ = 1 + \tanh\left[\kappa\left(x - x_0 - 4\kappa^2 t + \varepsilon_+\right)\right] \tag{46}$$

to ensure that the soliton $q(x, t)$ varies as $q(x - 4\kappa^2 t)$.

To illustrate the present soliton solution, it is convenient to define a new function $r(x, t)$ in terms of the following relation:

$$(1 + q)^{-1/2} = 1 - r \tag{47}$$

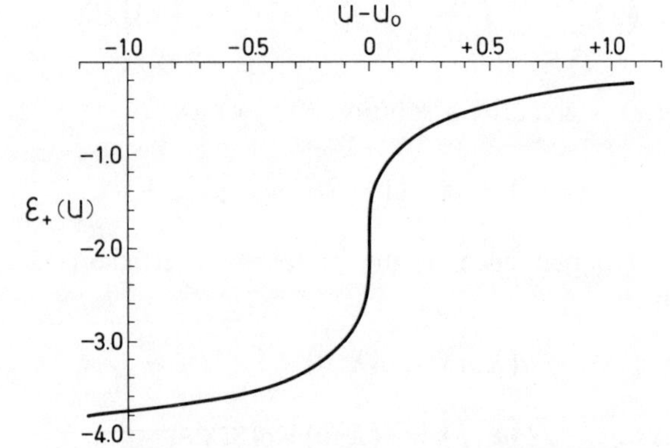

Figure 3. The curve of $\varepsilon_+(u)$ for $\kappa = -\frac{1}{2}$.

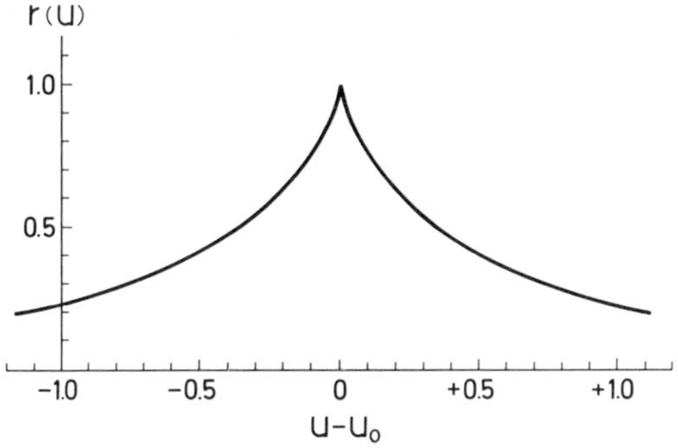

Figure 4. The cusp soliton $r(u)$ for $\kappa = -\frac{1}{2}$.

Equation (29) is then transformed to

$$\frac{\partial}{\partial t}r + (1-r)^3\frac{\partial^3}{\partial x^3}r = 0 \tag{48}$$

for which the one-soliton solution is given by

$$r(x,t) = \operatorname{sech}^2\left[\kappa\left(x - x_0 - 4\kappa^2 t + \varepsilon_+\right)\right] \tag{49}$$

together with Eq. (46). Figure 3 shows explicitly the function $\varepsilon_+(u)$ determined from Eq. (46) for an arbitrary value of $\kappa = -\frac{1}{2}$. Substituting the value of $\varepsilon_+(u)$ into Eq. (49), we obtain the explicit form of the one-soliton solution $r(x,t)$ as shown in Fig. 4. We propose to call this singular type of soliton a cusp soliton.

6. NONLINEAR TRANSVERSE OSCILLATION OF ELASTIC BEAMS

Although these new integrable nonlinear evolution equations have novel features of mathematical interest, it is worthwhile to identify physical problems for which they are relevant. We can see immediately that Eq. (25) could be the key equation for physical problems in which the curvature of deformation has crucial effects.

One such problem is the nonlinear transverse oscillation of elastic beams under tension.[18] We can write down the equations of motion of the small element AB (Fig. 5) as

$$\rho A\frac{\partial^2}{\partial t^2}y = \frac{\partial}{\partial x}S \tag{50a}$$

$$0 = \frac{\partial}{\partial x}M + P\frac{\partial}{\partial x}y + S \tag{50b}$$

where ρ is the density of material, and A is the cross-sectional area; S is the resultant stress parallel to the y-axis, and P is the end thrust parallel to the x-axis. As long as we consider a uniform elastic beam, P is constant.

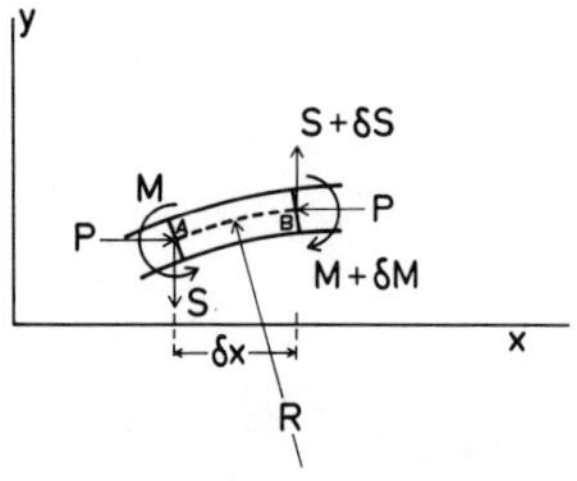

Figure 5. Transverse displacement of elastic beam under the end thrust.

The bending momentum M is given by

$$M = \frac{EI}{R} = EI \frac{1}{\left\{1 + (\partial y/\partial x)^2\right\}^{3/2}} \frac{\partial^2 y}{\partial x^2} \tag{51}$$

where E is Young's modulus, R represents the radius of curvature of bending beam, and I is the moment of inertia of the cross section of beam. When the beam is subject to a tension $P = -\mathfrak{T}$, we obtain the following nonlinear partial differential equation from Eqs. (50a), (50b), and (51).

$$\frac{\partial^2}{\partial t^2} y - \frac{\mathfrak{T}}{\rho A} \frac{\partial^2}{\partial x^2} y + \frac{EI}{\rho A} \frac{\partial^2}{\partial x^2} \left\{ \frac{\partial^2 y/\partial x^2}{\left[1 + (\partial y/\partial x)^2\right]^{3/2}} \right\} = 0 \tag{52}$$

Since we are dealing with the uniform elastic beam, there is no dynamical nonlinear effect, but we have taken fully account of the geometrical nonlinear effect.

Introducing dimensionless variables X, Y, and T by

$$x = A^{1/2} X \tag{53a}$$

$$y = A^{1/2} Y \tag{53b}$$

$$t = \frac{A^{1/2}}{\lambda} T \tag{53c}$$

where $\lambda = (\mathfrak{T}/\rho A)^{1/2}$ is the linear wave velocity, and defining the stretched coordinates

$$\xi = X + T \tag{54a}$$

$$\tau = \varepsilon T \tag{54b}$$

with the dimensionless parameter ε given by

$$\varepsilon = \frac{EI}{2\mathfrak{T} A} \tag{55}$$

we can reduce Eq. (52) through the first order of ε,

$$\frac{\partial}{\partial \tau} \frac{\partial}{\partial \xi} Y + \frac{\partial^2}{\partial \xi^2} \left\{ \frac{\partial^2 Y/\partial \xi^2}{\left[1 + (\partial Y/\partial \xi)^2\right]^{3/2}} \right\} = 0 \tag{56}$$

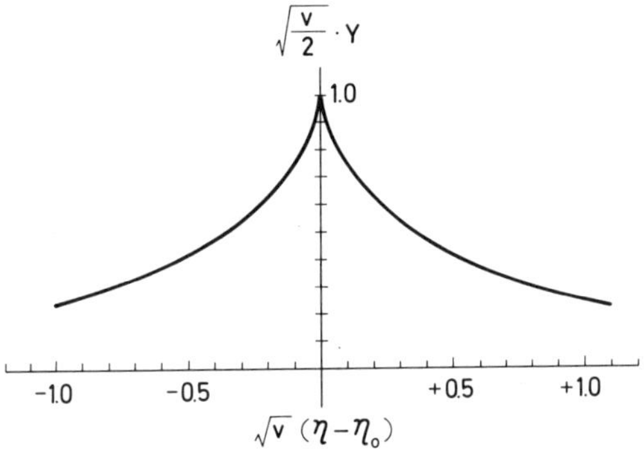

Figure 6. The subsonic solitary wave $Y(\eta)$ given by Eq. (64).

Defining

$$q(\xi, \tau) = \frac{\partial Y}{\partial \xi} \tag{57}$$

we see immediately that Eq. (56) is nothing but the second type of new integrable nonlinear evolution equation [i.e., Eq. (25)].

Although Eq. (56) is integrable by the inverse scattering transformation, we can also examine traveling solutions that depend on the single variable η

$$\eta = \xi \pm v\tau \qquad v > 0 \tag{58}$$

allowing ourselves to examine periodic solutions. Carrying out the integrations, we obtain the expression

$$\pm\sqrt{v}\,(\eta - \eta_0) = -F(\psi, k) + 2E(\psi, k) \tag{59}$$

where the functions $F(\psi, k)$ and $E(\psi, k)$ are the first and the second kind of elliptic, respectively.

The argument ψ is defined by

$$\sin\psi = u = \left(1 - \frac{v}{4k^2} Y^2\right)^{1/2}, \tag{60}$$

while the modulus k is given by

$$k^2 = \tfrac{1}{2}(1 + C), \qquad 1 > C > -1, \tag{61}$$

where C is the integration constant. Taking the limit of $k^2 \to 1$, we investigate a localized solitary solution. Since we have

$$E(\psi, k \to 1) = \sin\psi, \tag{62}$$

$$F(\psi, k \to 1) = \operatorname{sech}^{-1}|\cos\psi|, \tag{63}$$

we can reduce Eq. (59) into the following expression

$$\pm\sqrt{v}\,(\eta-\eta_0) = -\operatorname{sech}^{-1}\left|\frac{\sqrt{v}}{2}Y\right| + 2\left(1-\frac{v}{4}Y^2\right)^{1/2} \tag{64}$$

In Fig. 6, we illustrate the solitary wave solution expressed by Eq. (64), which can be called a cusp soliton.

7. A LOOP SOLITON PROPAGATING ALONG A STRETCHED ROPE

Now, if the beam is flexible enough, it could be deformed into a shape of loop, of which upper half-portion takes the negative curvature. One can easily realize such a situation on a stretched rope. To describe such a case, we need to modify the expression for the curvature in the following way:

$$\begin{aligned}\frac{1}{R} \equiv \frac{d\theta}{ds} &= \frac{dx}{ds}\cdot\frac{d}{dx}(\tan^{-1}y_x)\\ &= \operatorname{sgn}\left(\frac{ds}{dx}\right)\cdot\left\{\left[1+y_x^2\right]^{-3/2}y_{xx}\right\}\end{aligned} \tag{65}$$

where $d\theta$ is an increment of the tangential angle θ at a point of the stretched rope, and ds is an increment of the length of arc.[19] We notice that the increment ds is positive definite, while the displacement dx changes its sign. Thus, the basic equation becomes

$$y_{tt} - y_{xx} + 2\varepsilon\operatorname{sgn}\left(\frac{ds}{dx}\right)\cdot\left\{\left[1+y_x^2\right]^{-3/2}y_{xx}\right\}_{xx} = 0 \tag{66}$$

Here, we regard the factor $\operatorname{sgn}(ds/dx)$ as an index to define the branches of the deformation having the opposite sign of the curvature. If the deformation is small, $\operatorname{sgn}(ds/dx)$ takes a definite sign, and Eq. (66) is reduced to Eq. (65).

Aiming to describe nonlinear propagation of a deformation in one direction along the rope, we introduce stretched coordinates

$$\xi = x + t \tag{67a}$$

$$\tau = \varepsilon t \tag{67b}$$

Thus, through first order in ε, Eq. (66) reduces to

$$\frac{\partial}{\partial\tau}y_\xi + \operatorname{sgn}\left(\frac{ds}{dx}\right)\frac{\partial^2}{\partial\xi^2}\left\{\left[1+y_\xi^2\right]^{-3/2}y_{\xi\xi}\right\} = 0 \tag{68}$$

which is a modified version of the second type of integrable equation derived by us.[18] Modifying our scheme of the inverse scattering transformation, we present a one-soliton solution of Eq. (68) under the boundary condition of $y \to 0$ and $y_\xi \to 0$ as $|\xi| \to \infty$.

Regarding the factor $\text{sgn}(ds/dx)$ as an index to define the branches of deformation, we can write down an inverse scattering transformation for Eq. (68) as follows: the eigenvalue problem is

$$v_{1\xi} + i\lambda v_1 = \lambda y_\xi v_2 \tag{69a}$$

$$v_{2\xi} - i\lambda v_2 = -\lambda y_\xi v_1 \tag{69b}$$

and the temporal evolution of the eigenfunctions is determined by

$$v_{1\xi} = Av_1 + Bv_2 \tag{70a}$$

$$v_{2\tau} = Cv_1 - Av_2 \tag{70b}$$

where

$$A = \text{sgn}\left(\frac{ds}{d\xi}\right)(-4i\lambda^3 Y^{-1}) \tag{71a}$$

$$B = \text{sgn}\left(\frac{ds}{d\xi}\right)\left[4y_\xi\lambda^3 Y^{-1} + 2iy_{\xi\xi}\lambda^2 Y^{-3} - \left(y_{\xi\xi}Y^{-3}\right)_\xi\lambda\right] \tag{71b}$$

$$C = \text{sgn}\left(\frac{ds}{d\xi}\right)\left[-4y_\xi\lambda^3 Y^{-1} + 2iy_{\xi\xi}\lambda^2 Y^{-3} - \left(y_{\xi\xi}Y^{-3}\right)_\xi\lambda\right] \tag{71c}$$

with the abbreviation of $Y = (1 + y_\xi^2)^{1/2}$. The integrability condition for Eqs. (69) and (70) gives rise to Eq. (68), provided $\lambda_\tau = 0$.

Following the same analysis as for the first kind of our nonlinear evolution equation,[17] the Gelfand–Levitan equation can be obtained for the inverse scattering transformation defined by Eqs. (69) and (70) with Eqs. (71). It takes exactly the same form as given in Ref. 17. The only necessary modifications are in the expressions of the conserved densities. The first two conserved densities are

$$\sigma_{-1} = 1 - \text{sgn}\left(\frac{ds}{dx}\right)Y \tag{72a}$$

$$\sigma_0 = \frac{y_{\xi\xi}(1 - \text{sgn}(ds/dx)Y)}{2y_\xi Y^2} \tag{72b}$$

Analytic properties of the Jost functions for large $|\lambda|$ are, for example,

$$\psi = \begin{pmatrix} \dfrac{i\sigma_{-1}}{y_\xi} \\ 1 \end{pmatrix} \exp\{i\lambda(x + \varepsilon_+) - \mu_+\} + O\left(\frac{1}{\lambda}\right) \tag{73}$$

where

$$\varepsilon_+(x) = \int_x^\infty \sigma_{-1}\, dx \tag{74a}$$

$$\mu_+(x) = \int_x^\infty \sigma_0\, dx \tag{74b}$$

The one-soliton solution of Eq. (68) arises out of a bound state located at $\lambda = i\eta$ ($\eta > 0$). Solving the Gelfand–Levitan equation, we obtain

$$y(\xi, \tau) = \eta^{-1}\operatorname{sech}(\delta(\xi, \tau)) \tag{75a}$$

where

$$\delta = 2\eta(u + \varepsilon_+(u)) - 2\delta_0 \tag{75b}$$

$$u = \xi - 2\eta^2\tau \tag{75c}$$

The quantity $2\delta_0$ is an initial phase. To ensure that the one-soliton solution Eq. (75a) is indeed a traveling soliton, we obtain the relation

$$\eta\varepsilon_+(u) = \tanh[2\eta(u + \varepsilon_+(u))] - 1 \tag{76}$$

Referring to Eq. (67) we notice that this one soliton is moving toward the negative direction with subsonic velocity.

As discussed in Section 5, the function $\varepsilon_+(u)$ is determined by solving Eq. (76) numerically. Figure 7 plots the function $\eta\varepsilon_+(u)$ against $2\eta u$. Substituting the value of $\eta\varepsilon_+(u)$ into Eq. (75b), we obtain explicitly the shape of the soliton solution of Eq. (75a). Indeed, we observe that it takes the form of a loop (Fig. 8), which propagates along the stretched rope.

Finally, we notice that if Eq. (68) is expressed in terms of the variables tangential angle θ, and arc length s, it becomes

$$\frac{\partial}{\partial t}\theta + \cos\theta \frac{\partial}{\partial s}\left(\sec\theta \frac{\partial^2}{\partial s^2}\theta\right) = 0 \tag{77}$$

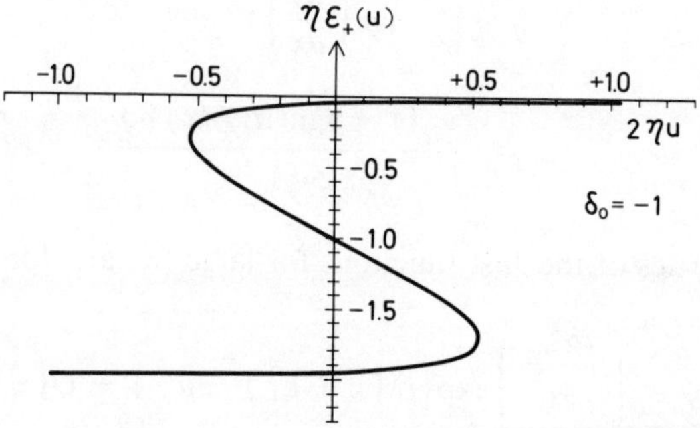

Figure 7. The function $\eta\varepsilon_+(u)$ plotted against the moving coordinate $2\eta u$.

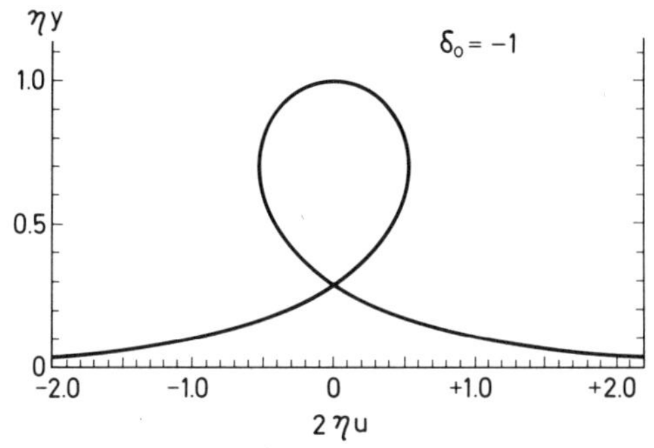

Figure 8. The loop soliton ηy plotted against the moving coordinate $2\eta u$.

which is further reduced to the equation of Euler's elastica

$$\frac{d^2}{ds^2}\theta + a^2 \sin\theta = 0 \tag{78}$$

if the solution is time independent. The solution of Eq. (78) has been given by Love,[20] with the limiting case being the loop shape shown in Fig. 8.

8. CONCLUDING REMARKS

We have surveyed some recent results of our investigation of nonlinear evolution equations. Generalization of the inverse scattering transformation has demonstrated that the linear superposition of two schemes of the inverse scattering transformation solves equation with the sum of two nonlinear terms, each corresponding to one of the schemes of the inverse scattering transformation. We need to explore the extent to which such a superposition works for general nonlinear evolution equation.

In this connection, it is interesting to recall the work of Hirota,[21] who shows that the equation

$$iq_t + \rho q_{xx} + i\sigma q_{xxx} + \delta |q|^2 q + i3\alpha |q|^2 q_x = 0 \tag{79}$$

has N-soliton solutions, provided the condition

$$\alpha\rho = \sigma\delta \tag{80}$$

holds. Equation (79) is a superposition of the nonlinear Schrödinger equation and the complex modified Korteweg–de Vries equation. As can be seen from the analysis of Lamb,[22] Eq. (79) illustrates that the superposition of some nonlinear terms can be linearized by an inverse scattering scheme, which, in this case, is the AKNS scheme. In Section 3 we show that the superposition of different schemes of the inverse scattering transformation can be used to rigorously integrate an evolution equation. It is also interesting to note that Chen, Lee, and Liu[23] have shown the integrability of Eq. (79) for the special case of $\sigma = 0$ and $\delta = 0$, which violates the condition given by Eq. (80).

Studies of these new integrable equations have revealed a challenging problem for the inverse scattering transformation: construction of singular potentials. In particular, we have shown explicitly how to construct exotic solitons such as the cusp soliton and the loop soliton through the inverse scattering transformation. Since these exotic solitons should be awarded citizenship in the soliton empire, the concept of a soliton has been extended far beyond what was expected when Zabusky and Kruskal[3] coined the term in conjunction with the Korteweg–de Vries equation.

Speaking of the future of soliton science, we mention recent progress in the study of the dynamics of two-dimensional solitary vortices. Hasegawa, Mima, and their co-workers have shown that nonlinear behavior of the drift wave is described by:[24,25]

$$\frac{\partial}{\partial t}(\nabla^2\phi - \phi) + v^*\frac{\partial}{\partial x}\phi - \left(\frac{\partial}{\partial y}\phi\frac{\partial}{\partial x} - \frac{\partial}{\partial x}\phi\frac{\partial}{\partial y}\right)\nabla^2\phi = 0 \tag{81}$$

where ϕ is the electrostatic potential, ∇^2 the two-dimensional Laplacian, and v^* the drift velocity in the y-direction due to a density gradient. Equation (81), often called the Hasegawa–Mima equation in plasma physics, also describes the propagation of Rossby waves in an atmospheric pressure system.[26,27] An exact solitary vortex solution has been obtained by Larichev and Reznik,[28] and numerical studies of dynamical properties of this solitary vortex solution have been carried out by Makino, Kamimura, and Taniuti.[29] Indeed, they have numerically confirmed that these vortex solutions behave as stable two-dimensional solitons. This development in the study of multidimensional nonlinear equation brings to mind the early stage of analysis of the Korteweg–de Vries equation. We can expect that in the coming decade or two multidimensional solitons will be fully understood and this new horizon of nonlinear science will be put on a firm mathematical basis.

ACKNOWLEDGMENTS

We gratefully acknowledge the patient support provided by the Institute of Plasma Physics, Nagoya University. We are obliged to many participants of the symposia and workshops on nonlinear wave phenomena held at the Institute during the past several years, in particular to Profs. T. Taniuti, M. Toda, and N. Yajima for their precious discussions concerning our studies.

REFERENCES

1. R. H. G. Helleman, in *Fundamental Problems in Statistical Mechanics*, Vol. 5, E. G. D. Cohen, Ed. (North Holland, Amsterdam and New York, 1980), p. 165.
2. E. Fermi, J. R. Pasta, and S. M. Ulam, Los Alamos Scientific Laboratory Report LA-1940 (1955); also in *Collected Works of Enrico Fermi*, Vol. II, (University of Chicago Press, Chicago, 1965) p. 978.
3. N. J. Zabusky and M. D. Kruskal, *Phys. Rev. Lett.* **15**, 240 (1965).
4. M. Toda, *J. Phys. Soc. Jpn.* **22**, 431 (1967); see also *Theory of Nonlinear Lattice*, Springer Series in Solid-State Sciences, Vol. 20 (Springer-Verlag, New York, 1980).
5. For instance, see *Proceedings of the 1976 Tucson Conference on the Theory and Applications of Solitons*, H. Flaschka and D. W. McLaughlin, Eds., in *Rocky Mt. J. Math.* **8**:1 and 2 (1978).
6. C. S. Gardner and G. M. Morikawa, Courant Institute of Mathematics and Science Report NYO-9082 (1969).
7. T. Taniuti, *Prog. Theor. Phys. Suppl.* **55**, 1 (1974).
8. A. Rogister, *Phys. Fluids* **14**, 2733 (1971).
9. K. Mio, T. Ogino, K. Minami, and S. Takeda, *J. Phys. Soc. Jpn.* **41**, 265 (1976).
10. M. J. Ablowitz, D. J. Kaup, A. C. Newell, and H. Segur, *Stud. Appl. Math.* **53**, 249 (1974).
11. M. Wadati, H. Sanuki, K. Konno, and Y. H. Ichikawa, *Proceedings of the 1976 Tucson Conference on the Theory and Application of Solitons*, H. Flaschka and D. W. McLaughlin, Eds., in *Rocky Mt. Math. J.* **8**, 323 (1978).
12. Y. H. Ichikawa, K. Konno, M. Wadati, and H. Sanuki, *J. Phys. Soc. Jpn.* **48**, 279 (1980).
13. D. J. Kaup and A. C. Newell, *J. Math. Phys.* **19**, 798 (1978).
14. M. Wadati, K. Konno, and Y. H. Ichikawa, *J. Phys. Soc. Jpn.* **46**, 1965 (1979).
15. M. Wadati, K. Konno, and Y. H. Ichikawa, *J. Phys. Soc. Jpn.* **47**, 1698 (1979).
16. M. D. Kruskal, *Lect. Notes Phys.* **38**, 310 (1975).
17. M. Wadati, Y. H. Ichikawa, and T. Shimizu, *Prog. Theor. Phys.* **64**, 1959 (1980).
18. Y. H. Ichikawa, K. Konno, and M. Wadati, *J. Phys. Soc. Jpn.* **50**, 1799 (1981).
19. K. Konno, Y. H. Ichikawa, and M. Wadati, *J. Phys. Soc. Jpn.* **50**, 1025 (1981).
20. A. E. H. Love, *A Treatise on the Mathematical Theory of Elasticity* (Cambridge University Press, Cambridge, England, 1927).
21. R. Hirota, *J. Math. Phys.* **14**, 805 (1973).
22. G. L. Lamb, *J. Math. Phys.* **18**, 1964 (1977).
23. H. H. Chen, Y. C. Lee, and C. S. Liu, *Phys. Scripta* **20**, 490 (1979).
24. A. Hasegawa and K. Mima, *Phys. Rev. Lett.* **39**, 205 (1977); *Phys. Fluids* **21**, 87 (1978).
25. A. Hasegawa, C. G. Maclennan, and Y. Kodama, *Phys. Fluids* **22**, 2122 (1979).
26. H. J. Stewart, *Q. Appl. Math.* **1**, 262 (1943).
27. G. K. Morikawa, *J. Meteorol.* **17**, 148 (1960).
28. V. D. Larichev and G. K. Reznik, *Polymode News* **19**, 3 (1976).
29. M. Makino, T. Kamimura, and T. Taniuti, *J. Phys. Soc. Jpn.* **50**, 980 (1981).

ON DAVYDOV'S α-HELIX SOLITONS

J. M. HYMAN AND D. W. McLAUGHLIN
Research Group T-7
Los Alamos Scientific Laboratory
Los Alamos, New Mexico

A. C. SCOTT
Department of Electrical and Computer Engineering
University of Wisconsin
Madison, Wisconsin

Abstract. *Numerical studies of Davydov's nonlinear dynamic model for the α-helix protein confirm his prediction of soliton formation. These solitons are robust, localized, dynamic entities that couple molecular (amide-I) vibrations to longitudinal sound waves; they may provide an efficient mechanism for energy transport in biological systems. Both the numerical studies and analytical computations show a threshold level of nonlinearity below which solitons will not form. A rough estimate indicates that this nonlinearity has the required order of magnitude.*

1. INTRODUCTION

"How can energy be transmitted in biological systems?" This basic question was discussed in depth at a meeting of the New York Academy of Sciences in 1973, amid talk of a "crisis in bioenergetics."[1] A central issue in the "crisis" is that the attractive mechanism of energy transduction via excited molecular vibrations is presumed (on the basis of a *linear* dynamic analysis) to have an unacceptably short lifetime; but a potential resolution of this objection has been proposed by Davydov.[2] He suggests that the *nonlinear* character of interatomic forces (e.g., the hydrogen bond) can lead to the formation of robust solitary waves (often called "solitons"[3]), which exhibit greatly increased radiative lifetimes and a correspondingly increased ability to transport energy over large distances.

As a specific context for the development of his idea, Davydov has concentrated on the α-helix protein and has chosen the relatively isolated amide-I (or C=O "stretch") vibration of the peptide group as the main "basket" in which energy is carried. According to a linear analysis, energy transported by this means should spread out from the effects of dispersion and should rapidly become disorganized and lost as a source for biological mechanisms. But in the nonlinear analysis of Davydov, propagation of amide-I vibrations is retroactively coupled to longitudinal sound waves of the α-helix, and the coupled excitation propagates as a localized and dynamically self-sufficient entity called a soliton. The amide-I vibrations generate longitudinal sound waves that in turn provide a potential well that prevents vibrational dispersion; thus the soliton holds itself together.

For such a coupled excitation to be viable, certain "threshold" conditions must be satisfied. The nonlinear coupling between amide-I vibrations and nonlinear sound waves must be sufficiently strong and the amide-I vibrations must be energetic enough for the retroactive interaction to "take hold." Below this threshold, a soliton cannot form and the dynamic behavior will be essentially linear. Above threshold the soliton is a possible mechanism for lossless energy transduction.

This chapter reports on a numerical study of Davydov's fundamental equations that confirms his analytical results. A sharp threshold between linear (dispersive) behavior and nonlinear (soliton) formation is clearly seen, and this threshold is related to fundamental physical parameters describing the α-helix protein. In Section 2 we describe for the general reader the basis for these numerical computations. To this end each term in Davydov's model is physically described with reference to the basic atomic structure. Section 3 displays our main numerical observations with emphasis on their physical significance. Finally we summarize our results and discuss some important open questions in this new area of *nonlinear biomolecular dynamics*. All mathematical discussions are presented in appendices, not because we feel that these are unimportant but to make the scientific logic of Davydov's theory as clear and as widely understandable as we can. This theory may, after, all, help to resolve the "crisis in bioenergetics."

2. DAVYDOV'S MODEL FOR α-HELIX PROTEIN

The atomic structure of α-helix protein is sketched (as a stereogram) in Fig. 1. The basic helix follows the sequence:

etc. —N—C—C—N—C—C—N— etc.

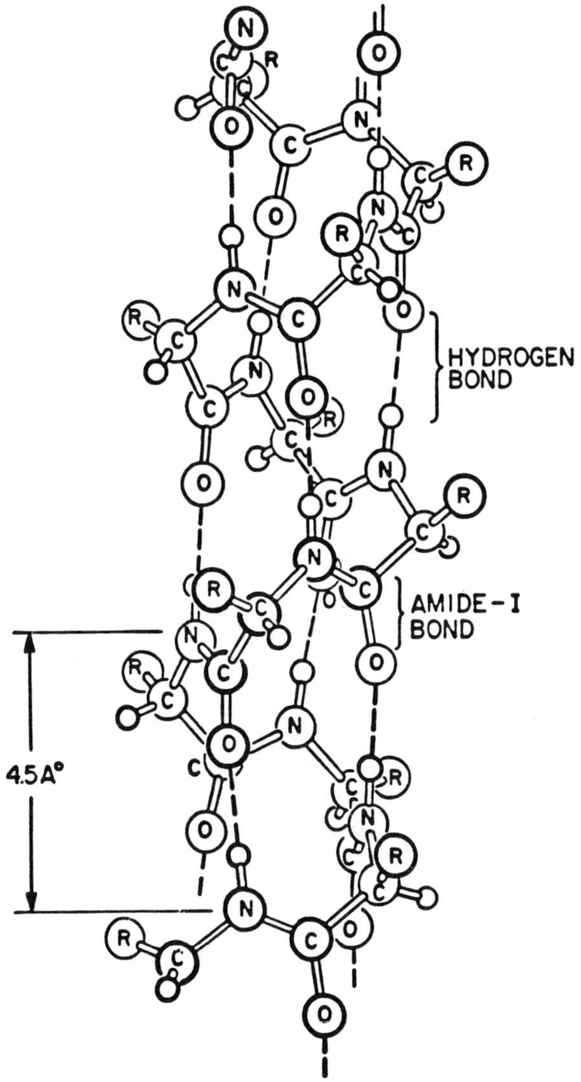

Figure 1. Stereogram of α-helix protein.

with a pitch of 5.4 Å. Superimposed on this basic structure are three "spines," which are almost longitudinal with the sequence:

etc. ———N—C=O———N—C=O———N—C=O——— etc.

where "O=C" represents the locus of the amide-I vibration and "O———N" is the longitudinal hydrogen bond that holds the structure in its helical form. Davydov's equations describe propagation along these three spines of amide-I bond energy and longitudinal sound waves. Nonlinearity of the hydrogen bond leads to coupling of these two propagating systems and, if certain threshold conditions are satisfied, the formation of a soliton.

Let us begin by considering the equations that Davydov has derived to describe propagation along the three spines. From[4] these are:

$$\begin{aligned} i\hbar\frac{da_{n\alpha}}{dt} = {} & [\mathcal{E}_0 + W + \chi_1(\beta_{n+1,\alpha} - \beta_{n-1,\alpha})]a_{n\alpha} \\ & -J(a_{n-1,\alpha} + a_{n+1,\alpha}) + L(a_{n,\alpha+1} + a_{n,\alpha-1}) \\ & +\chi_2[\beta_{n+1,\alpha}a_{n+1,\alpha} - \beta_{n-1,\alpha}a_{n-1,\alpha} - \beta_{n\alpha}(a_{n+1,\alpha} - a_{n-1,\alpha})] \end{aligned} \tag{1}$$

$$\begin{aligned} M\frac{d^2\beta_{n\alpha}}{dt^2} - w(\beta_{n+1,\alpha} - 2\beta_{n\alpha} + \beta_{n-1\alpha}) = {} & \chi_1(|a_{n+1,\alpha}|^2 - |a_{n-1,\alpha}|^2) \\ & +\chi_2[a^*_{n\alpha}(a_{n+1,\alpha} - a_{n-1,\alpha}) \\ & +(a^*_{n+1,\alpha} - a^*_{n-1,\alpha})a_{n\alpha}] \end{aligned} \tag{2}$$

$$W \equiv \frac{1}{2}\sum_{n,\alpha}\left[M\left(\frac{d\beta_{n\alpha}}{dt}\right)^2 + w(\beta_{n\alpha} - \beta_{n-1,\alpha})^2\right] \tag{3}$$

Broadly speaking, Eq. (1) describes the propagation of amide-I vibrations via dipole–dipole interactions and Eq. (2) represents the propagation of longitudinal sound. The total longitudinal sound energy is defined in Eq. (3). Each term, when individually considered, is quite plausible.

2.1. Subscripts

There are two subscripts to the dynamical variables, n and α. These run over the ranges:

$$n = -1, 0, 1, 2, \ldots, n_{\text{max}}$$

$$\alpha = 1, 2, 3$$

Thus n specifies a particular unit cell along a spine and α chooses a particular spine.

2.2. Bond Occupation Amplitude $A_{n\alpha}$

Consider Eq. (1) with the nonlinear coefficients χ_1 and χ_2, the dipole–dipole coupling coefficients J and L, and the sound energy W set equal to zero. Then we have

$$i\hbar\frac{da_{n\alpha}}{dt} \doteq \mathcal{E}_0 a_{n\alpha} \tag{1'}$$

In this equation

$$|a_{n\alpha}|^2 = a_{n\alpha}a_{n\alpha}^*$$

represents the probability of finding a quantum of bond energy $\mathcal{E}_0$ at unit cell n on spine α. If for the sum of such probabilities we have

$$\sum_{n,\alpha} |a_{n\alpha}|^2 = 1$$

a single quantum of amide-I bond energy is present on the helix.

Equation (1′) is the quantum dynamical description of a simple oscillator. It says that the magnitude of $a_{n\alpha}$ remains constant and its phase progresses linearly with time as

$$a_{n\alpha}(t) = a_{n\alpha}(0)\exp\left(\frac{-i\mathcal{E}_0 t}{\hbar}\right)$$

From Ref. 5 $\mathcal{E}_0 \doteq 1650\ \text{cm}^{-1}$ in spectrographic units,* thus

$$\mathcal{E}_0 = .205\ \text{eV}$$

$$= 0.328 \times 10^{-19}\ \text{J}$$

2.3. Longitudinal Displacement $\beta_{n\alpha}$

Consider Eq. (2) with the nonlinear coefficients χ_1 and χ_2 set equal to zero. Then

$$M\frac{d^2\beta_{n\alpha}}{dt^2} - w(\beta_{n+1,\alpha} - 2\beta_{n\alpha} + \beta_{n-1,\alpha}) = 0 \qquad (2')$$

This is a linear equation for longitudinal sound propation on the helix, where $\beta_{n\alpha}$ is the displacement of unit cell n on spine α from its equilibrium position and M is the mass of (see Fig. 1)

$$2\text{C} + \text{O} + \text{N} + \text{H} + \text{R}$$

For the computations reported here, we (quite arbitrarily) take R to be CH_3. Thus

$$\text{M} = 70 \times \text{mass of proton}$$

$$= 1.17 \times 10^{-25}\ \text{kg}$$

The parameter w in Eq. (2′) gives the linear restoring force per unit of

*One electron-volt (eV) = $8065.5\ \text{cm}^{-1} = 1.602 \times 10^{-19}$ joule (J).

hydrogen bond stretching. From Ref. 6 a somewhat similar bond is said to have a force constant of 0.76 mdyne/Å.* Thus we take

$$w = 76 \text{ N/m}$$

From Eq. (2′) the longitudinal sound speed is $[w/M]^{1/2}$ times the longitudinal distance between unit cells. Since the pitch of the helix is 5.4 Å, corresponding to 3.6 spines, the length of a single unit cell along one spine is 4.5 Å. Thus

$$\text{sound speed} = 1.15 \times 10^{4} \text{ m/sec}$$

2.4. Dipole–Dipole Coupling

If Eq. (1) is considered in the approximation that the sound energy W and the nonlinear coefficients χ_1 and χ_2 are zero, it can be written in the form

$$i\hbar \frac{da_{n\alpha}}{dt} + J(a_{n-1,\alpha} - 2a_{n\alpha} + a_{n-1,\alpha}) - \mathcal{E}_0 a_{n\alpha} = -2Ja_{n\alpha} + L(a_{n,\alpha+1} + a_{n,\alpha-1}) \tag{1''}$$

These terms with coefficients J and L represent the effects of dipole–dipole couplings between the amide-I vibrations. The particular form presented in Eq. (1″) emphasizes that the effect of the "J-term" is to provide a mechanism for longitudinal propagation of bond energy. Indeed if the left-hand side of Eq. (1″) were zero, it would be satisfied by a plane wave of probability amplitude propagating in a dispersive manner.

The "J-term" represents dipole–dipole coupling between a particular amide-I bond and its next neighbors in the longitudinal direction. The "L-term" represents a corresponding coupling to lateral neighbors. Fortunately for our numerical studies, the values for these coupling coefficients have been calculated (and checked for their effects on infrared spectra) as[5]:

$$J = 7.8 \text{ cm}^{-1} = 1.55 \times 10^{-22} \text{ J}$$

and

$$L = 12.4 \text{ cm}^{-1} = 2.46 \times 10^{-22} \text{ J}$$

2.5. Nonlinear Coefficients χ_1 and χ_2

The nonlinear coefficients χ_1 and χ_2 represent anharmonicity in the longitudinal hydrogen bonds. Their effect is to provide nonlinear coupling between the longitudinal sound waves Eq. (2′) and dispersive propagation of amide-I bond

*One dyne = 10^{-5} newton (N).

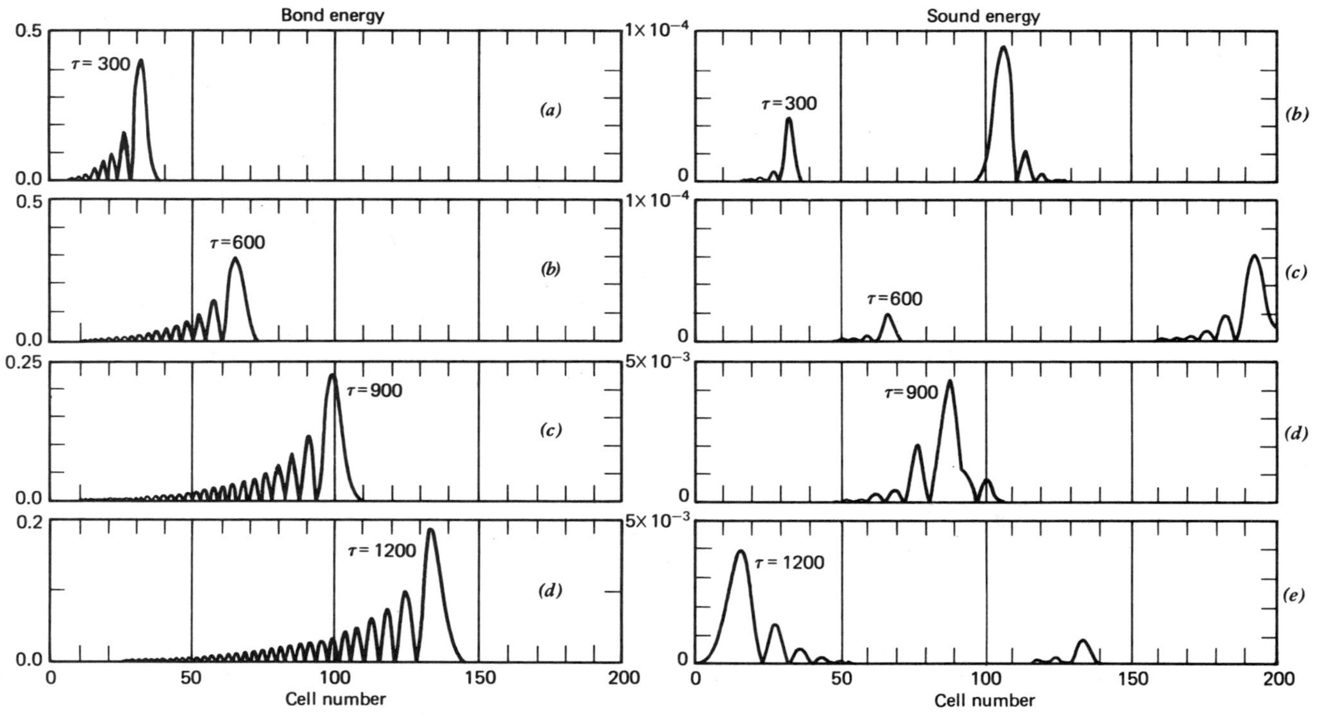

Figure 2. Symmetric three-spine excitation at $\chi = 10^{-11}$ N.

and

$$U(2) \equiv \sum_{\alpha} \left[\left(\frac{dB_{n\alpha}}{d\tau} \right)^2 + (B_{n+1,\alpha} - B_{n,\alpha})^2 \right]$$

Consider first the result for $\tau = 300$ (Fig. 2*a*). The bond energy $U(1)$ has dispersed somewhat and is moving away from the point of initiation at a normalized speed of about 0.1. The longitudinal sound energy $U(2)$ consists of two distinct components: a "fast" component traveling at the limiting sound speed and therefore found at $n = 100$, and a "slow" component that is locked to the bond energy. Does the interaction of bond energy and slow sound constitute of soliton? Figures 2*a* ($\tau = 300$), 2*b* ($\tau = 600$), 2*c* ($\tau = 900$), and 2*d* ($\tau = 1200$), indicate that the bond energy does *not* settle into the hyperbolic secant shape that characterizes a soliton.[2] On the contrary, it continues to disperse until at $\tau = 1500$ (Fig. 2*f*) it has spread itself over half the molecule.

To see how the bond energy dispersion at $\chi = 10^{-11}$ N differs from linear dispersion, turn to Fig. 3, where the computation is repeated for the case $\chi = 0$. Note that the linear bond dispersions at $\tau = 600$ (Fig. 3*a*) and at $\tau = 900$ (Fig. 3*b*) are *identical* to those in Figs. 2*c* and 2*d*. Thus we must conclude that nonlinear coupling between amide-I energy and sound energy plays no role in the computations of Fig. 2. The threshold level has not been attained; solitons have not formed.

If the nonlinearity parameter is raised an order of magnitude, to the level $\chi = 10^{-10}$ N, the dynamic behavior of the bond energy is strikingly different. As Fig. 4 clearly shows, it no longer disperses but propagates along the helix with a fixed shape and a normalized velocity of 0.132. In this case the level of

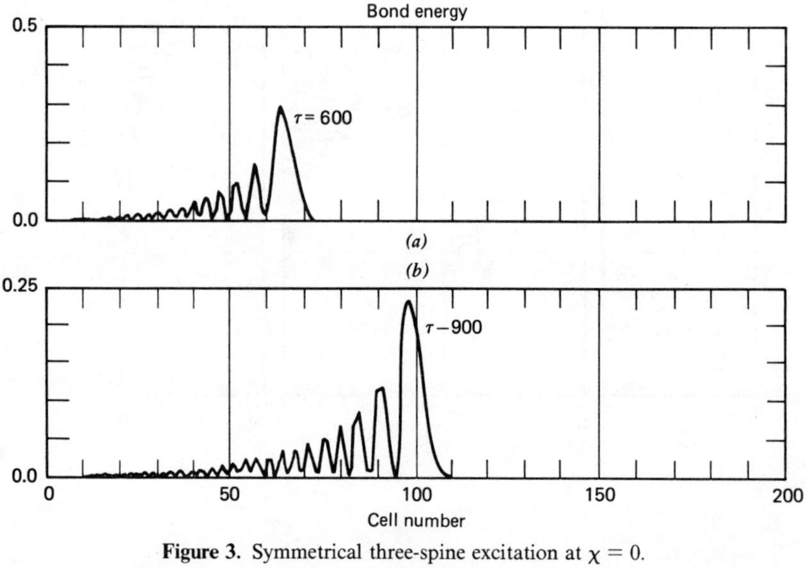

Figure 3. Symmetrical three-spine excitation at $\chi = 0$.

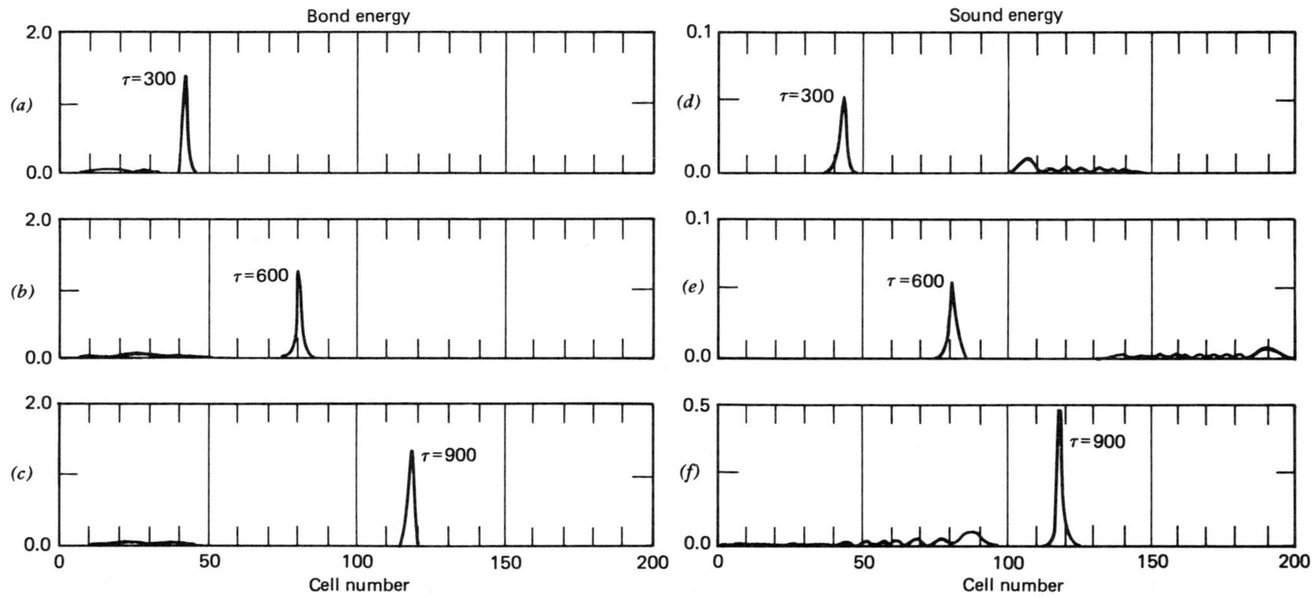

Figure 4. Symmetrical three-spine excitation at $\chi = 10^{-10}$ N.

amide-I bond excitation and the nonlinearity are great enough to permit the sound energy to form a "potential well," which holds the bond energy together. The threshold level is exceeded and Davydov's soliton is observed.

How sharp is this threshold? To answer this we present expanded plots of bond energy for $\chi = 3 \times 10^{-11}$ N (Fig. 5) and $\chi = 5 \times 10^{-11}$ N (Fig. 6). Now we see that the soliton is just beginning to form at the lower level and is well developed at the upper level. The detailed computations in Appendix A indicate that the threshold is inversely proportional to both χ and the energy of bond excitation. Thus if N is the number of amide-I quanta introduced onto a single spine,

$$\chi N > 3 \times 10^{-11}\ \mathrm{N} \tag{6}$$

should be a useful threshold condition for the formation of Davydov's solitons.

To this point our computations have been entirely for symmetric initial conditions; a quantum of the amide-I vibration has been placed on each of the three spines. Now we begin with a single quantum on a single spine. Results for $\chi = 10^{-10}$ N are displayed in Fig. 7. Comparison of Fig. 7*f* with Fig. 4*d* shows that a soliton has indeed developed, but its speed is about 85% of that observed in the symmetrical case. Furthermore the radiation of longitudinal sound from the soliton is much stronger with single-spine excitation. This sound radiation appears to be related to slow relaxation into a more symmetric

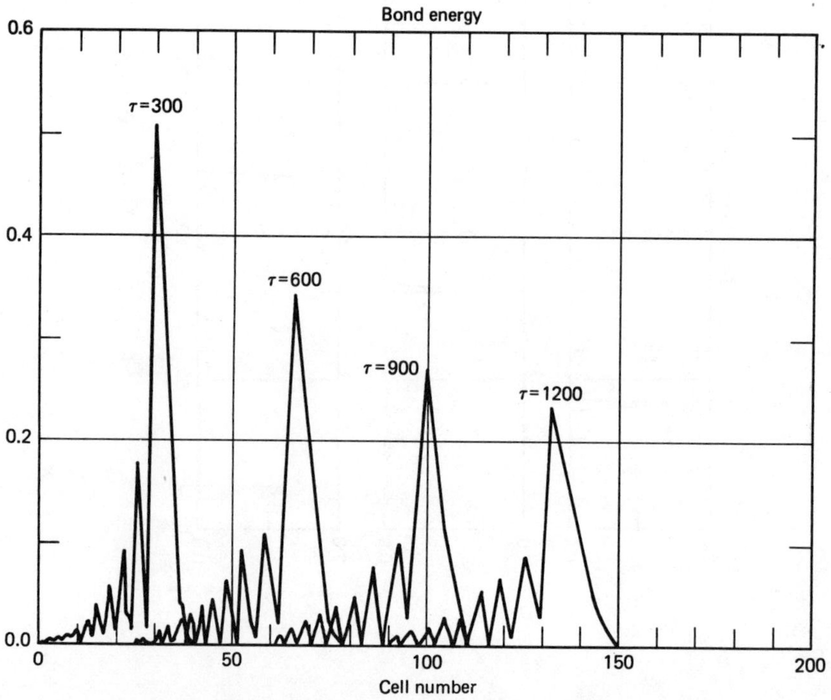

Figure 5. Symmetrical three-spine excitation at $\chi = 3 \times 10^{-11}$ N.

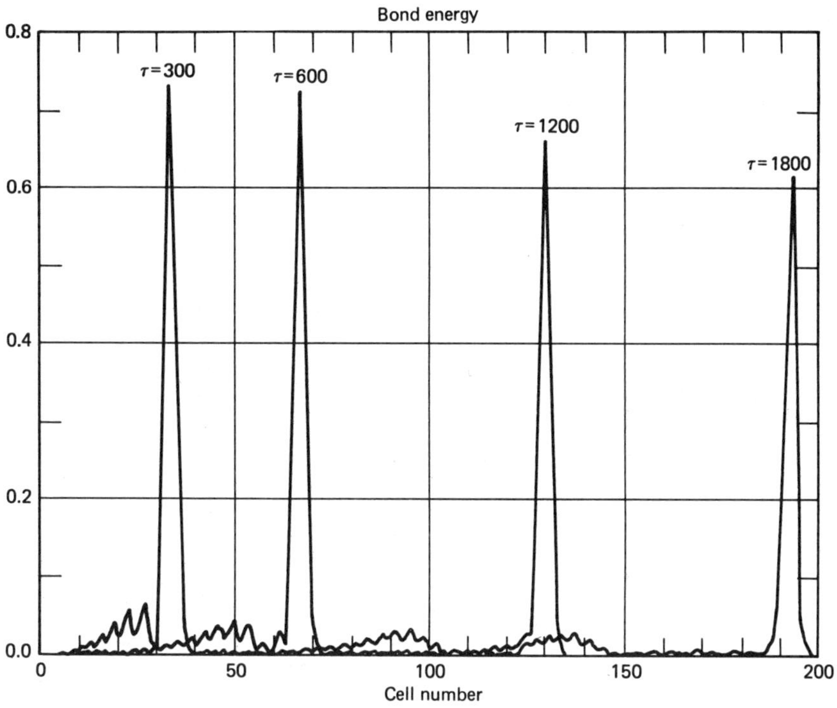

Figure 6. Symmetrical three-spine excitation at $\chi = 5 \times 10^{-11}$ N.

configuration. To illustrate this effect Fig. 8 plots the peak bond energy in the excited spine and that in one of the unexcited spines as a function τ.

Finally we consider the influence of single-spine excitation upon the threshold for soliton formation. For $\chi = 5 \times 10^{-11}$ N the appropriate data are presented in Fig. 9, which shows that a soliton does not develop; rather, the packet of bond energy emits "bursts" of longitudinal sound. If, however, the nonlinear parameter is raised to $\chi = 7 \times 10^{-11}$ N, the data of Fig. 10 clearly show the development of a soliton.

4. MECHANICAL BENDING OF THE HELIX

Davydov has suggested that one effect of soliton propagation along an α-helix might be to cause a mechanical bend (kink?) of the helix.[2] Such an effect would not appear for the symmetric (or three-spine) solitons described in Section 3 because each spine would be elongated equally. However the system composed of Eqs. (1‴) and (2″) can also support an antisymmetric soliton for which

$$A_{n1} = -A_{n2}$$

$$A_{n3} = 0$$

Since the longitudinal sound is insensitive to the phase of $A_{n\alpha'}$ this mode would elongate spines 1 and 2 but would leave spine 3 unchanged.

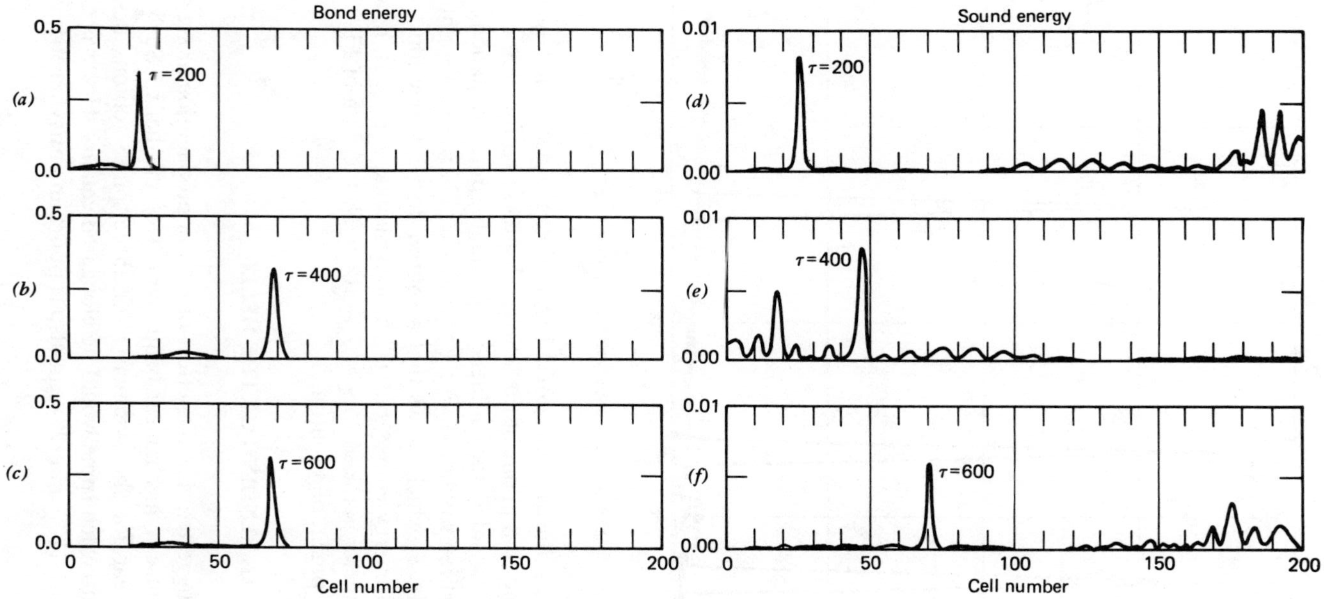

Figure 7. Single-spine excitation at $\chi = 10^{-10}$ N.

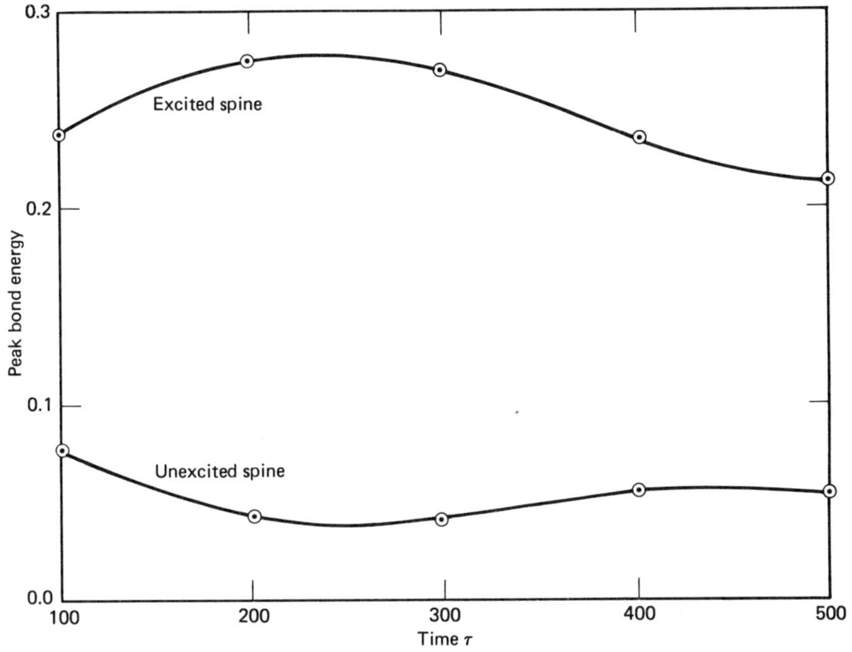

Figure 8. Bond energy in excited spine and unexpected spine versus time for single-spine excitation with $\chi = 10^{-10}$ N.

The amount of such elongation is readily computed from the results developed in Appendix A. From Eq. (A2) we have

$$\rho = 0.264 \frac{10^{10}\chi}{1 - s^2} |A|^2$$

and the total elongation is

$$\int \rho \, d\xi = 0.264 \frac{(10^{10}\chi)}{1 - s^2} \int A^2 \, d\xi$$

$$= 0.264 \frac{(10^{10}\chi)}{1 - s^2} \qquad (7)$$

where N is the total number of amide-I quanta being carried along a single spine by the soliton and s is the soliton speed.

Since the radius of α-helix is 2.8 Å, the turning radius for the bend is $(1 + \sqrt{3}/2)2.8 = 5.23$ Å. Also the right-hand side of Eq. (7) is in units of 0.1 Å [see Eq. (5b)]. Thus the angle (θ) of the bend will be

$$\theta = \tan^{-1}\left[0.00505 N \frac{10^{10}\chi}{1 - s^2}\right]$$

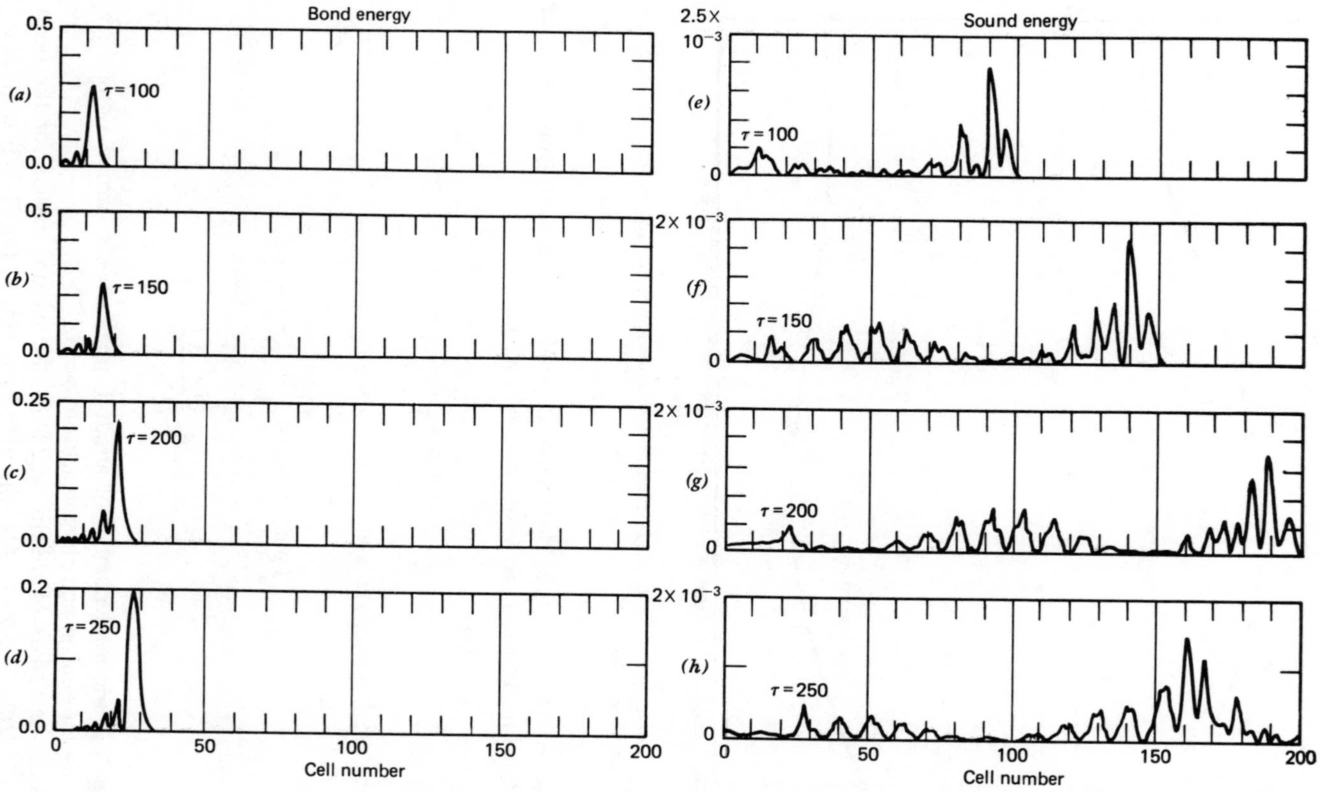

Figure 9. Single-spine excitation at $\chi = 5 \times 10^{-11}$ N.

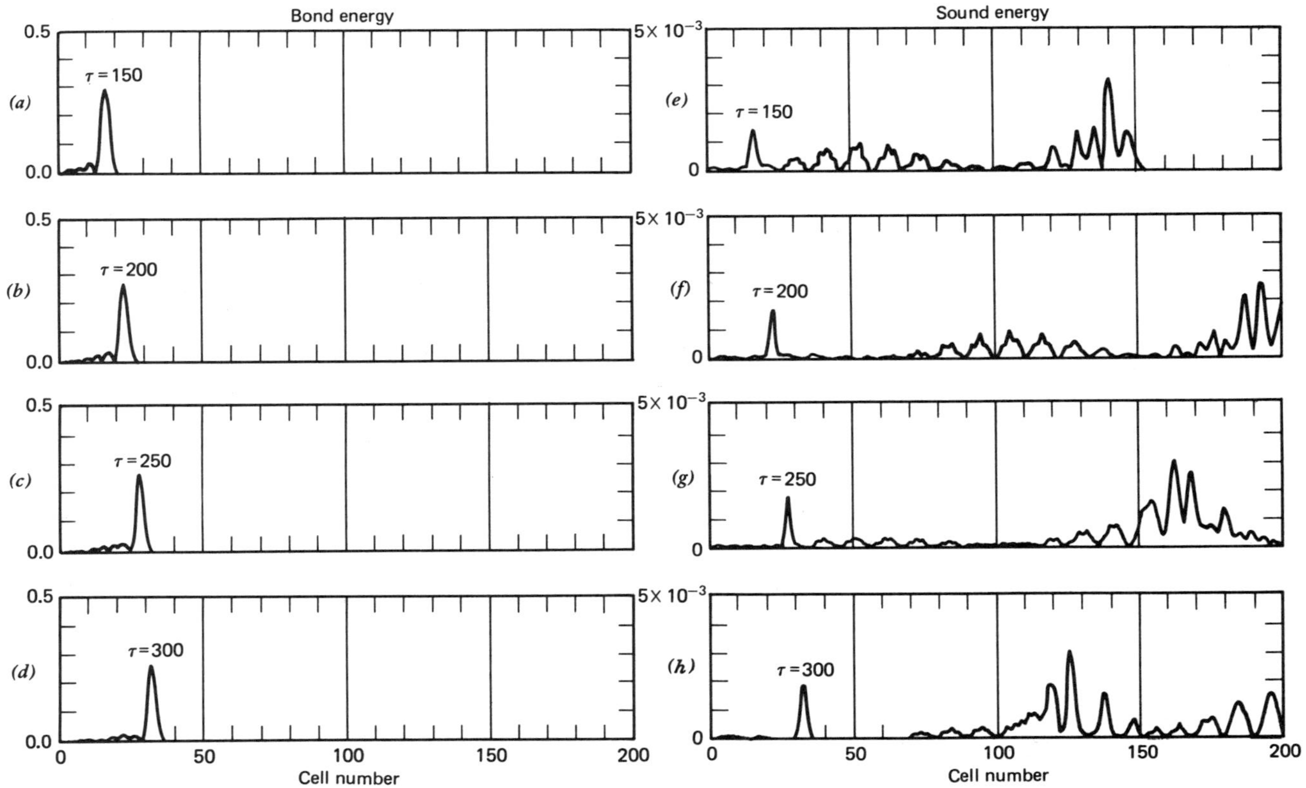

Figure 10. Single-spine excitation at $\chi = 7 \times 10^{-11}$ N.

To a good approximation, therefore, the angle of bend will be

$$\theta \doteq 0.29N(10^{10}\chi) \text{ deg} \tag{8}$$

The same angle of bend would be observed for N quanta propagating along a single spine. However, as we noted in connection with the single-spine computations presented in Figs. 7 and 9, such an excitation tends to relax into the symmetric mode.

5. SUMMARY OF RESULTS

5.1. Threshold for Soliton Formation

Our numerical and analytical studies show that with symmetrical (three-spine) excitation of the initial cell, the threshold level of nonlinearity for soliton formation is

$$\chi > \frac{3}{N} \times 10^{-11} \text{ N}$$

where N is the number of amide-I quanta introduced onto each spine. Comparing this result with our order estimates

$$\chi > 2 - 6 \times 10^{-11} \text{ N}$$

for the nonlinear parameter, we see a possibility of soliton formation with a single quantum on each spine. As the number of quanta introduced becomes larger, the likelihood of soliton formation increases. In this connection it is important to note that the 0.5 eV released in each event of ATP hydrolysis is more than enough to introduce two quanta into an amide-I bond.

5.2. Soliton Speed

From both numerical computations and analytical calculations we find the soliton speed near threshold to be almost equal to the group velocity of a linear pulse below threshold. From our numerical computations it is 0.11 of longitudinal sound speed or

$$\text{soliton speed} \sim 1.26 \times 10^{3} \text{ m/sec}$$

Thus the time required for a soliton to traverse 1000 Å (about the length of a typical myosin molecule in striated muscle) is about 80 psec.

5.3. Mechanical Bending

The mechanical bending of the α-helix under symmetrical excitation is zero, but if the soliton is antisymmetric (in the sense defined by Davydov) or if all the bond energy is confined to a single spine, the bend angle is approximately $\theta = 0.45N(10^{10}\chi)$. This could be a significant effect for several amide-I quanta in the soliton and a nonlinear parameter somewhat larger than the values estimated in Appendix B.

6. OPEN QUESTIONS

In an exploratory study such as this it is as important to indicate what we have *not* shown as it is to itemize our results. Davydov has clearly changed the question posed at the beginning of this chapter to: "Is biological energy transmitted by solitons?" A definitive answer is not yet available, however. Indeed it is rather exciting to await future scientific developments that should indicate whether the *real* level of nonlinearity in a α-helix is approximately that estimated in Appendix B (indicating that solitons should easily form at the single quantum level), or substantially smaller (indicating that several quanta must participate to form a soliton). We feel that the following questions should be given high priority.

1. *Additional Numerical Studies.* The numerical studies presented here are not complete, and additional investigations should include the following: (*a*) a more careful study of relaxation from single-spine excitation, (*b*) more general initial conditions, to initialize more than one cell and to excite Davydov's antisymmetric mode, (*c*) inclusion of additional dipole–dipole coupling terms from Ref. 5, (*d*) augmentation of Davydov's equations to include additional degrees of freedom, and (*e*) study of soliton propagation through a nonuniform α-helix.

2. *The Level of Anharmonicity.* Every effort should be given to obtain better experimental measurements and theoretical estimates of the anharmonicity (χ) in the hydrogen bonds of α-helix protein. We have not found really satisfactory estimates from the literature[7] and present the order estimates of Appendix B as a rough guide. But we are not biochemists (nor chemists, even) so relevant information may be lying about. The level of anharmonicity is the most important fact in nonlinear biomolecular dynamics.

ACKNOWLEDGMENTS

It is a pleasure to thank Profs. A. S. Davydov, C. Sandorfy, and D. E. Green for stimulating private discussions of this problem and Prof. M. Sundaralingam for several interesting comments on the structure of α-helix protein.

APPENDIX A: THE INITIAL VALUE PROBLEM IN LINEAR AND NONLINEAR LIMITS

The original difference–differential equations, Eqs. (1‴) and (2″), are approximated as partial differential equations and studied analytically. In the linear limit, a Fourier transform solution is discussed. In the nonlinear limit, the inverse scattering transform[8] is used to find the threshold for soliton creation and soliton speed.

From the numerical results presented above, it is evident that several unit cells participate in the structure of each soliton. Thus, as Davydov et al. have shown,[4] Eqs. (1‴) and (2″) can be approximated as

$$i\frac{\partial A}{\partial \tau} + 0.058\frac{\partial^2 A}{\partial \xi^2} - F(\tau)A \doteq 0.744(10^{10}\chi)\rho A \tag{A1}$$

$$\frac{\partial^2 \rho}{\partial \tau^2} - \frac{\partial^2 \rho}{\partial \xi^2} \doteq +0.264(10^{10}\chi)\frac{\partial^2}{\partial \xi^2}|A|^2 \tag{A2}$$

where ξ is a continuous variable approximating the longitudinal index n. Also

$$\rho \equiv -\frac{\partial B}{\partial \xi}$$

and

$$F(\tau) \equiv 0.068 + 1.41\sum_{\alpha}\int\left[\left(\frac{\partial B_\alpha}{\partial \tau}\right)^2 + \left(\frac{\partial B_\alpha}{\partial \xi}\right)^2\right]d\xi$$

The term $F(\tau)A$ can be eliminated from Eq. (A1) by adjusting the phase of A as

$$A = \mathcal{A}\exp\left[-i\int^{\tau}F(\tau')\,d\tau'\right]$$

The numerical results also show that soliton speed is slow compared with sound speed. Thus Eq. (A2) becomes approximately

$$\rho = \frac{0.264(10^{10}\chi)}{1-s^2}|\mathcal{A}|^2 \tag{A3}$$

where s is the wave speed. With these approximations, Eq. (A1) takes the form

$$i\frac{\partial \mathcal{A}}{\partial \tau} + 0.058\frac{\partial^2 \mathcal{A}}{\partial \xi^2} = -\frac{0.196(10^{10}\chi)^2}{1-s^2}|\mathcal{A}|^2\mathcal{A} \tag{A4}$$

This is the "nonlinear Schrödinger equation," which has been exactly solved by Zakharov and Shabat for arbitrary initial conditions.[9]

We are interested in the initial conditions (Fig. A1)

$$\begin{aligned} A &= \frac{N}{p} \quad \text{for} \quad 0 < \xi < p \\ &= -\frac{N}{p} \quad \text{for} \quad -p < \xi < 0 \\ &= 0 \quad \text{for} \quad |\xi| > p \end{aligned}$$

These initial conditions deserve a word of explanation (Fig. A2). In our numerical computations a certain number (N) of amide-I quanta were put

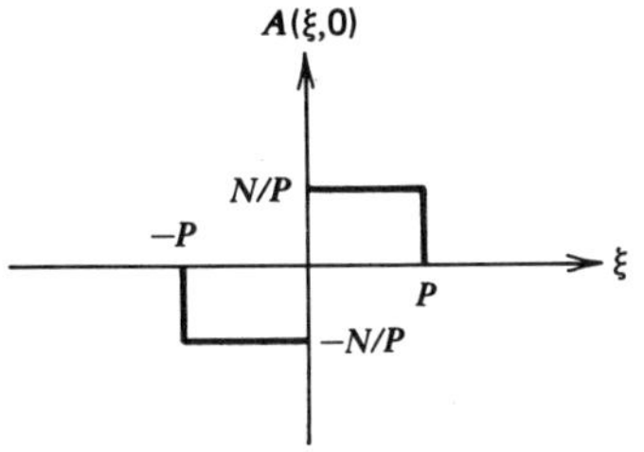

Figure A1. Initial conditions.

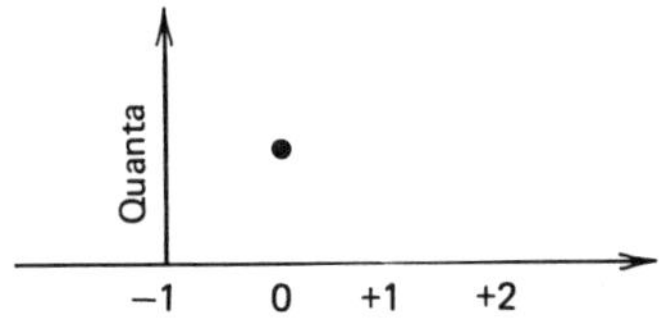

Figure A2.

onto the $n = 0$ bond at time $\tau = 0$. All other amide-I bonds were without energy at $\tau = 0$ and the energy at $n = -1$ was maintained at the value zero throughout the computations. An antisymmetrical form is chosen for the full line $(-\infty < \xi < +\infty)$ to maintain the boundary condition of zero at the origin $(n = -1)$. We have no precise value for p, but it should be approximately 2.

For analysis it is convenient to normalize Eq. (A4) by writing

$$\mathcal{A} = 3.19 \frac{(1 - s^2)^{1/2}}{(10^{10}\chi)} \phi$$

$$\xi = 0.241x$$

$$\tau = t$$

Then Eq. (A4) takes the form*

$$i\phi_t + \phi_{xx} = -2|\phi|^2\phi \qquad \text{(A4')}$$

with the initial conditions

$$\phi = \frac{(10^{10}\chi)N}{3.19p(1 - s^2)^{1/2}} \quad \text{for} \quad 0 < x < p/0.241$$

$$= -\frac{(10^{10}\chi)N}{3.19p(1 - s^2)^{1/2}} \quad \text{for} \quad -p/0.241 < x < 0$$

$$= 0 \quad \text{for} \quad |x| > p/0.241$$

*Subscripts are used to indicate partial differentiation.

A.1. Linear Limit

When the amplitude in Eq. (A4′) is small enough so the nonlinear term ($|\phi|^2\phi$) can be neglected, it becomes simply $i\phi_t + \phi_{xx} = 0$ with the Fourier transform solution

$$\phi(x,t) \approx \int_{-\infty}^{\infty} \frac{\sin^2(k\tilde{p}/2)}{(k\tilde{p}/2)} \exp[i(kx - k^2 t)]\, dk \tag{A5}$$

where

$$\tilde{p} \equiv \frac{p}{0.241}$$

The integrand in Eq. (A5) takes its maximum value when k is the root of $\tan(k\tilde{p}/2) = k\tilde{p}$ or

$$k_{\max} = \frac{2.331\ldots}{\tilde{p}}$$

and the corresponding group velocity [where the phase $(kx - k^2 t)$ is stationary] is $2k_{\max}$. Thus the linear pulse velocity (in unnormalized units that correspond to the numerical computations) is

$$\text{linear pulse velocity} = \frac{0.271}{p} \tag{A6}$$

For $p = 2$ this implies a linear pulse velocity of 0.135 whereas from the numerical computations displayed in Fig. 3 we find a velocity of 0.11.

The amplitude in Eq. (A5) should fall asymptotically as $t^{-1/2}$, indicating that the maximum bond energy in the linear limit should fall as $1/\tau$.[10] This in turn implies that the bond energy must "spread out" over a length of the α-helix that is proportional to τ. Such an effect is observed in the data of Figs. 2 and 5.

A.2. Soliton Limit

We now use the analytical tools of the inverse scattering transform method to find the threshold for soliton formation and its corresponding velocity. Readers are forewarned that this discussion will be completely unintelligible unless they have some working knowledge of the inverse scattering transform method. Those who do not should merely note that it is a generalization of the Fourier transform method.[8]

For the Zakharov–Shabat linear scattering operator[9]

$$i\begin{bmatrix} \partial_x & -\phi \\ -\phi & -\partial_x \end{bmatrix}\begin{bmatrix} \psi_1 \\ \psi_2 \end{bmatrix} = \gamma\begin{bmatrix} \psi_1 \\ \psi_2 \end{bmatrix}$$

and the initial conditions listed in Eq. (A4′), we assume asymptotic scattering amplitudes at $t = 0$ to be

$$\begin{bmatrix} \psi_1 \\ \psi_2 \end{bmatrix} = \begin{bmatrix} 1 \\ 0 \end{bmatrix} \exp(-i\gamma x) \qquad \text{for} \quad x < -p$$

and

$$\begin{bmatrix} \psi_1 \\ \psi_2 \end{bmatrix} = a(\gamma) \begin{bmatrix} 1 \\ 0 \end{bmatrix} \exp(-i\gamma x) + b(\gamma) \begin{bmatrix} 0 \\ 1 \end{bmatrix} \exp(+i\gamma x) \qquad \text{for } x > p$$

Then we find

$$a(\gamma) = \exp(2i\gamma\tilde{p}) \left[\left(\cos m\tilde{p} - \frac{i\gamma}{m} \sin m\tilde{p} \right)^2 + \frac{K^2}{(m\tilde{p})^2} \sin^2 m\tilde{p} \right] \tag{A7}$$

where

$$m^2 \equiv \gamma^2 + \frac{K^2}{\tilde{p}^2}$$

$$\tilde{p} \equiv \frac{p}{0.241}$$

$$K \equiv \frac{(10^{10}\chi)N}{0.241 \times 3.19(1 - s^2)^{1/2}}$$

Solitons correspond to zeros of Eq. (A7) that lie in the upper half of the γ-plane. For such a zero at $\gamma = \gamma_r + i\gamma_i$, the corresponding soliton has[8]

$$\text{speed} = 4\gamma_r$$

$$\text{amplitude} = 2\gamma_i$$

For K small (i.e., as $\chi \to 0$), $a(\gamma)$ has no upper half-plane zeros. Thus the threshold for soliton formation occurs when the first zero Eq. (A7) crosses the real axis of the γ-plane. Since a zero of Eq. (A7) implies

$$\cot(m\tilde{p}) = i\left(\frac{\gamma\tilde{p} \pm K}{m\tilde{p}} \right)$$

a real axis zero can only occur where $\cot(m\tilde{p}) = 0$ or at

$$\pm\gamma_r\tilde{p} = K = \frac{\pi}{2\sqrt{2}}$$

For units that correspond to the numerical computations, the threshold condition is for $s = 0.11$)

$$\chi N > 7.64 \times 10^{-11}\ \mathrm{N} \tag{A8}$$

This threshold level is higher than that in Eq. (6).

The soliton velocity at threshold (again in units that correspond to the numerical computations) is

$$\text{soliton velocity} = \frac{0.258}{p} \tag{A9}$$

Comparison with Eq. (A6) shows that the soliton velocity at threshold should be quite close (i.e., within 5%) to the linear pulse velocity. This is confirmed by the numerical data of Figs. 5 and 6 (particularly 5*c* and 6*d*).

We note finally that solitary wave solutions of the set composed of Eqs. (A1) and (A2) have been studied rather extensively in plasma physics, where they are called "Langmuir solitons." Reference 11 is a particularly lucid introduction to this work.

APPENDIX B: ORDER ESTIMATES OF HYDROGEN BOND NONLINEARITY

This appendix obtains order of magnitude estimates for the level of anharmonicity to be expected in longitudinal vibrations of α-helix protein. The hydrogen bond is assumed to have the anharmonic potential (about the minimum at x_0)

$$U(x) = \tfrac{1}{2}wx^2 + ax^3 \tag{B1}$$

Nonlinearity enters Davydov's Hamiltonian formalism as an "interaction" term

$$H_{\text{int}} = \chi B^\dagger B x \tag{B2}$$

where $B^\dagger B$ gives the number of quanta in the amide-I bond. If the restoring force of this bond is taken to be K N/m, the bond extension is

$$B^\dagger B \hbar\omega_0 = \tfrac{1}{2}Kx^2 \tag{B3}$$

where

$$\hbar\omega_0 = \mathcal{E}_0 = 0.328 \times 10^{-19}\ \mathrm{J}$$

the quantum energy of an amide-I vibration. From Eqs. (B1)–(B3)

$$\chi = \frac{2\hbar^2 a}{\mathcal{E}_0 M_r} \tag{B4}$$

where the "reduced mass" of amide-I is

$$M_r = \tfrac{48}{7} \times \text{mass of proton}$$

$$= 1.15 \times 10^{-26}\ \text{kg}$$

To estimate the parameter a in Eq. (B1) we note, from Pauling et al.,[12] that the binding energy of a hydrogen bond in α-helix protein is about 8 kcal/mole or

$$\Delta U = 5.5 \times 10^{-20}\ \text{J}$$

But from Eq. (B1) we find (Fig. B1) that $\Delta U = w^3/54a^2$ so

$$a = \frac{w^{3/2}}{\sqrt{54\,\Delta U}}$$

$$= 38.3 \times 10^{10}\ \text{N/m}^2$$

Thus from Eq. (B4) we obtain the order estimate

$$\chi \sim 2 \times 10^{-11}\ \text{N}$$

We expect the readers to be as suspicious of this estimate as we are; thus we turn to Ref. 7. There the potential Eq. (B1) is written in the form

$$U = \tfrac{1}{2}kq^2 + k_3q^3 \qquad \text{(B1')}$$

where

$$q \equiv \frac{x}{\lambda}$$

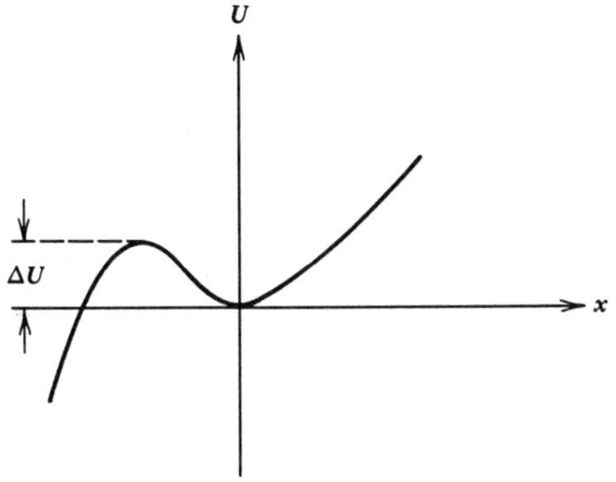

Figure B1. Plot of $\Delta U = w^3/54a^2$.

is a normalized space variable and

$$\lambda \equiv \left[\frac{\hbar}{\omega\mu}\right]^{1/2}$$

in which ω is the radian frequency and μ the reduced mass of the resulting oscillation. Evidently

$$w = \frac{k}{\lambda^2}$$

$$a = \frac{k_3}{\lambda^3}$$

Sandorfy[7] states: "Model calculations...have shown that k_3 must not exceed 7 or 8% of k...for [second-order perturbation theory] to be valid. According to our experience this is probably fulfilled for weak hydrogen bonds with ΔU values not higher than about 3 or 4 kcal/mole" (p. 617). Since *our* ΔU is taken to be 8 kcal/mole, it seems reasonable to assume $k_3/k = 0.15$ and to calculate

$$a = \frac{k_3}{k}\,\frac{w^{5/4}\mu^{1/4}}{\hbar^{1/2}}$$

With μ the reduced mass of an O———N,

$$\mu = \tfrac{112}{15} \times \text{mass of proton}$$

we find

$$a = 10.96 \times 10^{11}\ \text{N/m}^2$$

and therefore

$$\chi \sim 6.3 \times 10^{-11}\ \text{N}$$

APPENDIX C: NOTES ON THE NUMERICAL CODE

Equations (1‴) and (2″) are solved along three spines ($\alpha = 1, 2, 3$), each containing 201 unit cells ($n_{\max} = 200$). Thus, at each unit cell there are three first-order complex ordinary differential equations Eq. (1‴) and three second-order real ordinary differential equations Eq. (2″). These equations are split into a coupled system of 12 first-order real equations at each unit cell to give a total of 2412 ordinary differential equations.

The differential equations were then solved on a CDC 7600 computer using a numerical method of lines code, called PDE1D.[13] This code integrated the equations with a third-order Adams–Bashford–Moulton PECE method.[14] The time step was chosen to approximate the solution within a relative error of 0.0001 per unit time interval. For these calculations the time step ranged between 0.15 and 0.5.

The accuracy of the time integration was also checked by rerunning some of the calculations with a much smaller time step. No significant differences were found when these more accurate solutions were compared with the previous calculations.

An independent check on the overall accuracy of each calculation was made by monitoring the total probability

$$P = \sum_{n=0}^{200} \sum_{\alpha=1}^{3} A_{n\alpha}^2$$

during the calculation. This probability should remain invariant as $A_{n\alpha}$ evolves according to Eqs. (1‴) and (2″). This check remained constant within a few percentage points of its initial value in all the calculations presented here.

APPENDIX D: COMPUTER FILM NOTICE

A computer film illustrating the dynamic effects discussed in this chapter, and in particular the soliton threshold, has been prepared by J. C. Eilbeck, Department of Mathematics, Heriot–Watt University, Edinburgh, Scotland. Under the title "Davydov Solitons," it is available from Swift Film Productions, 1, Wool Road, Wimbledon, London, SW20 OHN, UK at a price of £55.00.

REFERENCES

1. D. E. Green, "Mechanism of Energy Transduction in Biological Systems," *Science* **181**, 583–584 (1973). See also *Ann. N.Y. Acad. Sci.* **227** (1974) and, particularly, the "General Discussion" on pp. 108–115.
2. A. S. Davydov, "Solitons in Molecular Systems with Applications in Biology," *Phys. Scripta 20* (1979) 387–394. See also *Phys. Stat. Sol.* **75**, 735–742 (1976) and *Stud. Biophys.* **62**, 1–8 (1977).
3. Since the term "soliton" was coined by Zabusky and Kruskal [*Phys. Rev. Lett.* **15**, 240–243 (1965)], thousands of papers have appeared in a wide variety of research areas. Several recent books include: R. Hermann, *The Geometric Theory of Non-linear Differential Equations* (Math Sciences Press, Brookline, MA, 1977); *Solitons in Action*, K. Lonngren and A. Scott, Eds. (Academic Press, New York, 1978); *Solitons in Condensed Matter Physics*, A. R. Bishop and T. Schneider, Eds., Springer-Verlag, New York, (1979); *Solitons*, P. J. Caudry and R. K. Bullough, Eds. (Springer-Verlag, New York, to appear); G. L. Lamb, Jr., *Elements of Soliton Theory* (Wiley-Interscience, New York, to appear).
4. A. S. Davydov, A. A. Eremko, and A. I. Seraienko, "Solitonui *b* α-Spiralnui Belkovuix Molekulax," *Ukr. Fiz. Zh.* **23**, 983–993 (1978).

5. N. A. Nevskaya and Yu. N. Chirgadze, "Infrared Spectra and Resonance Interactions of Amide-I and -II Vibrations of α-Helix," *Biopolymers* **15**, 637–648 (1976).

6. W. C. Hamilton and J. A. Ibers, *Hydrogen Bonding in Solids*, (Benjamin, New York, 1968, p. 164.

7. For recent surveys of anharmonicity in hydrogen bonds see: P. Schuster, G. Zundel, and C. Sandorfy, *The Hydrogen Bond*, Vols. 1–3. (North-Holland, New York, Of particular interest are: Chapter 2 by Schuster, "Energy Surfaces for Hydrogen Bonded Systems," and Chapter 13 by Sandorfy, "Anharmonicity and Hydrogen Bonding."

8. M. J. Ablowitz, D. J. Kaup, A. C. Newell, and H. Segur, "The Inverse Scattering Transform–Fourier Analysis for Nonlinear Problems," *Stud. Appl. Math.* **53**, 249–315 (1974).

9. V. E. Zakharov and A. B. Shabat, "Exact Theory of Two-Dimensional Self-Focusing and One-Dimensional Self Modulation of Waves in Nonlinear Media," *Sov. Phys.* [*J. Exp. Theor. Phys.* **34**, 62–69 (1972)].

10. See G. B. Whitham, *Linear and Nonlinear Waves* (Wiley-Interscience, New York, 1974), pp. 371–374, for a clear discussion of asymptotic behavior.

11. J. Gibbons, S. G. Thornhill, M. J. Wardrop, and D. ter Haar, "On the Theory of Langmuir Solitons," *J. Plasma Phys.* **17**, part 2, 153–170 (1977).

12. L. Pauling, R. B. Corey, and H. R. Branson, "The Structure of Proteins: Two Hydrogen-Bonded Helical Configurations of the Polypeptide Chain," *Proc. Natl. Acad. Sci.* **37**, 205–211 (1951).

13. D. Durack and J. M. Hyman, "PDEL1B, A Library for the Numerical Solution of Partial Differential Equations," in *Advances in Computer Methods for PDE-III*, R. Vichnevetsky and R. S. Stapleman, Eds. (IMACS, 1979), p. 43.

14. L. S. Shampine and M. K. Gordon, *Computer Solution of Ordinary Differential Equations* (Freeman, San Francisco, 1975).

Part 4

BEAM–BEAM INTERACTION

EXPERIMENTAL OBSERVATIONS AND THEORETICAL MODELS FOR BEAM–BEAM PHENOMENA

S. KHEIFETS
Stanford Linear Accelerator Center
Stanford University
Stanford, California

It is impossible to grasp a boundlessness. . . .

Kosma Prutkov

Abstract. *The phenomenology and theory of beam blowup due to dynamical instabilities is reviewed for a number of high energy colliding beam experiments.*

1. INTRODUCTION: PHENOMENOLOGY OF THE BEAM–BEAM INTERACTION

The beam–beam interaction in storage rings exhibits all the characteristics of nonintegrable dynamical systems. Here one finds all kinds of resonances, closed orbits, stable and unstable fixed points, stochastic layers, chaotic behavior, diffusion, and so on. The storage ring itself is an expensive device; nevertheless, once constructed and put into operation it presents a good opportunity to study experimentally the long-time behavior of both conservative (proton machines) and nonconservative (electron machines) dynamical

---

Work supported by the U.S. Department of Energy, contract DE-AC03-76SF00515.

Kosma Prutkov is the nickname for a group of Russian humorists of the nineteenth century (A. Tolstoi, *Three Brothers Zhemchuzhnikov*).

systems—the number of bunch–bunch interactions routinely reaches values of 10^{10}–10^{11} and could be increased by decreasing the beam current. At the same time the beam–beam interaction puts practical limits for the yield of the storage ring. This phenomenon not only determines the design value of main storage ring parameters (luminosity, space charge parameters, beam current), but also in fact prevents many of the existing storage rings from achieving design parameters. Hence, the problem has great practical importance along with its enormous theoretical interest.

This chapter is a brief overview of the problem. Experimental observations, including the latest available results from the PEP storage ring at Stanford are discussed here together with the different theoretical and computational models suggested for understanding of the beam–beam phenomena.

The bibliography on the subject includes hundreds of titles and is far beyond the space and time available. The few references listed here are given only for the use of readers who want to go deeper into the field. Some of them describe the most recent work performed on the beam–beam phenomena. Some work is reviewed in Refs. 1, 2 (see especially Refs. 6 and 7), and 5. For more extensive version of this chapter containing more details, see Ref. 8.

1.1. Single-Particle Dynamics

The presence in the machine of the second counterrotating beam has significant influence on the dynamics of a particle. The force a particle sees from an oncoming bunch is $\simeq 2\gamma^2$ times larger than the force from the other particles in the same bunch. In ultrarelativistic machines the latter effect is negligibly small. In most cases the action of the beam–beam force can be described as a δ-function-like change in particle transverse momenta, the magnitude of which are functions of its transverse displacements x and y at the interaction point.

For the three-Gaussian bunch "the kicks" $\Delta x'$ in the horizontal and $\Delta y'$ in the vertical directions, respectively, can be found from the corresponding potential function in the following integral form[9]:

$$\Delta x' = -G\left(\frac{x}{a}\right)\int_0^1 d\tau\, e^{-\tau Q(x,y)}[p + \tau(1-p)]^{-3/2} \tag{1.1}$$

$$\Delta y' = -G\left(\frac{y}{b}\right)\int_0^1 d\tau\, e^{-\tau Q(x,y)}[p + \tau(1-p)]^{-1/2} \tag{1.2}$$

where

$$Q(x,y) = \left(\frac{y}{b}\right)^2 + \left(\frac{x}{a}\right)^2[p + \tau(1-p)]^{-1} \tag{1.3}$$

$$G = \frac{2r_0N}{a\gamma} \tag{1.4}$$

r_0 is the classic electron radius ($= 2.82 \times 10^{-13}$ cm), N is the number of particles in the bunch, a and b are horizontal and vertical dispersions of the Gaussian distributions ($a = \sqrt{2}\,\sigma_x$, $b = \sqrt{2}\,\sigma_y$), respectively; and $p = b^2/a^2$ is the aspect ratio of the bunch. The signs in Eqs. (1.1) and (1.2) describe an attractive force. They should be reversed for a repulsive one.

The most striking features of the force [Eqs. (1.1) and (1.2)] are its nonlinear character and the additional coupling it produces in particle motion. The force is linear only for very small values of y/b and x/a where the charge is distributed more or less uniformly. It reaches a maximum at the intermediate values of x/a. At very large distances from the origin the force falls off as $1/(x^2+y^2)^{1/2}$. Figure 1.1 illustrates the kick $\Delta y'$ as a function of y/b for elliptical bunch cross section ($p = 0.25$).

For very small values of $x/a \ll 1$ and $y/b \ll 1$, Eqs. (1.1) and (1.2) give:

$$\Delta x' = -\frac{Gx}{\sigma_x + \sigma_y} \qquad (1.1')$$

$$\Delta y' = -\frac{Gy}{\sigma_x + \sigma_y} \qquad (1.2')$$

In this approximation beam–beam interaction is linear and decoupled. The action of the opposite bunch can be described as an additional thin lens focusing (for attractive force) in both planes. Such a lens can be included into the lattice to find the linear incoherent tune shift $\Delta\nu$ and the linear change of amplitude β-function. For one interaction place one gets, for example, for the y-plane:

$$\cos 2\pi(\nu_y + \Delta\nu_y) = \cos 2\pi\nu_y - 2\xi_y(\sin 2\pi\nu_y) \qquad (1.5)$$

$$\beta_y = \frac{\beta_{0y}(\sin 2\pi\nu_y)}{\sin 2\pi(\nu_y + \Delta\nu_y)} \qquad (1.6)$$

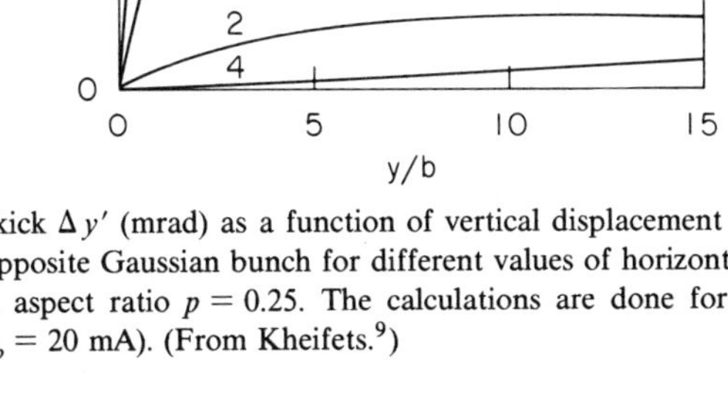

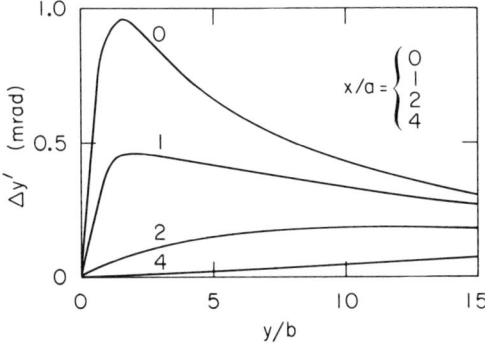

Figure 1.1. Vertical kick $\Delta y'$ (mrad) as a function of vertical displacement y in units of vertical dispersion b of the opposite Gaussian bunch for different values of horizontal displacement x/a. Elliptical beam with aspect ratio $p = 0.25$. The calculations are done for PETRA parameters (current per bunch $I_b = 20$ mA). (From Kheifets.[9])

wherein appears the famous space charge parameter[10–13]:

$$\xi_y = \frac{\beta_{0y} N r_0}{2\pi\gamma\sigma_y(\sigma_x + \sigma_y)} \tag{1.7}$$

An analogous expression holds for ξ_x:

$$\xi_x = \frac{\beta_{0x} N r_0}{2\pi\gamma\sigma_x(\sigma_x + \sigma_y)} \tag{1.8}$$

In these expressions β_{0x}, β_{0y} mean the values of the corresponding unperturbed β-functions at the interaction point; ν_x, ν_y are the corresponding unperturbed tunes of the machine.

These simple expressions hold only for a small portion of particles. The forces are strongly nonlinear for particles with $x/a \sim 1$ and $y/b \sim 1$. The nonlinear character of the beam–beam force brings up a whole set of nonlinear resonances.

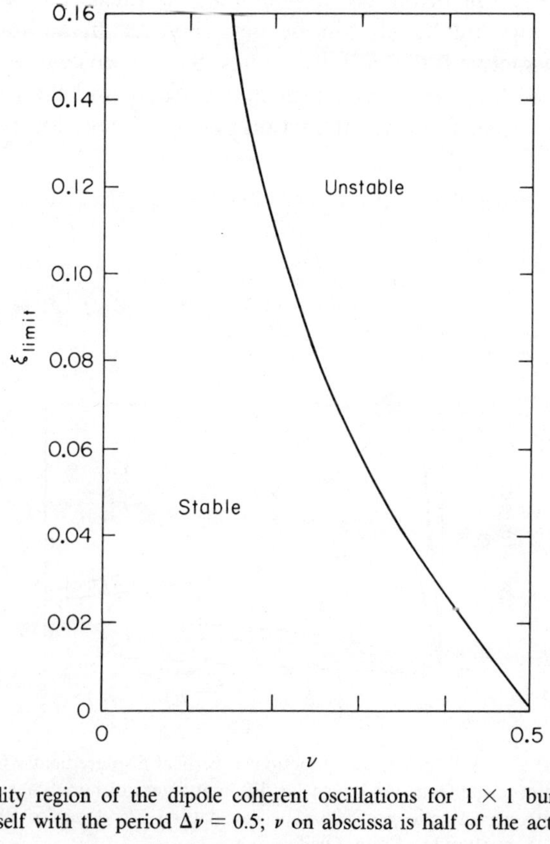

Figure 1.2. Stability region of the dipole coherent oscillations for 1×1 bunch collision. The picture repeats itself with the period $\Delta\nu = 0.5$; ν on abscissa is half of the actual machine tune. (From Chao[7].)

In addition to purely transverse resonances, resonances due to coupling of transverse and longitudinal motion can appear (and play an important role).[14–18]

1.2. Coherent Dynamics

Let us consider coherent effects in bunch motion brought about by the beam–beam interaction.

In linear approximation the beam–beam force induces dipole oscillations of bunches as a whole. The oscillations of different bunches are coupled to each other and can be described by their normal modes. The eigenfrequency of each

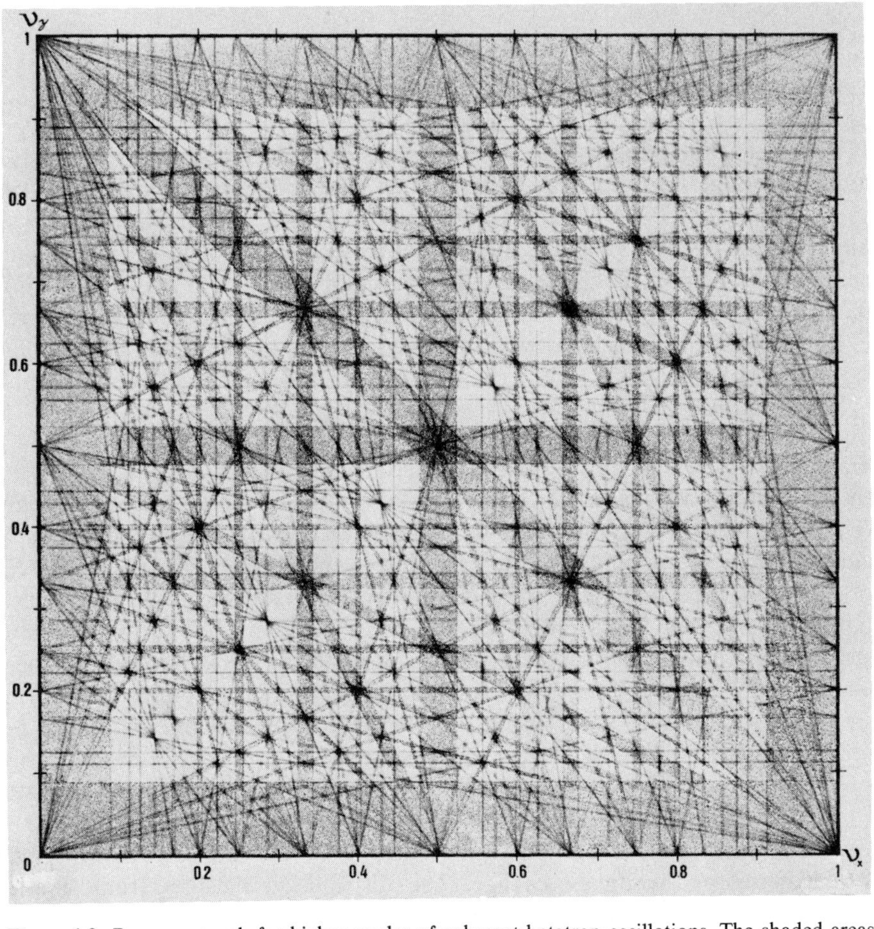

Figure 1.3. Resonance web for higher modes of coherent betatron oscillations. The shaded areas around the resonance lines show the widths of the resonance due to beam–beam interaction. The space charge parameter $\xi = 0.05$. Only the resonances are shown, the widths of which are greater than 10^{-3}. The largest available working zones are located along the difference coupling resonance $\nu_x = \nu_y$. (From Chau and Potaux.[21])

mode should be chosen outside the stopband of all resonances. Figure 1.2 presents the stability regions for dipole oscillations of bunches for the case of collisions of 1×1 bunches. The coherent space charge parameter is twice as big as the incoherent one, since the beam–beam force seen by a center of a bunch displaced by y is the same as that for a single particle displaced by $2y$ in the incoherent motion.

The nonlinear terms in the beam–beam force can also cause higher multipole coherent oscillations. The analysis of these oscillations is much more complicated and involves difficult and cumbersome calculations.[19–21] Figure 1.3 illustrates the widths of a few first resonances of coherent oscillations excited by the beam–beam interaction in the system with charge compensation.[21]

1.3. Beam Blowup

It is useful to describe briefly what happens when beams collide. Normally, during storing the beams are separated at interaction points by means of special electrostatic fields (at all other places of the machine circumference bunches pass at different times and do not disturb each other). When enough current is stored in each beam, the separation field is switched out and particles start to interact, producing desirable (as well as undesirable) effects.

The behavior of bunches now completely depends on the value of a current stored in both of them.

1.3.1. Weak Beam–Strong Beam Incoherent Instability. Suppose first that the number of particles in bunch "1" (weak beam) is much less than that in the opposite bunch "2" (strong beam). In this case there exist a critical value of the current and correspondingly a critical value of the space charge parameter ξ_y of the strong beam. Below it there are hardly any visible beam–beam effects, but as soon as the threshold is reached the vertical size of the weak beam suddenly increases significantly (beam blowup.) At the same time the horizontal size of the beam does not change significantly. The strong beam shows very small change, if any. Typical values for the crucial space charge parameter ξ_y range around 0.03–0.05. The space charge parameter of the weak beam, being small because of its current, becomes still less because of increase in size. Figure 1.4 illustrates beam blowup in the experiment at SPEAR.[22]

Along with the change in the bunch size, the particle distribution also experiences significant change. The distribution deviates from simple Gaussian shape; the tails become much more populated.

Figure 1.5 presents the results of vertical bunch profile measurements[23] on SPEAR for electrons and positrons before and after the separation was switched out.

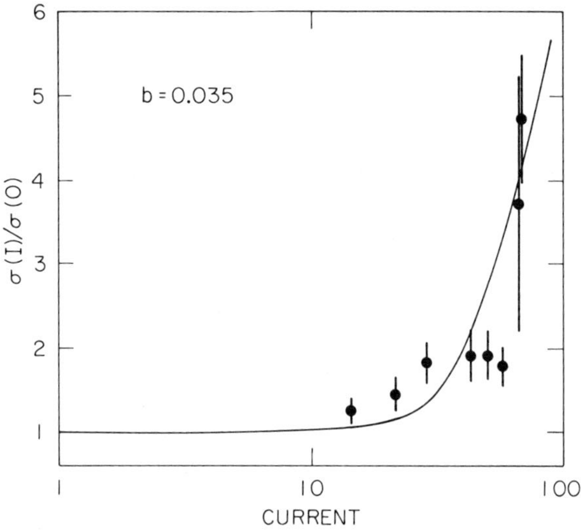

Figure 1.4. Beam blowup in the weak beam–strong beam case as measured on SPEAR.[22] The solid line represents calculations according to diffusion phenomenological model (see Section 3.3). The parameter η used as abscissa is defined as $\eta = 2Bf\tau(2\pi\xi_y)^2$.

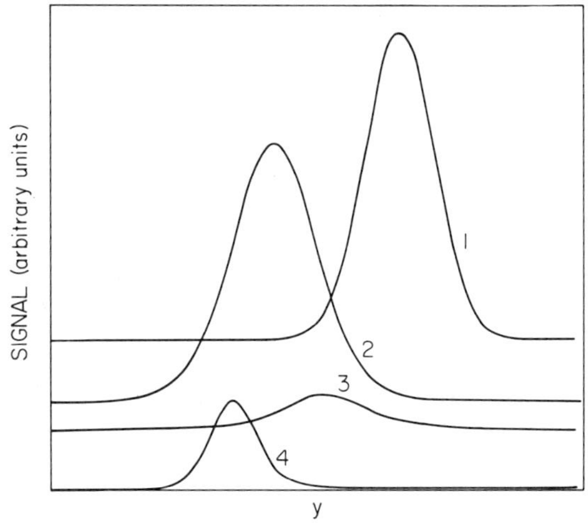

Figure 1.5. Vertical beam profile measurements on SPEAR.[23] Curves correspond to the following conditions:

N	Bunch	Current (mA) I_+	Current (mA) I_-	Condition
1	Electron	1.70	6.8	Beams are separated
2	Electron	1.70	6.8	Beams collide
3	Positron	1.75	6.9	Beams collide
4	Positron	1.75	7.0	Beams are separated

The question of the rise time of the weak beam–strong beam instability is difficult and still not resolved. The attempts to measure it at SPEAR showed a value much smaller than the damping time. This fact was confirmed also on PETRA.

1.3.2. Strong Beam–Strong Beam Incoherent Instability. The usual mode of operation of a storage ring is one of maintaining the currents in both beams as equal as is practically possible. The main reason is of course the desire to achieve the maximum luminosity of the storage ring. If the current of one beam were less than that in another, the weaker beam would blow up at much lower current magnitude, reducing the luminosity.

The strong beam–strong beam instability exhibits many common characteristics with the weak–strong case. One observes beam blowup (usually bigger for the beam with slightly smaller current in it and smaller for the stronger beam), redistribution of particles in the bunch, and increase of the population of its tails. The deviation from the Gaussian distribution is still more pronounced (see, e.g., Fig. 1.6[6]). The change in particle distributions makes it extremely difficult to simulate strong–strong case in a computer study, since forces depending on the distribution should be recalculated separately for each interaction.

On the other hand there is no clear answer to the question of the existence of a threshold. It seems that in some cases such a threshold exists[6]; in other cases the beam blows up starting from very small values of current.[24-26] The same holds for the tune dependence. In some cases there is marked dependence

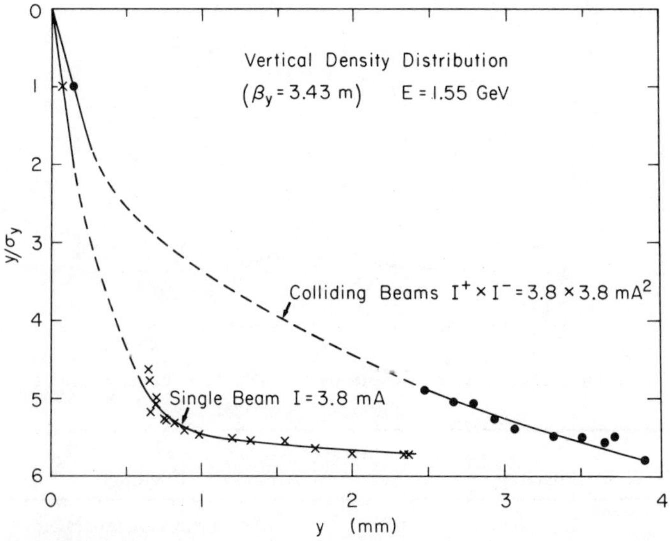

Figure 1.6. Beam blowup for the strong beam–strong beam case. The density distribution is measured[6] with the help of scrapers by observing the reduction of beam lifetime as a function of the scraper position y. For a Gaussian bunch the value y/σ_y calculated from the beam lifetime would be a linear function of y.

of the beam blowup limits on the values of the tunes.[26,27] For the high beam intensity only a small island of stability can be found on the tune diagram. As an example,[26] Fig. 1.7 shows the regions of stability for the ACO and DCI storage rings. At the same time, in some cases strong dependence on tune is not observed.[24] An interesting situation is observed with the storage ring SPEAR. In the old version of the machine (SPEAR 1) such dependence has been observed.[6] The beam–beam limit was raised by a factor of five by decreasing the vertical tune. Further decrease was impossible because there was strong two-beam resonance $\nu_x - 2\nu_y = 5$ (forbidden for the twofold symmetry of the machine with one beam). This effect disappeared in the modified machine (SPEAR 2).

The fate of the blown-up beam strongly depends on the beam current and the energy. At the maximum currents and energies for a given storage ring the

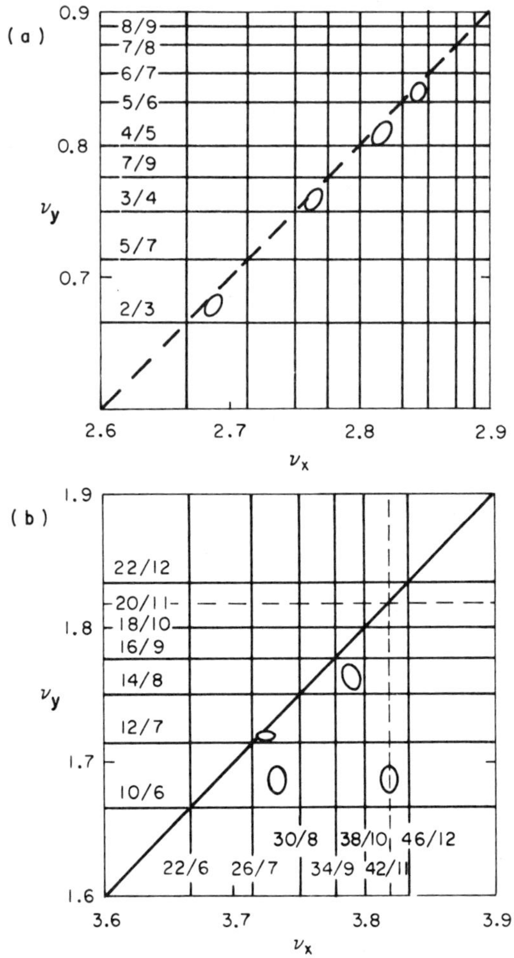

Figure 1.7. Operation zones found[26] on the storage rings ACO (a) and DCI (b).

lifetime of the beam is reduced drastically and beam is usually lost in very short time. At smaller currents or energies the lifetime of the blown-up beam seems to be unaffected and equal to the lifetime of the unblown beam.[28] This can be used to restore normal conditions, by using strong dependence of the blowup on the machine tunes.

Near the blowup limit, but below it, the beams are highly sensitive to all kinds of perturbation. Even a small change of conditions usually causes beam blowup.

The dependence of the effect on the number of bunches is also still not understood completely. The intensity of the beam–beam interaction increases with the increase of the number of bunches. In principle this should lead to a decrease in the luminosity at the given interaction place and to a decrease of the beam–beam limt. There are observations showing indeed such dependence[28,29]; but there are also others[30] that can be interpreted as exhibiting maximum luminosity independent of the number of bunches.

1.3.3. Flip-Flop Effect. A remarkable controlled beam blowup is discovered on SPEAR[29] at certain conditions; by changing the difference of the phase of two accelerating cavities, it is possible to blow up one (flip) or the other (flop) of two beams. The flip-flop effect is apparent only near the beam–beam limit at a given energy. There is also pronounced influence on the effect of the magnitude of the horizontal dispersion function at the interaction point. Other parameters of the ring (chromaticities, sextupole strengths, bunch length, tunes, and coupling) seem to be important or not depending on whether they influence the separation of the bunches or the horizontal dispersion at the interaction point of the machine.

The flip-flop effect is not yet completely understood. Nevertheless the effect is successfully employed to balance the heights of the opposite bunches at the maximum current, hence maximizing the luminosity of the ring.

Preliminary observations[31] show the existence of the flip-flop effect also in PEP.

2. EXPERIMENTAL OBSERVATIONS

The comprehensive review of the experimental results obtained with different storage rings is difficult not only because of wide spectra of different types, and quite different parameters of the storage rings (energy, space charge parameters, currents, number of bunches, etc.), but mainly because of completely different conditions of the measurements and the different ways of interpreting results.

Even for a given storage ring the results strongly depend on the tunes, energy, quality of closed orbits correction, chromaticity, transverse offsets of the bunches at the interaction point, asymmetry of the ring (differences in phase advance between the interaction points), aspect ratio of the bunch, and so on, almost ad infinitum.

1. *Investigation of Beam–Beam Limitations.* Measurements of this kind are done during special machine physics runs. The main goal of these measurements is to achieve the maximum possible luminosity for given parameters by increasing the currents to the point at which the lifetime of the beam starts to decrease sharply. To maximize the luminosity of the ring, both currents are usually maintained at much the same value. One tries to do the same with the vertical size of the beam. At least at SPEAR this condition was met by means of adjustment of the phase between the radio frequency cavities positioned symmetrically around the interaction point.[29] Experimental data obtained in this situation should be more sensitive to the particle distribution at large amplitudes (to the tails of distribution) rather than to the distribution in the core of the beam.

2. *Investigation of Storage Ring Performance.* Measurements of this kind are usually done during high energy physics runs in a parasitic mode. Maximum luminosity is achieved under a restrained condition of the beam lifetime being unaffected by beam–beam phenomena or by demand to have reasonable background in experimental devices. These measurements should be more sensitive to the distribution in the core of the beam.

It is important also to distinguish the regions of parameters below and above the blowup limit. The boundary between these two regions is not always sharp and pronounced. Since the functional behavior of relevant parameters is quite different in these two regions, one should be specific when talking about their values and dependencies on energy, current, bunch number, and the like.

An experimental fact observed on all the machines is that the horizontal size of the bunch is not influenced by the beam–beam interaction[6,30] with accuracy $\lesssim 10\%$.

2.1. SPEAR Dependence on Energy

Recently a set of new measurements of the maximum luminosity and the beam current versus machine energy was undertaken by H. Wiedemann.[32] The range of energy variation was 0.6–3.7 GeV and is much wider than in all previous experiments. The data are taken during the special runs of the SPEAR

Table 1. Dependence of SPEAR Parameters on the Particle Energy E (GeV)[a]

f	k	q	Comment
$\mathcal{L}_{max}$	0.033	6.6	In 10^{30} cm^{-2} sec^{-1}
i_{max}	1.2	3.6	In milliamperes
σ_y/σ_x	0.5	−1.0	—
ξ_x	0.022	0.87	—
ξ_y	0.011	2.3	—

[a] The fit is done[32] by a function $f = kE^q$.

dedicated to machine physics. Much work was done to adjust all the machine parameters to achieve maximum luminosity. Special attention was paid to balancing the vertical sizes of electron and positron bunches to avoid the loss of the luminosity due to the blowup effect.

The fit by a power law to recent data seems to give slopes, especially for the vertical space charge parameter, quite different from those in previous measurements.[6] The difference may be attributed to the fact that the energy range in the work[6] was much narrower ($\approx$ 1.2–2.5 GeV). Table 1 summarizes the results of fitting to these measured and calculated data.

2.2. SPEAR Dependence on the Beam Current

Table 2 summarizes the data picked up from SPEAR logbooks by M. Cornacchia.[33] The data were mostly taken during regular physics runs of the machine. The fits to the data taken at high energy physics run are recalculated. Instead of fitting data by the least-squares method, the maximum luminosity was fitted.

Table 2. Dependence of SPEAR Parameters on the Beam Current *i* (mA)[a]

f	E (GeV)	k	q	β_y	Comment
$\mathcal{L}_{max}$	1.5	0.030	1.95		High energy physics runs; in 10^{30} cm^{-1} sec^{-1}
	2.5	0.046	1.55		
	3.7	0.054	1.45		
$\mathcal{L}_{max}$	1.95	0.052	1.41	10 cm	Machine physics runs; in 10^{30} cm^{-1} sec^{-1}
	1.95		1.45	20 cm	
σ_y			0.59		
σ_x			0		
ξ_y	2.4		0.33		

Source. Data from SPEAR logbooks, summarized by Cornaccia.[33]
[a]The fit is done by a function $f = ki^q$.

Table 3. Dependence of ADONE Parameters on the Particle Energy *E* (GeV)[a]

f	k	q	Comment
$\mathcal{L}_{max}$	0.64	7	in 10^{30} cm^{-2} sec^{-1}
$\xi_x \simeq \xi_y$	0.068	1.57	—
			Strong beam–weak beam
i_{max} (mA)	105	4.34	Three bunches
	42.4	4.12	One bunch

Source. S. Tazzari.[28]
[a]The fit is done by a function $f = kE^q$.

2.3. ADONE

Table 3 summarizes the dependencies of the maximum luminosity and the beam current versus energy taken from the report by S. Tazzari.[28] The space charge parameters of this machine were kept approximately equal. The fit for the space charge parameters is derived from the calculated values plotted in the work.[28] The number of bunches B in ADONE can be changed, and this was done. The data taken with one and three bunches do not contradict the assumption

$$\xi_y \sim \frac{1}{\sqrt{B}}$$

2.4. PETRA

The data from the measured specific luminosity LB/i^2 during high energy physics experiments were fitted with the help of the blowup function σ_y assumed[24] to behave according to the following:

$$\sigma_y^2 = \sigma_0^2 + \left(\frac{ai}{\sigma_y}\right)^2 \tag{2.1}$$

Here σ_0 is the value of σ_y at zero current i and a is a parameter. From the data taken at different energies, a is found to be:

$$a = \frac{\text{const}}{E^4} \tag{2.2}$$

The values of aspect ratio of the beam are estimated to be of the order of several percent at all energies.

2.5. PEP

Because of the sixfold symmetry of the machine, the operation of PEP is performed in 1×1 or 3×3 bunch modes, and the beam–beam data obtained in these modes differ. The reason for this difference is not yet understood.[34] Figures 2.1–2.3 present the results[35] of measurements of the luminosity, the specific luminosity L/BI_B^2, and the vertical tune shift $\Delta\nu$ for different bunch currents. The data are taken with the particle energy 14.5 GeV. It is clearly seen that the initial slope of $L(I_B)$ for 1×1 mode agrees with the I_B^2 law (constant cross section of a bunch). The function saturates at higher current values indicating the cross-sectional increase (blowup). At the same time the luminosity curve for the 3×3 mode does not seem to follow the I_B^2 law even for very small current. This can be seen more clearly from the specific luminosity data. The 3×3 data show beam blowup starting from the lowest available current data.

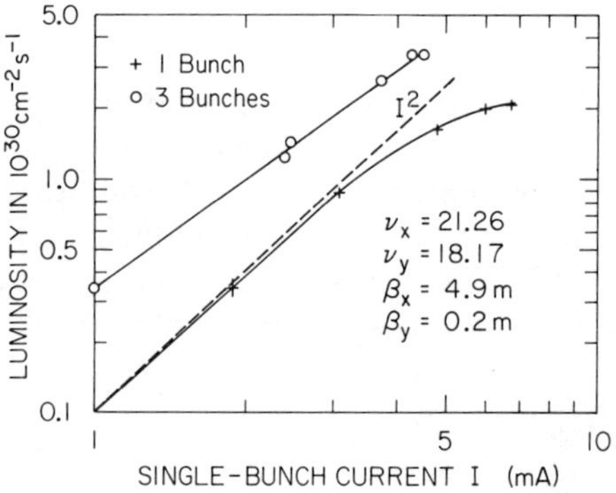

Figure 2.1. Results of luminosity measurements[35] on PEP for two collision modes, 1×1 and 3×3 bunches. The deviation from the I^2 law starts for the 1×1 mode at bunch current $\simeq 4$ mA; for the 3×3 mode there is no threshold.

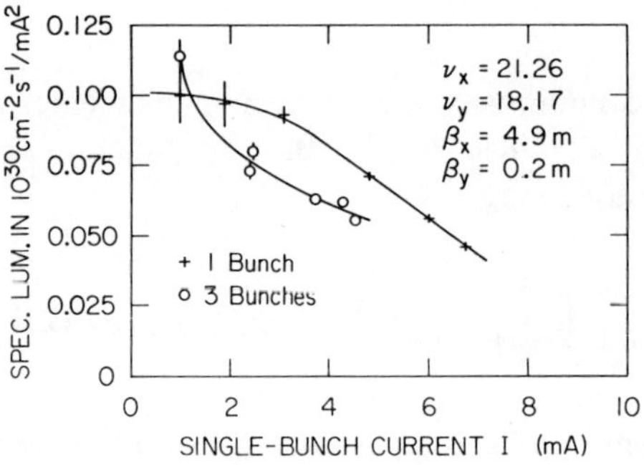

Figure 2.2. Specific luminosity ($L_{sp} = L/BI_B^2$) for PEP.[35]

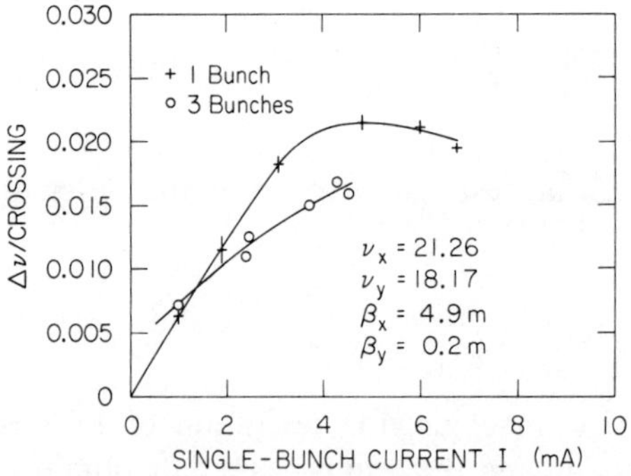

Figure 2.3. Space charge parameter ξ_y (tune shift) for PEP.[35]

Unfortunately, there are not enough data to fulfill an analysis similar to one for PETRA. More data for different energies are needed for the comparison between two storage rings.

2.6. CESR

Luminosity measurements and observations of the related beam–beam effect is done[36] on the CESR storage ring at the energy 5.5 GeV (the design energy of the ring is 8 GeV). Maximum luminosity of 3×10^{30} cm^{-2} sec^{-1} has been achieved, which corresponds to a maximum vertical tune shift of 0.035. Luminosity curves versus current show i^2 dependence up to $i \lesssim 4$ mA for

Table 4. Some Experimental Observations

Machines	E (GeV)	Vertical Tune, ν_y	β_y (cm)	Bunches per Beam, B	Scaling Laws, E $I(E)$	$\mathcal{L}(E)$	$S^*(E)$	$\xi(E)$	ξ_{max} (10^{-2})
ACO	0.54	0.829	400	1–2	$\gamma^{3.5}$	γ^5	γ^2	$\gamma^{0.5}$	3.1
VEPP-2M	0.65	3.086	5.6	1		$\gamma^4-\gamma^6$			4.0[a]
VEPP-4	1.84	9.1	15.0			γ^4			1.5
ADONE	1.5	3.05	340	3	$\gamma^{4.5}$	γ^7	γ^2	$\gamma^{1.5}$	6.6
DCI	1.85	1.79	200	1	γ^1	γ^2	γ^0	γ^0	4.1
DORIS	5.0	5.18	30	360	γ^3	γ^4	γ^2	γ^0	1.0
SPEAR	2.0	5.17	10	1	$\gamma^{3.6}$	$\gamma^{6.7}$		$\gamma^{2.4}$	4.5

[a] Data extracted from contribution.[76]

different lattice configurations. Above this value one or both beams flip and the saturation is seen clearly. It is found that small changes in tune can lead to large changes in specific luminosity and/or lifetime of the beam.

2.7. Low Energy Machines

The summary of experimental observations on several other machines can be found in Table 4, taken from Ref. 37. Representative samples of data for SPEAR and for VEPP-4 are taken from Refs. 32 and 38, respectively. Columns 2–5 give the maximum energy, the vertical tune ν_y, the vertical β-function at interaction point β_y, and the number of bunches per beam. The next four columns deal with scaling laws for maximum value of the beam strength parameter ξ_{max} achieved.

2.8. Intersecting Storage Rings (ISR)

To give a feeling of the order of magnitude of relevant parameters for proton bunched beams, I include here the results of the beam–beam study[39] undertaken on the intersecting storage ring at the European Organization for Nuclear Research (CERN). Although for the electron storage ring the beam–beam shift tune is of the order of 0.03–0.05, the maximum achieved value for the same parameter for the proton storage ring is 3×10^{-3}. This value is obtained for well-adjusted vertical position of two beams at the interaction point. When beams were off-center by approximately one standard deviation, dramatic blowup was observed. The rate of the vertical blowup increases substantially, also by the noise in the bunching radio frequency system.

For continuous beams the beam–beam limit was observed[40] to be in the range 0.01–0.02. The difference between this result and the results for bunched collisions is tentatively attributed to the fact that for continuous beams the motion is essentially one-dimensional (since the vertical tune shift is much larger than the horizontal one, while there are no synchrotron oscillations). For one-dimensional motion one can expect the presence of a stable region below the "stochastic limit."

3. THEORETICAL MODELS AND TECHNIQUES

At present there is no satisfactory theory to account for the beam–beam phenomena. The complexity of the physical object—being, essentially, extremely hot plasma very far from equilibrium—makes the analysis very difficult. The situation is aggravated also by the highly nonlinear character of the beam force, the presence of other nonlinearities in the equation of motion, the influence of noise of different origins, and by nonuniformity in space (aperture of the machine) and time (damping).

There are many approaches to the problem and many theoretical and computational techniques being used in the attempt to attack the problem. Many of these give a qualitative explanation to some observed facts, but it is hardly surprising that none is able to comprehensively describe the phenomenon as a whole, nor to give ways and means to calculate the relevant parameters.

Computational technique is hardly in better shape. Until now all computer simulations have been restricted to particle tracking. In most cases the results of tracking are difficult to interpret in terms of real machine parameters. In better cases, these results seem to be satisfactory to their authors and can be applied only to the particular machine for which they are designed. There is also one practically important restriction of the tracking method. If, as I believe to be true, the nonlinear elements (sextupoles) of the machine are important for correct description of the beam–beam phenomena, the capacity of modern computers is insufficient to do the job, at least for the future big machines, since computational time becomes prohibitive. This is true to an even greater extent for proton machines. Although for electron storage rings it is believed that only several damping times are essential to get to the stationary state ($\sim 10^3$–10^4 collisions), for the proton ring such time should be the lifetime of the beam ($\sim 10^{10}$–10^{11} collisions), which is far beyond the ability of any computer.

In subsequent sections, I describe very superficially some different theoretical models, to inform readers of the main underlying ideas and obtained results. I do not supply details of sometimes very complicated mathematics.

I have restricted myself to the beam–beam interactions in a storage ring. There is a related subject, that of the beam–beam interaction in a linear collider. Some interesting theoretical and computational work[41–45] has been done in this area. This work helps us to look into the beam–beam phenomena, studying them in a self-consistent manner; but since it lies outside the area of long-time prediction, and mainly because no experimental observations have been done yet, I do not include the subject in this chapter.

3.1. Single-Resonance Models

I follow here Ref. 7, where one can find more details as well as some references to other papers on this subject.

The foundation of the model contains two main assumptions:

1. There is only one set of integers q, p, n satisfying the condition $q\nu_x + p\nu_y \simeq n$, which dominates the motion of a single particle interacting with a fixed opposite bunch. In principle there are many other sets of numbers p, q satisfying this condition. The rationale for ignoring them is that the influence (width) of the corresponding resonance drops the faster with the increase of the particle amplitude, the more quickly the numbers p and q increase.
2. All the fast oscillating terms in the equation of motion can be ignored (smooth approximation). Formally that goal is achieved by substituting the Hamiltonian by its average over unperturbed phase trajectory. Such averaging changes the volume of the phase space, which can lead to some errors. Besides that, this approach does not take into account the existence of stochastic regions and possible Arnol'd diffusion.

Two such models are discussed. In the first one all parameters are considered to be constant (static model). The second model allows slow (adiabatic) change of the parameters due to coupling to longitudinal motion (dynamic model).

3.1.1. Static Model. Consider transverse motion of a single particle in the presence of a fixed oposite bunch (weak beam–strong beam case, neglecting synchrotron oscillations).

The Hamiltonian of this problem is:

$$H = \tfrac{1}{2}\left(px^2 + K_x(s)x^2\right) + \tfrac{1}{2}\left(py^2 + K_y(s)y^2\right) + U(x, y)\varepsilon(s) \quad (3.1)$$

where $K_x(s)$ and $K_y(s)$ are respectively, the lattice horizontal and vertical focusing functions of the longitudinal coordinate s. Potential of the beam–beam interaction $U(x, y)$ should be chosen in such a way as to give correct "kicks" $\Delta x' = -\partial U/\partial x$ and $\Delta y' = \partial U/\partial y$ [see Eqs. (1.1) and (1.2)]. The periodic function $\varepsilon(s)$ describes the longitudinal dependence of the beam–beam force. For a short bunch it can be approximated by the δ-function periodic in s. In the vicinity of the single resonance $q\nu_x + p\nu_y = n$ (assumption 1) the Hamiltonian has rapidly oscillating and slowly changing terms. The averaging procedure (assumption 2) throws the rapidly oscillating terms away. If the particle distribution in the strong bunch is Gaussian and its action can be described as a δ-function kick in time, the remaining Hamiltonian depends on five dimensionless parameters,

$$\xi_x, \xi_y, \nu_x, \nu_y, \frac{\sigma_y}{\sigma_x}$$

The tune shift $\Delta\nu$ and the width of resonance $\delta\nu$ as functions of particle amplitude can be calculated. For the one-dimensional case ($q = 0$) the result is:

$$\Delta\nu(\alpha) = \xi e^{-\alpha}[I_0(\alpha) + I_0'(\alpha)] \tag{3.2}$$

$$\delta\nu(\alpha) = 8\xi \sum_s \left| \frac{e^{-\alpha}\left[(1+2\alpha)I_{sp/2}(\alpha) + 2\alpha I'_{sp/2}(\alpha)\right]}{s^2p^2 - 1} \right| \tag{3.3}$$

where $\sqrt{2\alpha}$ is the normalized amplitude J in units of the strong bunch vertical dispersion

$$\alpha = \frac{J\beta}{2\sigma^2} \tag{3.4}$$

and I_m is the modified Bessel function of mth order. Qualitative behaviors of $\Delta\nu$ and $\delta\nu$ are shown in Fig. 3.1.

Fast decay of the resonance width with the amplitude according to the static model causes the motion to be stable with respect to beam–beam interaction, in striking contradiction to experimental observations. This discrepancy seems to be overcome by dynamic models.

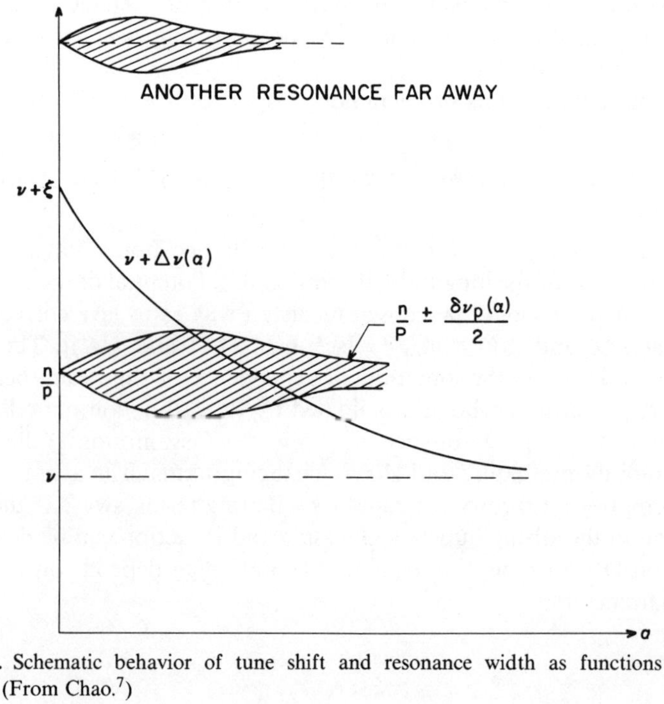

Figure 3.1. Schematic behavior of tune shift and resonance width as functions of particle amplitude. (From Chao.[7])

3.1.2. Dynamic Models: Trapping. The static model does not provide a mechanism for transporting particles from small to large amplitudes. Such a mechanism seems to be necessary to provide for particle losses and lifetime decrease. There are several models[46–48] with time-variable parameters, such as ν or ξ, that lead to an instability. The change of the parameters may arise from different sources. It may be coupling to longitudinal motion, in which case the tune ν may be modulated by synchrotron frequency ν_s, or it may be some noise in the system that is causing the tune diffusion. The change of the tune makes assumption 1 (single resonance) still less sound than for the static model.

In the trapping model[46] the tune of the particle is adiabatically changed in time close to a single resonance value n/p. These changes caused the resonance islands (the separatrices) to move. At a certain rate of the time change it is possible that an island will trap a particle, bringing it from one phase trajectory to another. The island's movement is completely analogous to the autophasing in longitudinal motion in a synchrotron, where a slow increase of the magnetic field causes the separatrix to move toward bigger mean energy. Synchrotron oscillations of the particle inside the separatrix then follow its movement, exhibiting trapping. As we know, the speed of the magnetic field change should not exceed a certain limit, which is determined by the dependence of the revolution frequency on energy. At this value the size of the separatrix becomes zero. In the same manner the speed of the tune change in the trapping model should not exceed a certain limit, which in turn is determined by the amplitude dependence of the tune, at which the width of the island tends to zero.

To find the rate of particle transport toward larger amplitudes in the trapping model, one needs an evaluation of the trapping probability. It is assumed that this probability is proportional to the volume of the island, being small for small amplitudes. It then reaches its maximum at a dimensionless amplitude of the order of magnitude 1–3 (depending on the order of the particular resonance), then drops toward larger amplitude again (the decrease is due to both the decrease in resonance width and the distance to resonance). The net effect is the difference in the number of particles brought from the small to large amplitude (proportional to particle density inside the bunch) and the number of particles brought back from larger to smaller amplitude (proportional to particle density in the tails of distribution).

3.1.3. Method of Lie Transformation. The application of the transfer map method[49–51] to the analysis of the beam–beam interaction is not discussed here. The discussion of this method can be found in Ref. 8.

3.2. Many-Resonance Models: Stochastic Limit

The assumption made in single-resonance models does not appear to be very realistic. A picture of the simultaneous action of many resonances together seems to be closer to reality. Insofar as the strength of the beam–beam interaction is small enough, the separation between the different resonances (at least for low order resonances) is greater than their widths. In this case the Kolmogorov–Arnol'd–Moser (KAM) theorem is valid and the motion of almost all particles with different initial conditions is stable, deterministic, and reversible in time. The trajectories of such particles are only slightly perturbed by the presence of resonances. There are exponentially thin layers in the phase space (near the separatrices associated with resonances), where particles that happen to be there as a result of initial conditions behave erratically. In one-dimensional systems the KAM surfaces are closed and provide stability even for a particle inside such a layer. In multidimensional systems the layers associated with different resonances can intersect each other, making it possible for the particle in the layer to drift from one resonance to another and eventually away to the machine aperture (so-called Arnol'd diffusion). This would not practically reduce the bunch lifetime unless there were no external noise in the equations of motion, which can occasionally bring more and more particles inside the stochastic layer.

The situation changes drastically with increase of the strength parameter. At a certain value of this parameter the widths of the next resonances become big enough to touch each other. The KAM surface breaks down, and the motion becomes stochastic for a substantial portion of the phase space (initial conditions). This resonance overlapping situation is believed to create the beam–beam limit and is a content of the Chirikov criterion for stability.

In the following, I describe very briefly (for completeness only) the most important ideas developed along this line. The subject links the problem of beam–beam interaction to the more general problem of long-time behavior of the nonintegrable Hamiltonian systems and exhibits some basic concepts of nonlinear Hamiltonian dynamics: resonance, closed orbits, stochasticity, and Arnol'd diffusion. More comprehensive description and more details can be found elsewhere in Part 4 and in the literature (see, e.g., Refs. 52–54, and references therein).

3.2.1. Estimate of the Beam–Beam Limit According to the Chirikov Criterion. Let us evaluate the beam–beam limit according to the Chirikov criterion.[55] I follow here the derivation of the work.[56] Practically the same treatment is used by Bountis.[53] One finds an interesting approach[57] expressing the Chirikov criterion in terms of a turbulence limit of a flow in the phase space.

Consider one-dimensional motion of a particle in the presence of the second bunch. The Hamiltonian of the motion in action-angle variables is

$$H = \nu J + \sum_{m,\, k=-\infty}^{\infty} \varepsilon_{km}(J)\, e^{i(m\phi - k\theta)} \tag{3.5}$$

Term $\varepsilon_{00}(J)$ determines the dependence of the oscillation frequency on amplitude:

$$\alpha = \frac{d\omega}{dJ} = \frac{d^2\varepsilon_{00}}{dJ^2} \tag{3.6}$$

For δ-function-like perturbations in θ, ε_{km} does not depend on k. If ε_m is small enough, each of the other terms in Eq. (3.5) produces a resonance at the frequency

$$\omega_{km} = \frac{k}{m} = \nu + \alpha J_{km} \tag{3.7}$$

The last equation in Eq. (3.7) determines the resonant amplitude J_{km}.

In phase space a resonance is surrounded by a separatrix having the width

$$\delta J_{km} = 4\left[\frac{2\varepsilon_m(J_{km})}{\alpha}\right]^{1/2} \tag{3.8}$$

The same value in terms of frequency is

$$\delta\omega_{km} = \alpha\,\Delta J_{km} = 4\left[2\alpha\varepsilon_m(J_{km})\right]^{1/2} \tag{3.9}$$

The resonance separation around the resonance of the order m is $\Delta\omega_{km} \simeq 1/m$. Now the Chirikov criterion can be formulated quantitatively if one finds the sum of the ratio of the widths to the separations in a unit frequency interval. If one introduces the notation

$$\frac{1}{\lambda} = \sum_m m\left[2\alpha\varepsilon_m(J_{km})\right]^{1/2} \tag{3.10}$$

the Chirikov stability criterion becomes

$$\lambda > 4 \tag{3.11}$$

The numerical investigation of the validity of this criterion on the model called the standard mapping[58] shows that Eq. (3.11) overestimates stability (actual loss of stability appears for a smaller value of ε).

The same seems to be true for the beam–beam stability limit, at least in the simple one-dimensional model with the round Gaussian strong bunch. Equation (3.11) gives in this case, for the space charge parameter,

$$\varepsilon < \tfrac{1}{8} \tag{3.12}$$

which is clearly three to four times better than the experimental observations.

It is common thinking that the limit will look more realistic for bunched beams if one takes into account the existence of additional synchrobetatron resonances arising because of coupling of the transverse motion to the longitudinal one. The overlap of synchrobetatron resonances has been studied by Israilev,[59] who showed that indeed the beam–beam limit can be reached at $\xi \approx 0.04$. For unbunched proton beams the more severe limit is believed to be connected with Arnol'd diffusion—in the absence of damping the needed lifetime can be achieved only below the stochastic limit when resonance overlapping has not developed to full scale.

3.2.2. Study of Nonlinear Equations of Motion. In a completely different approach to the beam–beam problem, the stability limit is sought from the investigation of the nonlinear Hill's equation[60,61] or corresponding finite difference equations.[62]

Again the one-dimensional problem is investigated. Consider two first-difference mappings

$$x_{t+1} = y_t \tag{3.13}$$

$$y_{t+1} = -x_t + 2y_t \cos 2\pi\nu + \xi F(y_t)\frac{\sin 2\pi\nu}{\nu} \tag{3.14}$$

describing the vertical motion in the presence of the δ-function beam–beam force $\xi F(y)$, where ξ is the strength parameter. For the piecewise linear beam–beam force

$$F(y) = \begin{cases} y & \text{for} \quad |y| < \sqrt{\pi}/2 \\ \dfrac{\sqrt{\pi}}{2} & \text{for} \quad y > \sqrt{\pi}/2 \\ -\dfrac{\sqrt{\pi}}{2} & \text{for} \quad y < -\sqrt{\pi}/2 \end{cases} \tag{3.15}$$

it appeared possible to construct an invariant curve $y(x)$ in the phase plane, that is, the curve invariant in respect to the nonlinear map under consideration. It contains a stable region of the motion for all times. For each given value of ν only one such curve is found to exist. The stability limit is determined from the following expression:

$$\Delta\nu = \frac{\xi}{4\pi\nu} = \frac{\cos 2\pi\nu(1 + \cos 2\pi\nu)}{\pi(1 + 2\cos 2\pi\nu)(-\sin 2\pi\nu)} \tag{3.16}$$

For the ISABELLE storage ring this gives $\delta\nu \approx 0.03$. This value seems again to be too optimistic.

3.3. Diffusion Model

We have discussed attempts to describe the beam–beam limit as a combined effect of a single resonance and a diffusionlike change of its parameters. Yet when there are many rather strong resonances inside the tune change region of the particle, the motion of the particle can become stochastic even in the absence of special noise sources. That circumstance makes it plausible to try to consider the beam–beam interaction in the limit where there is no correlation between the results of different collisions and between the phases of a particle at different interaction points.[22,63–66] Such considerations do not pretend to constitute a rigorous theory but rather a phenomenological model that helps to make parametrization of the experimental data in a suitable way and to derive some scaling laws by means of a few fitting parameters. The behavior of these fitting parameters is not described by a theory and should be taken from comparison with experiment.

3.3.1. Beam Blowup. At each interaction the vertical coordinate y and the angle in the vertical plane y' are changed as follows:

$$\Delta y = 0 \tag{3.17}$$

$$\Delta y' = 2\pi\xi_y \frac{\sigma_0}{\beta_y} K_b \phi_b(u) \tag{3.18}$$

where $b = (\sigma_y/\sigma_x)/[1 - (\sigma_y/\sigma_x)^2]^{1/2}$, $u = y/\sigma_0$, and $K_b\phi_b$ is a function describing the electromagnetic force of the opposite bunch. For Gaussian distribution[22]

$$K_b = \left[\frac{(1+b^2)^{1/2} + b}{(1+b^2)^{1/2} - b}\right]^{1/2} \tag{3.19}$$

$$\phi_b(u) = u\int_0^1 \frac{dw}{[w+b^2]^{1/2}} e^{-wu^2} \tag{3.20}$$

We know that at least the linear part of the force cannot cause the stochasticity. It can be considered to be an additional focusing force, hence should be included in the regular part of particle motion. Probably the same is true also for some nonlinear parts of the force.

That is why for the purpose of calculating beam blowup as a consequence of a diffusionlike process we should consider not all the force $\phi_b(u)$, but only some nonlinear part of it $\tilde{\phi}_b(u)$. The means of obtaining $\tilde{\phi}_b$ from ϕ_b is not clear and should be considered here only as a way to introduce in the theory a phenomenological fitting parameter.

It is reasonable to assume that for particles that behave erratically there is a complete mixing of phases within the bunch, and in the long run each particle can be expected to acquire any value of coordinate y. In this case the beam blowup can be found by averaging the value $\theta^2 = (\Delta y')^2$ over the distribution function[67]

$$\sigma_y^2 = \sigma_0^2 + fB\tau\beta_y^2\theta_{\text{rms}}^2 \tag{3.21}$$

where θ_{rms} is the effective root-mean-square scattering angle of a particle in the vertical plane, produced by the opposite bunch, f is the revolution frequency, and τ is damping time of vertical oscillations.[68] We get

$$\sigma_y^2 = \sigma_0^2 + \frac{2\pi^2 e^2 \tau\beta_y^2 h^2 \sigma_0^2 i^2}{fBE^2\sigma_y^2\sigma_x^2(1 + \sigma_y/\sigma_x)^2} \tag{3.22}$$

First of all we see here exactly the same formula [Eq. (2.1)] that was postulated in Ref. 24. Comparing Eqs. (3.32) and (2.1), we find

$$a = \frac{\pi e\beta_y h\sigma_0}{E\sigma_x(1 + \sigma_y/\sigma_x)}\left(\frac{2\tau}{fB}\right)^{1/2} \tag{3.23}$$

3.3.2. Scaling Laws. Equations (3.22) and (3.23) contain only one unknown parameter h. Let us consider it to be a phenomenological parameter that should be determined from experimental data. One way to do this is to use PETRA results[24] in Eq. (2.2). It is easy to see that to satisfy an E^{-4} decrease for the value a, we need the following dependence of h on energy:

$$h \sim E^{-3/2} \tag{3.24}$$

Since we are interested now in maximum values of luminosity and current, we derive from Eq. (2.1) that asymptotically at large current i, $\sigma_y^4 \simeq a^2 i^2$ or

$$\sigma_y \sim \frac{\sqrt{i}}{E^2} \tag{3.25}$$

The maximum possible value of σ_y limited by particle losses and beam lifetime should be some constant, which can be written as $(A_y)^{1/2}$, where A_y is an effective vertical acceptance of the storage ring. Let us see now what consequences follow from these assumptions.

1. *Dependence on Energy.* Consider first what happens when the limitation arises from the beam lifetime. Assuming $\sigma_y = \text{const}$ in Eq. (3.25), we immediately get

$$i_{\text{max}} \sim E^4 \tag{3.26}$$

With the help of this expression we also get the following scaling laws (note that for the electron storage ring $\sigma_x \sim E$):

$$L_{\max} \sim E^7 \tag{3.27}$$

$$\xi_{y_{\max}} \sim E^2 \tag{3.28}$$

$$\xi_{x_{\max}} \sim E \tag{3.29}$$

$$\frac{\sigma_y}{\sigma_x} \sim \frac{1}{E} \tag{3.30}$$

2. *Dependence on Current.* Let us now turn to experiments in which beam lifetime limit has not been reached yet. At a given energy one gets from the same expressions:

$$\sigma_x \sigma_y \sim i^{1/2} \tag{3.31}$$

$$\xi_{y_{\max}} \sim i^{1/2} \tag{3.32}$$

$$L_{\max} \sim i^{3/2} \tag{3.33}$$

Table 5. The Power q in the Power Law $f(E) \sim E^q$

	Experiment					
Parameter	SPEAR	ADONE	PETRA	Model	Equation	Comment
h				$-\frac{3}{2}$	(3.24)	
$\mathfrak{L}_{\max}$	6.6	7		7	(3.27)	
$i_{\max}$	3.6	4.5		4	(3.26)	Strong–strong
$i_{\max}$		4.12 4.34		5	(3.35)	Weak–strong
ξ_y	2.3	1.5		2	(3.28)	
ξ_x	0.9			1	(3.29)	
σ_y/σ_x	−1			−1	(3.30)	
a			−4	−4	(2.2)	

Table 6. The Power q in the Power Law $f(i) \sim i^q$

	Experiment				
Parameter	SPEAR	ADONE	PETRA	Model	Equation
$\mathfrak{L}_{\max}$	1.4			1.5	(3.33)
$\mathfrak{L}_{sp_{\max}}$			−0.5	−0.5	
$\sigma_x\sigma_y$	0.6			0.5	(3.31)
ξ_y	0.4			0.5	(3.32)

Tables 5 and 6 summarize the theoretical and experimental values for different parameters relevant for the beam–beam interaction. Keeping in mind the number of assumptions and the approximations made, the agreement seems to be astonishingly good.

3.4. Computational Models

In this section I discuss some computer simulations of the beam–beam interaction. All such simulations are performed by tracing a set of particles with different initial conditions through many interactions. Motion between the interactions is assumed to be linear (i.e., simple rotation in phase space with constant amplitude). Hence, the nonlinearities of the lattice are neglected.

The main attraction of the computational method for the investigation of beam–beam phenomena (as well as many other complicated objects) is that this method presents a unique possibility to see the behavior of a sample of particles when the solution of the equations governing their motion is unknown. The finite capability of a computer forces us to neglect some features of the motion, thus making the method as approximate as any other. The approximation can be more or less physically sound, but it is always there. We see that the computational method is always applied to a certain physical model in the same way as the analytical method is. To some extent both methods can be considered to complement each other, since the analytical method is usually applicable only if there is a small parameter in the model. The computational method usually causes trouble in the presence of a small parameter.

There is also a substantial disadvantage in using the computational technique. It demands much skill to extract from the results of computations any general natural law—if it is possible at all. On the other hand, this method is sometimes the only one that is available.

Out of many simulations done,[54–69, 70–74] I have selected only two recent examples, which quite substantially represent the bright and dark sides of all computational efforts. I omit here one of the most interesting, Tennyson's model,[69] since it is described in this volume.

Piwinski's model[18] deals with the weak beam–strong beam interaction. All three degrees of freedom are considered here. Damping and quantum fluctuations of radiation are also taken into account.

Charge distribution of the strong beam was assumed to be Gaussian. The "kicks" $\Delta x'$ and $\Delta y'$ are given in the integral form [Eqs. (1.1) and (1.2)]. The integrals were tabulated for the grid of 75×200 points with the distances of $0.2\sigma_x$ and $0.2\sigma_y$ between them. The exact kick experienced by a particle by the passage through the strong beam was interpolated for each collision.

The tracking is performed for PETRA parameters $\sigma_x/\sigma_y = \beta_x/\beta_y = 15$, $\sigma_s/\beta_y = 0.1$, $\nu_x = 25.2$, $\nu_s = 0.07$, and different ν_y. Three different particle energies (7, 11.3, and 17.9 GeV) and three different numbers of bunches (one, two, and four) are studied. The initial coordinates of particles were uniformly distributed in phase and Gaussian in amplitude. The typical number of particles is 125. They were traced for up to 12 damping times.

Special attention is paid to possible asymmetries and distortions in the machine. In particular, differences of betatron phase advances between interaction points (including the change of this value due to synchrotron oscillation for a particle) and spurious horizontal dispersion function are investigated and show drastic influence on the beam blowup. Small vertical dispersion at the interaction point as well as linear coupling of horizontal and vertical oscillations due to skew quadrupoles show no effect on the blowup.

The results of simulation clearly demonstrate the appearance of many additional resonances arising because of disturbances of the storage ring and synchrotron oscillations. The resonance augmentation of y_{rms} is seen even for the smallest space charge parameter studied ($\xi = 0.02$), and is markedly stronger when ξ is increased to 0.06.

Talman's computational model[74,75] is the only one I know of that is designed to study the strong beam–strong beam effect. Many particles (~ 100 in each of the two beams) are tracked for many turns (~ 3000, which corresponds to approximately three damping times τ) in six-dimensional phase space. Radiation damping, quantum fluctuations, and vertical coupling are taken into account.

For a period of 0.3τ one bunch is held rigid. From the tracking results for the particles of the second bunch, smooth distribution and the horizontal and vertical fields of this bunch are calculated. Then the roles of the two bunches are reversed.

The possibility of particle loss is incorporated by installing a mask (typically set at $\pm 10\sigma_x$ and $\pm 10\sigma_y$ for corresponding deviations). In this way the lifetime of the bunch can be found. For the conditions giving a "good" lifetime, luminosity is calculated from the equilibrium particle distributions. It is found that the regions of "bad" lifetime correlate with the positions of the resonances of the nonlinear equation of vertical motion parametrically "pumped" by the horizontal oscillations.

A comparison between the experimental observations and the results of computations shows good agreement both in the dependence of $L(i)$ and in the values of i_{max} beyond which the lifetime is too short (Fig. 3.2).

The sources of the beam blowup as seen by the authors of all three computational studies are discerned quite differently. It is noise in the first,[69] machine distortions in the second[73] and strong coupling to horizontal motion in the third.[74] The same difference holds in respect to a mechanism of the blowup. It is a single resonance in the first, simultaneous action of many resonances in the second, and parametric amplification of the vertical oscillations by horizontal motion in the third.

Such controversy is hardly surprising and might point to the existence of all three (and maybe even more) reasons for the beam instability.

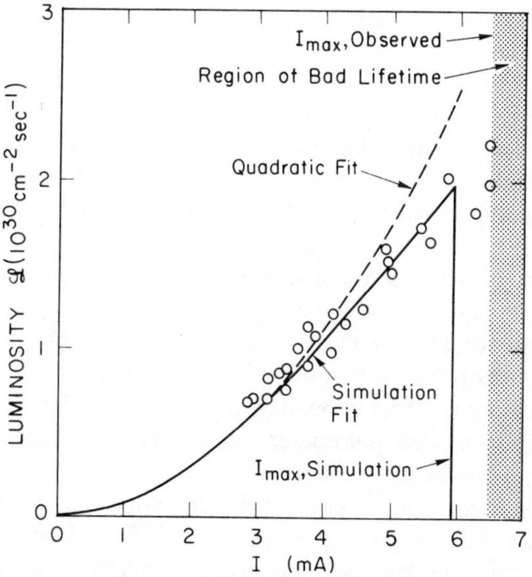

Figure 3.2. Comparison of the experimental observations at CESR and the results of computation according to Talman's model. (From Peggs and Talman.[74])

4. CONCLUSION

Considering the state of the beam–beam problem should one be frustrated or excited? The answer is of course subjective.

Personally, I am more exhilarated—the physics with and of the storage ring is progressing rapidly, after all—rather than disappointed by all the failures to properly describe the beam–beam phenomena.

ACKNOWLEDGMENTS

This chapter would never appear without the encouragement and advise of Dr. M. Month.

REFERENCES

1. *Proceedings of the Beam-Beam Interaction Seminar*, SLAC PUB 2624 (Stanford Linear Accelerator Center, Stanford, CA, May 22–23, 1980).
2. *Nonlinear Dynamics and the Beam–Beam Interaction*, American Institute of Physics Conference Proceedings, Vol. 57, M. Month and J. Herrera, Eds. (AIP, New York, 1979).
3. *Proceedings of the Eleventh International Conference on High Energy Accelerators*, Geneva (1980).
4. *Proceedings of the Tenth International Conference on High Energy Accelerators*, Serpukhov (1977).

5. F. Amman, *IEEE Trans. Nucl. Sci.* **NS-20**:3, 858 (1973).
6. H. Wiedemann, in Ref. 2, p. 84.
7. A. Chao, in Ref. 2, p. 42.
8. S. Kheifets, SLAC PUB 2700 PEP NOTE 346 March 1981 (A). (Stanford Linear Accelerator Center, Stanford, CA, 1981)
9. S. Kheifets, PETRA-KURZMITTEILUNG N113 (Deutsches Elektronen-Synchrotron, Hamburg, West Germany); (1977).
10. F. Amman and D. Ritson, in *International Conference on High Energy Accelerators* (Brookhaven National Laboratory, Upton, NY, 1961), p. 471.
11. E. Courant, *IEEE Trans. Nuclear Sci.* **NS-12**:3, 550 (1965).
12. D. Ritson and J. Rees, SLAC-TN-65-39 (Stanford Linear Accelerator Center, Stanford, CA, 1965).
13. M. Bassetti, *Fifth International Conference on High Energy Accelerators* (Frascati, 1965), p. 708.
14. Y. Orlov, *Sov. Phys. JETP* **5**:1, 45 (1957).
15. K. Robinson, CEA-54 (Cambridge Electron Accelerator, Cambridge, Mass.) (1958).
16. R. Sundelin, *IEEE Trans. Nucl. Sci.* **NS-26**:3, 3604 (1979).
17. N. Vinokurov et al., in Ref. 4, p. 254.
18. A. Piwinski, in Ref. 3, p. 638.
19. Y. Derbenev, SLAC Translation 151 (Stanford Linear Accelerator Center, Stanford, CA, 1973).
20. N. N. Chau and D. Potaux, Orsay, Rapport Technique 5-74 (1974).
21. N. N. Chau and D. Potaux, Orsay, Rapport Technique 2-75 (1975).
22. S. Kheifets, *IEEE Trans. Nucl. Sci.* **NS-26**:3, 3615 (1979).
23. S. Kheifets, M. Donald, and P. Morton (unpublished).
24. G. Voss, "First and Only Partial Analysis of Space Charge Effects in Petra," DESY, M-VM-79/6 (Deutsches Elektronen-Synchrotron, Hamburg, West Germany) (1979).
25. D. Degéle et al., DESY 80/10 (Deusches Elektronen-Synchrotron, Hamburg, West Germany) (1980); also, D. Degéle, in Ref. 3, p. 16.
26. H. Zyngier, in Ref. 2, p. 136.
27. J. LeDuff, in Ref. 1, p. 80.
28. S. Tazzari, in Ref. 2, p. 128.
29. M. Donald and J. Paterson, *IEEE Trans. Nucl. Sci.* **NS-26**:3, 3580 (1979).
30. G. Voss, in Ref. 3, p. 748.
31. M. Donald, private communication.
32. H. Wiedemann, in Ref. 3, p. 744.
33. M. Cornaccia, in Ref. 2, p. 99.
34. J. Paterson, in Ref. 3, p. 7.
35. J. Rees, private communication.
36. The CESR Operations Group (Cornell Electron Storage Ring, Cornell University, Ithaca, NY), in Ref. 3, p. 26.
37. P. Marin, in Ref. 3, p. 742.
38. The VEPP-4 Group (Vstrechnie Electron-Positronii Puchki, Nuclear Physics Institute, Novosibirsk, USSR), in Ref. 3, p. 38.
39. A. Hoffman and B. Zotter, in Ref. 3, p. 713.
40. B. Zotter, in Ref. 4, Vol. II, p. 23.

41. H. Uhm and C. Liu, *Phys. Rev. Lett.* **43**, 914 (1979).
42. S. Kheifets and A. Chao, AATF/79/13, PEP-Note-325 (Stanford Linear Accelerator Center, Stanford, CA) (1979).
43. R. Hollebeek, in Ref. 1, p. 165.
44. A. Garren, in Ref. 3, p. 725.
45. R. Sah, in Ref. 3, p. 736.
46. M. Month, *IEEE Trans. Nucl. Sci.* **NS-22**:3, 1376 (1975).
47. H. Hereward, CERN/ISR-DI/72-26 (European Organization for Nuclear Research, Geneva, Switzerland) (1972).
48. J. LeDuff, CERN/ISR-AS/74-53 (European Organization for Nuclear Research, Geneva, Switzerland) (1974).
49. A. Dragt and O. Jakubowicz, in Ref. 1, p. 205.
50. J. Cary, LBL-6350 (Lawrence Berkeley Laboratory, Berkeley, CA, 1970).
51. A. Dragt and J. Finn, *J. Appl. Math.* **17**:12, 2215 (1976).
52. J. Greene, in Ref. 1, p. 235.
53. T. Bountis, in Ref. 1, p. 248.
54. C. Eminhizer, in Ref. 1, p. 273.
55. B. Chirikov, *Phys. Rep.* **52**:5, 265 (1979).
56. E. Courant, in Ref. 3, p. 763.
57. L. Teng, *IEEE Trans. Nucl. Sci.* **NS-20**:3, 843 (1973).
58. J. Greene, *J. Math. Phys.* **20**, 1183 (1979).
59. F. Israilev, "Nearly Linear Mappings and Their Applications," submitted to *Physica D* (1980).
60. P. Vuillermot, in Ref. 1, p. 312.
61. T. Bountis and E. Coutsias, in Ref. 2, p. 311.
62. R. Hellemann, in Ref. 2, p. 236.
63. E. Keil, PEP-Note-59 (Stanford Linear Accelerator Center, Stanford, CA) (1973).
64. L. Teng, in Ref. 1, p. 99.
65. S. Kheifets, in Ref. 1, p. 40.
66. A. Ruggiero, "The Theory of Nonlinear Systems in Presence of Noise (the Beam–Beam Effect)," Fermilab (Fermi National Laboratory, Batavia, Ill.) (1980).
67. J. Rees, private communication.
68. M. Sands, SLAC-121 (Stanford Linear Accelerator Center, Stanford, CA, November 1970).
69. J. Tennyson, in Ref. 1, p. 1.
70. E. Close, in Ref. 2, p. 210.
71. J. Herrera, M. Month, and R. Peierls, in Ref. 2, p. 202.
72. D. Neuffer and A. Ruggiero, in Ref. 1, p. 332.
73. A. Piwinski, in Ref. 3, p. 754.
74. S. Peggs and R. Talman, in Ref. 3, p. 754.
75. S. Peggs, "Some Aspects of Machine Physics in the Cornell Electron Storage Ring," Ph.D. Thesis, Cornell University, Ithaca, NY (May 1981).
76. I. Vasserman et al., Proceedings of the 5th ALLUNION Conference on Accelerators, Dubna, USSR, 1976), p. 252.

RESONANCE STREAMING IN ELECTRON–POSITRON COLLIDING BEAM SYSTEMS

JEFFREY L. TENNYSON
Electronic Research Laboratory
University of California
Berkeley, California

Abstract. *The two dimensional Hamiltonian describing betatron oscillation is subjected to an infinite sequence of delta function impulses as a model for beam–beam interactions. Theoretical analysis of both the primary and higher order phase space resonances determines the conditions for radial and vertical beam blowup.*

1. INTRODUCTION

The luminosities of electron–positron colliding beam machines, such as those presently operating at Stanford (PEP) and Hamburg (PETRA), are limited by the perturbing effects of the "beam–beam interaction" (the electromagnetic force experienced by a particle of one beam as it passes through a bunch of the opposing beam).[1,2] Efforts to increase these luminosity limits have been severely hampered by lack of knowledge concerning the precise mechanism by which the beam–beam interaction imposes these limits. Until recently, attempts to understand the phenomenon resulted in theories that were either inadequate to explain the observations or not precise enough to be useful. Such studies, although largely inconclusive, nevertheless contributed to what is now a fairly comprehensive picture of the various dynamical processes involved in this elusive effect.

If radiation effects are neglected, the motion of a single stored particle can be approximately determined using perturbation theory. The single-particle system is modeled by a two-dimensional linear oscillator that is subject to a small time-dependent perturbation (the beam–beam interaction). The dynamics of such "near-integrable" systems have been reviewed in detail by Chirikov[3] and can be calculated with surprising precision. These systems are characterized by a "stochasticity threshold," a critical perturbation strength below which the oscillator amplitudes exhibit bounded oscillations and above which they diffuse chaotically. For the colliding beam machines, this means that there exists some critical strength of the beam–beam interaction (and correspondingly, a critical current of the opposing beam) above which the normal modes of the transverse oscillation exchange energy in a statistically random way. Such a diffusion could easily result in beam enlargement and perhaps even particle loss (as observed at the beam–beam limit). Consequently, it has long been suspected that the beam–beam limit results from intrinsic stochasticity in the the conservative (Hamiltonian) motion.[4]

Although the stochasticity theory has some rather attractive features, the near-integrable system approach also suggests a second mechanism that is equally appealing. This involves the enhancement of non-Hamiltonian processes, such as diffusion or dissipation, by isolated nonlinear resonances. Because the resonances must be isolated, the system is predominantly regular (at least on some appropriate time scale) and therefore operating at currents below the stochasticity threshold. A number of different enhancement effects have been investigated. These can be divided into two groups: effects involving enhancement of motion *across* nonlinear resonances[5] and those involving enhancement of motion *along* nonlinear resonances.[6–9]

The systematic application of these theories to the beam–beam interaction has not been easy, primarily because of the complexity of the system (the large number of independent variables). Notable progress has been made in Novosibirsk (by Izrailev et al.) with the stochasticity theory.[10,11] These studies have shown that a slow and sufficiently deep modulation of the beam–beam interaction strength (e.g., that experienced by a particle with large amplitude synchrotron oscillations) reduces the analytic stochasticity threshold to values comparable to those actually observed. Although these successes lend considerable credence to the stochasticity theory of the beam–beam limit, a number of details remain unexplained. Among these are the existence of "tails" in the cross-sectional profile of the beam (SPEAR) and a significant dependence of the beam–beam interaction on damping time.[12,13] Furthermore, a number of observations (see, e.g., Ref. 14) seem to indicate that beam enlargement is not a threshold effect at all, but a smoothly varying function of the beam current.

This chapter applies the theory of enhanced transport along nonlinear resonances[9] to a simple model of the beam–beam interaction. This particular mechanism was first observed in a two-dimensional simulation[15] of a flat beam device that included radiation effects but not synchrotron modulation. The simulation produced both beam enlargement and particle loss at currents comparable to those observed on real machines. It also exhibited both the "tails" observed at SPEAR, and a marked dependence of the critical current on energy. The mechanism responsible for this behavior was found to have nothing to do with stochasticity. It involved instead the trapping of particles in certain sum resonances and the rapid transport of these particles along the resonances to large vertical amplitudes. The transport (referred to here as *resonance streaming*), was predominantly nonrandom, and was found to be driven primarily by the radiation damping.

It now appears quite likely that either intrinsic stochasticity or resonance streaming (and possibly both) are responsible for the beam–beam limit. A detailed comparison of the two processes and their compatibility with the observations is made in Ref. 16.

This chapter presents only the principal features of the resonance streaming theory and its application to a simple "flat beam" model of single-particle motion in a colliding beam machine. Although the analysis is specific to electron–positron devices with flat beams, the procedure is applicable, with only slight modifications, to any colliding beam machine.

A description of a simple Hamiltonian model is given in Section 2. Section 3 describes briefly the theory of resonance streaming and provides derivations of the expressions used in the analysis. The results of the analysis are presented in Section 4. These results show that above a certain order, only sum resonances effect the beam, and that certain specific sum resonances are more important than others. They also show that the number of dangerous sum resonances increases rapidly with the damping time, necessitating a drop in the "linear tune shift" for low energy beams. A discussion of the results appears in Section 5.

2. DESCRIPTION OF THE MODEL

As a particle travels around the ring, it oscillates about an ideal trajectory called the "design orbit" (for details on e^+–e^- storage rings see Sands[17]). These oscillations, called betatron oscillations, are approximately linear with normal modes in the vertical and radial directions. Without the beam–beam interaction, the single-particle system may be approximately represented by an integrable Hamiltonian function of the form

$$H(x, p_x, z, p_z, t) = \frac{p_x^2}{2} + \omega_x^2(t)\frac{x^2}{2} + \frac{p_z^2}{2} + \omega_z^2(t)\frac{z^2}{2} \tag{1}$$

The canonically conjugate variables p_x, x and p_z, z are the radial and vertical momenta and displacements from the design orbit. The time dependence in the frequencies comes from the nonuniform distribution of quadrupole magnets around the ring (where the dependence on position along the design orbit has been converted to dependence on elapsed time via the constant speed of light c).

When there are only two bunches in the storage ring, each particle collides with its opposing bunch twice per revolution at diametrically opposed interaction points. It is convenient to choose as the unit of time, the travel time of the particle between interaction points (half the revolution period). Assuming that the functions $\omega_x(t)$ and $\omega_z(t)$ are periodic with period one. Eq. (1) may be solved,[18] and the solution expressed in the form

$$x(t) = \alpha\beta_x^{1/2}(t)\cos(\psi_x) \tag{2}$$

$$\dot{\psi}_x(t) = \frac{1}{\beta_x(t)} \tag{3}$$

and similarly for z. The "beta function" β_x is a property of the magnet lattice as a whole, but depends on the longitudinal position along the design orbit. Equation (3) can be integrated between interaction points to give a mapping

$$x_{n+1} = x_n\cos(\pi\nu_x) + x'_n\beta_x\sin(\pi\nu_x) \tag{4}$$

$$x'_{n+1} = -x_n\beta_x^{-1}\sin(\pi\nu_x) + x'_n\cos(\pi\nu_x)$$

The constants ν_x and ν_z (called the "betatron tunes") are properties of the magnet lattice. The variable x' is defined by

$$x' = -\alpha\beta_x^{-1/2}(t)\sin(\psi_x) \tag{5}$$

and is related to the momentum p_x by

$$x' = p_x + \frac{\alpha}{\beta}x \tag{6}$$

where α is another lattice-dependent function of the longitudinal position (see the reference in note 18 for definitions of these functions). To an observer who can see the particle only as it passes an interaction point, the particle appears to be oscillating harmonically about the design orbit with frequencies

$$\bar{\omega}_x = \pi\nu_x \tag{7}$$

$$\bar{\omega}_z = \pi\nu_z$$

However, the local frequencies at the interaction points are actually

$$\omega_x^* = \frac{1}{\beta_x(t=n)} \tag{8}$$

$$\omega_z^* = \frac{1}{\beta_z(t=n)}$$

where n is an integer. By making ω^* larger than $\bar{\omega}$, the machine designers are able to compress the beam's cross section as the bunch passes the interaction points.

When the beam–beam interaction is added to the system, it appears in Eq. (1) as a series of δ functions in time.

$$H = H_0 + \sum_{m=-\infty}^{\infty} \delta(t-m)V(x,z) \tag{9}$$

Since the motion between interaction regions is known [Eq. (4)], it is necessary to investigate only the motion on the infinitesimal time intervals during which the collisions take place. For this purpose, it is useful to introduce a new "local" Hamiltonian function

$$H(x,x',z,z',t) = \frac{z'^2}{2} + \omega_z^{*2}\frac{z^2}{2} + \frac{x'^2}{2} + \omega_x^{*2}\frac{x^2}{2} + \sum_{m=-\infty}^{\infty} \delta(t-m)V(x,z) \tag{10}$$

with the restriction $m- < t < m+$ where m is an integer. The replacement of p_x and p_z by x' and z' is justified because the jump in p_x due to the beam–beam interaction is equal to the jump in x',

$$\Delta p_x = \Delta x' \tag{11}$$

as can be seen from Eq. (6). When the colliding beams are precisely aligned, and there is no dispersion at the interaction point, the potential $V(x,z)$ is an odd function of both x and z. In this case, the $\Delta x'$ and $\Delta z'$ in an e^+-e^- machine always advance the phases ψ_x and ψ_z of the betatron oscillation. The resulting changes in the betatron tunes for small amplitude oscillations are called the "linear tune shifts" and are easily derived from Eq. (10):

$$\xi_z = \frac{1}{4\pi\omega_z^*}\left.\frac{\partial^2 V}{\partial z^2}\right|_{z,x=0} \tag{12}$$

$$\xi_x = \frac{1}{4\pi\omega_x^*}\left.\frac{\partial^2 V}{\partial x^2}\right|_{z,x=0}$$

The two linear tune shifts can be adjusted independently by varying the local frequencies ω^*.

When the beams are flat (the ratio of vertical to radial width is very small $\sigma_z/\sigma_x \ll 1$), the Hamiltonian function Eq. (10) may be split into vertical and radial parts.[15]

$$H_z(z', z, t) = \frac{z'^2}{2} + \omega_z^{*2}(t)\frac{z^2}{2} + \sum_{m=-\infty}^{\infty} \delta(t-m)V(z, x(t)) \tag{13}$$

$$H_x(x', x, t) = \frac{x'^2}{2} + \omega_x^{*2}(t)\frac{x^2}{2} + \sum_{m=-\infty}^{\infty} \delta(t-m)W(x) \tag{14}$$

Here the solution $x(t)$ for Eq. (14) serves as an explicit time dependence in Eq. (13). The system of vertical oscillations is then one dimensional with time dependence. The time dependence is quasi-periodic if the solution for the radial motion Eq. (14) has a discrete frequency spectrum (at the linear tune shifts of interest, this condition is satisfied for most initial conditions). In this case, Eq. (13) is a "near-integrable system" and exhibits both regular and stochastic motion (for a general discussion, see Ref. 19 or 20).

It is useful at this point to switch to action-angle variables. The transformed local Hamiltonian function for the vertical motion becomes

$$H_z(I_z, \theta_z, t) = \omega_z^* I_z + \sum_{m=-\infty}^{\infty} \delta(t-m) f(I_z, \theta_z, I_x(t), \theta_x(t)) \tag{15}$$

where

$$I_z = \frac{1}{\omega_z^*}\frac{z'^2}{2} + \omega_z^*\frac{z^2}{2} \tag{16}$$

$$\theta_z = \tan^{-1}\left(\frac{1}{\omega_z^*}\frac{z'}{z}\right)$$

and I_x, θ_x are similarly defined. The next step is to subject the perturbation sum to Fourier transformation:

$$H_z = \omega_z^* I_z + \langle V\rangle_{\theta_x,\theta_z}(I_x, I_z)$$
$$+ \sum_{k=1}^{\infty}\sum_{n=-\infty}^{\infty} F_k(I_x, I_z)\cos(m_{kx}\theta_x + m_{kz}\theta_z + 2\pi n t) \tag{17}$$

where the m_{kx}, m_{kz}, $n = 0$ term has been removed from the sum and expressed explicitly. This term, which is just the average of V over the angles θ_x and θ_z, depends only on the actions I_x, I_z. If I_x is considered a constant, this term by itself does not destroy the integrability of the system. It does, however, introduce a nonlinearity. If $\langle V\rangle(I_z)$ is small compared to $\omega_z^* I_z$, the dependence of the oscillator frequency on amplitude is given approximately by

$$\omega_z(I_z I_x) \approx \bar{\omega}_z + \frac{\partial}{\partial I_z}\langle V\rangle_{\theta_x,\theta_z}\bigg|_{I_z, I_x} \tag{18}$$

(the validity of this formula is questionable on low order resonances). When

I_x, I_z are small, the second term on the right-hand side of Eq. (18) gives the linear tune shift,

$$\xi_z \approx \frac{1}{2\pi} \frac{\partial}{\partial I_z} \langle V \rangle_{\theta_x, \theta_z} \bigg|_{I_z, I_x = 0} \qquad \text{for} \quad \xi_z \ll 1 \tag{19}$$

The potential energy function for vertical motion in a flat beam may be roughly approximated by[15]:

$$V(x, z) = 8\pi e^{-x^2/2} \left(1 + \frac{z^2}{2}\right)^{1/2} \tag{20}$$

where x and z are normalized to the beam dimensions σ_x and σ_z. Inverting the definitions, Eq. (16) yields

$$x = \left(I_x \frac{2}{\omega_x^*}\right)^{1/2} \cos(\theta_x) \tag{21}$$

$$z = \left(I_z \frac{2}{\omega_z^*}\right)^{1/2} \cos(\theta_z)$$

Equation (20), when used with Eq. (21), can be shown to satisfy both Eqs. (12) and (19), when ξ_z is small and $I_x = 0$. From Eq. (21), the amplitude A_z of the oscillator is related to the action (to zero order) by

$$I_z = \frac{\omega_z^*}{2} A_z^2 \tag{22}$$

The radial motion is treated in a similar way. The radial potential is approximated with[15]:

$$W(x) = -10.22\pi \left(1 + \frac{x^2}{5.11}\right)^{-1} \tag{23}$$

With equal β-functions $\beta_x = \beta_z$, and a bi-Gaussian current distribution, the tune shift ratio is equal to the inverse of the aspect ratio $\xi_z/\xi_x = \sigma_x/\sigma_z$. Since this is infinite for flat beams, a factor σ_x/σ_z has been included in Eq. (23), and ω_x^* in Eq. (14) should be reinterpreted accordingly. Note that the radial tune shift does not depend on I_z and that averages over θ_z in Eqs. (18) and (19) are now unnecessary.

Using a computer to numerically evaluate the nonlinear tune shifts given by Eq. (18), it is possible to plot the amplitude contours of the flat beam in frequency space. This is shown in Fig. 1, where the linear tune shifts are equal

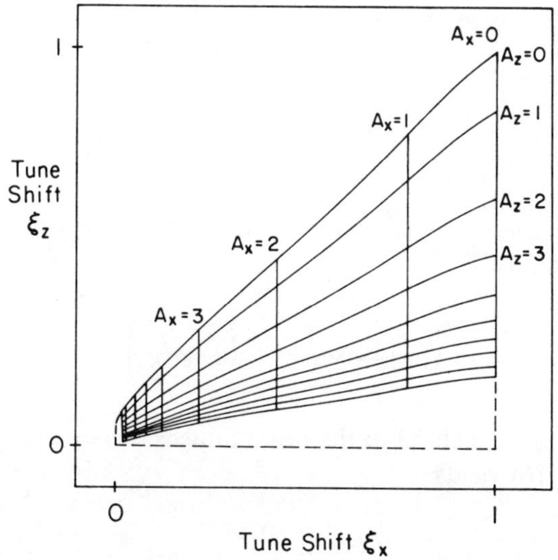

Figure 1. Amplitude contours of the flat beam in the tune shift space. The amplitudes A_x and A_z are normalized to the beam width and height σ_x and σ_z. A computer was used to calculate the tune shifts at various different amplitudes of the betatron oscillation. The model used is that described by Eqs. (20) and (23). The coordinates may be independently scaled to give the proper linear tune shifts.

and assumed small compared to one. The tune shifts are maximal at $(A_x, A_z) = (0, 0)$ and go to zero at $(A_x, A_z) = (\infty, \infty)$.

Also of interest in the frequency space are the various resonance lines. The resonances are defined by the conditions for stationary phase in the cosine terms of Eq. (17). Using the approximations

$$\dot{\theta}_x = \omega_x \approx \pi \nu_x \tag{24}$$

$$\dot{\theta}_z = \omega_z \approx \pi \nu_z$$

the resonances are defined by the linear equations

$$m_{kx}\nu_x + m_{kz}\nu_z = 2n \tag{25}$$

where m_{kx}, m_{kz}, n is any combination of positive or negative integers. Resonances of fourth order and less are shown in Fig. 2.

To see which resonances affect the beam, it is necessary to superimpose Fig. 1 onto Fig. 2. This is shown in Fig. 3, where the working tunes are $\nu_x = 5.24$, $\nu_z = 5.18$, and the tune shifts are $\xi_x = \xi_z = 0.06$ (close to the actual values at SPEAR). The intuitive appeal of this figure can be improved by changing the coordinates and metric. Figure 4 plots the resonance lines in a Euclidean amplitude space. The density of resonance lines is greatest in the regions of

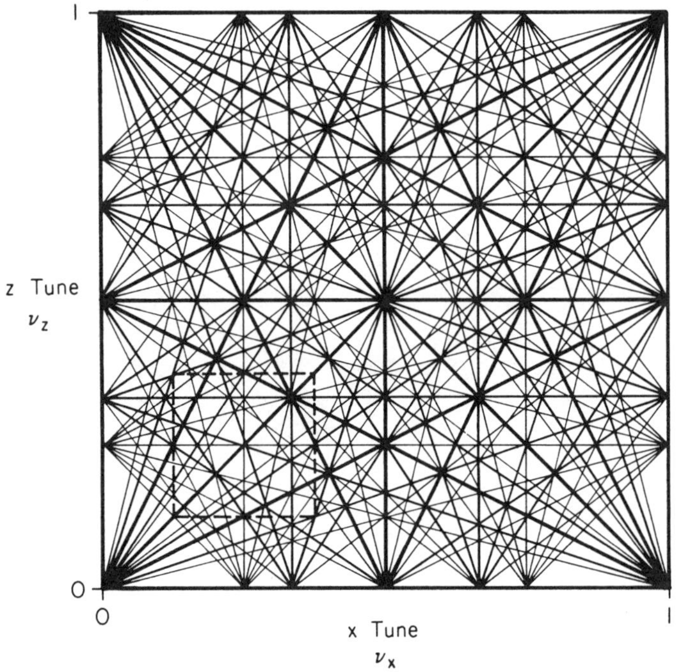

Figure 2. Resonance lines in tune space. Resonances of order four and less are shown in the tune space. The resonance pattern is periodic in both ν_x and ν_z with period one. The resonances are defined by the equations $m_{kx}\nu_x + m_{kz}\nu_z + n = 0$. The area inside the dashed line is expanded in Fig. 3.

strongest nonlinearity. Resonances that extend out to large A_z become steeper as A_z increases (because of the falling nonlinearity).

In trying to understand the beam–beam limit, the microscopic motion of individual particles in the amplitude space is of paramount interest. Equations (13) and (14), with Hamilton's equations, describe one component of this motion. At realistic tune shifts, using the simple models Eqs. (20) and (23), this component is characterized by regular motion for all but a very small fraction of initial conditions. It is important to note that a number of potentially significant effects have been ignored:

1. Contributions to the discrete spectrum from sources other than the beam–beam interaction (e.g., sextupole and moments of higher order in the magnet lattice).
2. Small errors in magnet alignment and other effects that destroy the strict symmetries of the beam–beam interaction Eqs. (20) and (23), and of the ring with respect to the interaction points.
3. Synchrotron modulation effects.

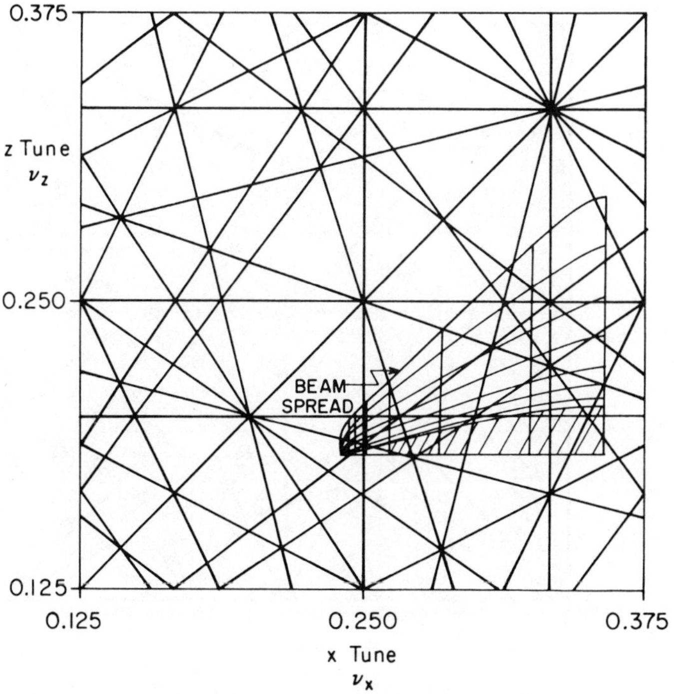

Figure 3. Amplitude contours of the flat beam in the tune space. With working tunes of $\nu_x = 5.24$, $\nu_z = 5.18$ and linear tune shifts of $\xi_x = 0.06$, $\xi_z = 0.06$, the beam is distributed within the triangle shown. Several resonances are seen to intersect the beam. This figure corresponds to the dashed outline in Fig. 2.

There is, in fact, very little justification for ignoring any of these effects. Their neglect should be considered a matter of theoretical convenience at this stage.

A fourth effect, synchrotron radiation, is not ignored. Unlike the three effects mentioned above, the influence of synchrotron radiation cannot be included as a simple addition to the discrete perturbation spectrum in Eq. (17). It is useful to think of the classical effect (radiation damping) and the quantum effect (quantum fluctuations) as distinct non-Hamiltonian processes in the amplitude space. The equations of motion for the damping are

$$\dot{A}_x = \frac{-A_x}{T_D} \tag{26}$$

$$\dot{A}_z = \frac{-A_z}{T_D}$$

where T_D is the damping time. The quantum fluctuations are represented by a uniform, isotropic diffusion on the amplitude space

$$D^{ij} = \langle \Delta A^i \, \Delta A^j \rangle = D_0 \delta^{ij} \tag{27}$$

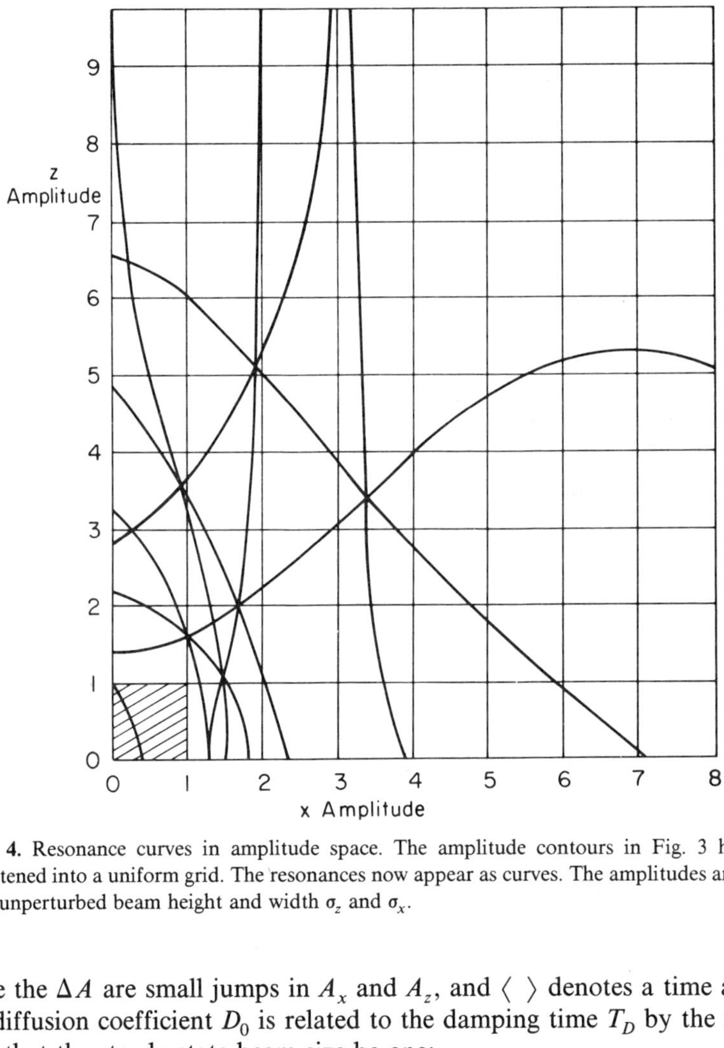

Figure 4. Resonance curves in amplitude space. The amplitude contours in Fig. 3 have been straightened into a uniform grid. The resonances now appear as curves. The amplitudes are in units of the unperturbed beam height and width σ_z and σ_x.

where the ΔA are small jumps in A_x and A_z, and $\langle\ \rangle$ denotes a time average. The diffusion coefficient D_0 is related to the damping time T_D by the requirement that the steady state beam size be one:

$$D_0 = \frac{1}{T_D} \tag{28}$$

In summary, then, this model for single-particle motion in a colliding beam machine consists of three distinct processes on the amplitude space of the betatron oscillation.

1. The projection of a Hamiltonian phase flow onto the amplitude space.
2. A dissipation due to radiation damping.
3. A diffusion due to quantum fluctuations.

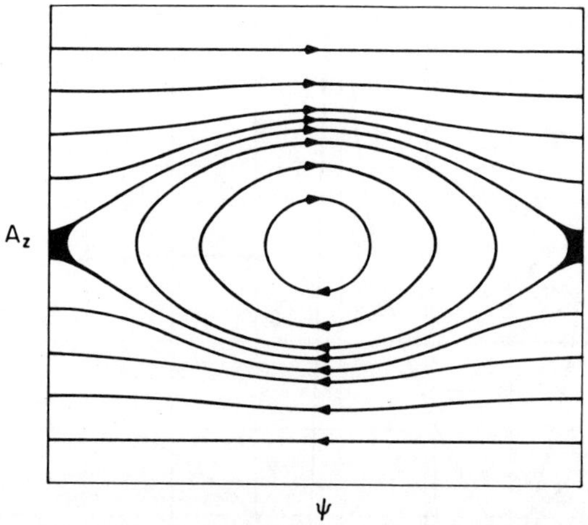

Figure 5. Resonance libration. Eleven different particle trajectories are projected, approximately, onto the two-dimensional phase space defined by the vertical amplitude and the resonance phase (the true projections are not perfectly closed curves). The trajectories inside the separatrix belong to resonance particles, all of which have the same average amplitude $\langle A_z \rangle$.

Although none of these effects, independently, is sufficient to produce the beam–beam limit, Section 3 demonstrates that radiation damping, under certain conditions, can drive a fast migration of particles along the lines of nonlinear resonance shown in Fig. 4.

3. RESONANCE STREAMING

The simplified model of the vertical betatron motion consists of a linear oscillator, a small time-dependent (Hamiltonian) perturbation, a weak dissipation, and a weak diffusion. At realistic linear tune shifts, ξ_x, $\xi_z \approx 0.05$ or less, the Hamiltonian function Eq. (17), and Hamilton's equations, induce bounded vertical oscillations in the amplitude space:

$$\dot{I}_z = -\frac{\partial H}{\partial \theta_z} = \sum_{k,n} m_{kz} F_k(I_x, I_z) \sin(m_{kx}\theta_x + m_{kz}\theta_z + 2\pi n t) \tag{29}$$

The phases associated with the different terms in Eq. (29) rotate, for the most part, at fairly constant frequencies (determined by ω_x, ω_z). If, however, a particular term is resonant, that term is, by definition, phase locked. In this case, instead of rotating, the phase oscillates about zero. This phase oscillation (henceforth referred to as "resonant libration"), is qualitatively distinct from the rotational behavior of the nonresonant terms. This difference is illustrated in Fig. 5, where approximate trajectories associated with a number of initial conditions are plotted in the phase space. The coordinates here are A_z (vertical amplitude) and

$$\psi_k = m_{kx}\theta_x + m_{kz}\theta_z + 2\pi nt \tag{30}$$

(the kth resonance phase). The average position of a particle is called the "oscillation center" of that particle. All the resonant trajectories in Fig. 5 have the same oscillation center. This picture is associated with a particular fixed value of A_x. If A_x is changed, the position of the resonant oscillation center will, in general, also change. This is illustrated in Fig. 6, where sections corresponding to different values of A_x are stacked together to show the change in the resonance position with A_x. Comparing Fig. 6 and Fig. 4, it appears that the resonance curves might more aptly be described as "tubes." These tubes follow the resonance curves of Fig. 4, but have a finite width (defined by the maximum width of the separatrix).

When the resonant libration of a particle is projected onto the amplitude space, it appears as a straight line segment centered on the resonance. If such a particle is instantaneously displaced to a different libration curve at a different A_x (Fig. 7), the new libration curve will also be a line segment centered on the resonance. The actual ("classical") displacement of the particle in amplitude is then somewhat different from the displacement of the oscillation center. In

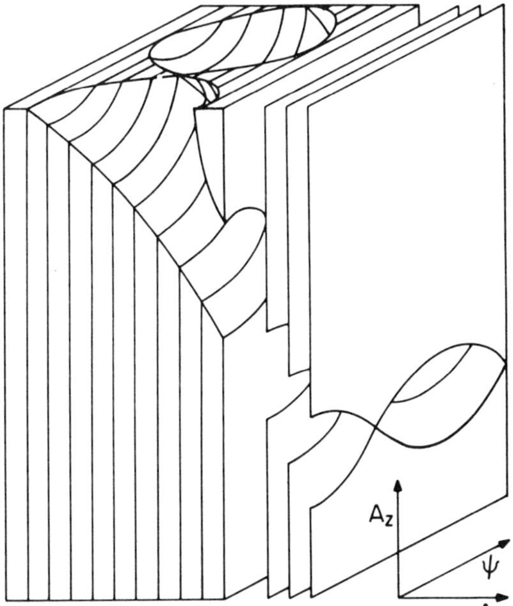

Figure 6. A stack of many sections. The location of the resonance center in A_z depends on the value of A_x. This dependence is illustrated by forming a stack of sections, each one of which is associated with a particular value of A_x. The resonances look somewhat like tubes.

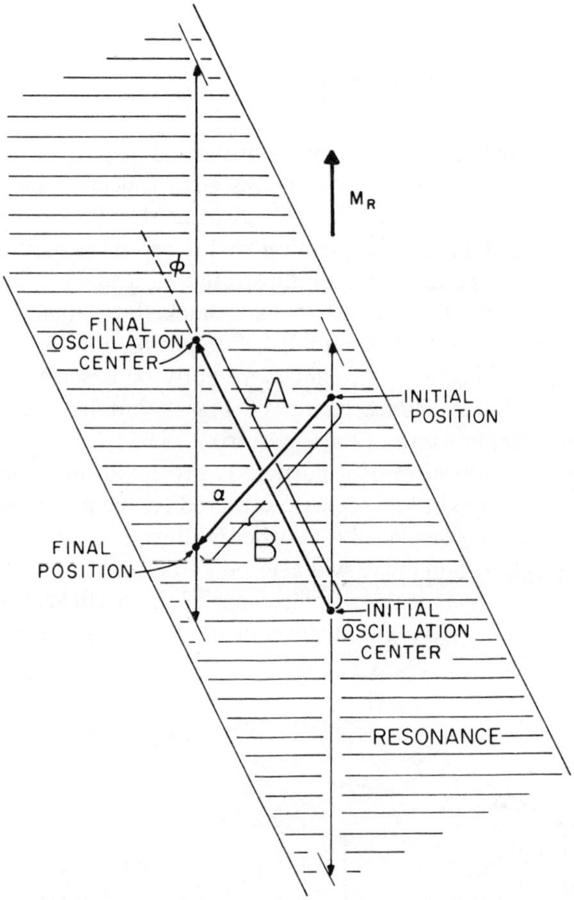

Figure 7. Oscillation center displacement inside a resonance. A particle is originally on a libration curve inside the resonance. The projection of its trajectory is approximately a vertical line segment in the amplitude space, centered on the resonance. When the particle is instantaneously displaced from point **a** to a point **b** on another libration curve, its oscillation center moves up the resonance, from point **A** to point **B**. The oscillation center displacement in A_z is larger in magnitude than the actual displacement in A_z and of opposite sign from it.

particular, both the directions and the magnitudes of the two jumps are different. Since both dissipation and diffusion may be constructed from a series of infinitesimal classical displacements of this type, it is clear that the diffusive and dissipative motions of the resonant oscillation centers may be quite different from those of the nonresonant oscillation centers. In Fig. 7, the magnitude of the resonant oscillation center displacement is related to the magnitude of the classical displacement by

$$(\mathbf{A} - \mathbf{B}) = (\mathbf{a} - \mathbf{b})\sin(\alpha)\csc(\phi) \tag{31}$$

Since the beam size is determined by a balance between the dissipation and the diffusion, it follows that the dissipative process is dominant outside the beam, at $\sigma_x, \sigma_z > 1$, whereas diffusion dominates inside. Since the phenomena of particular interest here are those that appear between the beam and the wall, it is natural to first examine resonant oscillation center transport in the presence of dissipation only. The classical dissipation Eq. (26) always pulls the particle toward the origin in the amplitude space. But if the particle happens to be in a steep sum resonance (one with a large negative slope), the oscillation center will move up the resonance, rather than down to the origin (Fig. 8). Furthermore, the rate at which the particle moves up the resonance is approximately proportional to the absolute value of the slope of the resonance. For such a resonant particle, the radiation damping is not just neutralized, it is actually reversed and amplified. It is easy to see that such a mechanism does not occur

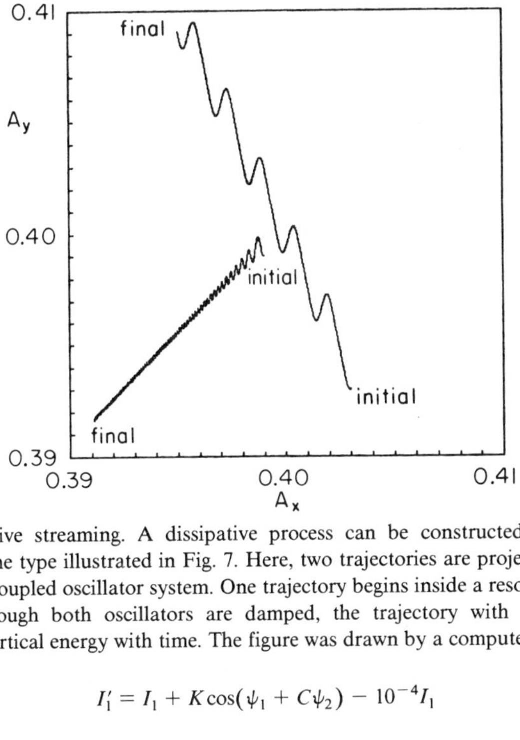

Figure 8. Dissipative streaming. A dissipative process can be constructed from infinitesimal displacements of the type illustrated in Fig. 7. Here, two trajectories are projected onto the action space of a simple coupled oscillator system. One trajectory begins inside a resonance and the other just outside. Although both oscillators are damped, the trajectory with the resonant initial conditions gains vertical energy with time. The figure was drawn by a computer using the mapping

$$I_1' = I_1 + K\cos(\psi_1 + C\psi_2) - 10^{-4}I_1$$

$$\psi_1' = \psi_1 + P2\pi I_1$$

$$I_2' = I_2 + CK\cos(\psi_1 + C\psi_2) - 10^{-4}I_2$$

$$\psi_2' = \psi_2 + 2\pi I_2,$$

where $K = 10^{-5}$; $P = 10$; $C = 5$. The resonance shown is defined by $PI_1 + CI_2 = 5$.

in a difference resonance and that it cannot increase the horizontal amplitude of a particle.

It should be noted that parametric resonances in the radial direction have been neglected here. The resonant libration associated with these resonances (which are precisely vertical), is horizontal in the amplitude space. These resonances do not, therefore, contribute to the streaming effects.

Although there are an infinite number of very steep sum resonances, only a finite number of them are dangerous. This is because the validity of the oscillation center concept depends on the particle being inside the resonance for at least one period of the resonant libration. It turns out that the higher the order of the resonance, the longer its period of libration and the smaller its width. Therefore, resonances above a certain order are invisible to the particles. They are "drowned out in the noise."

Given the width of a resonance, its libration frequency, and the amplitude drift velocity of a particle due to radiation damping, the resonance may be classified into one of three regimes:

1. *The Oscillation Center Regime.* Particles remain inside the resonance for a time greater than one libration period.
2. *The Plateau Regime.* Particles remain in the resonance for a time less than a libration period, but the velocity $\dot{A}_H$ induced by the Hamiltonian motion is greater in magnitude than that induced by the dissipation $\dot{A}_D$. So the particle sees a resonant "kick" as it passes through the resonance.
3. *The Classical Regime.* The velocity $\dot{A}_D$ is greater than $\dot{A}_H$. In this case, the presence of the resonance is inconsequential, since the dissipation pulls the particle across the resonance before the coherent kicks ΔA_H associated with the beam–beam interaction can accumulate.

Since both oscillation center and plateau regime resonances can potentially effect the beam shape and lifetime, these resonances are referred to as "active" resonances. The regime to which a particular resonance belongs depends on the amplitudes A_x, A_z of the particle inside the resonance (resonance width, libration frequency, and drift velocity all depend on amplitude). Therefore, for each point A_x, A_z in the amplitude space, it is possible to draw a tune diagram (Fig. 2) with only the active sum resonances represented (a few of the very lowest order difference resonances should be included, since these are dangerous because of their large size alone). To identify the active resonances at a given amplitude A_x, A_z, the beam–beam interaction model Eq. (17), must be dissected term by term. A simple analytic procedure is entailed.

Aside from the particular analytic model used, the independent variables are the two amplitudes A_x, A_z, the two tune shifts ξ_x, ξ_z, the damping time T_D, and the resonance integers (m_x, m_z). When a particular resonance is examined at a point A_x, A_z, it is assumed that the resonance in question intersects the point A_x, A_z in the amplitude space. This assumption amounts to a constraint on the betatron tunes ν_x, ν_z. The tunes are constrained to be such that the resonance m_x, m_z, n intersects the amplitude A_x, A_z when the linear tune shifts are ξ_x, ξ_z (the integer n is arbitrary).

The first step in the analysis is to subject the perturbation to Fourier transformation, as shown in Eq. (17), and to calculate the elements of the nonlinearity matrix. These are, from Eq. (18),

$$\Lambda_{ij} = \frac{\partial^2}{\partial I_i \, \partial I_j} \langle V \rangle_{\theta_x, \theta_z} \Bigg|_{I_x, I_z} \tag{32}$$

Note that from the flat beam approximation, $\Lambda_{xz} = 0$.

Using Eq. (9) and the normalized interaction potentials Eqs. (20) and (23), the tune shifts are related to ω_x^* and ω_z^* by

$$\xi_z = \frac{1}{\omega_z^*} \tag{33}$$

$$\xi_x = \frac{1}{\omega_x^*}$$

Since action coordinates are related to amplitude coordinates via Eq. (22), the elements of the Jacobian matrix are

$$J_x^x = \frac{\partial A_x}{\partial I_x} = \frac{\xi_x}{A_x} \tag{34}$$

$$J_z^z = \frac{\xi_z}{A_z} \qquad J_z^x = J_x^z = 0$$

The next step is to examine the individual resonances.

In amplitude space, the resonance appears as a curve that intersects the point A_x, A_z. This curve is defined by the resonance condition

$$m^i \omega_i = 2\pi n \tag{35}$$

If the metric is taken to be Euclidean with respect to amplitude coordinates, the components of a vector normal to the resonance at A_x, A_z are given by

$$n^j = \delta^{jk} \frac{\partial}{\partial A_k} \left(m^i \omega_i \right) = \delta^{jk} m^i \Lambda_{li} (J^{-1})_k^l \tag{36}$$

where δ^{jk} is the Kronecker delta. The vector with components m_x, m_z in action coordinates is called the "resonance vector." In amplitude coordinates, these components are

$$r^i = m^j J_j^i \tag{37}$$

Normally, the resonance vector points in the direction of resonant libration. But because the flat beam approximation involves an infinite normalization factor in Eq. (14), the situation is different here. In this case, the resonant libration is always vertical. The "effective nonlinearity" in the vertical direction is defined to be

$$\Lambda = |m_z \Lambda_{zz} m_z| = |n_z r_z| \tag{38}$$

The libration width can then be calculated using the standard formula (see, e.g., Ref. 3, pp. 278 ff),

$$w = 4|r_z|\left(\frac{F}{\Lambda}\right)^{1/2} \tag{39}$$

where F is the Fourier amplitude, Eq. (17). The small amplitude frequency of resonant libration is given by

$$f = \sqrt{F\Lambda} \tag{40}$$

The tangent to the resonance curve at A_x, A_z makes an angle relative to the vertical of

$$\phi = \sin^{-1}\left(\frac{\Lambda}{|\mathbf{n}|\,|r_z|}\right) \tag{41}$$

The "enhancement factor" gives the ratio of the streaming speed up the resonance to the classical horizontal drift speed

$$E = \csc(\phi) = \frac{|\mathbf{n}|\,|r_z|}{\Lambda} \tag{42}$$

The width of the resonance is then

$$\Delta - w\sin(\phi) \tag{43}$$

$$\Delta = 4\frac{f}{|\mathbf{n}|}$$

The resonance may now be classified using the computed values of F, Λ, n_i, r_z, and E. The drift velocity associated with the dissipation is

$$\dot{A}_i = -\frac{A_i}{T_D} \tag{44}$$

The time it takes for the particle to drift across a thin resonance at A_x, A_z is

$$t_D = \frac{|\mathbf{n}|}{(\dot{\mathbf{A}} \cdot \mathbf{n})} \Delta \tag{45}$$

The libration period is

$$t_L = \frac{2\pi}{f} \tag{46}$$

The oscillation center regime is then defined by

$$t_L < t_D$$

or

$$\left| \frac{4F\Lambda}{A_i n_i} \right| > \frac{2\pi}{T_D} \tag{47}$$

The rms speed of libration in the amplitude space may be estimated using the small amplitude frequency and the resonance width

$$\mathbf{A}_H = \frac{w}{2} \sin(ft) \tag{48}$$

$$\dot{\mathbf{A}}_{H(\mathrm{rms})} = \frac{wf}{\sqrt{8}}$$

The classical regime is then

$$|\dot{\mathbf{A}}_D| > \dot{\mathbf{A}}_{H(\mathrm{rms})}$$

where $|\dot{\mathbf{A}}_D|$ is the drift speed from Eq. (44). This may be written

$$\frac{|\mathbf{A}|}{T_D} > \sqrt{2}\,|\mathbf{r}|\,F \frac{\Lambda}{|\mathbf{r}|\,|\mathbf{n}|} \sin(\phi)$$

or

$$\sqrt{2}\,T_D F E \frac{\Lambda}{|\mathbf{n}|\,|\mathbf{A}|} < 1 \tag{49}$$

The intermediate, or plateau regime, corresponds to the case in which neither Eq. (47) nor Eq. (49) holds:

$$\frac{E\pi}{\sqrt{2}} < \left| \sqrt{2}\,T_D F E \frac{\Lambda}{(n_i A_i)} \right| < \left| \sqrt{2}\,T_D F E \frac{\Lambda}{|\mathbf{n}|\,|\mathbf{A}|} \right| < 1 \tag{50}$$

A similar analysis may be done for the diffusion. The oscillation center, plateau, and classical regimes are defined differently when the external process is diffusive rather than dissipative. These definitions are (see Ref. 9):

1. *Oscillation Center*:

$$\frac{8T_D(F\Lambda)^{3/2}}{2\pi|\mathbf{n}|^2} > 1 \tag{51}$$

2. *Classical*:

$$\frac{8T_D(F\Lambda)^{3/2}}{2\pi|\mathbf{n}|^2} < \frac{\sqrt{2}\,\Lambda}{|\mathbf{n}|\,|\mathbf{r}_z|\pi} = \frac{\sqrt{2}}{E\pi} \tag{52}$$

3. *Plateau*:

$$\frac{E\pi}{\sqrt{2}} > \frac{4E(F\Lambda)^{3/2}}{\sqrt{2}\,|\mathbf{n}|^2} > 1 \tag{53}$$

It is convenient to define as the "streaming threshold" the quantity

$$S_T = \frac{E\pi}{\sqrt{2}} \tag{54}$$

which appears on the left-hand sides of Eqs. (50) and (53). Then by calculating the other three quantities appearing in Eqs. (50) and (53), it is possible to not only see which regime the resonance is in, but also how far it is into that regime. These streaming parameters are labeled

$$P_1 = \left|\frac{\sqrt{2}\,T_D E F\Lambda}{N_i A^i}\right| \tag{55}$$

$$P_2 = \left|P_1 \frac{(\mathbf{n}\cdot\mathbf{A})}{|\mathbf{n}|\,|\mathbf{A}|}\right|$$

$$P_3 = \left|\frac{4E(F\Lambda)^{3/2}}{\sqrt{2}\,|\mathbf{n}|^2}\right|$$

4. RESONANCE ANALYSIS

The objective here is to examine each of the resonances in the tune space individually and to answer the following questions, using real numbers:

1. Is the resonance a potential threat to the beam? If the resonance is in the classical regime for all amplitudes below the aperture value, it cannot affect either the size or the lifetime of the beam.

Table 1. Computer Analysis of Resonance Streaming Regimes[a]

tsx	tsz	td	ax	az
.060	.060	5.0e+04	2.00	5.00

m1	m2	st	p1	p2	p3	delta	f	e	type
0	2	7.5	*******	*******	*******	5.06	8.47e-02	3.4	3
0	4	7.5	*******	*******	*******	1.87	6.26e-02	3.4	3
2	2	30.4	*******	*******	*******	0.83	5.66e-02	13.7	3
2	-2	16.2	*******	*******	*******	1.56	5.66e-02	7.3	3
0	6	7.5	*******	854.70	539.45	0.99	4.96e-02	3.4	3
2	4	18.8	766.06	*******	575.32	0.49	4.18e-02	8.5	3
4	-2	39.3	428.72	*******	506.23	0.31	2.77e-02	17.7	3
4	2	53.5	314.78	*******	506.23	0.23	2.77e-02	24.1	3
2	-4	5.0	*******	*******	575.32	1.87	4.18e-02	2.2	3
0	8	7.5	399.57	401.24	253.24	0.59	3.92e-02	3.4	2
2	6	15.0	212.58	477.07	240.80	0.33	3.31e-02	6.8	2
6	-2	62.4	17.80	244.51	82.67	0.08	1.12e-02	28.1	1
4	4	30.4	55.93	315.27	138.19	0.15	2.05e-02	13.7	2
4	-4	16.2	104.81	574.38	138.19	0.28	2.05e-02	7.3	2
6	2	76.7	14.50	207.64	82.67	0.07	1.12e-02	34.5	1
2	-6	2.3	*******	301.68	240.80	2.14	3.31e-02	1.0	2
0	10	7.5	124.60	198.75	125.44	0.37	3.09e-02	3.4	2
2	8	13.1	67.83	215.95	113.05	0.22	2.62e-02	5.9	2
8	-2	85.6	0.56	29.40	10.21	0.02	3.94e-03	38.5	0
4	6	22.7	16.56	125.58	57.84	0.11	1.62e-02	10,2	1
6	-4	27.7	4.04	76.29	22.57	0.07	8.28e-03	12.5	1
6	4	41.9	2.67	53.72	22.57	0.04	8.28e-03	18.9	1
4	-6	8.6	43.57	483.89	57.84	0.28	1.62e-02	3.9	2
8	2	99.8	0.48	26.04	10.21	0.02	3.94e-03	44.9	0
2	-8	2.6	343.48	114.46	113.05	1.13	2.62e-02	1.2	2
0	12	7.5	41.54	101.55	64.10	0.24	2.42e-02	3.4	2
2	10	12.0	23.17	104.20	56.00	0.15	2.06e-02	5.4	2
10	-2	108.7	0.01	2.87	1.01	0.01	1.24e-03	49.0	0
4	8	18.8	5.55	56.77	27.15	0.08	1.28e-02	8.5	1
8	-4	39.3	0.12	8.75	2.79	0.02	2.91e-03	17.7	0
6	6	30.4	0.82	21.55	9.45	0.03	6.56e-03	13.7	0
6	-6	16.2	1.53	39.26	9.45	0.06	6.56e-03	7.3	1
8	4	53.5	0.09	6.80	2.79	0.01	2.91e-03	24.1	0
4	-8	5.0	20.97	335.43	27.15	0.29	1.28e-02	2.2	2
10	2	123.0	0.01	2.61	1.01	0.00	1.24e-03	55.4	0
2	-10	3.3	83.14	62.51	56.00	0.55	2.06e-02	1.5	2
0	14	7.5	14.48	52.95	33.42	0.16	1.88e-02	3.4	2
2	12	11.2	8.25	52.20	28.61	0.11	1.61e-02	5.0	1
12	-2	131.9	0.00	0.24	0.08	0.00	3.57e-04	59.4	0
4	10	16.5	1.97	27.29	13.45	0.05	1.01e-02	7.4	1
10	-4	50.8	0.00	0.84	0.28	0.00	9.16e-04	22.9	0
6	8	24.6	0.28	9.77	4.43	0.02	5.19e-03	11.1	0
8	-6	23.9	0.05	4.12	1.17	0.01	2.31e-03	10.7	0
8	6	38.1	0.03	2.74	1.17	0.01	2.31e-03	17.1	0
6	-8	10.5	0.66	26.64	4.43	0.06	5.19e-03	4.7	0
10	4	65.1	0.00	0.69	0.28	0.00	9.16e-04	29.3	0
4	-10	3.1	10.55	32.89	13.45	0.29	1.01e-02	1.4	2
12	2	146.1	0.00	0.22	0.08	0.00	3.57e-04	65.8	0

[a] The values of the input parameters are listed at the top of the output. Resonances are classical if either P_2 or $P_3 < 1$ (type 0), in the oscillation center regime if both P_1 and $P_3 > S_T$ (type 2), and dangerously large if $Ew > A_z/2$ (type 3). All resonances not in one of these catagories are in the plateau regime and are labeled type 1.

2. Is the resonance capable of causing streaming out of the beam? In other words, is it a sum resonance in either the oscillation center or the plateau regime? If so, it should be avoided.

3. Is the resonance so large that it can seriously disrupt the beam, even without streaming or stochasticity effects? Then it must be avoided with plenty of room to spare.

No attempt is made here to calculate the actual lifetime of the beam or the actual blowup. Such a calculation appears to be very difficult and would require a careful treatment of synchrotron modulation effects (thus far, only particles with negligible phase oscillation amplitudes have been considered).

In principle, it is possible to dispense with the analytic expressions for the potentials and numerically integrate over an analytic (bi-Gaussian) density distribution. In practice, however, the additional integration makes the analysis prohibitively long, even for the fastest computers. The following results, then, are derived for the flat beam model, Eqs. (20) and (23).

A detailed analysis of the dependence on ξ_x, ξ_z, A_x, A_z, and T_D would be very lengthy and inappropriate for this chapter. Therefore, the representative values $\xi_x = \xi_z = 0.06$, $A_x = 2.$, $A_z = 5.$ are held fixed throughout the analysis. Only the one parameter T_D is varied.

A typical resonance analysis is shown in Table 1 for fixed values of ξ_x, ξ_z, A_x, A_z, and T_D. The resonances are analyzed one by one, starting with the lowest order and counting up systematically to higher orders. The resonances are labeled by the resonance integers m_x, m_z ($m1$, $m2$ in Table 1). The label n has been omitted because none of the results depend on n, but it should be remembered that for each resonance listed, there is actually a family of resonances in the tune diagram and that the size of this family is approximately proportional to the magnitude of the largest of the two integers m_x, m_z. The numbers on the left-hand side divide the active resonances into three types. In type 3 the resonance is dangerous by virtue of its size alone, regardless of its streaming potential. In type 2 the resonance is in the dissipative and diffusive oscillation center regimes. A resonance is type 1 if it is not type 3 or 2 and if it is in neither the dissipative nor the diffusive classical regime. If the resonance is in the classical regime for either process, it is type 0. The damping time shown here corresponds to an energy of 4.5 GeV at SPEAR or 15. GeV at PEP.

In Fig. 9, the "type 3" and active sum resonances are drawn in frequency space. If the beam is exactly symmetric with respect to the interaction points, and the beam–beam force is exactly symmetric with respect to $x, z = 0$, then Fig. 9 will be periodic in ν_x, ν_z with period one. The small triangle represents the beam with tune shifts ξ_x, $\xi_z = 0.06$. The most important feature of this picture is that it shows quite clearly that active sum resonances do intersect the amplitude space of a real beam. It is thus not only possible but quite likely that streaming effects are important in $e^+ - e^-$ colliding beam machines.

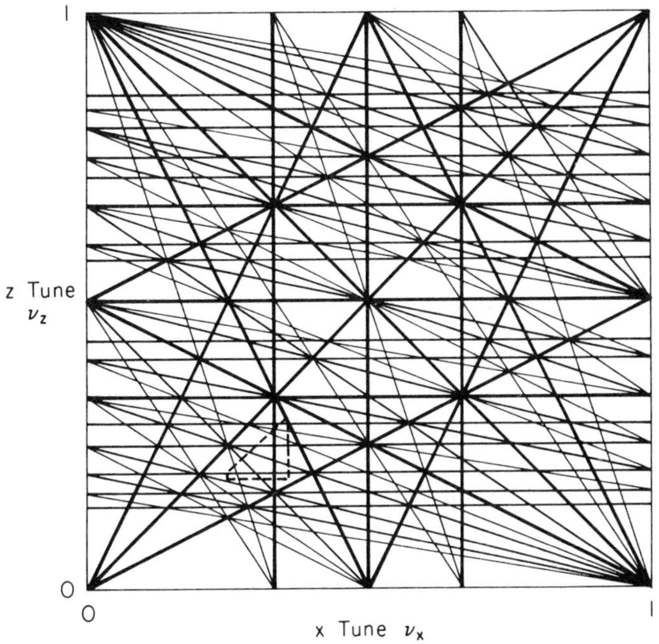

Figure 9. Active resonances in tune space. The type 3 and active sum resonances from Table 1 are plotted in the tune spaces.

As a specific example, consider the resonance passing through $A_x = 2.$ and $A_z = 5.$ in Fig. 4. From Fig. 3, this is the $10\nu_z = 2$ "parametric" resonance and in Table 1, it is labeled $m_1 = 0$, $m_2 = 10$. From Table 1, the angle of this resonance with respect to the vertical is $\csc^{-1} e = 17°$. The libration width is $\Delta A_z = \delta \times e = 1.26$, and the period of small amplitude libration is $P = 2\pi/f = 203$ collisions. Since $P1$, $P2$, and $P3$ are all much larger than S_T, the resonance is well inside both oscillation center regimes, and should be avoided if possible.

The dependence of the beam–beam limit on energy can be crudely estimated by plotting the number of active resonances versus the damping time (Fig. 10). The number plotted includes the multiplicity associated with each of the pairs m_x, m_z. As the energy is lowered, the number of active resonances increases. For a real machine, this means that to keep the beam lifetime a constant, the linear tune shift must decrease with the energy (reducing the tune shift has three effects: it shrinks the beam in frequency space, reduces the number of active resonances in the tune diagram, and weakens the streaming effects on those that remain). A drop in the beam lifetime with energy has been observed at SPEAR.[6]

5. CONCLUSION

Since it is possible to calculate the relative (and to some extent the absolute) dangers of individual resonances, the streaming theory of the beam–beam limit should be fairly easy to verify experimentally. If streaming effects do, in fact, prove to be important, attempts to minimize them can be raised from a purely empirical level to a theoretical and computational level where they are more likely to succeed.

The theory seems promising in a number of respects. For flat beams, it gives vertical but not radial blowup, the dependence on tune shift is smooth rather than sudden, it predicts a drop in tune shift for low energy beams, and it explains the presence of inflated tails on the density distribution.

In addition to the qualitative agreement, Section 3 showed that relatively high order resonances, resonances that have never been considered especially dangerous before, can significantly increase the vertical betatron amplitudes of a small class of particles. Furthermore, it was shown that at linear tune shifts comparable to those observed in actual machines, these active resonances are virtually unavoidable in the tune space.

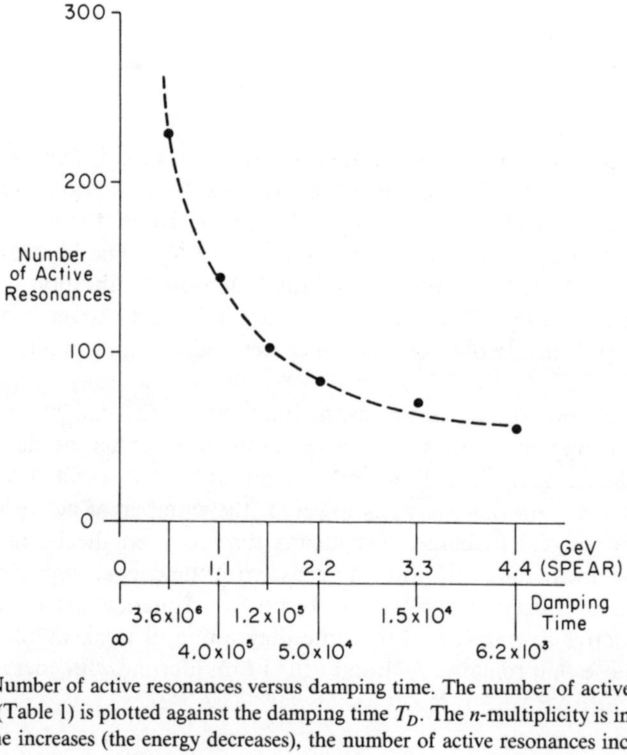

Figure 10. Number of active resonances versus damping time. The number of active resonances in the analysis (Table 1) is plotted against the damping time T_D. The n-multiplicity is included. As the damping time increases (the energy decreases), the number of active resonances increases rapidly.

ACKNOWLEDGMENTS

This work was supported by the Office of Naval Research, grant N00014-79-C-0674, the National Science Foundation grant ENG-78-26372, and the U.S. Department of Energy DE-AS03-76F00034-PA#DE-ATOE-76ET53059.

REFERENCES

1. M. Month and J. C. Herrera, Eds., *Nonlinear Dynamics and the Beam–Beam Interaction*, American Institute of Physics Conference Proceedings Series, Vol. 57, (AIP, New York, 1979).
2. *Proceedings of the Beam–Beam Interaction Seminar*, SLAC PUB 2624 (Stanford Linear Accelerator Center, Stanford, CA, 1980).
3. B. V. Chirikov, "A Universal Instability of Many Dimensional Oscillator Systems," *Phys. Rep.* **52**:5 (1979).
4. B. V. Chirikov, "Research Concerning the Theory of Nonlinear Resonance and Stochasticity," CERN Trans. 71-40, (1971), as referenced in Ref. 11.
5. D. Neuffer and A. Ruggerio, "Enhancement of Diffusion by a Nonlinear Force," in Ref. 2.
6. H. G. Hereward, "Diffusion in the Presence of Resonances," CERN/ISR-DI/ 72-26 (1972), as referenced in A. Chao, pp. 42 of Ref. 1.
7. A. Chao and M. Month, "Particle Trapping During Passage Through a High Order Nonlinear Resonance," *Nuclear Instrum. Meth.* **121**, 129–138 (1974).
8. B. V. Chirikov, "Stability of the Motion of a Charged Particle in a Magnetic Confinement System," *Sov. J. Plasma Phys.* **5**:4, 492 (1979).
9. J. Tennyson, "Resonance Enhancement of Transport Processes in Near Integrable Systems," to be published in *Physica D*.
10. F. M. Izrailev, S. I. Misnev, and G. M. Tumaikin, "Numerical Studies of Stochasticity Limit in Colliding Beams (One-Dimensional Model)," Novosibirsk Institute of Nuclear Physics, preprint 77-43 (1977).
11. F. M. Izrailev, "Nearly Linear Mappings and Their Applications," *Physica D*, **1D**:3, 243 (1980).
12. H. Wiedemann, "Experiments on the Beam–Beam Effect in e^+-e^- Storage Rings," in Ref. 1, pp. 84 ff.
13. H. Wiedemann, "Beam–Beam Effect and Luminosity in SPEAR," in Ref. 2, pp. 33 ff.
14. A. Piwinski, "Computer Simulations of the Beam–Beam Interaction and Measurements with PETRA," unpublished.
15. J. Tennyson, "A Simulation Study of the Beam–Beam Interaction at SPEAR," in Ref. 2, pp. 1 ff.
16. J. Tennyson and F. M. Izrailev, in preparation.
17. M. Sands, in *Physics with Intersecting Storage Rings*, B. Touscheck, Ed. (Academic Press, New York, 1971).
18. A stable solution only exists when certain conditions on the $\omega(t)$ are satisfied. These conditions are necessarily met in operating storage rings. See E. D. Courant and H. S. Snyder, "Theory of Alternating Gradient Synchrotrons," *Ann. Phys.* **3**, 1 (1958).
19. M. Berry, in *Topics in Nonlinear Mechanics*, S. Jorna, Ed. American Institute of Physics Conference Proceedings Series, Vol. 46 (AIP, New York, 1978).
20. J. Moser, *Stable and Unstable Motion in Dynamical Systems* (Princeton University Press, Princeton, NJ, 1973).

GLOBAL STABILITY IN A FOUR-DIMENSIONAL MAPPING MODEL OF COLLIDING CYLINDRICAL BEAMS

TASSOS C. BOUNTIS
Department of Mathematics and Computer Science
Clarkson College of Technology
Potsdam, New York

CHARLES R. EMINHIZER
Physical Dynamics
La Jolla, California

ROBERT H. G. HELLEMAN
Theoretical Physics Group
Twente University of Technology
Enschede
The Netherlands

Abstract. *A four-dimensional, highly nonlinear mapping model of the interaction between two colliding beams with cylindrically symmetric charge distribution is exactly reduced to a family of two-dimensional, area preserving mappings. This family is parametrized by the value J of an exact, angular momentum integral of the motion. As a consequence of this reduction, Arnol'd diffusion is precluded for this system and globally stable behavior in phase space is guaranteed for all time. The stability properties of certain low-period solutions of the reduced, two-dimen-*

sional mapping are studied, and large ranges of initial conditions are found for which the orbits remain forever within limited domains, a highly desirable situation for colliding beam accelerator facilities.

1. INTRODUCTION

In recent years there has been a significant increase in research on the accelerator dynamics of storage rings and in particular on the stability of colliding beams under the cumulative effect of the so-called beam–beam interaction.[1, 2] This concentrated effort has mainly grown out of the realization that exciting new results of the past two decades in nonlinear dynamics[3, 4] can shed new light on these highly nonlinear and *nonintegrable*[3, 4] (i.e., nonseparable) equations of the beam–beam interaction. For example, in this interaction very interesting questions arise concerning Arnol'd diffusion.[3, 4]

Accelerator storage rings are made to intersect[1,2] with each other so that two (e.g., proton) beams taking part in the experiment collide and hopefully produce new elementary particles. Since the verification of recent results of elementary particle physics calls for collision experiments at higher and higher energies (and often extremely small cross sections), the long-time stability of particle beams becomes of paramount importance. Now, to achieve a reasonable number of collisions (i.e., high enough "luminosity"), particles must remain in storage rings for as many as 10^{11} revolutions sometimes. This is long enough for even the subtlest nonlinear phenomena to take effect, leading to beam blowup and severe particle loss.

This chapter discusses two such colliding beams (made up of identical "bunches"), each having a Gaussian charge distribution that is cylindrically symmetric in the x, y-coordinates of a plane transverse to the beam. We study the effects of such a beam on a single particle of the other beam colliding "head on." Even if there is no actual collision taking place, the particle still experiences a nonlinear Lorentz force

$$\mathbf{E} \equiv (E_x, E_y) = (xF(r), yF(r)) \tag{1.1}$$

where

$$F(r) = \frac{1 - \exp(-r^2/2)}{r^2} \qquad r^2 \equiv x^2 + y^2 \tag{1.2}$$

because of the current in the other beam. Even with the beams crossing at small angles, the interaction is exceedingly short compared with the period of one revolution around the rings. Thus we may formulate the problem as a *four-dimensional mapping*,[4–8] or, two coupled second-difference equations for the horizontal and vertical deflections of the particle x_t, y_t ($t = 0, 1, 2, \ldots,$), respectively.

Our results indicate that in this cylindrical beam case, and for operating parameter values accessible to a variety of accelerators [at the Fermi Laboratory, the European Organization for Nuclear Research (CERN), the Brookhaven National Laboratory, e.g.],[1, 2] *there exist large ranges of phase space to which the solutions remain confined for all time*! In particular, we derive an *exact*, angular momentum *integral* of the system, J, and, using this integral, we reduce the problem to a family of area preserving mappings in the plane, parameterized by the value of J. Such mappings have been extensively studied in the literature and have similar properties to Poincaré maps of two-degree-of-freedom dynamical systems.[1–4, 10–13]

It is known that in conservative systems of two degrees of freedom, there often exist "nested" invariant tori in phase space that confine the orbits within them for all time![1–4] A "flat" beam, for example, where all motion occurs in the horizontal direction, trivially falls in this category.[7,9] For systems of more than two degrees of freedom (and higher than two-dimensional mappings) however, a "slow", so-called Arnol'd diffusion[1–4, 12] of the orbits throughout phase space is always possible. By explicitly reducing our problem to a two-dimensional area preserving mapping (see Section 2), we prove that *Arnol'd diffusion is impossible* in the model considered here.

In Section 3 we locate the fixed points and period-2 orbits of this reduced, two-dimensional nonlinear mapping and study their stability properties analytically. Varying the value of J, we also observe, in a series of numerically generated plots, that the stability (or instability) of these lowest period orbits affects large regions of initial conditions nearby. Thus we can suggest ranges of parameter values for which the undesirable effects of the beam–beam force (as, e.g., sharp decreases in luminosity[1, 2, 5]) can be minimized.

2. REDUCTION TO TWO DEGREES OF FREEDOM AND AN INTEGRAL OF MOTION

That the differential equations for the beam–beam effect can be reduced to difference equations is explained in Refs. 4, 7, and 8. For the cylindrically symmetric beam these difference equations are:[5, 6]

$$x_{t+1} = -x_{t-1} + x_t f(r_t) \qquad t = 0, 1, 2, \ldots \tag{2.1a}$$

$$y_{t+1} = -y_{t-1} + y_t f(r_t) \tag{2.1b}$$

where $r_t = (x_t^2 + y_t^2)^{1/2}$,

$$f(r_t) = 2C + \frac{2BS}{Q}\frac{1 - e^{-r_t^2/2}}{r_t^2} \tag{2.2}$$

Q is the "tune" of the machine,[1, 2] B is a measure of the interaction strength

and

$$C \equiv \cos 2\pi Q, \qquad S \equiv \sin 2\pi Q \tag{2.3}$$

The "tune" Q is the frequency of transverse (betatron) oscillations of the particle during one revolution and is, in general, different for horizontal and vertical oscillations (i.e., in general, $Q_x \neq Q_y$).[4, 7, 8] In a cylindrical beam model $Q_x = Q_y \equiv Q$ and as the results in this chapter suggest, this may be a wise choice to achieve beam stability.

Introducing polar coordinates

$$x_t \equiv r_t \cos\theta_t \qquad \text{and} \qquad y_t \equiv r_t \sin\theta_t \tag{2.4}$$

for the intersections of the particle with the x_t, y_t plane we rewrite Eqs. (2.1):

$$r_{t+1}\cos\theta_{t+1} = -r_{t-1}\cos\theta_{t-1} + r_t f(r_t)\cos\theta_t \tag{2.5a}$$

$$r_{t+1}\sin\theta_{t+1} = -r_{t-1}\sin\theta_{t-1} + r_t f(r_t)\sin\theta_t \tag{2.5b}$$

It is possible to remove the arbitrariness of an initial angle θ_0 by multiplying Eq. (2.5a) by $\cos\theta_t$ and Eq. (2.5b) by $\sin\theta_t$ and adding:

$$r_{t+1}\cos\Delta\theta_{t+1} = -r_{t-1}\cos\Delta\theta_t + r_t f(r_t) \tag{2.6a}$$

where

$$\Delta\theta_t \equiv \theta_t - \theta_{t-1} \tag{2.6b}$$

Squaring Eqs. (2.5a) and (2.5b) and adding, we obtain:

$$r_{t+1}^2 = r_{t-1}^2 - 2r_{t-1}r_t f(r_t)\cos\Delta\theta_t + r_t^2 f^2(r_t) \tag{2.7}$$

Thus we have been able to reduce the two difference equations, Eqs. (2.1), a *fourth-order* system, to a *third-order* system, Eqs. (2.6) and (2.7) in $\Delta\theta_t$, r_t, since only three initial conditions are required now: $\Delta\theta_0$, r_1, r_0. In fact, further reduction is possible: multiplying Eq. (2.5a) by $r_t\sin\theta_t$ and Eq. (2.5b) by $r_t\cos\theta_t$ and subtracting gives:

$$r_t r_{t+1}\sin\Delta\theta_{t+1} = r_t r_{t-1}\sin\Delta\theta_t$$

We have thus derived an *integral* of motion[14]

$$J \equiv r_t r_{t-1}\sin\Delta\theta_t = \text{const} \tag{2.8a}$$

which in x_t, y_t variables has the form

$$J = x_{t-1}y_t - x_t y_{t-1} = \text{const} \tag{2.8b}$$

and is reminiscent of the z-component of the angular momentum of a dynamical system $J_z = yp_x - xp_y$ (where p_x and p_y are linear momenta in the x- and y-directions, respectively). The result above is not entirely unexpected in view of the cylindrical symmetry of the system.

Using Eq. (2.8a) we could eliminate $\cos \Delta\theta_t$ from Eq. (2.7) and thus arrive at a *single second-difference* equation in r_t. We prefer, instead, to introduce new variables and work with a different two-dimensional *reduced mapping* derived in the next section. The main point is, however, that the four-dimensional problem has been reduced to a two-dimensional mapping that is like the Poincaré map of a two-degree-of-freedom dynamical system.[1-4] Thus, as mentioned in Section 1, no Arnol'd diffusion can exist, and global stability for all time is, in general, guaranteed.[1, 2, 4, 12]

3. THE REDUCED MAPPING

We derive now a two-dimensional mapping equivalent to the four-dimensional system of Eqs. (2.1), obtain its lowest order periodic orbits, and discuss their stability properties. One way to achieve this reduction is as follows: define a new variable

$$p_t \equiv r_t^2 \qquad t = 0, 1, 2, \ldots \tag{3.1}$$

and rewrite Eq. (2.7) as

$$p_{t+1} = p_{t-1} - 2r_t r_{t-1} f(p_t) \cos \Delta\theta_t + p_t f^2(p_t) \tag{3.2}$$

where $f(p_t)$ is the same as (2.2) with r_t^2 replaced by p_t. Multiplying Eq. (2.6) by r_t and using the integral J of Eq. (2.8a), Eqs. (2.6) and (3.2) become

$$p_{t+1} = p_{t-1} - 2f(p_t) J \cot \Delta\theta_t + p_t f^2(p_t) \tag{3.3}$$

$$J \cot \Delta\theta_{t+1} = -J \cot \Delta\theta_t + p_t f(p_t) \tag{3.4}$$

Introducing the variable

$$z_t \equiv J \cot \Delta\theta_t \tag{3.5}$$

and using Eq. (2.8) again, we rewrite Eqs. (3.3) and (3.4) after some algebra as a mapping of the (p, z) plane onto itself:

$$p_{t+1} = \frac{1}{p_t}\left(z_{t+1}^2 + J^2\right) \tag{3.6a}$$

$$t = 0, 1, 2, \ldots$$

$$z_{t+1} = -z_t + p_t f(p_t) \tag{3.6b}$$

[cf. Eq. (2.2)], which we refer to as the reduced mapping of the problem.

To establish the equivalence of the reduced mapping Eq. (3.6) with the original system Eq. (2.1), thereby also establishing uniqueness of solutions of Eq. (3.6), we argue as follows: given a value of J and a point (p_0, z_0) we immediately deduce r_0 from Eq. (3.1) and cot $\Delta\theta_0$ from Eq. (3.5). If $J > 0$ (< 0), Eq. (2.8) implies $0 < \Delta\theta_0 < \pi$ ($\pi < \Delta\theta_0 < 2\pi$), which uniquely determines $\sin\Delta\theta_0$, $\cos\Delta\theta_0$. Solving Eq. (2.8) for $r_1 = J/r_0 \sin\Delta\theta_0$, we may substitute $r_1, r_0, \sin\Delta\theta_0, \cos\Delta\theta_0$ on the right-hand sides of Eqs. (2.6) and (2.7)—which are equivalent to Eq. (2.1)—evaluate $r_2, \cos\Delta\theta_1$ uniquely, and keep going. Alternatively, we pick a θ_0, find $\theta_1 = \Delta\theta_1 + \theta_0$, and using our r_0 and r_1 and Eq. (2.3), connect (p_0, z_0) with (x_0, x_1, y_0, y_1).

It is easy to establish that the uncoupled case Eq. (2.1) with, say, $y_0 = y_1 = 0$, corresponds in the present formulation to $\sin\Delta\theta_t = 0$ and $J = 0$. It appears difficult, however, to obtain it as a $J \to 0$ limit of Eq. (3.6), mainly because at $J = 0$ there is a sign ambiguity $\cos\Delta\theta_t = \pm 1$ in Eqs. (2.6) and (2.7). Of course, the uncoupled case, Eq. (2.1a) with $y_t = 0$ for all t, is itself a two-dimensional mapping and has been analyzed elsewhere in the more convenient x_t, x_{t-1} variables.[7]

The reduced mapping Eq. (3.6) is not *locally* area preserving: letting $p_t = \hat{p}_t + \Delta p_t$ and $z_t = \hat{z}_t + \Delta z_t$ in Eq. (3.6) and keeping terms up to order $\Delta p_t, \Delta z_t$ we find

$$\begin{pmatrix} \Delta p_{t+1} \\ \Delta z_{t+1} \end{pmatrix} = \mathbf{M}_t \begin{pmatrix} \Delta p_t \\ \Delta z_t \end{pmatrix} \tag{3.7a}$$

with

$$\mathbf{M}_t \equiv \begin{pmatrix} -\left(\dfrac{\hat{p}_{t+1}}{\hat{p}_t}\right) + 2\left(\dfrac{\hat{z}_{t+1}}{\hat{p}_t}\right)[f(\hat{p}_t) + \hat{p}_t f'(\hat{p}_t)] & \dfrac{-2\hat{z}_{t+1}}{\hat{p}_t} \\ f(\hat{p}_t) + \hat{p}_t f'(\hat{p}_t) & -1 \end{pmatrix} \tag{3.7b}$$

Since

$$\det \mathbf{M}_t = \frac{\hat{p}_{t+1}}{\hat{p}_t} \tag{3.8}$$

we conclude that Eq. (3.6) does not preserve areas,[3, 4] in general, going from one point to the next in the p_t, z_t plane. It is easy to show, however, that Eq. (3.6) *does* preserve areas in the $\ln p_t, z_t$ plane (which is the plane of Figs. 1–9), since, as we see from Eq. (3.8), it has an invariant in the p_t, z_t variables

$$\iint p_{t+1}^{-1}\, dp_{t+1}\, dz_{t+1} = \iint p_t^{-1}\, dp_t\, dz_t$$

and its properties are topologically equivalent to those of an area preserving mapping.

Moreover, the reduced mapping does preserve areas upon "return" to each point $(\hat{p}_t, \hat{z}_t)$ belonging to a periodic orbit of period m, say

$$\hat{p}_t = \hat{p}_{t+m} \qquad \hat{z}_t = \hat{z}_{t+m} \tag{3.9}$$

since the determinant of the linearized "*return*" mapping

$$\mathbf{M}^{(m)} \equiv \prod_{t=1}^{m} \mathbf{M}_t \tag{3.10}$$

is one:

$$\det \mathbf{M}^{(m)} = \prod_{t=1}^{m} \det \mathbf{M}_t = \frac{\hat{p}_{t+1}}{\hat{p}_t} \cdot \frac{\hat{p}_{t+2}}{\hat{p}_{t+1}} \cdots \frac{\hat{p}_{t+m}}{\hat{p}_{t+m-1}} = 1 \tag{3.11}$$

[cf. Eqs. (3.8)–(3.10)]. Thus every point of an m-periodic orbit can be viewed as a fixed point of an area preserving mapping: the mth power of the reduced mapping. Hence near every such (stable) point Kolomogorov–Arnol'd–Moser (KAM) theory holds and the Poincaré–Birkhoff theorem applies.[3, 4] In view of

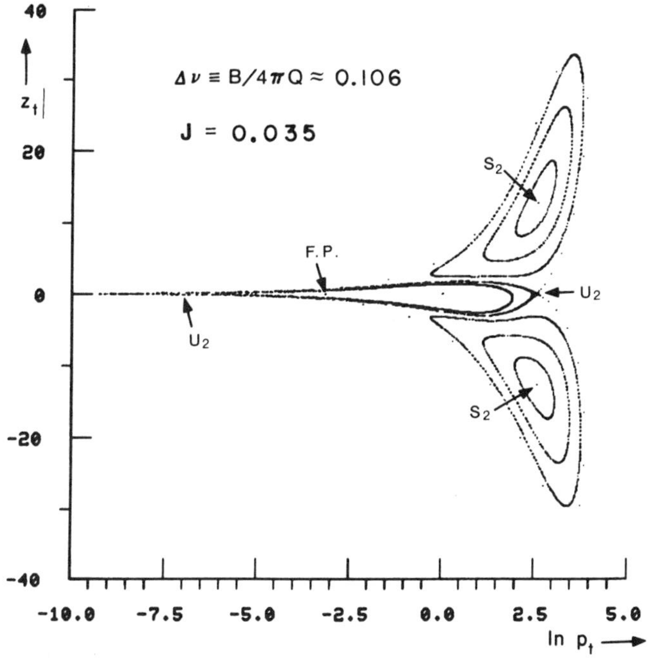

Figure 1. Orbits of the reduced mapping Eq. (3.6) near its fixed point (F.P.) and period-2 solutions (S_1, U_2) in the ln p_t, z_t plane for $J = 0.035$. In all cases $B = 5.02$ and $Q = 22.6/6$, giving a "tune shift" $\Delta\nu \equiv B/4\pi Q = 0.106$ (see the discussion in Section 3 and Tables 1 and 2).

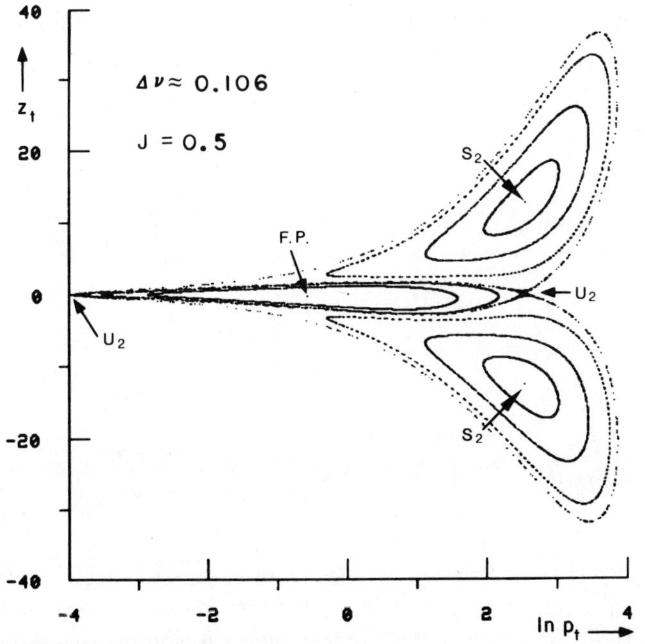

Figure 2. Orbits of the reduced mapping Eq. (3.6) near F.P. and period-2 solutions in the p_t, z_t plane for $j = 0.5$; B and Q as in Fig. 1.

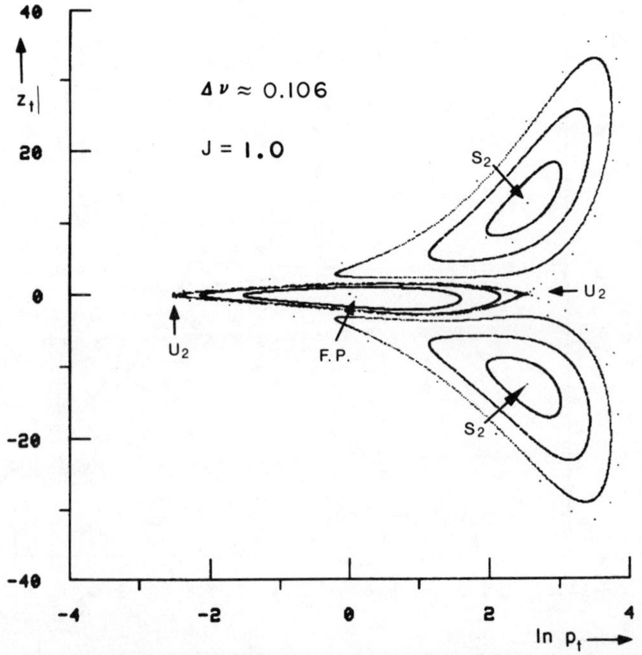

Figure 3. Orbits of the reduced mapping Eq. (3.6) near F.P. and period-2 solutions in the ln p_t, z_t plane for $J = 1.0$; B and Q as in Fig. 1.

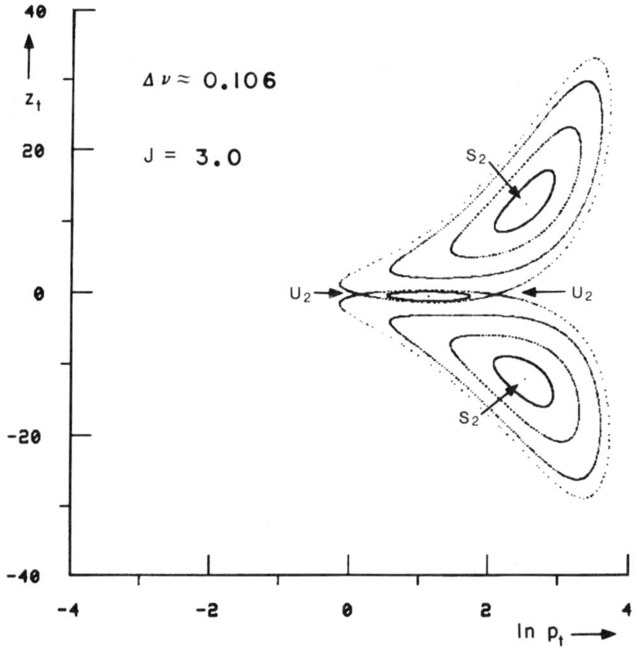

Figure 4. Orbits of the reduced mapping Eq. (3.6) near F.P. and period-2 solutions in the ln p_t, z_t plane for $J = 3.0$; B and Q as in Fig. 1.

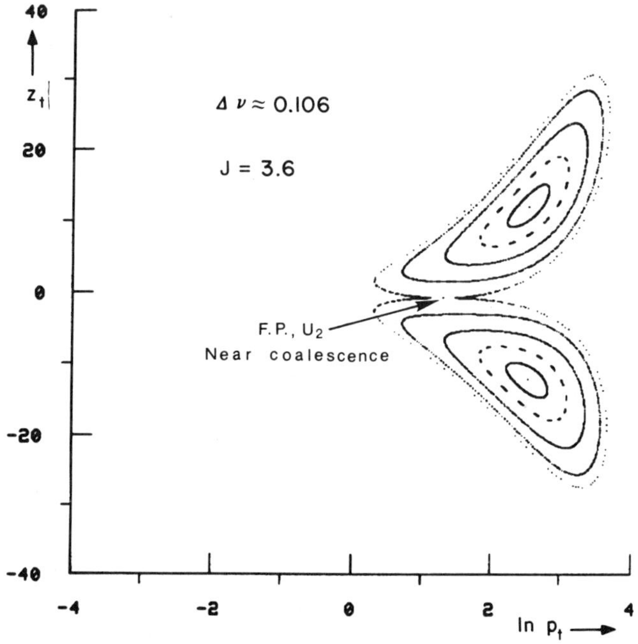

Figure 5. Orbits of the reduced mapping Eq. (3.6) near F.P. and period-2 solutions in the ln $\mathbf{p}_t$, z_t plane for $J = 3.6$; B and Q as in Fig. 1.

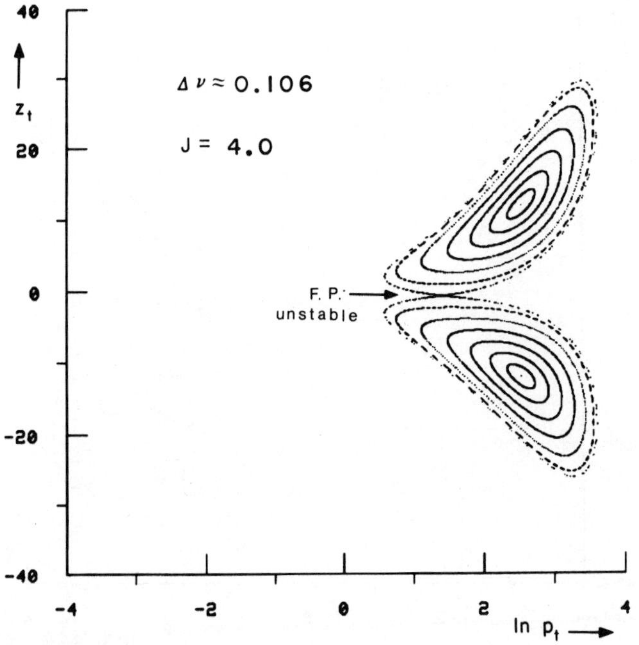

Figure 6. Orbits of the reduced mapping Eq. (3.6) near F.P. and the S_2 orbit in the ln p_t, z_t plane for $J = 4.0$; B and Q as in Fig. 1.

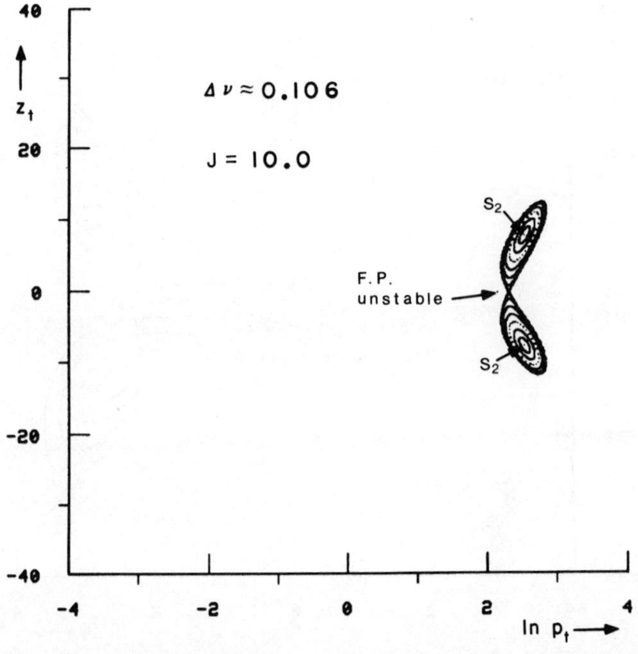

Figure 7. Orbits of the reduced mapping Eq. (3.6) near F.P. and the S_2 orbit in the ln p_t, z_t plane for $J = 10.0$; B and Q as in Fig. 1.

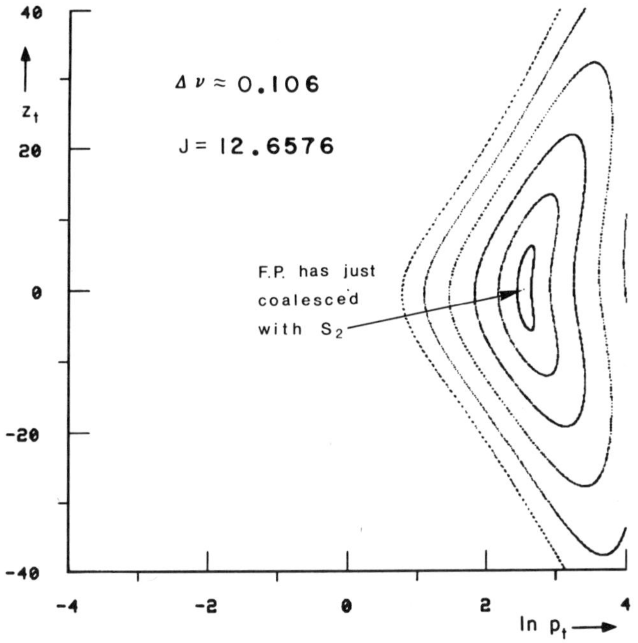

Figure 8. Orbits of the reduced mapping Eq. (3.6) near the stable F.P. in the ln p_t, z_t plane for $J = 12.6576$; B and Q as in Fig. 1.

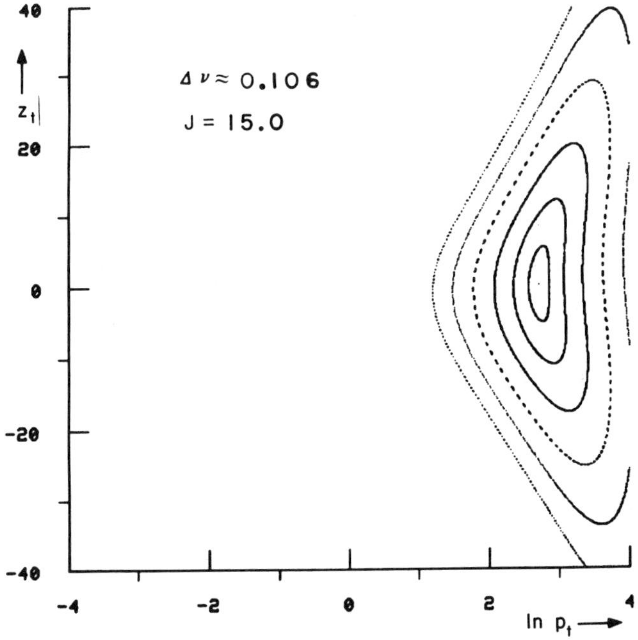

Figure 9. Orbits of the reduced mapping Eq. (3.6) near the stable F.P. in the ln p_t, z_t plane for $J = 15.0$; B and Q as in Fig. 1.

Eq. (3.11) the eigenvalues of $\mathbf{M}^{(m)}\lambda_1, \lambda_2$ are either:[10–13]

1. Complex conjugates with $|\lambda_1| = |\lambda_2| = 1$ and the periodic orbit is *stable* with "islands"[3, 4] around it in the p_t, z_t plane.
2. Real, with $|\lambda_1| > 1, |\lambda_2| < 1$ and the periodic orbit is *unstable* with two contracting and two expanding "eigendirections"[3, 4] originating from each point.

Judging from the behavior of the iterates of Eq. (3.6) near an unstable periodic orbit (see U_2 in Fig. 1–3) we expect that the reduced mapping is not integrable.[3, 4] For such mappings, most analytical stability results require the numerical computation of periodic orbits.[10, 11] As Figs. 1–9 indicate, for parameter values B, Q, and so on, of practical interest, the general behavior of the solutions of Eq. (2.1) is determined by the fixed points and the period-2 orbits of the reduced mapping, to which we now turn.

4. FIXED POINT AND PERIOD-2 ORBITS

For period-1 orbits of Eq. (3.6) we have

$$\hat{p}_t = \hat{p}_{t+1} \equiv p \qquad \hat{z}_{t+1} = \hat{z}_t \equiv z \qquad \text{for all } t \tag{3.12}$$

[cf. Eq. (3.9)], where p is a solution of

$$p^2 = J^2 + \left[Cp + \frac{BS}{Q}(1 - e^{-p/2})\right]^2 \tag{3.13}$$

and z is given by

$$z = Cp + \frac{BS}{Q}(1 - e^{-p/2}) \tag{3.14}$$

[cf. Eqs. (3.6) and (2.2)]. It is easy to see that for $C > 0$, $S < 0$ (and $p \geqslant 0$), Eq. (3.13) has a unique root, which we compute numerically for several values of J (see Table 1).

The stability properties of (p, z) are deduced directly from the trace of $\mathbf{M}^{(1)}$:

$$\operatorname{Tr} \mathbf{M}^{(1)} = -2 + \frac{2z}{p}\left(2C + \frac{BS}{Q}e^{-p/2}\right) \tag{3.15}$$

[cf. Eqs. (3.7) and (2.2)]. In the limit $J \to 0$ $(p, z) \to (0, 0)$ which is a stable fixed point of the uncoupled case.[7] As J increases, p increases monotonically, while z attains a minimum at $J_1 \cong 3.62088$, where the fixed point turns *unstable*

Table 1. Fixed Point Eq. (3.6): $B = 5.02$, $Q = 22.6/6$

J:	0.035	0.5	2.0	$J_1 = 3.62088$	4.0	10.0	$J_2 = 12.65758$	15.0
p:	0.04195	0.56732	2.10057	3.69379	4.065508	10.003668	12.65758	15.00197
z:	−0.023126	−.26805	−.642183	−0.73028	−0.726882	−0.27086	0.0	0.243423
Tr **M**: (1)	−.79946	−1.25439	−1.8443	−2.0	−2.01268	−2.01084	−2.0	−1.99324

(see Figs. 4–6). In fact, at that value of $J = J_1$, (p, z) coalesces with an unstable orbit of period 2 (see below). For $J > J_1$, z increases until $z = 0$ at $J = J_2 = 12.65758$, where (p, z) coalesces with a stable orbit of period 2 and turns stable again! (See Fig. 8.) For $J > J_2$, $z \sim Cp$, $p \sim J/|S|$ [cf. Eqs. (3.13) and (3.14)] and (p, z) remains always stable, since

$$-2 < \operatorname{Tr} \mathbf{M}^{(1)} < -2 + 4C^2 \tag{3.16}$$

The stability condition for the origin[7, 9] $(J = p = z = 0)$ can also be obtained from the results above:

$$z \sim Cp + \frac{BS}{2Q} p \qquad \text{as} \quad p \to 0 \tag{3.17}$$

[cf. Eq. (3.14)], and combining Eqs. (3.7) and (3.15) we find for stability

$$|\operatorname{Tr} \mathbf{M}^{(1)}| = \left| -2 + 4\left(C + \frac{BS}{2Q}\right)^2 \right| < 2 \tag{3.18}$$

or

$$\left| C + \frac{BS}{2Q} \right| < 1 \qquad J = p = 0$$

Condition (3.18) can also be derived directly from Eqs. (2.1) by linearizing about $x_t = y_t = 0$ [in which case, Eqs. (2.1), in fact, completely uncouple].

The other interesting feature of Figs. 1–4 consists of the two orbits of period 2, one *stable* (marked by S_2) and one *unstable* (marked by U_2), to which we now turn. In particular, the unstable orbit U_2 is especially interesting because there are significant "chaotic" regions around it that limit the extent of the stable region about the fixed point discussed above. From Eq. (3.6) we find that the coordinates of a period-2 orbit $(p_0, z_0) \leftrightarrow (p_1, z_1)$ satisfy

$$p_1 p_0 = z_0^2 + J^2 \tag{3.19}$$

$$Cp_0 + \frac{BS}{Q}(1 - e^{-p_0/2}) = Cp_1 + \frac{BS}{Q}(1 - e^{-p_1/2}) \tag{3.20}$$

$$z_1^2 = z_0^2 \tag{3.21}$$

and for $z_1 = z_0$

$$z_0 = Cp_0 + \frac{BS}{Q}(1 - e^{-p_0/2}) \tag{3.22}$$

(for the $z_1 = -z_0$ case, see below).

It turns out that the solution with $z_1 = z_0$ [cf. Eq. (3.21)] yields the unstable orbit U_2, whereas the one with $z_1 = -z_0$ gives the stable orbit S_2 (Figs. 1–4). Numerically solving Eqs. (3.19)–(3.22), for example, we list in Table 2 the coordinates of the unstable period-2 orbit U_2 for several values of J.

Table 2. Unstable Period 2: $B = 5.02$, $Q = 22.6/6$

J:	0.035	0.5	1.0	2.0	3.0	3.5	3.6	3.62088
p_0:	9.68×10^{-4}	0.01993	0.08196	0.37207	1.1372	2.3175	3.0698	3.69379
p_1:	12.657	12.5505	12.2249	10.8437	8.09696	5.4779	4.390156	3.69379
$z_1 = z_0$:	-5.4×10^{-5}	−0.01106	−0.04465	−0.18611	−0.45595	−0.66717	−0.71896	−0.73028
$\mathrm{Tr}\,\mathbf{M}^{(2)}$:	2.4615	2.45415	2.4164	2.33915	2.1671	2.03767	2.00672	2.00

The stability properties of these orbits are deduced from the trace of the "return" matrix $\mathbf{M}^{(2)}$ [cf. Eq. (3.10)]. In the case of the stable orbit $S_2(p_1 = p_0, z_1 = -z_0)$ we find a simple expression

$$S_2: \qquad \operatorname{Tr}\mathbf{M}^{(2)} = 2 - 4\frac{z_0^2}{p_0^2}\left(2C + \frac{BS}{Q}e^{-p_0/2}\right)^2 \tag{3.23}$$

where p_0, z_0 satisfy

$$Cp_0 + \frac{BS}{Q}(1 - e^{-p_0/2}) = 0 \tag{3.24}$$

and

$$z_0 = \left(p_0^2 - J^2\right)^{1/2} \tag{3.25}$$

From Eq. (3.24) we find with e.g. $B = 5.02$ and $Q = 22.6/6$ that $p_0 \cong 12.65758$, and using this value of p_0 and Eq. (3.25) in Eq. (3.23) gives

$$S_2: \qquad \operatorname{Tr}\mathbf{M}^{(2)} = 2 - 0.826768\left[1 - \left(\frac{J}{12.65758}\right)^2\right] \tag{3.26}$$

Thus this orbit is stable for $0 \leqslant J \leqslant J_2 = 12.65758$, at which point it coalesces with the fixed point and ceases to exist, while the fixed point becomes stable (see Table 1 and Figs. 8 and 9).

The analysis of the unstable period-2 orbit U_2 is slightly more complicated, but the full story is told by Table 2 and Figs. 1–5. In fact, U_2 is a "short-lived" orbit, since at $J = J_1 = 3.62088$ it coalesces with the fixed point (p, z) and is lost forever.

5. DISCUSSION

We have applied the concepts and methods of nonlinear dynamics to a model of the beam–beam interaction problem of high energy accelerators. In the case of beams with cylindrically symmetric charge distribution, we have derived an exact integral that enables us to reduce the original four-dimensional problem to a two-dimensional area preserving mapping. Such a reduction is indeed a rare occurrence in nonlinear systems and has quite beneficial consequences for the stability properties of the particle beams. In view of the potential usefulness of our results, we conclude with some remarks that may be of practical interest to researchers in accelerator physics.

In the accelerator field the relevant parameters are the two frequencies in the x, y-directions Q_x and Q_y, and the so called tune shifts $\Delta\nu_x \equiv B/4\pi Q_x$ and $\Delta\nu_y \equiv B/4\pi Q_y$. Our results suggest that taking $Q_x = Q_y = Q$ (and $\Delta\nu_x = \Delta\nu_y = \Delta\nu$) would enhance the stability properties of colliding beams in real-life machines, since the existence of the angular momentum integral reduces the dimensionality of the problem and precludes particle (Arnol'd) diffusion for all time.

Moreover, to ensure minimal size "excursions" of particle orbits in the x, y-plane, one should choose initial conditions close to the stable, lowest period solutions of the problem. As Figs. 1–5 indicate (for one "tune shift" value of $\Delta\nu \cong 0.106$), this can best be accomplished near the fixed point for $0 < J \lesssim 3.5$. Finally, orbits are also well behaved near the stable-2 orbit S_2 (for all J), and the stable fixed point for $J \gtrsim 12.6576$, but at radial distances $r_t \gtrsim 3.5577$ away from the origin of the x, y-plane.

ACKOWLEDGMENTS

We thank Dr. A. Ruggiero for informing us of similar results obtained independently by his group at Fermilab (see Ref. 14). This study was partially supported by U.S. Department of Energy grant DE-AC03-77ER01538.

REFERENCES

1. *Nonlinear Dynamics and the Beam–Beam Interaction*, M. Month and J. Herrera, Eds. American Institute of Physics Conference Proceedings Series, Vol. 57 (AIP, New York, 1979).
2. T. Bountis, in *Proceedings of the Beam–Beam Introduction Seminar*, SLAC Pub 2624 CONF-8005102 (Stanford Linear Accelerator Center, Stanford, CA, (October, 1980).
3. M. V. Berry, in *Topics in Nonlinear Dynamics*, S. Jorna, Ed. American Institute of Physics Conference Proceedings Series, Vol. 46 (AIP, New York, 1978), p. 16. See also other articles in that volume.
4. R. H. G. Helleman, "Self-Generated Chaotic Behavior in Nonlinear Mechanics," in *Fundamental Problems in Statistical Mechanics*, Vol. 5, E. G. D. Cohen, Ed. (North Holland, Amsterdam and New York 1981), pp. 166–233. It includes many references. See also A. J. Lichtenberg and M. A. Lieberman, "Regular and Stochastic Motion," to appear (Springer Verlag, 1982).
5. J. C. Herrera, in Ref. 1, p. 29.
6. C. R. Eminhizer, in same volume as Ref. 2, p. 273.
7. R. H. G. Helleman, in Ref. 1, p. 236.
8. F. M. Izrailev, G. M. Tumaikin, and I. B. Vasserman, "Stochasticity Limit in Colliding Beams on the Main Coupling Resonance," preprint 79-74, (Institute of Nuclear Physics, 630090, Novosibirsk, USSR, 1979).
9. F. M. Izrailev, "Nearly Linear Mappings and Their Applications," *Physica D* **1**, 243–266. (1980).
10. T. Bountis and R. H. G. Helleman, "On the Stability of Periodic Orbits of Two-Dimensional Mappings," *J. Math. Phys.* **22**, 1867 (1981).
11. J. M. Greene, *J. Math. Phys.* **20**:6, 1183 (1979); see also *J. Math. Phys.* **9**, 760 (1968).
12. B. V. Chirikov, "Universal Instability of Many-Dimensional Oscillator Systems," *Phys. Rep.* **52**:5 (May 1979).
13. J. Ford, in *Fundamental Problems in Statistical Mechanics*, Vol. 3, E. G. D. Cohen, Ed. (North Holland, Amsterdam and New York, 1975).
14. A. G. Ruggiero, Lecture given at Workshop on the Beam–Beam Effect, in the Workshop on "Long-Time Prediction in Dynamics," Austin, Texas (March 1981). See also D. Neuffer, A. Riddiford, and A. Ruggiero, Fermilab Report FN-333 8000.000 (April 1981).

STATISTICAL DESCRIPTION OF THE CHIRIKOV–TAYLOR MODEL IN THE PRESENCE OF NOISE

A. B. RECHESTER
Plasma Fusion Center, Massachusetts Institute of Technology, Cambridge, Massachusetts

M. N. ROSENBLUTH
Institute for Fusion Studies, University of Texas at Austin, Austin, Texas

R. B. WHITE AND C. F. F. KARNEY
Plasma Physics Laboratory, Princeton University, Princeton, New Jersey

Abstract. *A review of recent analytical and numerical results concerning the Chirikov–Taylor model is given. It is shown that the presence of noise makes the statistical description of this system unique. The form of the diffusion coefficient is quite different depending on the nature of the dynamical orbits (integrable, stochastic, or accelerator). We have found that in the presence of noise, dynamical averaging, performed numerically, and statistical averaging, performed analytically with the path-integral method, are the same. Some unsolved problems are presented and discussed.*

1. INTRODUCTION

Statistical description of systems far from thermal equilibrium constitutes one of the major unsolved problems of modern physics. The fundamental question in this area is how to derive the statistical properties of such systems based on the equations of motion. This is usually quite a difficult question to answer because it involves a description of dynamical systems for very long times. So far, studies of this kind have been done only for simple systems with a very few degrees of freedom. The purpose of the present chapter is to give a review of recent results related to the Chirikov–Taylor (C–T) model, also called standard mapping,[1, 2] in the presence of Gaussian noise.

The C–T model is described by the equations

$$x_{t+1} = x_t + y_{t+1}, \tag{1}$$

$$y_{t+1} = y_t - \varepsilon \sin x_t. \tag{2}$$

Here $t = 0, 1, 2, \ldots, T$ corresponds to a discrete time and $0 \leqslant x \leqslant 2\pi$ is a phase variable. The variable y corresponds to energy,[1] magnetic moment,[3] velocity,[4, 5], or other physical property, depending on the physical context. Consider, for example, the motion of a charged particle in a uniform magnetic field B in a cyclotron. Let us apply a periodic potential $U \sin \omega t$ within a small cross section (see Fig. 1). The equations that describe the changes of particle energy W and phase of the wave ϕ every time it passes through the acceleration region can be written in the form[1]

$$W_{i+1} = W_i + eU \sin \phi_i, \tag{3}$$

$$\phi_{i+1} = \phi_i + 2\pi \frac{\omega W_{i+1}}{eBc}. \tag{4}$$

Here e is the charge, and $\omega_c = eBc/W$ is the cyclotron frequency of the relativistic particle. Equations (3) and (4) can be rewritten in dimensionless form (1) and (2) with a parameter $\varepsilon = 2\pi\omega U/cB$. This method is used to heat particles: above stochastic threshold $\varepsilon \gtrsim 1$ the average energy of particles grows as the $\sqrt{t}$.

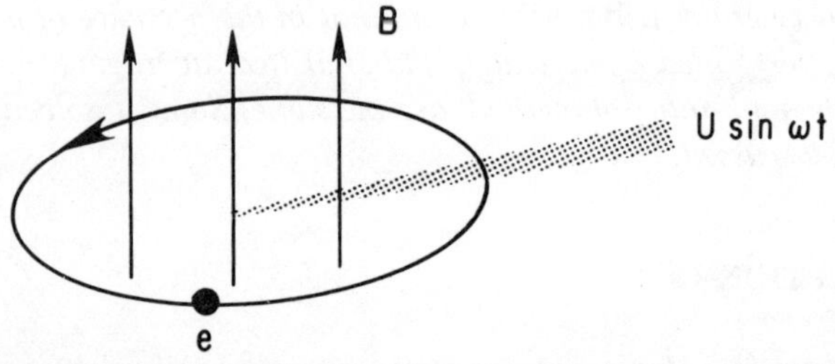

Figure 1. An electron trajectory in a cyclotron. A periodic potential $U \sin \omega t$ is applied in the shaded sector.

Equations (1) and (2) are area preserving. They can be rewritten in Hamiltonian form

$$\frac{dx}{dt} = y, \tag{5}$$

$$\frac{dy}{dt} = -\varepsilon \sum_{n=-\infty}^{\infty} \sin(x - 2\pi n t). \tag{6}$$

Equations (5) and (6) can be used as a simplified model to study, for example, the motion of charged particles in the field of electrostatic plane waves[4, 5] or the behavior of drift orbits in tokamaks in the presence of drift waves.[6]

2. DYNAMICS, DETERMINISTIC ORBITS

First we will review the behavior of a single orbit described by its coordinates x_t, y_t. A few orbits have been mapped in the phase plane xy on Fig. 2, taken from the paper of J. M. Greene.[7] There are two types of orbits: integrable

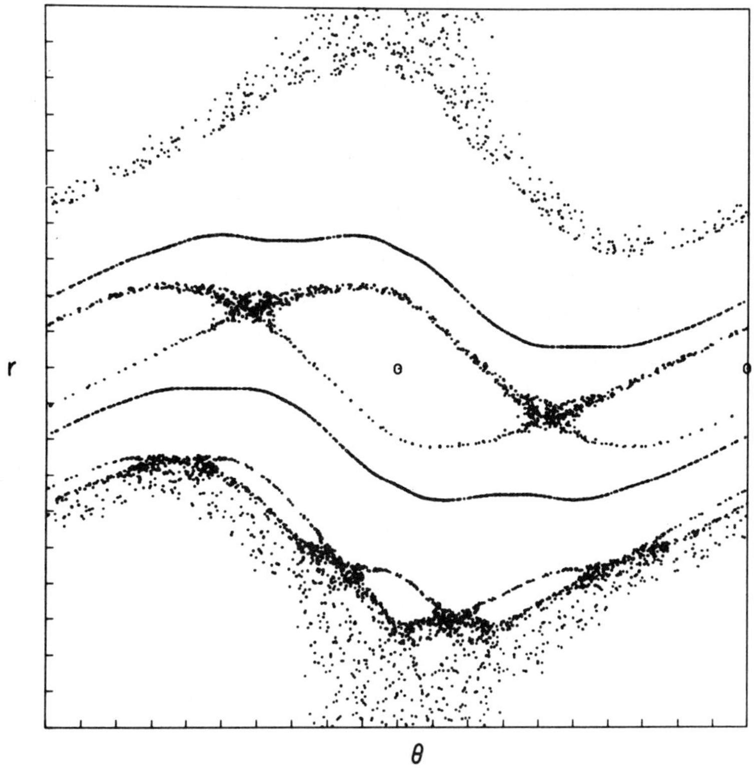

Figure 2. Mapping of orbits for the C–T model, $\varepsilon = 0.97 < \varepsilon_c$ (Ref. 7).

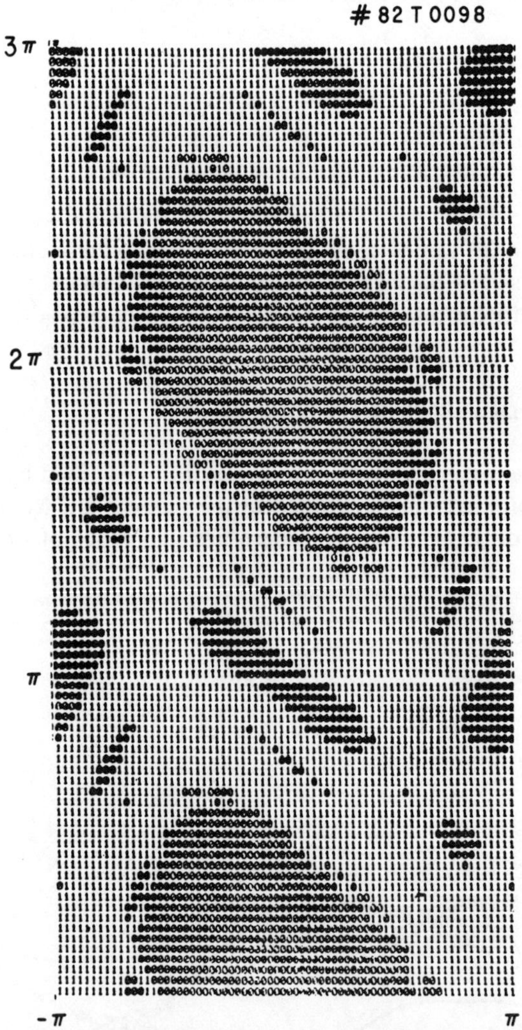

Figure 3. The area covered by a single orbit in the C–T map, $\varepsilon = 1.28$.

orbits, which stay on smooth curves called invariant curves or KAM surfaces, and stochastic orbits, which are area filling. The quantitative measure of stochasticity of an orbit is the maximum Liapunov number λ.[1, 2] It defines the rate of exponential divergence of nearby orbits. For stochastic orbits $\lambda > 0$, and for integrable orbits $\lambda = 0$. In the case $\varepsilon \ll 1$ most orbits are integrable, while for $\varepsilon \gg 1$ most orbits are stochastic. For values of $\varepsilon \leqslant \varepsilon_c = 0.9716\ldots$ there exist global invariant curves that divide the phase plane into upper and lower regions as is shown in Fig. 2.[7] These curves are periodic in the y direction because Eqs. (1), and (2) do not change with the transformation $y \to y \pm 2\pi$, $x \to x$. Obviously the existence of such curves bounds the excursion of the orbits in the y direction. At the value of $\varepsilon = \varepsilon_c$ the last of these curves is destroyed, and for some orbits the motion in the y direction becomes unbounded. In Fig. 3 we have plotted the area covered by one such orbit for $\varepsilon = 1.28$.[4] We can define quantitatively the diffusion coefficient for a single orbit by the equation

$$D_{\text{orb}} = \lim_{s\to\infty} \lim_{T\to\infty} \frac{1}{T} \sum_{t=0}^{T} \frac{1}{2s} (y_{t+s} - y_t)^2. \tag{7}$$

This type of averaging is called dynamic averaging. It requires following the orbit for very long times. We have calculated D_{orb} numerically using Eq. (7) and found that it changes significantly depending on the orbit chosen.[8] For example, if we take an orbit in the dark area of Fig. 3 we obviously get $D_{\text{orb}} \equiv 0$. But for the orbit located in the light area D_{orb} is finite.

Another interesting property of the C–T model is the existence of acceleration orbits.[2] For these orbits y_t increases approximately linearly with time $y_t \approx at$, where the parameter a corresponds to acceleration when y has the meaning of velocity. Acceleration orbits are integrable. In order to find corresponding KAM curves we will reduce the phase space in the y direction to the region $-\pi \leqslant y \leqslant \pi$. A number of orbits have been mapped on Fig. 4 in the

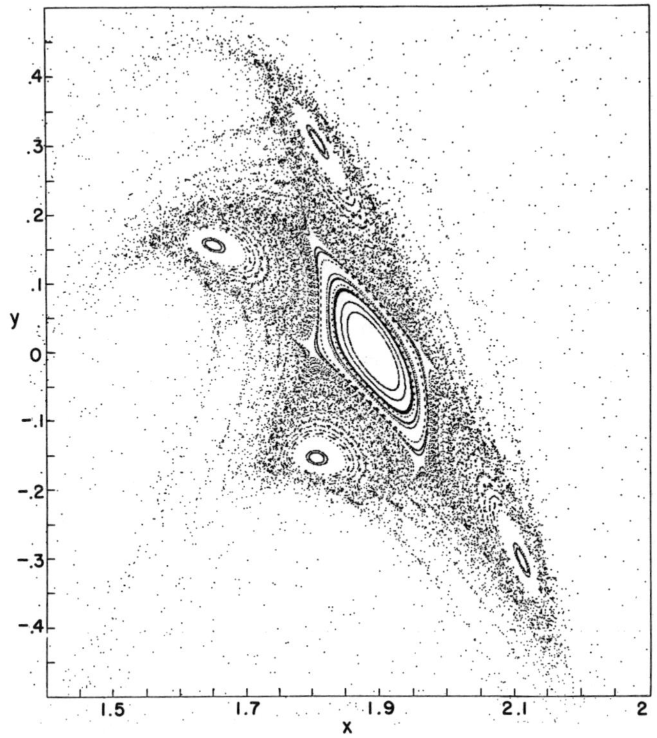

Figure 4. Mapping of accelerator orbits for C–T model, $\varepsilon = 6.615$.

reduced phase space around the stable elliptic point at $x = \cos^{-1}(\varepsilon/4)$, $y = 0$, and $\varepsilon = 6.615$. The large island around this first-order elliptic point is called an accelerator island. There are also four surrounding islands associated with stable elliptic points of fourth order. This whole structure moves in the original phase space with acceleration 2π, and there is a local rotation of the orbits around the elliptic points. Obviously acceleration orbits have $D_{\text{orb}} = \infty$. There are also apparently stochastic orbits filling the regions between the islands. We have calculated numerically[8] one of these orbits and found the diffusion enhanced over the value for stochastic orbits not near accelerator islands by two orders of magnitude. There is also a symmetric accelerator region near $y = 0$ and $x = -\cos^{-1}(\varepsilon/4)$ with acceleration $a = -2\pi$. The stability condition[2] for the elliptic point of the first order is $(2\pi n)^2 \leqslant \varepsilon^2 \leqslant 16 + (2\pi n)^2$. The case shown in Fig. 4 is for $n = 1$. The acceleration is $\pm 2\pi n$. The size of the acceleration island is a very complicated function of ε (Ref. 8), but the upper bound on its size is given by $\Delta x \approx \Delta y \approx 1/\varepsilon$.[2]

3. THE EFFECT OF NOISE

On the other hand there are always some collisions between particles or other noisy processes present in physical systems. Even when they are very small, over a very long time they make a significant contribution to the particle motion. We model them as random variables δx_t or δy_t added to the right-hand side of Eqs. (1) and (2) and assume that they have a Gaussian distribution

$$P(\delta x) = \frac{1}{\sqrt{2\pi\sigma}} \exp\left[-\frac{(\delta x)^2}{2\sigma}\right].$$

In the presence of noise calculations based on Eq. (7) we show that D is a unique function of ε independent of initial conditions even for a very weak noise. In Fig. 5 we have presented the results of such computations in the case of δx noise present with variance $\sigma = 10^{-3}$. Obviously noise destroys the memory of initial conditions and allows particles to diffuse from one deterministic orbit to another. In the presence of noise all regions in phase space become accessible to the particle. In the case of a deterministic orbit this is not true even in the case $\varepsilon \gg 1$.

To describe the particle motion in the presence of noise we need to introduce the concept of the probability of an orbit. Consider a particle that starts at $x_0 y_0$ at time $t = 0$. Due to the presence of the noise, there is a certain probability for finding the particle at position $x_t y_t$ at time t. For a given orbit, defined by $\{x_t, y_t\}$, $t = 0, 1, \ldots, T$, there exists a probability $P(\{x_t, y_t\})$ that the particle will traverse this orbit. The method of constructing the probability function P is called the path-integral method. It was introduced by Smolokowsky[9] and Einstein[10] to treat Brownian motion. Later Feynman used it to describe quantum mechanical wave functions in terms of classical orbits.[11]

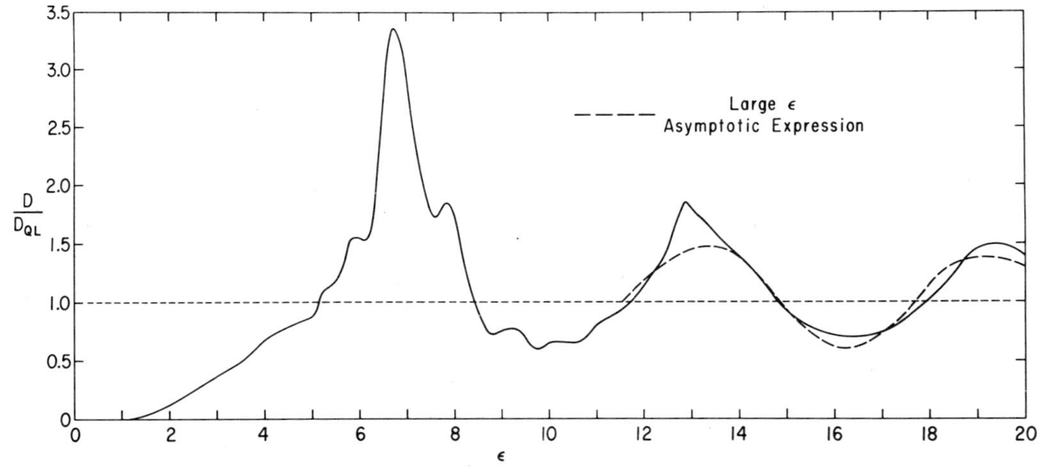

Figure 5. Diffusion in the C–T map. Here $\sigma = 10^{-3}$. The asymptotic curve is from Eq. (10).

For any function of the orbit we define the statistical averaging over the probability P given by

$$\langle f(\{x_t y_t\})\rangle = \prod_{t=1}^{T} \int_0^{2\pi} dx_t \int_{-\infty}^{\infty} dy_t f(\{x_t, y_t\}) P(\{x_t, y_t\}). \tag{8}$$

In this paper we will be particularly concerned with the calculation of the diffusion coefficient, defined by

$$D = \lim_{T\to\infty} \frac{\langle (y_T - y_0)^2 \rangle}{2T}. \tag{9}$$

The diffusion $D(\varepsilon, \sigma)$ is a complicated function of the two parameters ε, σ (see Fig. 5). This complexity is due to the great richness of the dynamical orbits, especially in the region $\varepsilon \approx 1$. However, for large or small ε we were able to explicitly evaluate Eq. (9) and find analytic expressions for the diffusion coefficient D.[5, 12] We will describe now the results of these calculations. Consider first the case $\varepsilon \gg 1$. This is the case when most orbits are stochastic. The first leading term of the asymptotic expansion of D in powers of Bessel functions $J_l(\varepsilon) \sim 1/\sqrt{\varepsilon}$ are given by

$$D = \frac{\varepsilon^2}{2}\left[\tfrac{1}{2} - J_2(\varepsilon)e^{-\sigma} - J_1^2(\varepsilon)e^{-\sigma} + J_2^2(\varepsilon)e^{-2\sigma} + J_3^2(\varepsilon)e^{-3\sigma} + \cdots\right]. \tag{10}$$

Here we have considered only the case of noise in x, δx. In Fig. 6 is shown the comparison of this expression with results obtained by numerically advancing the map. The reason for the good agreement is due to the rapid decay of phase correlations. The term $D_{QL} = \varepsilon^2/4$ corresponds to a random phase approximation often made in quasilinear theory. The Bessel terms result from residual phase correlations of the orbits. Define the phase correlation function

$$\langle \sin x_{t+\tau} \sin x_t \rangle, \tag{11}$$

then the $J_2(\varepsilon)$ term arises from $\tau = 1$, and the J_1^2, J_3^2 terms come from $\tau = 2$ and J_2^2 from $\tau = 3$.[8]

Now consider the case $\varepsilon \ll 1$. In this case most orbits are integrable. In this limit D takes the form

$$D = \frac{\varepsilon^2}{4} \tanh\left(\frac{\sigma}{2}\right). \tag{12}$$

Comparison with numerical results is shown in Fig. 7. In order to calculate Eq. (12), phase correlation functions Eq. (11) must be retained with $1 \ll \tau \lesssim 1/\sigma$. Thus the phase correlations decay only due to the presence of the noise, as

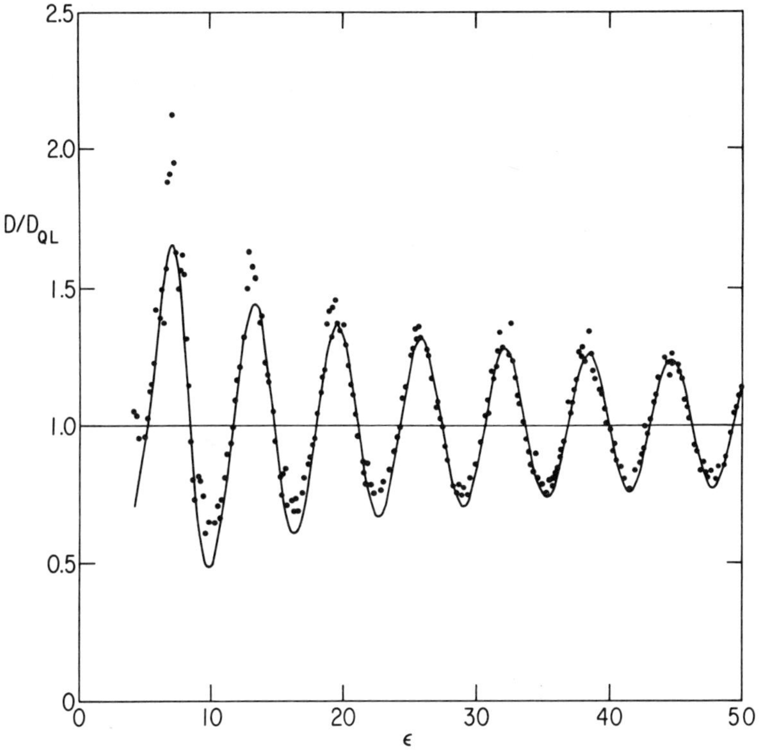

Figure 6. Diffusion in the C–T map. The points are from numerically advancing the map.

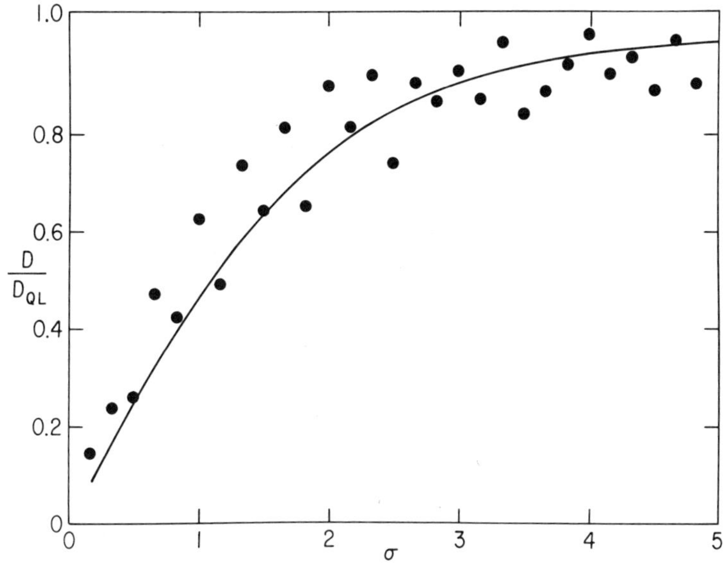

Figure 7. Diffusion in the C–T map as a function of noise parameter for small ε. Here $\varepsilon = 0.06$.

expected for integrable orbits. Equations (10) and (12) can easily be generalized to include noise δy.[8]

It is interesting to consider the limit of $\sigma \to 0$. In the stochastic regime there is a definite limit

$$D(\varepsilon) = \lim_{\sigma \to 0} D(\varepsilon, \sigma), \tag{13}$$

whereas in the integrable case,

$$\lim_{\sigma \to 0} D(\varepsilon, \sigma) = 0, \tag{14}$$

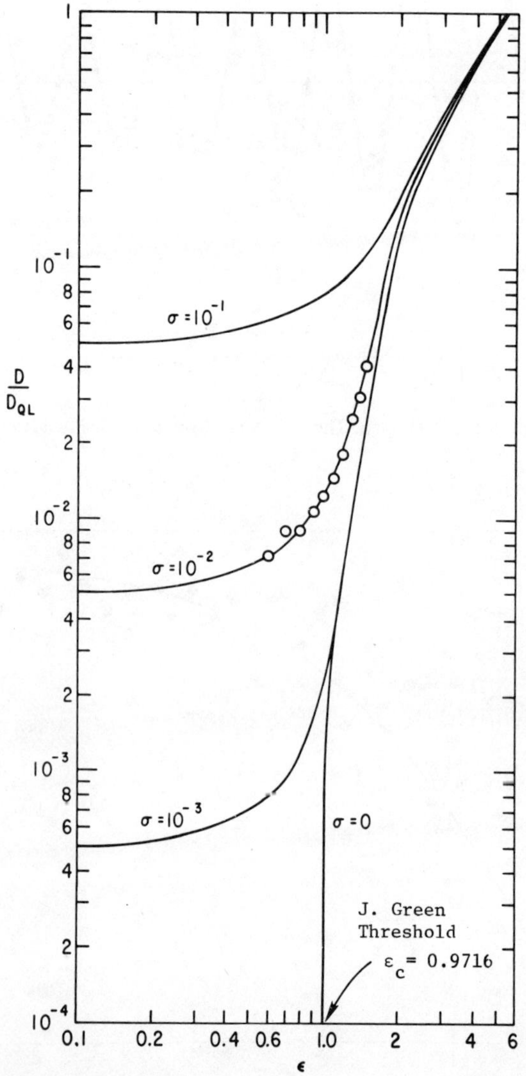

Figure 8. Diffusion in the C–T map for various σ near stochastic threshold.

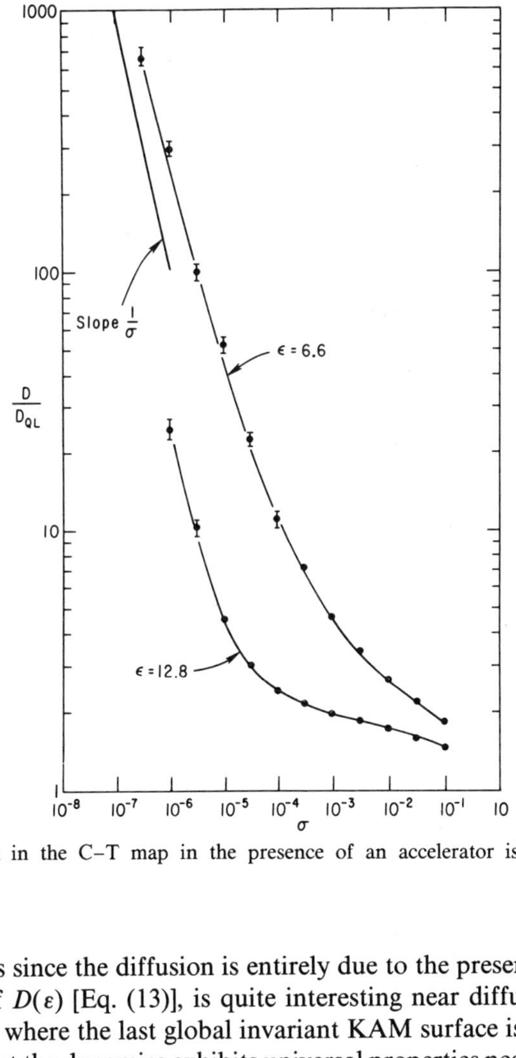

Figure 9. Diffusion in the C–T map in the presence of an accelerator island. For small σ, $D \sim 1/\sigma$.

which is obvious since the diffusion is entirely due to the presence of the noise. The behavior of $D(\varepsilon)$ [Eq. (13)], is quite interesting near diffusivity threshold given by $\varepsilon = \varepsilon_c$, where the last global invariant KAM surface is destroyed. The work[13] shows that the dynamics exhibits universal properties near this threshold, that is, is independent of the periodic function of x in Eq. (2). Our numerical results, shown in Fig. 8, indicate that D has the form

$$D = A(\varepsilon - \varepsilon_c)^{\alpha},$$

and we conjecture that α is a critical exponent.

To conclude this section we consider the effect of noise on accelerator orbits. The effect of the accelerator orbits on diffusion can be estimated as follows. As we saw in Section 2, the acceleration is $a = 2\pi n \approx \varepsilon$. The size of the island is bounded by Δx, $\Delta y \lesssim 1/\varepsilon$. In the presence of noise an orbit will diffuse outside an accelerator island in a time $\tau_{\text{is}} \approx (\Delta x)^2/\sigma$. The time spent outside the island structure is given by $\tau_{\text{out}} \approx 1/\sigma$. The diffusion can be estimated as

$$D_{\text{is}} \lesssim \frac{(\tau_{\text{is}} a)^2}{\tau_{\text{out}}} \lesssim \frac{1}{\varepsilon^2 \sigma}.$$

Thus $D_{\text{is}}/D_{QL} \lesssim 1/\varepsilon^4\sigma$.

The numerical confirmation of this scaling for accelerator islands with $n = 1, 2$ is shown in Fig. 9. We have plotted the ratio of the total diffusion D to D_{QL} for $\varepsilon = 6.6$ and $\varepsilon = 12.8$ where first-order acceleration islands exist. One can see that in the limit of very small σ the scaling $D \approx 1/\sigma$ is verified. We can also conclude that the relative importance of even the largest accelerator island is small unless $\sigma \lesssim 10^{-4}$. Such a small value of σ is necessary in order for an orbit to spend sufficient time in the small island structure. We were not able to calculate D analytically in this case. The main difficulty is that very long time phase correlations must be included.

4. CONCLUSION

The path-integral method proved to be very successful for the statistical description of the Chirikov–Taylor model in the presence of noise.[5, 8, 12] Recently this method has been applied to the variety of problems described by one- and two-dimensional mappings.[14–18] It is also applicable to nonlinear differential equations because they can be written in the form of mappings. To apply the path-integral method to such problems it is necessary to evaluate contributions from many time steps. In general there is no systematic way of doing this, but we have managed to analyze the case of the C–T model in the nearly integrable limit $\varepsilon \ll 1$. An important remaining problem is the development of analytic methods capable of treating similar but more difficult problems such as behavior near the stochastic threshold and in the presence of accelerator orbits.

ACKNOWLEDGMENT

This work was supported by U.S. Department of Energy Contract No. DE-AC02-76-CH03073 and DE-AS02-78 ET 53074.

REFERENCES

1. G. M. Zaslavsky and B. V. Chirikov, *Usp. Fiz. Nauk.* **14**, 195 (1972) [*Sov. Phys. Usp.* **14**, 549 (1972)].
2. B. V. Chirikov, *Phys. Rep.* **52**, 263 (1979).
3. M. N. Rosenbluth, *Phys. Rev. Lett.* **29**, 408 (1972).

4. T. H. Stix, in *Proceedings of the Joint Varenna-Grenoble International Symposium on Heating in Toroidal Plasmas*, Grenoble, France, 3–7 July 1978, edited by T. Consoli and P. Caldirola (Pergamon, Elmsford, N.Y., 1979).
5. A. B. Rechester and R. B. White, *Phys. Rev. Lett.* **44**, 1586 (1980).
6. A. B. Rechester and M. N. Rosenbluth, *Phys. Rev. Lett.* **40**, 38 (1978); also A. B. Rechester, M. N. Rosenbluth, and R. B. White, *J. Phys. Paris* **41**, C3-351 (1980).
7. J. M. Greene, *J. Math. Phys.* **20**, 1183 (1979).
8. C. F. F. Karney, A. B. Rechester, and R. B. White, *Physica* **4D**, 425 (1982).
9. M. V. Smolokowsky, *Ann. Phys. Paris* **25**, 205 (1908).
10. A. Einstein, *Z. Elektrochem.* **14**, 235 (1908).
11. R. P. Feynman and A. R. Hibbs, *Quantum Mechanics and Path Integrals* (McGraw Hill, New York, 1965).
12. A. B. Rechester, M. N. Rosenbluth, and R. B. White, *Phys. Rev.* **A23**, 2664 (1981).
13. J. M. Greene, R. S. Mackay, F. Vivaldi, and M. J. Feigenbaum, *Physica* **3D**, 530 (1981).
14. R. H. Cohen and G. Rowlands, *Phys. Fluids* **24**, 2295 (1981).
15. J. R. Cary, J. D. Meiss, and A. Bhatta-Charjee, *Phys. Rev.* **A24**, 2744 (1981).
16. R. V. Jensen and C. R. Oberman, *Phys. Rev. Lett.* **46**, 1547 (1981).
17. T. M. Antensen and E. Ott, *Phys. Fluids* **24**, 1635 (1981).
18. A. B. Rechester and R. B. White, Princeton Plasma Physics Laboratory Report PPPL-1821 (1981).

AUTHOR INDEX

SUBJECT INDEX